Lecture Notes in Computer Science 1412

Edited by G. Goos, J. Hartmanis and J. van Leeuwen

W0107516

Springer-Verlag Berlin Heidelberg GmbH

Robert E. Bixby   E. Andrew Boyd
Roger Z. Ríos-Mercado (Eds.)

Integer Programming and Combinatorial Optimization

6th International IPCO Conference
Houston, Texas, June 22-24, 1998
Proceedings

 Springer

Series Editors

Gerhard Goos, Karlsruhe University, Germany
Juris Hartmanis, Cornell University, NY, USA
Jan van Leeuwen, Utrecht University, The Netherlands

Volume Editors

Robert E. Bixby
Department of Computational and Applied Mathematics, Rice University
6020 Annapolis, Houston, TX 77005, USA
E-mail: bixby@rice.edu

E. Andrew Boyd
PROS Strategic Solutions
3223 Smith Street, Houston, TX 77006, USA
E-mail: boyd@prosx.com

Roger Z. Ríos-Mercado
Department of Industrial, Engineering,Texas A&M University
1000 Country Place Dr. Apt. 69, Houston, TX 77079, USA
E-mail: roger@hpc.uh.edu

Cataloging-in-Publication data applied for

Die Deutsche Bibliothek - CIP-Einheitsaufnahme

Integer programming and combinatorial optimization :
proceedings / 6th International IPCO Conference, Houston, Texas,
June 22 - 24, 1998. Robert E. Bixby ... (ed.). - Berlin ; Heidelberg ;
New York ; Barcelona ; Budapest ; Hong Kong ; London ; Milan ;
Paris ; Santa Clara ; Singapore ; Tokyo : Springer, 1998
    (Lecture notes in computer science ; Vol. 1412)

CR Subject Classification (1991): G.1.6, G.2.1-2, F.2.2

ISSN 0302-9743
ISBN 978-3-540-64590-0   ISBN 978-3-540-69346-8 (eBook)
DOI 10.1007/978-3-540-69346-8

Originally published by Springer-Verlag Berlin Heidelberg New York in 1998 .

Typesetting: Camera-ready by author
SPIN 10637207   06/3142 – 5 4 3 2 1 0   Printed on acid-free paper

Preface

This volume contains the papers selected for presentation at IPCO VI, the Sixth International Conference on Integer Programming and Combinatorial Optimization, held in Houston, Texas, USA, June 22–24, 1998. The IPCO series of conferences highlights recent developments in theory, computation, and applications of integer programming and combinatorial optimization.

These conferences are sponsored by the Mathematical Programming Society, and are held in the years in which no International Symosium on Mathematical Programming takes place. Earlier IPCO conferences were held in Waterloo (Canada) in May 1990; Pittsburgh (USA) in May 1992; Erice (Italy) in April 1993; Copenhagen (Denmark) in May 1995; and Vancouver (Canada) in June 1996.

The proceedings of IPCO IV (edited by Egon Balas and Jens Clausen in 1995) and IPCO V (edited by William Cunningham, Thomas McCormick, and Maurice Queyranne in 1996), were published by Springer-Verlag in the series Lecture Notes in Computer Science as Volumes 920 and 1084, respectively. The proceedings of the first three IPCO conferences were published by organizing institutions.

A total of 77 extended abstracts, mostly of an excellent quality, were initially submitted. Following the IPCO policy of having only one stream of sessions over a three day span, the Program Committee selected 32 papers. As a result, many outstanding papers could not be selected.

The papers included in this volume have not been refereed. It is expected that revised versions of these works will appear in scientific journals.

The Program Committee thanks all the authors of submitted extended abstracts and papers for their support of the IPCO conferences.

April 1998

Robert E. Bixby
E. Andrew Boyd
Roger Z. Ríos Mercado

Table of Contents

Integer Programming Applications

Integer Programming Computation

Network Flows

The Packing Property [*]

Gérard Cornuéjols[1], Bertrand Guenin[1], and François Margot[2]

[1] Graduate School of Industrial Administration
Carnegie Mellon University, Pittsburgh, PA 15213, USA
[2] Department of Mathematical Science
Michigan Technical University, Houghton, MI 49931, USA

Abstract. A clutter (V, E) *packs* if the smallest number of vertices needed to intersect all the edges (i.e. a transversal) is equal to the maximum number of pairwise disjoint edges (i.e. a matching). This terminology is due to Seymour 1977. A clutter is *minimally nonpacking* if it does not pack but all its minors pack. A 0,1 matrix is *minimally nonpacking* if it is the edge-vertex incidence matrix of a minimally nonpacking clutter. Minimally nonpacking matrices can be viewed as the counterpart for the set covering problem of minimally imperfect matrices for the set packing problem. This paper proves several properties of minimally nonpacking clutters and matrices.

1 Introduction

A *clutter* $\mathcal{C}$ is a pair $(V(\mathcal{C}), E(\mathcal{C}))$, where $V(\mathcal{C})$ is a finite set and $E(\mathcal{C}) = \{S_1, \ldots, S_m\}$ is a family of subsets of $V(\mathcal{C})$ with the property that $S_i \subseteq S_j$ implies $S_i = S_j$. The elements of $V(\mathcal{C})$ are the *vertices* of $\mathcal{C}$ and those of $E(\mathcal{C})$ are the *edges*. A *transversal* of $\mathcal{C}$ is a minimal subset of vertices that intersects all the edges. Let $\tau(\mathcal{C})$ denote the cardinality of a smallest transversal. A clutter $\mathcal{C}$ *packs* if there exist $\tau(\mathcal{C})$ pairwise disjoint edges.

For $j \in V(\mathcal{C})$, the *contraction* $\mathcal{C}/j$ and *deletion* $\mathcal{C} \setminus j$ are clutters defined as follows: both have $V(\mathcal{C}) - \{j\}$ as vertex set, $E(\mathcal{C}/j)$ is the set of minimal elements of $\{S - \{j\} : S \in E(\mathcal{C})\}$ and $E(\mathcal{C} \setminus j) = \{S : j \notin S \in E(\mathcal{C})\}$. Contractions and deletions of distinct vertices can be performed sequentially, and it is well known that the result does not depend on the order. A clutter $\mathcal{D}$ obtained from $\mathcal{C}$ by deleting $I_d \subseteq V(\mathcal{C})$ and contracting $I_c \subseteq V(\mathcal{C})$, where $I_c \cap I_d = \emptyset$ and $I_c \cup I_d \neq \emptyset$, is a *minor* of $\mathcal{C}$ and is denoted by $\mathcal{C} \setminus I_d / I_c$.

We say that a clutter $\mathcal{C}$ has the *packing property* if it packs and all its minors pack. A clutter is *minimally non packing* (mnp) if it does not pack but all its minors do. In this paper, we study mnp clutters.

[*] This work was supported in part by NSF grants DMI-9424348, DMS-9509581, ONR grant N00014-9710196, a William Larimer Mellon Fellowship, and the Swiss National Research Fund (FNRS).

R. E. Bixby, E. A. Boyd, and R. Z. Ríos-Mercado (Eds.): IPCO VI
LNCS 1412, pp. 1–8, 1998. © Springer–Verlag Berlin Heidelberg 1998

These concepts can be described equivalently in terms of 0,1 matrices. A 0,1 matrix A *packs* if the minimum number of columns needed to cover all the rows equals the maximum number of nonoverlapping rows, i.e.

$$\min\left\{\sum_1^n x_j : Ax \geq e, \; x \in \{0,1\}^n\right\}$$

$$= \max\left\{\sum_1^m y_i : yA \leq e, \; y \in \{0,1\}^m\right\}, \tag{1}$$

where e denotes a vector of appropriate dimension all of whose components are equal to 1. Obviously, dominating rows play no role in this definition (row $A_{i.}$ *dominates* row $A_{k.}$, $k \neq i$, if $A_{ij} \geq A_{kj}$ for all j), so we assume w.l.o.g. that A contains no such row. That is, we assume w.l.o.g. that A is the edge-vertex incidence matrix of a clutter. Since the statement "A packs" is invariant upon permutation of rows and permutation of columns, we denote by $A(\mathcal{C})$ any 0,1 matrix which is the edge-vertex incidence matrix of clutter $\mathcal{C}$. Observe that contracting $j \in V(\mathcal{C})$ corresponds to setting $x_j = 0$ in the set covering constraints $A(\mathcal{C})x \geq e$ (since, in $A(\mathcal{C}/j)$, column j is removed as well as the resulting dominating rows), and deleting j corresponds to setting $x_j = 1$ (since, in $A(\mathcal{C} \setminus j)$, column j is removed as well as all rows with a 1 in column j). The packing property for A requires that equation (1) holds for the matrix A itself and all its minors. This concept is dual to the concept of perfection (Berge [1]). Indeed, a 0,1 matrix is *perfect* if all its column submatrices A satisfy the equation

$$\max\left\{\sum_1^n x_j : Ax \leq e, \; x \in \{0,1\}^n\right\}$$

$$= \min\left\{\sum_1^m y_i : yA \geq e, \; y \in \{0,1\}^m\right\}.$$

This definition involves "column submatrices" instead of "minors" since setting a variable to 0 or 1 in the set packing constraints $Ax \leq e$ amounts to consider a column submatrix of A. Pursuing the analogy, mnp matrices are to the set covering problem what minimally imperfect matrices are to the set packing problem.

The 0,1 matrix A is *ideal* if the polyhedron $\{x \geq 0 : Ax \geq e\}$ is integral (Lehman [9]). If A is ideal, then so are all its minors [16]. The following result is a consequence of Lehman's work [10].

Theorem 1. *If A has the packing property, then A is ideal.*

The converse is not true, however. A famous example is the matrix Q_6 with 4 rows and 6 columns comprising all 0,1 column vectors with two 0's and two 1's. It is ideal but it does not pack. This is in contrast to Lovász's theorem [11] stating that A is perfect if and only if the polytope $\{x \geq 0 : Ax \leq e\}$ is integral.

The 0,1 matrix A has the *Max-Flow Min-Cut property* (or simply MFMC property) if the linear system $Ax \geq e$, $x \geq 0$ is totally dual integral (Seymour [16]). Specifically, let

$$\tau(A, w) = \min \left\{ wx : Ax \geq e, \ x \in \{0,1\}^n \right\},$$

$$\nu(A, w) = \max \left\{ \sum_1^m y_i : yA \leq w, \ y \in \{0,1\}^m \right\}.$$

A has the MFMC property if $\tau(A, w) = \nu(A, w)$ for all $w \in Z_+^n$. Setting $w_j = 0$ corresponds to deleting column j and setting $w_j = +\infty$ to contracting j. So, if A has the MFMC-property, then A has the packing property. Conforti and Cornuéjols [3] conjecture that the converse is also true.

Conjecture 1. A clutter has the packing property if and only if it has the MFMC property.

This conjecture for the packing property is the analog of the following version of Lovász's theorem [11]: A $0, 1$ matrix A is perfect if and only if the linear system $Ax \leq e$, $x \geq 0$ is totally dual integral.

In this paper, our first result is that this conjecture holds for diadic clutters. A clutter is *diadic* if its edges intersect its transversals in at most two vertices (Ding [6]). In fact, we show the stronger result:

Theorem 2. *A diadic clutter is ideal if and only if it has the MFMC property.*

A clutter is said to be *minimally non ideal* (mni) if it is not ideal but all its minors are ideal. Theorem 1 is equivalent to saying that mni clutters do not pack. Therefore mnp clutters fall into two distinct classes namely:

Remark 1. A minimally non packing clutter is either ideal or mni.

Next we consider ideal mnp clutters. Seymour [16] showed that Q_6 is the only ideal mnp clutter which is binary (a clutter is *binary* if its edges have an odd intersection with its transversals). Aside from Q_6, only one ideal mnp clutter was known prior to this work, due to Schrijver [14]. We construct an infinite family of such mnp clutters (see Appendix). The clutter Q_6, Schrijver's example and those in our infinite class all satisfy $\tau(\mathcal{C}) = 2$. Our next result is that all ideal mnp clutters with $\tau(\mathcal{C}) = 2$ share strong structural properties with Q_6.

Theorem 3. *Every ideal mnp clutter $\mathcal{C}$ with $\tau(\mathcal{C}) = 2$ has the Q_6 property, i.e. $A(\mathcal{C})$ has 4 rows such that every column restricted to this set of rows contains two 0's and two 1's and, furthermore, each of the 6 such possible 0,1 vectors occurs at least once.*

We make the following conjecture and we prove later that it implies Conjecture 1.

Conjecture 2. If $\mathcal{C}$ is an ideal mnp clutter, then $\tau(\mathcal{C}) = 2$.

The *blocker* $b(\mathcal{C})$ of a clutter $\mathcal{C}$ is the clutter with $V(\mathcal{C})$ as vertex set and the transversals of $\mathcal{C}$ as edge set. For $I_d, I_c \subseteq V(\mathcal{C})$ with $I_d \cap I_c = \emptyset$, it is well known and easy to derive that $b(\mathcal{C} \setminus I_d/I_c) = b(\mathcal{C})/I_d \setminus I_c$.

We now consider minimally non ideal mnp clutters. The clutter $\mathcal{J}_t$, for $t \geq 2$ integer, is given by $V(\mathcal{J}_t) = \{0, \dots, t\}$ and $E(\mathcal{J}_t) = \{1, \dots, t\}, \{0, 1\}, \{0, 2\}, \dots,$ $\{0, t\}$. Given a mni matrix A, let $\tilde{x}$ be any vertex of $\{x \geq 0 : Ax \geq e\}$ with fractional components. A maximal row submatrix $\bar{A}$ of A for which $\bar{A}\tilde{x} = e$ is called a *core* of A. The next result is due to Lehman [10] (see also Padberg [13], Seymour [17]).

Theorem 4. *Let A be a mni matrix, $B = b(A)$, $r = \tau(B)$ and $s = \tau(A)$. Then*

(i) *A (resp. B) has a unique core $\bar{A}$ (resp. $\bar{B}$).*
(ii) *$\bar{A}, \bar{B}$ are square matrices.*

Moreover, either $A = A(\mathcal{J}_t)$, $t \geq 2$, or the rows and columns of $\bar{A}$ can be permuted so that

(iii) *$\bar{A}\bar{B}^T = J + (rs - n)I$.*

Here J denotes a square matrix filled with ones and I the identity matrix. Only three cores with $rs = n + 2$ are known and none with $rs \geq n + 3$. Nevertheless Cornuéjols and Novick [5] have constructed more than one thousand mni matrices from a single core with $rs = n + 2$. An *odd hole* C_k^2 is a clutter with $k \geq 3$ odd, $V(C_k^2) = \{1, \dots k\}$ and $E(C_k^2) = \{\{1, 2\}, \{2, 3\}, \dots, \{k-1, k\}, \{k, 1\}\}$. Odd holes and their blockers are mni with $rs = n + 1$ and Luetolf and Margot [12] give dozens of additional examples of cores with $rs = n + 1$ and $n \leq 17$. We prove the following theorem.

Theorem 5. *Let $A \neq A(\mathcal{J}_t)$ be a mni matrix. If A is minimally non packing, then $rs = n + 1$.*

We conjecture that the condition $rs = n + 1$ is also sufficient.

Conjecture 3. Let $A \neq A(\mathcal{J}_t)$ be a mni matrix. Then A is minimally non packing if and only if $rs = n + 1$.

Using a computer program, we were able to verify this conjecture for all known mni matrices with $n \leq 14$.

A clutter is *minimally non MFMC* if it does not have the MFMC property but all its minors do. Conjecture 1 states that these are exactly the mnp clutters. Although we cannot prove this conjecture, the next proposition shows that a tight link exists between minimally non MFMC and mnp clutters. The clutter $\mathcal{D}$ obtained by *replicating* element $j \in V(\mathcal{C})$ of $\mathcal{C}$ is defined as follows: $V(\mathcal{D}) = V(\mathcal{C}) \cup \{j'\}$ where $j' \notin V(\mathcal{C})$, and

$$E(\mathcal{D}) = E(\mathcal{C}) \cup \{S - \{j\} \cup \{j'\} : j \in S \in E(\mathcal{C})\}.$$

Element j' is called a *replicate* of j. Let e_j denote the jth unit vector.

Remark 2. $\mathcal{D}$ packs if and only if $\tau(\mathcal{C}, e + e_j) = \nu(\mathcal{C}, e + e_j)$.

Proposition 1. *Let C be a minimally non MFMC clutter. We can construct a minimally non packing clutter $\mathcal{D}$ by replicating elements of $V(C)$.*

Proof. Let $w \in Z_+^n$ be chosen such that $\tau(C, w) > \nu(C, w)$ and $\tau(C, w') = \nu(C, w')$ for all $w' \in Z_+^n$ with $w' \leq w$ and $w'_j < w_j$ for at least one j. Note that $w_j > 0$ for all j, since otherwise some deletion minor of C does not have the MFMC property. Construct $\mathcal{D}$ by replicating $w_j - 1$ times every element $j \in V(C)$. By Remark 2, $\mathcal{D}$ does not pack. Let $\mathcal{D}' = \mathcal{D} \setminus I_d / I_c$ be any minor of $\mathcal{D}$. If j or one of its replicates is in I_c then we can assume that j and all its replicates are in I_c. Then $\mathcal{D}'$ is a replication of a minor C' of C/j. Since C' has the MFMC property, $\mathcal{D}'$ packs by Remark 2. Thus we can assume $I_c = \emptyset$. By the choice of w and Remark 2, if $I_d \neq \emptyset$ then $\mathcal{D}'$ packs. $\qquad\square$

Proposition 1 can be used to show that if every ideal mnp clutter C satisfies $\tau(C) = 2$ then the packing property and the MFMC property are the same.

Proposition 2. *Conjecture 2 implies Conjecture 1.*

Proof. Suppose there is a minimally non MFMC clutter C that packs. By Theorem 1, C is ideal. Then by Proposition 1, there is a mnp clutter $\mathcal{D}$ with a replicated element j. Furthermore, $\mathcal{D}$ is ideal. Using Conjecture 2, $2 = \tau(\mathcal{D}) \leq \tau(\mathcal{D}/j)$. Since $\mathcal{D}/j$ packs, there are sets $S_1, S_2 \in E(\mathcal{D})$ with $S_1 \cap S_2 = \{j\}$. Because j is replicated in $\mathcal{D}$, we have a set $S'_1 = S_1 \cup \{j'\} - \{j\}$. Remark that $j' \notin S_2$. But then $S'_1 \cap S_2 = \emptyset$, hence $\mathcal{D}$ packs, a contradiction. $\qquad\square$

Finally, we introduce a new class of clutters called weakly binary. They can be viewed as a generalization of binary and of balanced clutters. (A 0,1 matrix is *balanced* if it does not have $A(C_k^2)$ as a submatrix, $k \geq 3$ odd, where as above C_k^2 denotes an odd hole. See [4] for a survey of balanced matrices). We say that a clutter C has an odd hole C_k^2 if $A(C_k^2)$ is a submatrix of $A(C)$. An odd hole C_k^2 of C is said to have a *non intersecting set* if $\exists S \in E(C)$ such that $S \cap V(C_k^2) = \emptyset$. A clutter is *weakly binary* if, in C and all its minors, all odd holes have non intersecting sets.

Theorem 6. *Let C be weakly binary and minimally non MFMC. Then C is ideal.*

Note that, when C is binary, this theorem is an easy consequence of Seymour's theorem saying that a binary clutter has the MFMC property if and only if it does not have Q_6 as a minor [16]. Observe also that Theorem 6 together with Conjecture 2, Proposition 2, and Theorem 3, would imply that a weakly binary clutter has the MFMC property if and only if it does not contain a minor with the Q_6 property.

References

1. C. Berge. Färbung von Graphen deren sämtliche bzw. deren ungerade Kreize starr sind (Zusammenfassung), Wisenschaftliche Zeitschritch, Martin Luther Universität Halle-Wittenberg, Mathematisch-Naturwissenschaftliche Reihe, 114–115, 1960.
2. W. G. Bridges and H. J. Ryser. Combinatorial designs and related systems. *J. Algebra*, 13:432–446, 1969.
3. M. Conforti and G. Cornuéjols. Clutters that pack and the max-flow min-cut property: A conjecture. In W. R. Pulleyblank and F. B. Shepherd, editors, *The Fourth Bellairs Workshop on Combinatorial Optimization*, 1993.
4. M. Conforti, G. Cornuéjols, A. Kapoor, and K. Vušković. Balanced matrices. In J. R. Birge and K. G Murty, editors, *Math. Programming, State of the Art 1994*, pages 1–33, 1994.
5. G. Cornuéjols and B. Novick. Ideal 0, 1 matrices. *J. Comb. Theory Ser. B*, 60:145–157, 1994.
6. G. Ding. Clutters with $\tau_2 = 2\tau$. *Discrete Math.*, 115:141–152, 1993.
7. J. Edmonds and R. Giles. A min-max relation for submodular functions on graphs. *Annals of Discrete Math.*, 1:185–204, 1977.
8. B. Guenin. Packing and covering problems. Thesis proposal, GSIA, Carnegie Mellon University, 1997.
9. A. Lehman. On the width-length inequality. *Mathematical Programming*, 17:403–417, 1979.
10. A. Lehman. On the width-length inequality and degenerate projective planes. In W. Cook and P.D. Seymour, editors, *Polyhedral Combinatorics, DIMACS Series in Discrete Math. and Theoretical Computer Science, Vol. 1*, pages 101–105, 1990.
11. L. Lovász. Normal hypergraphs and the perfect graph conjecture. *Discrete Math.* 2:253–267, 1972.
12. C. Luetolf and F. Margot. A catalog of minimally nonideal matrices. *Mathematical Methods of Operations Research*, 1998. To appear.
13. M. W. Padberg. Lehman's forbidden minor characterization of ideal $0-1$ matrices. *Discrete Math.*, 111:409–420, 1993.
14. A. Schrijver. A counterexample to a conjecture of Edmonds and Giles. *Discrete Math.*, 32:213–214, 1980.
15. A. Schrijver. *Theory of Linear and Integer Programming*, Wiley, 1986.
16. P. D. Seymour. The matroids with the max-flow min-cut property. *J. Comb. Theory Ser. B*, 23:189–222, 1977.
17. P. D. Seymour. On Lehman's width-length characterization. In W. Cook and P. D. Seymour, editors, *Polyhedral Combinatorics, DIMACS Series in Discrete Math. and Theoretical Computer Science, Vol. 1*, pages 107–117, 1990.

Appendix

We construct ideal minimally non packing clutters $\mathcal{C}$ with $\tau(\mathcal{C}) = 2$. By Theorem 3, these clutters have the Q_6 property. Thus $V(\mathcal{C})$ can be partitioned into $I_1, \ldots, I_6$ and there exist edges $S_1, \ldots, S_4$ in $\mathcal{C}$ of the form:

$$S_1 = I_1 \cup I_3 \cup I_5, \qquad S_2 = I_1 \cup I_4 \cup I_6,$$
$$S_3 = I_2 \cup I_4 \cup I_5, \qquad S_4 = I_2 \cup I_3 \cup I_6.$$

Without loss of generality we can reorder the vertices in $V(\mathcal{C})$ so that elements in I_k preceed elements in I_p when $k < p$.

Given a set $\mathcal{P}$ of p elements, let $\mathcal{H}_p$ denote the $((2^p - 1) \times p)$ matrix whose rows are the characteristic vectors of the nonempty subsets of $\mathcal{P}$, and let $\mathcal{H}_p^*$ be its complement, i.e. $\mathcal{H}_p + \mathcal{H}_p^* = J$.

For each $r, t \geq 1$ let $|I_1| = |I_2| = r$, $|I_3| = |I_4| = t$ and $|I_5| = |I_6| = 1$. We call $Q_{r,t}$ the clutter corresponding to the matrix

$$
A(Q_{r,t}) =
\begin{array}{c}
\begin{array}{cccccc} I_1 & I_2 & I_3 & I_4 & I_5 & I_6 \end{array} \\
\left[
\begin{array}{cc|c|c|c|c}
\mathcal{H}_r & \mathcal{H}_r^* & J & \mathbf{0} & 1 & 0 \\
\mathcal{H}_r^* & \mathcal{H}_r & \mathbf{0} & J & 1 & 0 \\
J & \mathbf{0} & \mathcal{H}_t^* & \mathcal{H}_t & 0 & 1 \\
\mathbf{0} & J & \mathcal{H}_t & \mathcal{H}_t^* & 0 & 1
\end{array}
\right]
\end{array}
$$

where J denotes a matrix filled with ones. The rows are partitioned into four sets that we denote respectively by $T(3,5)$, $T(4,5)$, $T(1,6)$, $T(2,6)$. The indices k, l for a given family indicate that the set $I_k \cup I_l$ is contained is every element of the family. Note that the edge S_1 occurs in $T(3,5)$, S_2 in $T(1,6)$, S_3 in $T(4,5)$ and S_4 in $T(2,6)$.

Since $\mathcal{H}_1$ contains only one row, we have $Q_{1,1} = Q_6$ and $Q_{2,1}$ is given by

$$
A(Q_{2,1}) =
\left[
\begin{array}{cc|cc|cc|cc}
1 & 1 & 0 & 0 & 1 & 0 & 1 & 0 \\
1 & 0 & 0 & 1 & 1 & 0 & 1 & 0 \\
0 & 1 & 1 & 0 & 1 & 0 & 1 & 0 \\
\hline
0 & 0 & 1 & 1 & 0 & 1 & 1 & 0 \\
1 & 0 & 0 & 1 & 0 & 1 & 1 & 0 \\
0 & 1 & 1 & 0 & 0 & 1 & 1 & 0 \\
\hline
1 & 1 & 0 & 0 & 0 & 1 & 0 & 1 \\
0 & 0 & 1 & 1 & 1 & 0 & 0 & 1
\end{array}
\right]
\begin{array}{l}
\\ T(3,5) \\ \\ \\ T(4,5) \\ \\ T(1,6) \\ T(2,6)
\end{array}
$$

Proposition 3. *For all $r, t \geq 1$, the clutter $Q_{r,t}$ is ideal and minimally non packing.*

The clutter $\mathcal{D}$ obtained by *duplicating* element $j \in V(\mathcal{C})$ of $\mathcal{C}$ is defined by: $V(\mathcal{D}) = V(\mathcal{C}) \cup \{j'\}$ where $j' \notin V(\mathcal{C})$ and $E(\mathcal{D}) = \{S : j \notin S \in E(\mathcal{C})\} \cup \{S \cup \{j'\} : j \in S \in E(\mathcal{C})\}$. Let $\alpha(k)$ be the mapping defined by: $\alpha(1) = 2$, $\alpha(2) = 1$, $\alpha(3) = 4$, $\alpha(4) = 3$, $\alpha(5) = 6$, $\alpha(6) = 5$.

Suppose that, for $k \in \{1, .., 6\}$, we have that I_k contains a single element $j \in V(\mathcal{C})$. Then j belongs to exactly two of $S_1, \ldots, S_4$. These two edges are of the form $\{j\} \cup I_r \cup I_t$ and $\{j\} \cup I_{\alpha(r)} \cup I_{\alpha(t)}$. We can construct a new clutter $\mathcal{C} \otimes j$ by duplicating element j in $\mathcal{C}$ and including in $E(\mathcal{C} \otimes j)$ the edges:

$$\{j\} \cup I_{\alpha(j)} \cup I_r \cup I_t,$$

$$\{j'\} \cup I_{\alpha(j)} \cup I_{\alpha(r)} \cup I_{\alpha(t)}. \tag{2}$$

Since the $\otimes$ construction is commutative we denote by $\mathcal{C} \otimes \{k_1, \ldots, k_s\}$ the clutter $(\mathcal{C} \otimes k_1) \ldots \otimes k_s$. For Q_6, we have $I_1 = \{1\} = S_1 \cap S_2$ and $\{1\} \cup I_{\alpha(1)} \cup I_3 \cup I_5 = \{1, 2, 3, 5\}$ and finally $\{1'\} \cup I_{\alpha(1)} \cup I_{\alpha(3)} \cup I_{\alpha(5)} = \{1', 2, 4, 6\}$. Thus

$$
A(Q_6 \otimes 1) = \left[
\begin{array}{cc|c|cc|c|c}
1 & 1 & 0 & 1 & 0 & 1 & 0 \\
1 & 1 & 0 & 0 & 1 & 0 & 1 \\
0 & 0 & 1 & 0 & 1 & 1 & 0 \\
0 & 0 & 1 & 1 & 0 & 0 & 1 \\
\hline
1 & 0 & 1 & 1 & 0 & 1 & 0 \\
0 & 1 & 1 & 0 & 1 & 0 & 1
\end{array}
\right]
$$

Proposition 4. *Any clutter obtained from Q_6 and the $\otimes$ construction is ideal and minimally non packing.*

The clutter $Q_6 \otimes \{1, 3, 5\}$ was found by Schrijver [14] as a counterexample to a conjecture of Edmonds and Giles on dijoins. Prior to this work, Q_6 and $Q_6 \otimes \{1, 3, 5\}$ were the only known ideal mnp clutters. Eleven clutters can be obtained using Proposition 4. In fact it can be shown [8] that this proposition remains true if we replace Q_6 by $Q_{r,t}$. There are also examples that do not fit any of the above constructions, as shown by the following ideal mnp clutter.

$$
A(\mathcal{C}) = \left[
\begin{array}{cc|cc|cc|c}
1 & 1 & 0 & 0 & 1 & 0 & 1 & 0 \\
1 & 1 & 0 & 0 & 0 & 1 & 0 & 1 \\
0 & 0 & 1 & 1 & 1 & 0 & 0 & 1 \\
0 & 0 & 1 & 1 & 0 & 1 & 1 & 0 \\
\hline
1 & 0 & 1 & 1 & 1 & 0 & 1 & 0 \\
0 & 1 & 1 & 0 & 0 & 1 & 0 & 1 \\
\hline
0 & 1 & 1 & 0 & 1 & 0 & 0 & 1 \\
1 & 1 & 0 & 1 & 0 & 1 & 1 & 0
\end{array}
\right]
$$

A Characterization of Weakly Bipartite Graphs

Bertrand Guenin

Graduate School of Industrial Administration
Carnegie Mellon University, Pittsburgh, PA 15213, USA

Abstract. A labeled graph is said to be weakly bipartite if the clutter of its odd cycles is ideal. Seymour conjectured that a labeled graph is weakly bipartite if and only if it does not contain a minor called an odd K_5. An outline of the proof of this conjecture is given in this paper.

1 Introduction

Let $\mathcal{G} = (V, E)$ be a graph and $\Sigma \subseteq E$. Edges in Σ are called *odd* and edges in $E - \Sigma$ are called *even*. The pair $(\mathcal{G}, \Sigma)$ is called a *labeled graph*. Given a subgraph $\mathcal{H}$ of $\mathcal{G}$, $V(\mathcal{H})$ denotes the set of vertices of $\mathcal{H}$, and $E(\mathcal{H})$ the set of edges of $\mathcal{H}$. A subset $L \subseteq E(\mathcal{G})$ is *odd* (resp. *even*) if $|L \cap \Sigma|$ is odd (resp. even). A *cycle* of $\mathcal{G}$ is a connected subgraph of $\mathcal{G}$ with all degrees equal to two.

A labeled graph $(\mathcal{G}, \Sigma)$ is said to be *weakly bipartite* if the following polyhedron Q is integral (i.e. all its extreme points are integral):

$$Q = \left\{ x \in \Re_+^{|E|} : \sum_{i \in C} x_i \geq 1, \text{ for all odd cycles } C \text{ of } (\mathcal{G}, \Sigma) \right\} \qquad (1)$$

See Gerards [7] for a recent survey on weakly bipartite graphs and connexions with multicommodity flows. Particularly interesting is the case where $\Sigma = E(\mathcal{G})$. Let $\hat{x}$ be any $0, 1$ extreme point of Q. Then $\hat{x}$ is the incidence vector of a set of edges which intersect every odd cycle of $\mathcal{G}$. In other words $e - \hat{x}$ is the incidence vector of a bipartite subgraph of $\mathcal{G}$. Let $w \in \Re_+^{|E|}$ be weights for the edges of $\mathcal{G}$ and let $\bar{x}$ be a solution to

$$\min \left\{ wx : x \in Q \cap \{0, 1\}^{|E|} \right\}. \qquad (2)$$

Then $e - \bar{x}$ is a solution to the Weighted Max-Cut problem. This problem is known to be NP-Hard even in the unweighted case [10]. Note, weakly bipartite graphs are precisely those graphs for which the integrality constraints in (2) can be dropped.

Weakly bipartite graphs $\mathcal{G}$ with $\Sigma = E(\mathcal{G})$ were introduced by Grötschel and Pulleyblank [8]. They showed that the optimization problem $\min\{wx : x \in Q\}$ can be solved in polynomial time.

Barahona [1] proved that planar graphs are weakly bipartite. In fact, Fonlupt, Mahjoub and Uhry [5] showed that all graphs which are not contractible to K_5

R. E. Bixby, E. A. Boyd, and R. Z. Ríos-Mercado (Eds.): IPCO VI
LNCS 1412, pp. 9–22, 1998. © Springer–Verlag Berlin Heidelberg 1998

are weakly bipartite. This is closely related to an earlier result by Barahona [2] on the cut polytope. Note, this does not yield a characterization of weakly bipartite graphs. Consider the graph obtained from K_5 by replacing one edge by two consecutive edges. This graph is weakly bipartite and contractible to K_5.

Following Gerards [6], we will define a number of operations on labeled graphs, which maintain the weak bipartite property. Given $U \subseteq V(\mathcal{G})$, the cut $\{(u, v) : u \in U, v \notin U\}$ is denoted by $\delta(U)$. Given two sets S_1, S_2, the symmetric difference $(S_1 \cup S_2) - (S_1 \cap S_2)$ is denoted by $S_1 \triangle S_2$. The labeled graph obtained by replacing Σ by $\Sigma \triangle \delta(U)$ in $(\mathcal{G}, \Sigma)$ is called a *relabeling* of $(\mathcal{G}, \Sigma)$. Since $\delta(U)$ intersects every cycle an even number of times we readily obtain that:

Remark 1. $(\mathcal{G}, \Sigma)$ and $(\mathcal{G}, \Sigma \triangle \delta(U))$ have the same set of odd cycles.

$(\mathcal{G}, \Sigma) \setminus e$ denotes a labeled graph $(\mathcal{G}', \Sigma')$, where $\Sigma' = \Sigma - \{e\}$ and $\mathcal{G}'$ is obtained by removing edge e from $\mathcal{G}$. $(\mathcal{G}, \Sigma)/e$ denotes the labeled graph obtained as follows: (1) if e is odd (i.e. $e \in \Sigma$) then find a relabeling of $(\mathcal{G}, \Sigma)$ such that e is even, (2) contract edge e in $\mathcal{G}$ (remove edge e in $\mathcal{G}$ and identify both of its endpoints). $(\mathcal{G}, \Sigma) \setminus e$ is called a *deletion minor* and $(\mathcal{G}, \Sigma)/e$ a *contraction minor*. Let Q be the polyhedron associated with $(\mathcal{G}, \Sigma)$, see (1). It can be readily shown (see for example the introduction of [18]) that deleting edge e corresponds to projecting Q onto a lower subspace and contracting e corresponds to setting x_e to zero. A labeled graph $(\mathcal{H}, \theta)$ is called a minor of $(\mathcal{G}, \Sigma)$, if it can be obtained as a sequence of relabelings, deletions and contractions. It follows from Remark 1 and the above observations that:

Remark 2. If $(\mathcal{G}, \Sigma)$ is weakly bipartite then so are all its minors.

$(\mathcal{H}, \theta)$ is called a *proper minor* of $(\mathcal{G}, \Sigma)$ if it is a minor of $(\mathcal{G}, \Sigma)$ and $|E(\mathcal{H})| < |E(\mathcal{G})|$. An *odd* K_5, denoted by $\widetilde{K_5}$, is the complete graph on 5 vertices where all edges are labeled odd. For the polyhedron Q associated with $\widetilde{K_5}$, the 10 constraints corresponding to the triangles (the odd cycles of length three) define a fractional point $(\frac{1}{3}, \ldots, \frac{1}{3})$ of Q. Thus $\widetilde{K_5}$ is not weakly bipartite. Seymour [18],[19] predicted, as part of a more general conjecture on binary clutters (see Sec. 2.3) that:

Conjecture 1. $(\mathcal{G}, \Sigma)$ is weakly bipartite if and only if it has no $\widetilde{K_5}$ minor.

A labeled graph is said to be *minimally non weakly bipartite* if it is not weakly bipartite but all its proper minors are. An outline to the proof of the following theorem (which is equivalent to Conjecture 1) is given in this paper.

Theorem 1. *Every minimally non weakly bipartite graph is a relabeling of $\widetilde{K_5}$.*

Section 2 introduces some basic notions on clutters, before stating a theorem by Lehman [13] on minimally non ideal (mni) clutters. The section concludes by deriving key properties of binary mni clutters. Section 3 gives an outline of the proof of Theorem 1. A complete proof of this result can be found in [9].

2   Clutters

2.1   Elementary Definitions

A clutter $\mathcal{A}$ is a pair $(E(\mathcal{A}), \Omega(\mathcal{A}))$ where $E(\mathcal{A})$ is a finite set and $\Omega(\mathcal{A})$ is a family of subsets of $E(\mathcal{A})$, say $\{S_1, \ldots, S_m\}$, with the property that $S_i \subseteq S_j$ implies $S_i = S_j$. $M(\mathcal{A})$ denotes a $0, 1$ matrix whose rows are the incidence vectors of the elements of $\Omega(\mathcal{A})$. A clutter $\mathcal{A}$ is said to be *ideal* if the polyhedron $Q(\mathcal{A}) = \{x \geq 0 : M(\mathcal{A})x \geq e\}$ is integral. We say that $\mathcal{A}$ is the *clutter of odd cycles* of a labeled graph $(\mathcal{G}, \Sigma)$ if $E(\mathcal{A}) = E(\mathcal{G})$ and the elements of $\Omega(\mathcal{A})$ are the odd cycles of $(\mathcal{G}, \Sigma)$. Thus a labeled graph is weakly bipartite when the clutter of its odd cycles is ideal.

Given a clutter $\mathcal{A}$ and $i \in E(\mathcal{A})$, the *contraction* $\mathcal{A}/i$ and *deletion* $\mathcal{A} \setminus i$ are clutters defined as follows: $E(\mathcal{A}/i) = E(\mathcal{A} \setminus i) = E(\mathcal{A}) - \{i\}$, $\Omega(\mathcal{A}/i)$ is the set of inclusion-wise minimal elements of $\{S - \{i\} : S \in \Omega(\mathcal{A})\}$ and, $\Omega(\mathcal{A} \setminus i) = \{S : i \notin S \in \Omega(\mathcal{A})\}$. Contractions and deletions can be performed sequentially, and the result does not depend on the order. A clutter $\mathcal{B}$ obtained from $\mathcal{A}$ by a set of deletions I_d and a set of contractions I_c, where $I_c \cap I_d = \emptyset$ is called a *minor* of $\mathcal{A}$ and is denoted by $\mathcal{A} \setminus I_d / I_c$. We say that $\mathcal{B}$ is a *contraction minor* (resp. *deletion minor*) if $I_d = \emptyset$ (resp. $I_c = \emptyset$). The following is well known [13]:

Remark 3. If $\mathcal{A}$ is ideal then so are its minors.

We saw in Remark 1 that relabeling a labeled graph leaves the clutter of odd cycles unchanged. Contractions and deletions on a labeled graph (as defined in Sec. 1) are equivalent to the corresponding operations on the clutter of odd cycles.

Remark 4. Let $\mathcal{A}$ be the clutter of odd cycles of $(\mathcal{G}, \Sigma)$. Then

- $\mathcal{A}/e$ is the clutter of odd cycles of $(\mathcal{G}, \Sigma)/e$ and
- $\mathcal{A} \setminus e$ is the clutter of odd cycles of $(\mathcal{G}, \Sigma) \setminus e$.

Given a clutter $\mathcal{A}$, the clutter $b(\mathcal{A})$ is called the *blocker* of $\mathcal{A}$ and is defined as follows: $E(b(\mathcal{A})) = E(\mathcal{A})$ and $\Omega(b(\mathcal{A}))$ is the set of inclusion-wise minimal elements of $\{U : C \cap U \neq \emptyset, \forall C \in \Omega(\mathcal{A})\}$. It is well known that $b(\mathcal{A} \setminus I_c / I_d) = b(\mathcal{A})/I_c \setminus I_d$ and that $b(b(\mathcal{A})) = \mathcal{A}$ [17]. If $\mathcal{A}$ is the clutter of odd cycles of a labeled graph $(\mathcal{G}, \Sigma)$, then the elements of $\Omega(b(\mathcal{A}))$ are of the form $\delta(U) \triangle \Sigma$, where $\delta(U)$ is a cut of $\mathcal{G}$ [6].

2.2   Minimally Non Ideal Clutters

A clutter $\mathcal{A}$ is called *minimally non ideal* (mni) if it is not ideal but all its proper minors are ideal. Because of Remark 4, the clutter of odd cycles of a minimally non weakly bipartite labeled graph is mni. In this section we review properties of mni clutters.

The clutter $\mathcal{J}_t$, for $t \geq 2$ integer, is given by $E(\mathcal{J}_t) = \{0, \ldots, t\}$ and $\Omega(\mathcal{J}_t) = \{\{1, \ldots, t\}, \{0, 1\}, \{0, 2\}, \ldots, \{0, t\}\}$. The cardinality of the smallest element of

$\Omega\left(b(\mathcal{A})\right)$ is denoted by $\tau(\mathcal{A})$. In this section we consider the matrix representation $A = M(\mathcal{A})$ of a clutter $\mathcal{A}$. We say that a matrix $M(\mathcal{A})$ is mni when the clutter $\mathcal{A}$ is mni. The blocker of $b\left(M(\mathcal{A})\right)$ is the matrix $M\left(b(\mathcal{A})\right)$ and $\tau(\mathcal{A})$ is the smallest number of non-zero entries in any row of $b\left(M(\mathcal{A})\right)$.

Given a mni matrix A, let $\tilde{x}$ be any extreme point of $Q(A) = \{x \geq 0 : Ax \geq e\}$ with fractional components. A maximal row submatrix $\bar{A}$ of A for which $\bar{A}\tilde{x} = e$ is called a *core* of A. Two matrices are said to be *isomorphic* if one can be obtained from the other by a sequence of permutations of the rows and columns. The next result is by Lehman [13] (see also Padberg [16], Seymour [20]).

Theorem 2. *Let A be a mni matrix. Then $B = b(A)$ is mni. Let $r = \tau(B)$ and $s = \tau(A)$. Either A is isomorphic to $M(\mathcal{J}_t)$ or*

(i) *A (resp. B) has a unique core $\bar{A}$ (resp. $\bar{B}$).*
(ii) *$\bar{A}, \bar{B}$ are square matrices.*

Moreover, the rows and columns of $\bar{A}$ can be permuted so that

(iii) *$\bar{A}\bar{B}^T = J + (rs - n)I$, where $rs - n \geq 1$.*

Here J denotes a square matrix filled with ones and I the identity matrix. Also e is the vector of all ones, e_j is the jth unit vector, and $\bar{B}_{.j}$ denotes column j of $\bar{B}$. The following is a special case of a result of Bridges and Ryser [3]:

Theorem 3. *Let $\bar{A}, \bar{B}$ be matrices satisfying (ii),(iii) of Theorem 2.*

(i) *Columns and rows of $\bar{A}$ (resp. $\bar{B}$) have exactly r (resp. s) ones.*
(ii) *$\bar{A}\bar{B}^T = \bar{A}^T\bar{B}$*
(iii) *$\bar{A}^T\bar{B}_{.j} = e + (rs - n)e_j$.*
(iv) *Let j be the index of any column of A. Let $C_1, \ldots, C_s$ (resp. $U_1, \ldots, U_r$) be the characteristic sets of the rows of $\bar{A}$ (resp. $\bar{B}$) whose indices are given by the characteristic set of column j of $\bar{B}$ (resp. $\bar{A}$). Then $C_1, \ldots, C_s$ (resp. $U_1, \ldots, U_r$) intersect only in $\{j\}$ and exactly $q = rs - n + 1$ of these sets contain j.*

Note that in the last theorem, Property (ii) implies (iii) that in turn implies (iv). Because of Theorem 2 and Theorem 3(i) the fractional point $\tilde{x}$ must be $(\frac{1}{r}, \ldots, \frac{1}{r})$. The next remark follows from the fact that $\bar{A}$ is a maximal row submatrix of A for which $\bar{A}\tilde{x} = e$.

Remark 5. Rows of A which are not rows of $\bar{A}$ have at least $r + 1$ non-zero entries. Similarly, rows of B which are not in $\bar{B}$ have at least $s + 1$ ones.

Let $\mathcal{A} \neq \mathcal{J}_t$ be a mni clutter with $A = M(\mathcal{A})$. $\bar{A}$, with $M(\bar{\mathcal{A}}) = \bar{A}$, denotes the core of $\mathcal{A}$. Let $\mathcal{B}$ be the blocker of $\mathcal{A}$ and $\bar{B}$ the core of $\mathcal{B}$. Consider the element $C \in \Omega(\bar{A})$ (resp. $U \in \Omega(\bar{B})$) which corresponds to the ith row of $\bar{A}$ (resp. $\bar{B}$). By Theorem 2(iii), C intersects every element of $\Omega(\bar{B})$ exactly once except for U which is intersected $q = rs - n + 1 \geq 2$ times. We call U the *mate* of C. Thus every element of $\Omega(\bar{A})$ is paired with an element of $\Omega(\bar{B})$. Notice that Theorem 3(iv) implies the following result.

Remark 6. Let $\mathcal{A}$ be a mni clutter distinct from $\mathcal{J}_t$ and consider $C_1, C_2 \in \Omega(\bar{A})$ with $i \in C_1 \cap C_2$. The mates U_1, U_2 of C_1, C_2 satisfy $U_1 \cap U_2 \subseteq \{i\}$.

2.3   Binary Clutters

A clutter $\mathcal{A}$ is said to be binary if for any three sets $S_1, S_2, S_3 \in \Omega(\mathcal{A})$, the set $S_1 \triangle S_2 \triangle S_3$ contains a set of $\Omega(\mathcal{A})$. Lehman [11] showed (see also Seymour [17]):

Theorem 4. $\mathcal{A}$ *is binary if and only if for any* $C \in \Omega(\mathcal{A})$ *and* $U \in \Omega\left(b(\mathcal{A})\right)$ *we have* $|C \cap U|$ *odd.*

Thus in particular if $\mathcal{A}$ is binary then so is its blocker. The following is easy, see for example [6].

Proposition 1. *Let* $(\mathcal{G}, \Sigma)$ *be a labeled graph. Then the clutter of odd cycles of* $(\mathcal{G}, \Sigma)$ *is binary.*

2.4   Minimally Non Ideal Binary Clutters

Note that the blocker of $\mathcal{J}_t$ is $\mathcal{J}_t$ itself. We therefore have $\{1, 2\} \in E(\mathcal{J}_t)$ and $\{1, 2\} \in E\left(b(\mathcal{J}_t)\right)$. It follows by Theorem 4 that $\mathcal{J}_t$ is not binary. The clutter $\mathcal{F}_7$ is defined as follows: $E(\mathcal{F}_7) = \{1, \ldots, 7\}$ and

$$\Omega(\mathcal{F}_7) = \{\{1, 3, 5\}, \{1, 4, 6\}, \{2, 4, 5\}, \{2, 3, 6\}, \{1, 2, 7\}, \{3, 4, 7\}, \{5, 6, 7\}\}.$$

The clutter of odd cycles of K_5 is denoted by $\mathcal{O}_{K_5}$. Conjecture 1 is part of a more general conjecture by Seymour on minimally non ideal binary clutters. See [18] p. 200 and [19] (9.2), (11.2).

Conjecture 2. If $\mathcal{A}$ is a minimally non ideal binary clutter, then $\mathcal{A}$ is either $\mathcal{F}_7$, $\mathcal{O}_{K_5}$ or $b(\mathcal{O}_{K_5})$.

Since we can readily check that $\mathcal{F}_7$ and $b(\mathcal{O}_{K_5})$ are not clutters of odd cycles this conjecture implies Conjecture 1. Next are two results on mni binary clutters.

Proposition 2. *Let* $\mathcal{A}$ *be a mni binary clutter and* $C_1, C_2 \in \Omega(\bar{\mathcal{A}})$. *If* $C \subseteq C_1 \cup C_2$ *and* $C \in \Omega(\mathcal{A})$ *then either* $C = C_1$ *or* $C = C_2$.

Proof. Let r denote the cardinality of the elements of $\Omega(\bar{\mathcal{A}})$.

Case 1: $|C| = r$.
   By Remark 5, we have $C \in \Omega(\bar{\mathcal{A}})$. Let U be the mate of C and $q = |C \cap U| \geq 2$. By Theorem 4, q is odd so in particular $q \geq 3$. Since $C \subseteq C_1 \cup C_2$, we must have $|U \cap C_1| > 1$ or $|U \cap C_2| > 1$. This implies that U is the mate of C_1 or C_2, i.e. that $C = C_1$ or $C = C_2$.

Case 2: $|C| > r$.
   Let $t = |C_1 \cap C_2 \cap C|$. Since $C \subseteq C_1 \cup C_2$, it follows that

$$|C| = t + |(C_1 \triangle C_2) \cap C|. \tag{3}$$

For $T = C_1 \bigtriangleup C_2 \bigtriangleup C$, we have

$$
\begin{aligned}
|T| &= |(C_1 \cap C_2 \cap C) \cup [(C_1 \bigtriangleup C_2) - C]|, && C \subseteq C_1 \cup C_2 \\
&= |C_1 \cap C_2 \cap C| + |C_1 \bigtriangleup C_2| - |(C_1 \bigtriangleup C_2) \cap C| \\
&= t + |C_1 \bigtriangleup C_2| - (|C| - t), && \text{by (3)} \\
&= 2t + |C_1| + |C_2| - 2|C_1 \cap C_2| - |C| \\
&\leq |C_1| + |C_2| - |C|, && t \leq |C_1 \cap C_2| \\
&\leq 2r - (r+1), && C_1, C_2 \in \Omega(\bar{A})
\end{aligned}
$$

Since $\mathcal{A}$ is binary we have that T is equal to, or contains an element of $\Omega(\mathcal{A})$. But $|T| \leq r - 1$ which contradicts Theorem 3(i) and Remark 5. $\qquad \square$

Notice that for $\mathcal{O}_{K_5}$ the previous theorem simply says that given two triangles C_1, C_2 there is no odd cycle (distinct from C_1 and C_2) which is contained in the union of C_1 and C_2. It is worth mentioning that this is a property of mni *binary* clutters only. Indeed the property does not hold for odd holes or more generally for any circulant matrix with the consecutive one property. For a description of many classes of mni clutters see [4] and [14].

Proposition 3. *Let $\mathcal{A}$ be a mni binary clutter and $\mathcal{B}$ its blocker. For any $e \in E(\mathcal{A})$ there exist $C_1, C_2, C_3 \in \Omega(\bar{A})$ and $U_1, U_2, U_3 \in \Omega(\bar{B})$ such that*

(i) $C_1 \cap C_2 = C_1 \cap C_3 = C_2 \cap C_3 = \{e\}$
(ii) $U_1 \cap U_2 = U_1 \cap U_3 = U_2 \cap U_3 = \{e\}$
(iii) For all $i, j \in \{1, 2, 3\}$ we have:

$$
C_i \cap U_j = \{e\} \text{ if } i \neq j, \text{ and } |C_i \cap U_j| = q \geq 3, \text{ if } i = j.
$$

(iv) For all $e_i \in U_i$ and $e_j \in U_j$ with $i, j \in \{1, 2, 3\}$

$$
\exists C \in \Omega(\mathcal{A}) \text{ with } C \cap U_i = \{e_i\} \text{ and } C \cap U_j = \{e_j\}.
$$

Proof. Let r (resp. s) denote the cardinality of the elements of $\Omega(\bar{A})$ (resp. $\Omega(\bar{B})$).

(i) By Theorem 3(iv) there exist s sets $C_1, \ldots, C_s \in \Omega(\mathcal{A})$ such that $C_1 - \{e\}, \ldots, C_s - \{e\}$ are all disjoint. Moreover, exactly $q = rs - n + 1 \geq 2$ of these sets, say $C_1, \ldots, C_q$, contain e. Finally, by Theorem 4 $q \geq 3$.

(ii) Let U_i be the mate of C_i, where $i \in \{1, 2, 3\}$. We know $|U_i \cap C_i| > 1$ and for all $j \in \{1, \ldots, s\} - \{i\}$ we have $|U_i \cap C_j| = 1$. Since $C_1, \ldots, C_s$ only intersect in e and since by Theorem 3(i), $|U_i| = s$ it follows by counting that $e \in U_i$. Finally, by Remark 6 and the fact that $e \in C_1 \cap C_2 \cap C_3$, we obtain $U_i \cap U_j \subseteq \{e\}$ for all $i \neq j$ and $i, j \in \{1, 2, 3\}$.

(iii) Follows from (i),(ii) and the fact that U_i is the mate of C_j if and only if $i = j$.

(iv) Let $T = U_i \cup U_j - \{e_i, e_j\}$. Since $\mathcal{A}$ is binary so is its blocker $\mathcal{B}$. By Proposition 2, there is no $U \in \Omega(\mathcal{B})$ with $U \subseteq T$. Thus $E(\mathcal{A}) - T$ intersects every element of $\Omega(\mathcal{B})$. Since the blocker of the blocker is the original clutter, it follows that $E(\mathcal{A}) - T$ contains or is equal to, an element C of $\Omega(\mathcal{A})$. Since $C \cap U_i \neq \emptyset$ and $C \cap U_j \neq \emptyset$ we have by construction $C \cap U_i = \{e_i\}$ and $C \cap U_j = \{e_j\}$.

3   Outline of the Proof

3.1   From Binary Clutters to Labeled Graphs

Let $(\mathcal{G}, \Sigma)$ be a minimally non weakly bipartite graph. Let $\mathcal{A}$ be the clutter of odd cycles of $(\mathcal{G}, \Sigma)$ and $\mathcal{B}$ its blocker. As noted in Sec. 2.2, $\mathcal{A}$ (and thus $\mathcal{B}$) is mni. Let e be any edge of $E(\mathcal{A})$ and let $U_1, U_2, U_3 \in \Omega(\mathcal{B})$ be the sets defined in Proposition 3. We define

$$R = U_1 - \{e\} \qquad B = U_2 - \{e\} \qquad G = U_3 - \{e\} \tag{4}$$

and

$$W = E(\mathcal{A}) - (U_1 \cup U_2 \cup U_3) \cup \{e\}. \tag{5}$$

Note that by Proposition 3(ii), R, B, G and W form a partition of the edges of $\mathcal{G}$. It may be helpful to think of R, B, G, W as a coloring of the edges of $\mathcal{G}$.

Let C be any odd cycle of $(\mathcal{G}, \Sigma)$ with $e \notin C$. Since $C \in \Omega(\mathcal{A})$ we have $C \cap U_i \neq \emptyset$ for $i \in \{1, 2, 3\}$ it follows that $C \cap R \neq \emptyset, C \cap B \neq \emptyset, C \cap G \neq \emptyset$. Therefore, the minimally non weakly bipartite graph $(\mathcal{G}, \Sigma)$ satisfies the following property:

(P1) Every odd cycle C of $(\mathcal{G}, \Sigma)$ that does not contain e, has at least one edge in R, one in B, and one in G.

Consider edges $e_i \in R$ and $e_j \in B$ then $e_i \in U_1, e_j \in U_2$. Hence by Proposition 3(iv), there is an odd cycle $C \in \Omega(\mathcal{A})$ with $\{e_i\} = C \cap U_1 \supseteq C \cap R$ and $\{e_j\} = C \cap U_2 \supseteq C \cap B$. Therefore, $(\mathcal{G}, \Sigma)$ also satisfies:

(P2) For any $e_i \in R$ (resp. B, G) and $e_j \in B$ (resp. G, R) there is an odd cycle C of $(\mathcal{G}, \Sigma)$ with the following properties. C does not contain e, the only edge of C in R (resp. B, G) is e_i, and the only edge of C in B (resp. G, R) is e_j.

3.2   Building Blocks

Definition 1. *We say that a sequence of paths* $\mathcal{S} = [P_1, \ldots, P_t]$ *forms a circuit if each path* P_i *has endpoints* v_i, v_{i+1} *and* $v_{t+1} = v_1$. *We denote the set of edges* $\cup_{i=1}^{t} P_i$ *by* $E(\mathcal{S})$.

The next result is easy.

Lemma 1. *Let* $\mathcal{S} = [P_1, \ldots, P_t]$ *be a sequence of paths that form a circuit. If there is an odd number of odd paths in* $\mathcal{S}$ *then* $E(\mathcal{S})$ *contains an odd cycle.*

Given two vertices v_1 and v_2 of path P, the subpath of P between v_1 and v_2 is denoted by $P(v_1, v_2) = P(v_2, v_1)$. Given a set $S \subseteq E$ and $e \in E$, we denote the set $S - \{e\}$ by $S-e$. Let C be a cycle, v_1, v_2 two vertices of C and e an edge of C. Using the notation just defined we have that $C-e(v_1, v_2)$ defines a unique path. Next comes the first building block of the odd K_5.

Lemma 2. *Let $(\mathcal{G}, \Sigma)$ be a minimally non weakly bipartite graph with a partition of its edges as given in (4)-(5).*

(i) *There exist odd cycles $C_R \subseteq R \cup W, C_B \subseteq B \cup W$ and $C_G \subseteq G \cup W$ which intersect exactly in e. Moreover, $|C_R \cap R|, |C_B \cap B|, |C_G \cap G|$ are all even and non-zero.*

(ii) *C_R, C_B and C_G have only vertices w_1 and w_2 in common, where $e = (w_1, w_2)$.*

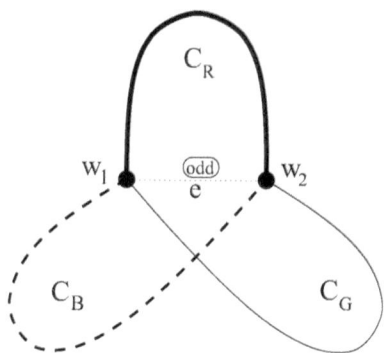

Fig. 1. Lemma 2. Bold solid lines represent paths in $R \cup W - e$, dashed lines paths in $B \cup W - e$, and thin solid lines paths in $G \cup W - e$.

Proof (of Lemma 2).

(i) Let us rename sets C_1, C_2, C_3 of Proposition 3 by C_R, C_B, C_G. We know from Proposition 3(i) that C_R, C_B and C_G intersect exactly in e. By Proposition 3(iii), $\emptyset = C_R \cap (U_2 \cup U_3 - \{e\}) = C_R \cap (B \cup G)$. Thus $C_R \subseteq R \cup W$. Also by Proposition 3(iii), $|C_R \cap R| = |C_R \cap (U_1 - \{e\})| = |C_R \cap U_1| - 1 = q - 1$ where $q \geq 3$ is odd. Identically we show, $C_B \subseteq B \cup W, C_G \subseteq G \cup W$ and $|C_B \cap B|, |C_G \cap G|$ both non-zero and even.

(ii) Suppose, for instance, C_R and C_B have a vertex t distinct from w_1 and w_2 in common. Let $P = C_R - e(w_1, t), P' = C_R - e(w_2, t)$ and $Q = C_B - e(w_1, t), Q' = C_B - e(w_2, t)$, see Fig. 2. Since we can relabel edges in $\delta(\{w_1\})$ and in $\delta(\{t\})$ we can assume w.l.o.g. that edge $e = (w_1, w_2)$ is odd and that paths P, P' are both even. If Q is odd then let $\mathcal{S} = [P, Q]$ otherwise let $\mathcal{S} = [\{e\}, Q, P']$. By Lemma 1, $E(\mathcal{S})$ contains an odd cycle, a contradiction to Proposition 2. □

Since C_R, C_B and C_G have only vertices w_1 and w_2 in common, we can relabel $(\mathcal{G}, \Sigma)$ so that e is the only odd edge in $C_R \cup C_B \cup C_G$. Let us now proceed to add 3 more paths P_R, P_B and P_G to our initial building block.

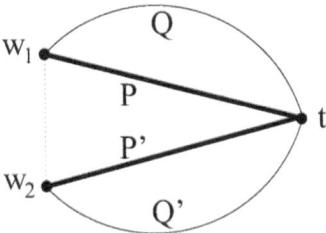

Fig. 2. Lemma 2(ii).

Lemma 3. *Let $(\mathcal{G}, \Sigma)$ be a minimally non weakly bipartite graph with a partition of its edges as given in (4)-(5). Suppose we also have odd cycles C_R, C_B and C_G as defined in Lemma 2 where e is the only odd edge in $C_R \cup C_B \cup C_G$.*

(i) *There is an odd path P_R (resp. P_B, P_G) between a vertex v_{BR} (resp. v_{RB}, v_{BG}) of C_B (resp. C_R, C_B) distinct from w_1, w_2, and a vertex v_{GR} (resp. v_{GB}, v_{RG}) of C_G (resp. C_G, C_R) distinct from w_1, w_2.*
(ii) *$P_R \subseteq R \cup W - e$, $P_B \subseteq B \cup W - e$ and $P_G \subseteq G \cup W - e$.*

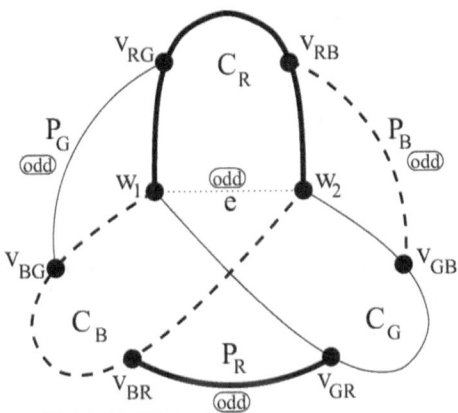

Fig. 3. Lemma 3.

Proof (of Lemma 3). By symmetry it is sufficient to show the result for path P_R. Since $|C_R \cap R| \geq 2$ there is an edge $e_B = (v_{BR}, v'_{BR}) \in B$ of C_B such that v_{BR} is distinct from w_1, w_2 and $C_B - e(w_1, v_{BR})$ contains exactly one edge in B namely e_B. Similarly, we have edge $e_G = (v_{GR}, v'_{GR}) \in C_G$ with v_{GR} distinct from w_1, w_2 and $C_G - e(w_1, v_{GR}) \cap G = \{e_G\}$.

By property (P2) there is an odd cycle C such that $C \cap B = \{e_B\}$ and $C \cap G = \{e_G\}$. The cycle C can be written as $\{e_B, e_G\} \cup P_R \cup P'_R$ where P_R and P'_R are paths included in $R \cup W - e$. Since C is odd we can assume w.l.o.g. that P_R is odd and P'_R is even.

Case 1: The endpoints of P_R are v'_{BR}, v_{GR} (resp. v_{BR}, v'_{GR}).
Then let $\mathcal{S} = [C_B{-}e(w_1, v'_{BR}), P_R, C_G{-}e(v_{GR}, w_1)]$. By Lemma 1, $E(\mathcal{S})$ contains an odd cycle but $e \notin E(\mathcal{S})$ and $E(\mathcal{S}) \cap B = \emptyset$, a contradiction to (P1).
Case 2: The endpoints of P_R are v'_{BR}, v'_{GR}.
Then let $\mathcal{S} = [C_B{-}e(w_1, v'_{BR}), P_R, C_G{-}e(v'_{GR}, w_1)]$. By Lemma 1, $E(\mathcal{S})$ contains an odd cycle but $e \notin E(\mathcal{S})$ and $E(\mathcal{S}) \cap B = E(\mathcal{S}) \cap G = \emptyset$, a contradiction to (P1).

Thus P_R has endpoints v_{BR}, v_{GR}. $\qquad\qquad\qquad\qquad\qquad\qquad\qquad$ □

An *internal vertex* of a path P is a vertex of P which is distinct from the endpoints of P. By choosing paths P_R, P_B, P_G carefully we can show the following additional property:

Lemma 4. *P_R, P_B and P_G have no internal vertices in common with C_R, C_B or C_G.*

Remark 7. Let $(\mathcal{G}, \Sigma)$ be a labeled graph with an odd path P where all internal vertices of P have degree two (in G). Then there is a sequence of relabeling and contractions that will replace P by a single odd edge, without changing the remainder of the graph.

Remark 8. Consider $(\mathcal{G}, \Sigma)$ as defined in Lemma 3. If P_R, P_B, P_G have no internal vertices in common, then $(\mathcal{G}, \Sigma)$ contains a $\widetilde{K_5}$ minor.

This is because in this case the following sequence of operations yields $\widetilde{K_5}$:

1. delete all edges which are not in C_R, C_B, C_G and not in P_R, P_B, P_G,
2. contract $C_R{-}e(v_{RG}, v_{RB}), C_B{-}e(v_{BR}, v_{BG}), C_G{-}e(v_{GR}, v_{GB})$,
3. relabel edges in $\delta(\{w_1, w_2\})$,
4. replace each odd path by a single odd edge (see Remark 7).

3.3   Intersections of Paths P_R, P_B, and P_G.

Because of Remark 8 we can assume at least two of the paths P_R, P_B and P_G must share an internal vertex. One of the main step of the proof of Theorem 1 is to show the following lemma (we give a simplified statement here) which describes how paths P_R, P_B and P_G must intersect.

Lemma 5. *A minimally non weakly bipartite graph $(\mathcal{G}, \Sigma)$ is either a relabeling of $\widetilde{K_5}$ or it contains a contraction minor $(\mathcal{H}, \Sigma_{\mathcal{H}})$ with the following properties (see Fig. 4):*

(i) There are odd paths P'_R, P'_B, P'_G of the form given in lemmas 3, 4 but with vertices v'_{BG}, v'_{GB} instead of v_{BG}, v_{GB}.
(ii) $P'_R \subseteq R, P'_B \subseteq B, P'_G \subseteq G$.

   (iii) P'_R and P'_B *(resp.* P'_G*) share an internal vertex* t_{RB} *(resp.* t_{RG}*).*
   (iv) $P'_B(t_{RB}, v'_{GB})$ *and* P'_G *share an internal vertex* t_{BG}*.*
   (v) P'_B *and* $P'_G(t_{RG}, v'_{BG})$ *share an internal vertex* t'_{BG}*.*
   (vi) Paths $P'_R(v_{BR}, t_{RB})$ *and* $P'_R(v_{GR}, t_{RG})$ *consist of a single edge.*
   (vii) P'_B *(resp.* P'_G*) has exactly one odd edge which is incident to* v'_{GB}
      (resp. v'_{BG}*).*
  (viii) $P'_R(v_{BR}, t_{RB}), P'_R(v_{GR}, t_{RG})$ *are even and* $P'_R(t_{RB}, t_{RG})$ *is odd.*
   (ix) No vertex is common to all three paths P_R, P_B *and* P_G*.*

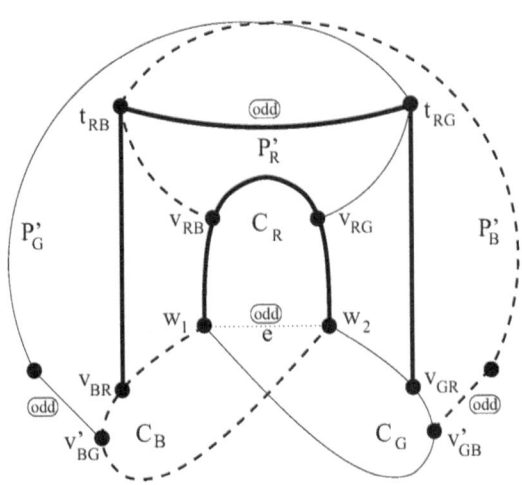

Fig. 4. Lemma 5. Graph $(\mathcal{H}, \Sigma_\mathcal{H})$.

We did not represent vertices t_{BG} (and t'_{BG}) in the previous figure. Let q denote the first vertex of P'_G, starting from v_{RG}, which is also a vertex of $P'_B(t_{RB}, v'_{GB})$ (see Fig. 4). By Lemma 5(iv) there exist such a vertex. Let q' be the first vertex of $P'_G(q, v_{RG})$, starting from q, which is either a vertex of P'_B or equal to v_{RG}. By Lemma 5(ix) q, q' are distinct from t_{RB}.

Definition 2. $(\mathcal{K}, \Sigma_\mathcal{K})$ *is the graph (see Fig. 5) obtained by deleting every edge of* $(\mathcal{H}, \Sigma_\mathcal{H})$ *which is not an edge of* C_R, C_B, C_G *or* $P'_R(v_{BR}, t_{RB})$, P'_B *and* $P'_G(q, q')$*.*

From Lemma 5 we can readily obtain the following properties:

Remark 9.

   (i) There are exactly two odd edges in $(\mathcal{K}, \Sigma_\mathcal{K})$, namely e and the edge of P'_B
      incident to v'_{GB}.
   (ii) Let S be the set of vertices $\{w_1, w_2, v_{BR}, v_{RB}, v'_{GB}, t_{RB}, q, q'\}$ shown in
      Fig. 5. v_{RB} and q' may denote the same vertex but all other vertices of S
      are distinct.
  (iii) S is the set of all vertices of $(\mathcal{K}, \Sigma_\mathcal{K})$ which have degree greater than two.

20   Bertrand Guenin

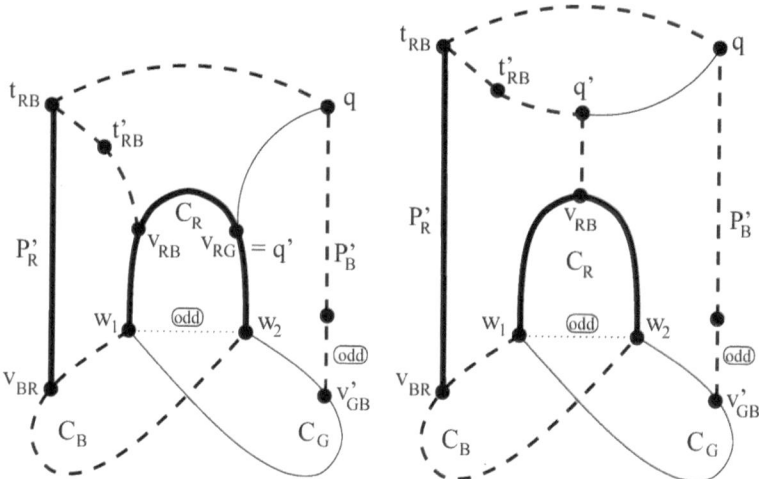

Fig. 5. Graph $(\mathcal{K}, \Sigma_\mathcal{K})$, Left $q' = v_{RG}$, right $q' \neq v_{RG}$

Definition 3. Let t_{RB} be the vertex of $(\mathcal{H}, \Sigma_\mathcal{H})$ defined in Lemma 5(iii). $\bar{e} = (t_{RB}, t'_{RB})$ denotes the edge of $P'_B(t_{RB}, v_{RB})$ incident to t_{RB} (see Fig 5). Note, t'_{RB} need not be distinct from v_{RB}.

Lemma 6. Let $(\mathcal{H}, \Sigma_\mathcal{H})$ be the graph defined in Lemma 5. There are three distinct odd paths F_1, F_2, F_3 from t_{RB} to t'_{RB}.

Proof. By applying Proposition 3(i) to edge $\bar{e}$ we obtain odd cycles L_1, L_2, L_3 of $(\mathcal{G}, \Sigma)$ which intersect exactly in $\bar{e}$. Let L'_1, L'_2, L'_3 be the corresponding odd cycles in $(\mathcal{H}, \Sigma_\mathcal{H})$ and F_i, for $i \in \{1, 2, 3\}$, denotes the path $L'_i - \bar{e}$. Since $\bar{e}$ is even and L'_i is odd we must have F_i odd as well. □

Lemma 7. Let $(\mathcal{K}, \Sigma_\mathcal{K})$ be the graph given in Definition 2 and let F_1, F_2, F_3 be the paths given in Lemma 6. Then F_1, F_2, F_3 all have an internal vertex in common with $(\mathcal{K}, \Sigma_\mathcal{K})$.

Proof. Suppose for a contradiction this is not the case and we have F_i with no internal vertices in common with $(\mathcal{K}, \Sigma_\mathcal{K})$, see Fig. 6. Consider the graph obtained from $(\mathcal{H}, \Sigma_\mathcal{H})$ by deleting $\bar{e} = (t_{RB}, t'_{RB})$ and all edges which are not edges of $(\mathcal{K}, \Sigma_\mathcal{K})$ or edges of F_i. The following sequence of operations yields $\widetilde{K_5}$:

1. relabel edges in $\delta(\{q\})$,
2. if $q' = v_{RG}$ then contract $C_R - e(v_{RG}, v_{RB})$ otherwise contract $P'_B(q', v_{RB})$,
3. relabel edges in $\delta(\{q\}), \delta(\{w_1, w_2\})$,
4. contract $P'_B(q, v'_{GB})$ and $C_B - e(t, v_{BR})$,
5. replace odd paths by odd edges, see Remark 7.

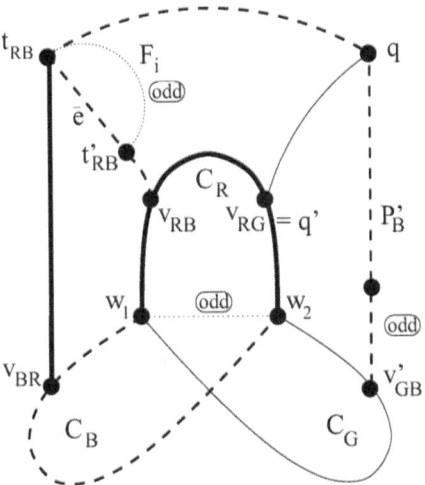

Fig. 6. Lemma 7. We represent the case where $q' = v_{RG}$ only.

Hence $(\mathcal{G}, \Sigma)$ contains $\widetilde{K_5}$ as a proper minor, a contradiction since $(\mathcal{G}, \Sigma)$ is minimally non weakly bipartite.   □

Let $(\bar{\mathcal{H}}, \Sigma_{\bar{\mathcal{H}}})$ be the graph obtained by deleting from $(\mathcal{H}, \Sigma_{\mathcal{H}})$ all the edges which are not edges of C_R, C_B, C_G and not edges of P'_R, P'_B and P'_G. Because of Lemma 7 we can define f_i, for $i = 1, 2, 3$, to be the first internal vertex of F_i, starting from t_{RB}, which is also a vertex of $(\bar{\mathcal{H}}, \Sigma_{\bar{\mathcal{H}}})$. By symmetry (see Fig. 4) there is a vertex a vertex t'_{RG} of $P'_G(t_{RG}, v_{RG})$ which is incident to t_{RG} and there are odd paths F'_1, F'_2, F'_3 between t_{RG} and t'_{RG}. As previously we can define f'_i, for $i = 1, 2, 3$, to be the first internal vertex of F'_i, starting from t_{RG}, which is also a vertex of $(\bar{\mathcal{H}}, \Sigma_{\bar{\mathcal{H}}})$.

The remainder of the proof is a case analysis which shows that for each possible set of vertices f_1, f_2, f_3 and f'_1, f'_2, f'_3 the graph $(\mathcal{H}, \Sigma_{\mathcal{H}})$ contains a $\widetilde{K_5}$ minor. In order to prove Lemma 5 and to make the case analysis tractable we first establish general results for labeled graphs with properties (P1) and (P2).

Acknowledgments. I am most indebted to Prof. Gérard Cornuéjols and Prof. François Margot for their help.

References

1. F. Barahona. On the complexity of max cut. Rapport de Recherche No. 186, Mathematiques Appliqués et Informatiques, Université Scientifique et Medicale de Grenoble, France, 1980.
2. F. Barahona. The max cut problem in graphs not contractible to K_5. *Oper. Res. Lett.*, 2:107–111, 1983.

3. W. G. Bridges and H. J. Ryser. Combinatorial designs and related systems. *J. Algebra*, 13:432–446, 1969.

4. G. Cornuéjols and B. Novick. Ideal 0, 1 matrices. *J. Comb. Theory Ser. B*, 60:145–157, 1994.

5. J. Fonlupt, A. R. Mahjoub, and J. P. Uhry. Composition of graphs and the bipartite subgraph polytope. Research Report No. 459, Laboratoire ARTEMIS (IMAG), Université de Grenoble, Grenoble, 1984.

6. A. M. H. Gerards. *Graphs and Polyhedra: Binary Spaces and Cutting Planes*. PhD thesis, Tilburg University, 1988.

7. A. M. H. Gerards. Multi-commodity flows and polyhedra. *CWI Quarterly*, 6(3), 1993.

8. M. Grötschel and W. R. Pulleyblank. Weakly bipartite graphs and the max-cut problem. *Operations Research Letters*, 1:23–27, 1981.

9. B. Guenin. A characterization of weakly bipartite graphs. Working paper, GSIA, Carnegie Mellon Univ., Pittsburgh, PA 15213, 1997.

10. R. M. Karp. Reducibility among combinatorial problems. In R. E. Miller and J. W. Thatcher, editors, *Complexity of Computer Computations*, pages 85–103. Plenum Press, New York, 1972.

11. A. Lehman. A solution of the Shannon switching game. *J. SIAM*, 12(4):687–725, 1964.

12. A. Lehman. On the width-length inequality, mimeographic notes. *Mathematical Programming*, 17:403–417, 1979.

13. A. Lehman. On the width-length inequality and degenerate projective planes. In W. Cook and P.D. Seymour, editors, *Polyhedral Combinatorics, DIMACS Series in Discrete Math. and Theoretical Computer Science, Vol. 1*, pages 101–105, 1990.

14. C. Luetolf and F. Margot. A catalog of minimally nonideal matrices. *Mathematical Methods of Operations Research*, 1998. To appear.

15. B. Novick and A. Sebö. On Combinatorial Properties of Binary Spaces. *IPCO Proceedings*, 1995.

16. M. W. Padberg. Lehman's forbidden minor characterization of ideal $0-1$ matrices. *Discrete Math.* 111:409–420, 1993.

17. P. D. Seymour. The forbidden minors of binary clutters. *J. of Combinatorial Theory B*, 22:356–360, 1976.

18. P. D. Seymour. The matroids with the max-flow min-cut property. *J. Comb. Theory Ser. B*, 23:189–222, 1977.

19. P. D. Seymour. Matroids and multicommodity flows. *European J. of Combinatorics*, 257–290, 1981.

20. P. D. Seymour. On Lehman's width-length characterization. In W. Cook, and P. D. Seymour, editors, *Polyhedral Combinatorics, DIMACS Series in Discrete Math. and Theoretical Computer Science, Vol. 1*, pages 107–117, 1990.

Bipartite Designs

Grigor Gasparyan

Yerevan State University
Yerevan-49, Armenia
Grisha@@ysu.am

Abstract. We investigate the solution set of the following matrix equation: $A^T B = J + diag(\mathbf{d})$, where A and B are $n \times n$ $\{0,1\}$ matrices, J is the matrix with all entries equal one and $\mathbf{d}$ is a full support vector. We prove that in some special cases (such as: both Ad^{-1} and Bd^{-1} have full supports, where $\mathbf{d}^{-1} = (d_1^{-1}, \ldots, d_n^{-1})^T$; both A and B have constant column sums; $\mathbf{d}^{-1} \cdot \mathbf{1} \neq -1$, and A has constant row sum etc.) these solutions have strong structural properties. We show how the results relate to design theory, and then apply the results to derive sharper characterizations of (α, ω)-graphs. We also deduce consequences for "minimal" polyhedra with $\{0,1\}$ vertices having non-$\{0,1\}$ constraints, and "minimal" polyhedra with $\{0,1\}$ constraints having non-$\{0,1\}$ vertices.

1   Introduction

Suppose we are given two $n \times n$ $\{0,1\}$ matrices A and B, and a full support vector $\mathbf{d}$. Let us call the pair of matrices (A, B) a *(bipartite)* $\mathbf{d}$*-design* if

$$A^T B = J + diag(\mathbf{d}),$$

where J is the matrix filled with ones. It seems difficult to say anything about the structure of the matrices A and B in such a general setting. But if $\mathbf{d} > \mathbf{0}$ then a surprising result of Lehman [7] asserts that either

$$A = B \cong \mathrm{DPP} \equiv \begin{bmatrix} 0 & 1 \\ 1 & I \end{bmatrix}$$

(then we call the pair (A, B) a DPP-*design*), or for some r and s:

$$AJ = JA = rJ; BJ = JB = sJ; A^T B = BA^T = J + (rs - n)I$$

(then we call the pair (A, B) an (r, s)-*design*). This result generalizes the earlier results of de Bruijn and Erdős [2] and Ryser [12], and it is one of the main arguments in the proof of Lehman's theorem on minimally non-ideal polyhedra [8].

In this paper we would like to investigate the $\mathbf{d}$-designs a bit more generally. Our main goal is *to find sufficient conditions which force a $\mathbf{d}$-design to become an (r, s)-design*. The following theorem summarizes our results in that direction:

R. E. Bixby, E. A. Boyd, and R. Z. Ríos-Mercado (Eds.): IPCO VI
LNCS 1412, pp. 23–36, 1998. © Springer–Verlag Berlin Heidelberg 1998

Theorem 1. *Let (A, B) be a* **d***-design. Then:*

1. *If $n > 3$ and both $A\mathbf{d}^{-1}$ and $B\mathbf{d}^{-1}$ are full support vectors, then (A, B) either is an (r, s)-design or a DPP-design;*
2. *If $AJ = rJ$ and $\mathbf{d}^{-1} \cdot \mathbf{1} \neq -1$, then (A, B) either is an (r, s)-design or $r = 2$ (and we characterize all the possible such designs);*
3. *If A and B have constant column sums, then (A, B) is an (r, s)-design.*

Our proof (see Section 7) uses widely the ideas of Lehman [8], Padberg [10], Seymour [15] and Sebő [14]. In Sections 4 and 5 we state and prove the two main ingredients of the proof: de Bruijn and Erdős Lemma [2] and a lemma from matrix theory, which contains, as a special case, our key argument from linear algebra due to Ryser [12], Lehman [7] and Padberg [10]. In Section 6 we discuss the applications of the lemma to the classical design theory. In Section 8 we apply Theorem 1 to get some new characterizations for (α, ω)-graphs. In Section 9 we use Theorem 1 to characterize fractional vertices and facets of certain types of minimally non-$\{0, 1\}$ polyhedra. Our characterizations imply Lehman's theorem on minimally non-ideal polyhedra [8], Padberg's theorem on minimally imperfect polytopes [11] and the part of Sebő [14], which correspond to nonsingular matrix equations. Matrix generalization of the singular case is going to be considered in the forthcoming paper [5]. In Sections 2 and 3 we give some preliminaries and discuss some basic examples.

2 Notations and Terminology

The following conventions and terminology is used throughout this article. We use lower case boldface letters to name vectors and upper case letters to name matrices. In particular, $\mathbf{1}$ and $\mathbf{0}$ denote vector of all one and the zero vector, respectively, and I and J denote the identity matrix and the all one matrix of suitable size. $\mathbf{e}_i$ denotes the ith unite vector and e_{ij} denotes the (i, j)th element of I. For a vector $\mathbf{x}$, x_i is its ith coordinate. Similarly, if A is a matrix then a_{ij} denotes its (i, j)th element, $\mathbf{a}_{i\cdot}$ denotes its ith row and $\mathbf{a}_{\cdot j}$ denotes its jth column. Also, $diag(\mathbf{d})$ denotes the diagonal matrix made up of the vector $\mathbf{d}$, and $\mathbf{d}^{-1} = (d_1^{-1}, \ldots, d_n^{-1})^T$.

We will not distinguish the matrix from the ordered set of its columns. A matrix with $\{0, 1\}$ entries we call a $\{0, 1\}$ matrix and a linear constraint we call a $\{0, 1\}$ constraint if it has $\{0, 1\}$ coefficients and $\{0, 1\}$ RHS. $A \cong B$ means that A can be obtained from B after permutation of its columns and rows.

If $\mathbf{x} = (x_1, \ldots, x_n)$, then $\mathbf{x}/i := (x_1, \ldots, x_{i-1}, x_{i+1}, \ldots, x_n)$ and if A is a $\{0, 1\}$ matrix, then

$$A/i := \{\mathbf{a}_{\cdot j}/i : \text{for each } j \text{ such that } a_{ij} = 0\}.$$

The dot product of column (row) vectors $\mathbf{a}$ and $\mathbf{b}$ is $\mathbf{a} \cdot \mathbf{b} := \mathbf{a}^T \mathbf{b}$ ($\mathbf{a} \cdot \mathbf{b} := \mathbf{a}\mathbf{b}^T$).

If $\mathcal{A}$ is a set system, then A denotes its (point-set) incidence matrix (and vice versa), and $\mathcal{A}^T$ denotes its dual set system, i.e. the (point -set) incidence matrix of $\mathcal{A}^T$ is congruent to A^T.

A pairwise balanced design (with index λ) is a pair $(V, \mathcal{B})$ where V is a set (the point set) and $\mathcal{B}$ is a family of subsets of V (the block set) such that any pair of points is in precisely λ blocks. A square design is a pairwize balanced design with equal number of points and blocks. A symmetric design is a square design with blocks of constant size. A projective plane is a symmetric design with index 1. If

$$A \cong \begin{bmatrix} 0 & 1 \\ 1 & I \end{bmatrix} \equiv \mathrm{DPP},$$

then $\mathcal{A}$ is called a degenerated projective plane.

Let A be an $n \times n$ $\{0,1\}$ matrix. $G(A)$ denotes a graph, vertices of which are the columns of A and two vertices are adjacent iff the dot product of the corresponding columns is not zero. We say that A is connected if $G(A)$ is connected.

If $\mathcal{I}$ and $\mathcal{J}$ are sets of indices, then $A[\mathcal{I}; \mathcal{J}]$ denotes the submatrix of A induced by the rows $\mathcal{I}$ and columns $\mathcal{J}$.

Definition 1. *We say that A is r-uniform (r-regular) if $JA = rJ(AJ = rJ)$. We say that A is totally r-regular if $JA = AJ = rJ$.*

Definition 2. *We say that an $n \times n$ $\{0,1\}$ matrix A is a DE-matrix (de Bruijn-Erdős's matrix) if for each pair i and j with $a_{ij} = 0$, the number of zeros in the ith row is equal to the number of zeros in the jth column.*

Definition 3. *Let (A, B) be a $\mathbf{d}$-design. If $1 + \sum_{i=1}^{n} d_i^{-1} = 0$ then we call such a design singular. The ith row of A (B) we call a $\mathbf{d}$-row if $\mathbf{a}_{i.}\mathbf{d}^{-1} = 0$ ($\mathbf{b}_{i.}\mathbf{d}^{-1} = 0$).*

Let P be a polyhedron. We say that P is *vertex* $\{0,1\}$ if all its vertices are $\{0,1\}$ vectors. We say that P is *facet* $\{0,1\}$ if it can be given with the help of $\{0,1\}$ constraints. We call P a $\{0,1\}$-*polyhedron* if it is both vertex $\{0,1\}$ and facet $\{0,1\}$. P/j denotes the orthogonal projection of P on the hyperplane $x_j = 0$, and $P \backslash j$ denotes the intersection of P with the hyperplane $x_j = 0$. The first operation is called *contraction* and the second one *deletion* of the coordinate j. A polyhedron P' is called a *minor* of P if it can be obtained from P by successively deletion or contraction one or several coordinates.

For an $m \times n$ $\{0,1\}$ matrix A, we denote by $P_\le(A) = \{\mathbf{x} \in \mathcal{R}^n : A\mathbf{x} \le 1; \mathbf{x} \ge \mathbf{0}\}$ the *set packing polytope* (SP-polytope) and by $P_\ge(A) = \{\mathbf{x} \in \mathcal{R}^n : A\mathbf{x} \ge 1; \mathbf{x} \ge \mathbf{0}\}$ the *set covering polyhedron* (SC-polyhedron) associated with A.

A $\{0,1\}$ SP-polytope is called *perfect,* and a $\{0,1\}$ SC-polyhedron is called *ideal.* A SP-polytope P is called *minimally imperfect* if it is not perfect, but all its minors are perfect. A *minimally non-ideal* polyhedron is defined similarly.

It is easy to see that $P_\ge(\mathrm{DPP})$ is a minimally non-ideal polyhedron.

For more information on polyhedral combinatorics we refer to Schrijver [13].

If $G = (V, E)$ is a graph, then $n = n(G)$ denotes the number of vertices of G; $\omega = \omega(G)$ denotes the cardinality of a maximum clique of G; $\alpha = \alpha(G)$ denotes the cardinality of a maximum stable set; and $\chi = \chi(G)$ denotes the chromatic number of G. A k-clique or k-stable set will mean a clique or stable set of size

k. A graph G is called *perfect* if $\chi(H) = \omega(H)$ for every induced subgraph H of G. A graph G is called *minimal imperfect* if it is not perfect, but all its proper induced subgraphs are perfect. G is called an (α, ω)-*graph (or partitionable)*, if $n = \alpha\omega + 1$, and $V(G)\backslash v$ can be partitioned both into ω-cliques and into α-stable sets, for every $v \in V(G)$ (here we assume that the empty graph and the clique are (α, ω)-graphs). Lovász [9] proved the following important theorem:

Theorem 2. *If G is minimal imperfect, then it is an (α, ω)-graph.*

Theorem 2 provides the only known coNP characterization of perfectness, and it is used by Padberg[11] to show further properties of minimal imperfect graphs. For more on (α, ω)-graphs we refer to Chvátal et al. [3].

3   Constructions Associated with Bipartite d-Designs: Some Basic Examples

There are several ways to associate a combinatorial structure to a **d**-design (A, B). A straightforward way to do it is to take two set systems $\mathcal{A} = \{A_1, \ldots, A_n\}$ and $\mathcal{B} = \{B_1, \ldots, B_n\}$ on some ground set $V = \{v_1, \ldots, v_n\}$ such that the matrices A and B are (point-set) incidence matrices of $\mathcal{A}$ and $\mathcal{B}$, respectively. Then the pair of set systems $(\mathcal{A}, \mathcal{B})$ have the following property: for each $1 \leq i, j \leq n$, $|A_i \cap B_j| = 1 + e_{ij}d_i$. We call such a pair of set systems a **d**-design. In particular, it was proved by Padberg [11] (see also [3]) that the pair of set systems of ω-cliques and α-stable sets of an (α, ω)-graph is an (α, ω)-design. Another interesting special case is when $\mathcal{A} = \mathcal{B}$. It was proved by de Bruijn and Erdős [2] that $(\mathcal{A}, \mathcal{A})$ is a **d**-design iff $\mathcal{A}^T$ is a (may by degenerated) projective plane.

A **d**-design can be characterized with the help of just one set system $\mathcal{A}^T$. Then the ith column of B can be interpreted as an incidence vector of a set subsystem of $\mathcal{A}^T$, which contains all the points except A_i by exactly once and A_i by exactly $d_i + 1$ times. We call such a set system a **d**-*hypergraph*. A **d**-hypergraph corresponding to a (r, s)-design we call an (r, s)-*hypergraph*. In particular, -1-hypergraph is a hypergraph having equal number of edges and vertices such that for each vertex v, $V\backslash v$ can be partitioned with the help of its edges. We will show that a $-\mathbf{1}$-hypergraph is an (α, ω)-hypergraph, which corresponds to the ω-clique hypergraph of some (α, ω)-graph.

An interesting special case is when $\mathcal{A}^T$ is 2-uniform, i.e. it is a graph. A nonsingular **d**-design (A, B) we call a G-design, if A is 2-regular. A graph G we call a **d**-graph if there exists a G-design (A, B) such that A is the (edge-vertex) incidence matrix of G. If G is an odd cycle, then we call (A, B) a C-design.

Let G be a **d**-graph. Then it is not difficult to show that, for each $1 \leq i \leq n$, $d_i = \pm 1$ (see Lemma 9). Denote by $G\backslash v$ (G/v) the graph obtained from G after deleting (duplicating) the vertex v. It is easy to see that, for each vertex v, either $G\backslash v$ or G/v has a perfect matching. Call such a graph matchable. The following lemma characterizes **d**-graphs:

Lemma 1. *G is a **d**-graph iff it is a connected graph with odd number of vertices and exactly one odd cycle such that the distance from each not degree two vertex to the cycle is even.*

Proof. We will prove by induction on the number of vertices. Suppose the theorem is true for the graphs having less than n vertices and G is a **d**-graph with n vertices. Then, clearly, G is connected, has equal number of edges and vertices, and odd number of vertices. Hence G has exactly one cycle, which is odd, as the (edge-vertex) incidence matrix of G is nonsingular. Furthermore, if v_1 is a leaf of G, and $v_1 v_2 \in E(G)$, then $G \backslash \{v_1; v_2\}$ is matchable. Indeed, for each $v \neq v_1, v_2$, the perfect matching of $G \backslash v$ (or G/v) must contain the edge $v_1 v_2$. Hence after deleting the edge $v_1 v_2$ from the matching, it will be a perfect matching for $G \backslash v$ (or G/v). It follows that either v_2 has degree 2, or it is a vertex of the cycle and has degree 3.

Let $v_3 \neq v_1$ such that $v_3 v_2 \in E$. Now if v_2 has degree two then $G \backslash \{v_1; v_2\}$ is a **d**-graph. Hence by induction hypothesis, the distance from each not degree two vertex $v \neq v_1; v_3$ to the cycle is even. If v_1 is the unique leaf of G nonadjacent to the cycle, then the degree of v_3 in $G \backslash \{v_1; v_2\}$ is one, hence the distances from v_1 and v_3 to the cycle are also even. If $v_4 \neq v_1$ is a leaf nonadjacent to the cycle and $v_4 v_5 \in E$ then, by induction hypothesis, $G \backslash \{v_4; v_5\}$ and $G \backslash \{v_1; v_2; v_4; v_5\}$ are a **d**-graphs, hence the distances from v_1 and v_3 to the cycle are again even.

If G has no leafs then it is an odd cycle and we have nothing to prove. Suppose G has leafs, but all of them are adjacent to the cycle. Then it is easy to see that G has exactly two leafs, the neighbors of which are adjacent vertices in the cycle. Denote by V_1 the set of vertices of G such that $G \backslash v$ has a perfect matching and by $V_2 = V \backslash V_1$. Then it is easy to see that G/v has a perfect matching iff $v \in V_2$ and $|V_1| = |V_2| + 1$. Now if (A, B) is a **d**-design corresponding to G, then $\mathbf{d}^{-1} \cdot \mathbf{1} = |V_1| - |V_2| = -1$, a contradiction.

The sufficiency of the condition is proved by similar arguments.

It is possible to associate a full dimensional simplex to a nonsingular **d**-design as follows. Denote by $\mathbf{y}$ the unique solution of $B^T \mathbf{y} = \mathbf{1}$, and by $P(A, B) = $ *convex hull* $\{\mathbf{a}_{.1}, \ldots, \mathbf{a}_{.n}, \mathbf{y}\}$. The facets of $P(A, B)$ containing the vertex $\mathbf{y}$ are $\mathbf{b}_{.j} \cdot \mathbf{x} = 1$. Hence $P(A, B)$ has at most one non-$\{0, 1\}$ facet, and at most one non-$\{0, 1\}$ vertex, which is not in that facet.

Conversely, suppose $\mathbf{y}$ is a non-degenerated vertex of a polyhedron P such that all its neighbors and the facets containing it are $\{0, 1\}$. Then we can associate a **d**-design (A, B) with $\mathbf{y}$ taking as columns of A the neighboring vertices of $\mathbf{y}$ and as columns of B the supports of the facets of P containing $\mathbf{y}$ such that the vertex corresponding to the ith column of A is not in the facet corresponding to the ith column of B. We call such a vertex a **d**-design vertex. Similarly, if F is a simplicial facet such that all its vertices and neighboring facets are $\{0, 1\}$, then we associate a **d**-design with F and call it a **d**-design facet.

4   De Bruijn-Erdős's Matrices

In this section we summarize the information about DE-matrices, which we use in this paper.

Lemma 2 ([2]). *Let A be an $n \times m$ $\{0,1\}$ matrix without all one columns. If $n \leq m$ and for each pair i and j with $a_{ij} = 0$, the number of zeros in the ith row is less or equal than the number of zeros in the jth column, then A is a DE-matrix.*

Proof. Denote by $w_j = 1/(n - 1 \cdot \mathbf{a}_{.j})$ and $\mathbf{w} = (w_1, \ldots, w_m)^T$. Then

$$(\mathbf{1}^T - \mathbf{a}_{i.})\mathbf{w} = \sum_{\substack{j \\ a_{ij}=0}} \frac{1}{n - 1 \cdot \mathbf{a}_{.j}} \leq \sum_{\substack{j \\ a_{ij}=0}} \frac{1}{m - \mathbf{a}_{i.}\mathbf{1}} = 1,$$

and $m = \mathbf{1}^T(J - A)\mathbf{w} \leq n$. Hence we should have equality throughout, i.e. $m = n$, and if $a_{ij} = 0$ then $\mathbf{a}_{i.}\mathbf{1} = \mathbf{1} \cdot \mathbf{a}_{.j}$.

Lemma 3. *If A is a DE-matrix, then it has an equal number of all one rows and columns.*

Proof. Delete all one rows and columns and apply Lemma 2.

Lemma 4. *Let A and A' be DE-matrices, where $\mathbf{a}_{.1} \neq \mathbf{a}'_{.1}$ and $\mathbf{a}_{.j} = \mathbf{a}'_{.j}$, $1 < j \leq n$. Then either A or A' has an all one column.*

Proof. Indeed, if say $a_{i1} = 1$ and $a'_{i1} = 0$, then $\mathbf{a}_{i.} = \mathbf{1}^T$. Hence by Lemma 3, A has also an all 1 column.

Lemma 5. *If B is a DE-matrix and $J - B$ is connected, then B is totally regular.*

Proof. If $(1 - \mathbf{b}_{.i}) \cdot (1 - \mathbf{b}_{.j}) \neq 0$, then there exists an index k such that $b_{ki} = b_{kj} = 0$, hence $\mathbf{1} \cdot \mathbf{b}_{.j} = \mathbf{b}_{k.}\mathbf{1} = \mathbf{1} \cdot \mathbf{b}_{.i}$. As $J - B$ is connected, it follows that $JB = sJ$, for some integer s. Since B has no all 1 column, and it is a DE-matrix, it cannot have an all 1 row. Hence B is totally s-regular.

The following result, which has been extracted by Sebő [14] from Lehman [8] (see also [15] and [10]), is one of our key arguments in the proof of Theorem 13. Denote by A^* the set of all the solutions of $A^T\mathbf{x} = \mathbf{1}$, and by A^{01} the set of all $\{0,1\}$ vectors in A^*.

Theorem 3 ([8,14]). *Let A be a nonsingular $\{0,1\}$ matrix without an all one column. If the vector $\mathbf{x} \in \mathbf{A}^*$ has full support, and for each i,*

$$\mathbf{x}/i \in \text{linear hull } (A/i)^{01},$$

then A is a DE-matrix.

Proof. Let $B_i \subseteq (A/i)^{01}$ be a matrix such that the equation $B_i \mathbf{y} = \mathbf{x}/i$ has a unique solution. As $\mathbf{x}/i$ has full support, B_i has no all zero row.

Suppose $a_{ij} = 0$, $\mathcal{L} = \{l : a_{lj} = 1\}$ and B_i' is the submatrix of B_i induced by the rows $\mathcal{L}$. Then all the columns of B_i' have exactly one 1. Thus we have:

$$\mathbf{a}_{i.}\mathbf{1} = n - \text{rk}(A/i) \geq \text{rk}(B_i) = \mathbf{1}^T B_i' \mathbf{1} \geq \mathbf{1} \cdot \mathbf{a}_{.j},$$

and we are done by Lemma 2.

The next important result of Lehman [8] will be used to prove Theorem 12.

Theorem 4 ([8]). *Let A be an $n \times m$ $\{0,1\}$ matrix having full row rank and no all one or all zero or equal columns. If the vector $\mathbf{x} \in \mathbf{A}^*$ has full support, and for each i, affine hull (A/i) can be given with the help of $\{0,1\}$ constraints, then A is a DE-matrix.*

Proof. Suppose $A' = \{\mathbf{a}_{.1}, \ldots, \mathbf{a}_{.n}\}$ is nonsingular. Then for each $1 \leq i \leq n$:

$$\mathbf{x}/i \in (A/i)^* \subseteq \text{linear hull } (A/i)^{01} \subseteq \text{linear hull } (A'/i)^{01},$$

hence by Theorem 3, A' is a DE-matrix.

If $n < m$, then there must exist $i \leq n$ and $j > n$ such that $A'' = \{A' \cup \mathbf{a}_{.j}\} \backslash \mathbf{a}_{.i}$ is again nonsingular. But then A'' also is a DE-matrix and by Lemma 4, A has an all 1 column, a contradiction.

5   A Lemma on Matrices

The following lemma contains our main argument from linear algebra. Though we need just a very special case of the lemma, we would like to state it in general form, for the sake of possible other applications.

Suppose $A, B, D \in \mathcal{R}^{n \times n}$; $U, W \in \mathcal{R}^{n \times m}$, where D is nonsingular and U (or W) has full column rank.

Lemma 6. $\det(D + UW^T) \neq 0$ *iff* $\det \Delta \neq 0$ *and then*

$$A^T B = D + UW^T \Leftrightarrow BD^{-1}A^T = I + X\Delta^{-1}Y^T,$$

where $\Delta = W^T D^{-1} U + I$; $X = BD^{-1}U$; $Y^T = W^T D^{-1} A^T$.

Proof. Denote by

$$F = \begin{bmatrix} \Delta^T & U^T \\ Y & A \end{bmatrix}; E = \begin{bmatrix} -\Delta & -W^T \\ X & B \end{bmatrix}; D_0 = \begin{bmatrix} -\Delta & 0 \\ 0 & D \end{bmatrix};$$

Now if $A^T B = D + UW^T$, then $A^T X = U\Delta$; $Y^T B = \Delta W^T$; and

$$Y^T X = W^T D^{-1} A^T B D^{-1} U = W^T D^{-1} (D + UW^T) D^{-1} U = \Delta^2 - \Delta.$$

It follows that $A^T B = D + UW^T \Leftrightarrow F^T E = D_0$. As $A^T BD^{-1}U = U\Delta$, and U has full column rank, the singularity of Δ implies the singularity of either A or B. If Δ is nonsingular, then both F and E are nonsingular, and

$$F^T E = D_0 \Leftrightarrow ED_0^{-1}F^T = I \Leftrightarrow BD^{-1}A^T = I + X\Delta^{-1}Y^T.$$

Since $A^T X\Delta^{-1} = U$ and $\Delta^{-1}Y^T B = W^T$, A and B are also nonsingular.

Notice that if the inverse of the matrix D is easy to compute, then Lemma 6 reduces the singularity test of the $n \times n$ matrix $D + UW^T$ to the singularity test of an $m \times m$ matrix Δ.

Taking $m = 1$, $U = W = \mathbf{1}$ and $D = diag(\mathbf{d})$ we get:

Lemma 7 ([7,10]). *If (A, B) is a nonsindular $\mathbf{d}$-design, then $BD^{-1}A^T = I + \delta^{-1}\mathbf{xy}^T$, where $\mathbf{x} = B\mathbf{d}^{-1}$, $\mathbf{y} = A\mathbf{d}^{-1}$, and $\delta = \mathbf{d}^{-1} \cdot \mathbf{1} + 1$.*

It follows that if $\mathbf{a}_{i.}$ is a $\mathbf{d}$-row of a nonsingular $\mathbf{d}$-design (A, B), then for each $k \leq n$, $\sum_{j=1}^n a_{ij}b_{kj}d_j^{-1} = e_{ik}$. On the other hand, if for some i and k, $\sum_{j=1}^n a_{ij}b_{kj}d_j^{-1} = e_{ik}$, or, in particular, $i \neq k$ and $\mathbf{a}_{i.} \cdot \mathbf{b}_{k.} = 0$, then either $\mathbf{a}_{i.}$ or $\mathbf{b}_{k.}$ is a $\mathbf{d}$-row.

The following simple lemma will also be useful in the sequel.

Lemma 8. *If (A, B) is a $\mathbf{d}$-design, then the columns of A and B are affinely independent. Moreover, if B is singular then $B\mathbf{d}^{-1} = \mathbf{0}$.*

6   Applications to the Block Design

Lemma 6 has some important consequences in the classical design theory. In particular, it implies that if $\mathbf{d}$ has full support, then $\lambda J + diag(\mathbf{d})$ is nonsingular iff $\lambda\mathbf{d}^{-1} \cdot \mathbf{1} \neq -1$. This simple fact, which can be easily proved directly, was used by Bose [1] (see also [12]) to prove the well-known Fisher's inequality asserting that *in a pairwize balanced design the number of blocks is greater or equal than the number of points.*

Theorem 5 ([12]). *[1] Let $\mathcal{A} = \{A_1, \dots, A_m\}$, where $A_1, \dots, A_m$ are distinct subsets of a set of n elements such that for each $i \neq j$, $1 \leq |A_i \cap A_j| = \lambda < n$. Then $m \leq n$.*

Proof. If one of the sets has λ elements, then all the other sets contain this one and are disjoint otherwise. It follows that $m \leq n$. Hence we may suppose that $d_i = |A_i| - \lambda > 0$. Then $A^T A = diag(\mathbf{d}) + \lambda J$. Since $\lambda\mathbf{d}^{-1} \cdot \mathbf{1} \neq -1$, $m \leq$ rk $A \leq n$.

Here is another useful consequence of Lemma 6:

Corollary 1. *Let $A, B \in \mathcal{R}^{n \times n}$. If $A^T B = \lambda J + (k - \lambda)I$, where $\lambda(1 - n) \neq k$, then*

$$A^T B = BA^T \Leftrightarrow JB = rJ; JA = sJ \Leftrightarrow BJ = rJ; AJ = sJ.$$

Proof. By Lemma 6 we have:

$$BA^T = \lambda(\lambda(n-1)+k)^{-1}BJA^T + (k-\lambda)I.$$

Hence $A^T B = BA^T \Leftrightarrow BJA^T = tJ$, for some $t \Leftrightarrow BJ = rJ; AJ = sJ$, where $rs = \lambda(n-1)+k \Leftrightarrow JB = rJ; JA = sJ$, where $rs = \lambda(n-1)+k$ (as $A^T B = \lambda J + (k-\lambda)I$).

Thus, we get the following well-known result of Ryser [12]:

Theorem 6 ([12]). *The dual of a symmetric design is again a symmetric design.*

Proof. Suppose $AA^T = (k-\lambda)I + \lambda J$. Then $JA = kJ$ and by Corollary 1, $AJ = kJ$ and $A^T A = AA^T = (k-\lambda)I + \lambda J$.

The following interesting fact also can be easily deduced from Lemma 6.

Theorem 7 ([12]). *In any square design, there exists a set incident to each given pair of points.*

7   Some Sufficient Conditions

The following theorem completely characterizes the **d**-designs (A, B), where $J - A$ is disconnected (the proof is omitted).

Theorem 8. *Let (A, B) be a **d**-design. If it is not a DPP-design and $J - A$ is disconnected, then the following cases are possible:*

1.
$$(A, B) \cong \left(\begin{bmatrix} 1\,0\,0 \\ 1\,1\,0 \\ 1\,0\,1 \end{bmatrix}, \begin{bmatrix} 1\,0\,0 \\ 1\,0\,1 \\ 1\,1\,0 \end{bmatrix} \right);$$

2.
$$(A, B) \cong \left(\begin{bmatrix} 1 & e_1^T \\ 1 & J - I \end{bmatrix}, \begin{bmatrix} 1 & 0 \\ e_1 & I \end{bmatrix} \right);$$

3.
$$(A, B) \cong \left(\begin{bmatrix} 0 & 1 & 1 \\ 1 & J-I & J \\ 1 & 0 & I \end{bmatrix}, \begin{bmatrix} 1\,0\,1 \\ 0\,I\,0 \\ 0\,0\,I \end{bmatrix} \right);$$

4.
$$(A, B) \cong \left(\begin{bmatrix} 1\,0\,1\,1 \\ 0\,0\,1\,1 \\ 1\,1\,0\,0 \\ 1\,1\,0\,1 \end{bmatrix}, \begin{bmatrix} 1\,1\,0\,0 \\ 0\,0\,0\,1 \\ 1\,0\,0\,0 \\ 0\,0\,1\,1 \end{bmatrix} \right).$$

Theorem 9. *If (A, B) is a **d**-design without **d**-rows, and $n > 3$, then either (A, B) is an (r, s)-design or it is a DPP-design.*

Proof. As both $A\mathbf{d}^{-1}$ and $B\mathbf{d}^{-1}$ have full supports, it follows from Lemma 8 that (A, B) is nonsingular. If either A or B has an all one column, then it is not difficult to show that $n \leq 3$. So suppose that A and B have no all one columns. Denote by $L^j = \{l : a_{lj} = 1\}$; $K^i = \{k : a_{ik} = 1\}$; and $B^{ij} = B[L^j; K^i]$. Let $a_{ij} = 0$. As $\mathbf{a}_{.j} \cdot \mathbf{b}_{.k} = 1$, for each $k \in K^i$, and $\mathbf{a}_{i.} \cdot \mathbf{b}_{l.} \geq 1$, for each $l \in L^j$ (by Lemma 7), it follows that each row of B^{ij} contains at least one 1 and each column of B^{ij} contains exactly one 1. Hence $|K^i| \geq |L^j|$, for each pair i and j such that $a_{ij} = 0$. Therefore, by Lemma 2, $|K^i| = |L^j|$, B^{ij} is a permutation matrix, and both A and B are DE-matrices. Moreover, if $a_{ij} = 0$ and $a_{lj} = 1$, then $\mathbf{a}_{i.} \cdot \mathbf{b}_{l.} = 1$. Suppose now that $\mathbf{a}_{i.} \geq \mathbf{a}_{l.}$. As by Lemma 3, A has no all one row, there exists a j such that $a_{ij} = a_{lj} = 0$. Since A is a DE-matrix, $\mathbf{a}_{i.} = \mathbf{a}_{l.}$, a contradiction. Hence $\mathbf{a}_{i.} \cdot \mathbf{b}_{l.} = 1$, for each $i \neq l$. That is $BA^T = J + diag(\mathbf{d}')$, for some vector $\mathbf{d}'$.

Now, it follows from Theorem 8 and Lemma 5 that either (A, B) is a DPP-design, or both A and B are totally regular. Suppose A is totally r-regular and B is totally s-regular. Then $rsJ = JA^TB = nJ + J diag(\mathbf{d})$, hence $d_1 = d_2 \cdots = d_n = rs - n$, and $A^TB = J + (sr - n)I$. Similarly, $BA^T = J + (sr - n)I = A^TB$.

Notice that in the proof we are using just a very special case of Theorem 8 (when both A and B are DE-matrices), which can be easily proved directly.

As a consequence we get the important result of Lehman on the structure of square minimally non-ideal matrices.

Corollary 2 ([7]). *If (A, B) is a **d**-design, where $\mathbf{d} \geq 1$, then either (A, B) is an (r, s)-design or it is a DPP-design.*

Here are two other useful consequences:

Corollary 3. *If $A^TB = J - I$, then (A, B) is an (α, ω)-design, where $n = \alpha\omega + 1$.*

Corollary 4. *If (A, B) is a **d**-design, where A and B both are uniform, then it is an (r, s)-design.*

Proof. If A is uniform, then the solution of $A^T\mathbf{x} = \mathbf{1}$ has full support. Hence $B\mathbf{d}^{-1}$ also has full support and B has no **d**-rows.

For our next result we need the following nice lemma of Sebő [14]:

Lemma 9 ([14]). *If (A, B) is a **d**-design, where A is r-regular, then, for each $j \leq n$, $d_j \equiv -n \pmod{r}$.*

The following theorem characterizes the nonsingular **d**-designs (A, B), where A is regular:

Theorem 10. *If (A, B) is a nonsingular **d**-design, where A is r-regular, then either it is an (r, s)-design or a G-design.*

Proof. Suppose (A, B) is not an (r, s)-design. Then by Lemma 9 and Theorem 9, for each j, either $d_j = -1$ or $d_j \geq r - 1$, and by Theorem 9, either A or B has a **d**-row. Consider two cases:

Case 1: $\mathbf{a}_{1.}$ is a **d**-row. Now, it follows from Lemma 7 that for each $i \neq 1$, either $\mathbf{a}_{1.} \cdot \mathbf{b}_{i.} = 0$ or $\mathbf{a}_{1.} \leq \mathbf{b}_{i.}$. As B has no equal columns, it follows that $r = 2$, $d_j = \pm 1$, hence (A, B) is a G-design.

Case 2: $\mathbf{b}_{1.}$ is a **d**-row having maximum number of ones.

Suppose $\mathbf{b}_{1.} = (1 \ldots 1, 0 \ldots 0)$, where $\mathbf{b}_{1.} \mathbf{1} = k$. Then we have that for each $i \neq 1$, either $\mathbf{a}_{i.} \cdot \mathbf{b}_{1.} = 0$ or $\mathbf{a}_{i.} \cdot \mathbf{b}_{1.} = r$. Moreover, $\mathbf{a}_{1.} \cdot \mathbf{b}_{1.} = r - 1$. It follows that $r \leq k < n$. Suppose $\mathbf{a}_{i.} \cdot \mathbf{b}_{1.} = r$ if $1 < i \leq l$, $\mathbf{a}_{i.} \cdot \mathbf{b}_{1.} = 0$ if $i \geq l$, and $\mathbf{a}_{1 k+1} = 1$. Denote by $A_1 = A[l+1 \ldots n; k+1 \ldots n]$, $B_1 = B[l+1 \ldots n; k+1 \ldots n]$. As A is nonsingular, $l \geq k$. On the other hand, $A_1^T B_1 = diag(d_{k+1} \ldots d_n) + J$, where $\sum_{j=k+1}^{n} d_j^{-1} = \sum_{j=1}^{n} d_j^{-1} \neq -1$, hence $k \geq l$. It follows that $k = l$ and A_1 is nonsingular. Hence the equation $A_1^T \mathbf{x} = \mathbf{1} - \mathbf{e}_1$ has a unique solution. As all the columns of $B' = B[k+1 \ldots n; 1 \ldots k]$ satisfy that equation, B' has an all one row. Suppose $\mathbf{b}_{p.}$ is the row of B corresponding to that row of B'. As $\sum_{j=1}^{n} a_{2j} b_{pj} d_j^{-1} = 0$, $\mathbf{b}_{p.}$ is a **d**-row having more ones than $\mathbf{b}_{1.}$, which is a contradiction.

Notice that only Lemma 8 and Theorem 8 yet contain some information about the structure of singular designs. The characterization of singular designs seems to be a more difficult problem, as Lemma 6 cannot be applied directly. In particular, it would be interesting to check whether there exist singular **d**-designs (A, B), where A is r-regular, and $r > 2$. A partial answer to this question is given in [14]. Here is another result on that direction. The prove is similar to the proof of Theorem 10.

Lemma 10. *If (A, B) is a **d**-design, where A is r-regular and $r > 1$, then A is connected.*

8 Sharper Characterizations for (α, ω)-Graphs

In this section we apply Theorem 9 to get some new, smaller sets of conditions characterizing (α, ω)-graphs. It is not difficult to deduce from Theorem 9 (see [3]) that G is an (α, ω)-graph iff it has a family of n cliques $\mathcal{A}$ and a family of n stable sets $\mathcal{B}$ such that $A^T B = J - I$. The following reformulation of this statement is a strengthening of a similar result of Hougardy and Gurvich [6].

Corollary 5. *If G has a family of $\leq n$ cliques covering all the edges of G such that for each vertex $v \in V$, $G \backslash v$ can be partitioned with the help of these cliques, then G is an (α, ω)-graph.*

The following theorem provides another characterization of (α, ω)-graphs.

Theorem 11 ([4]). *G is an (α, ω)-graph iff it has an α-stable set A_1 such that for each vertex $s \in A_1$ and stable set $S \subset V$; $\chi(G \backslash s) = \omega = \omega(G \backslash S)$.*

Proof. Let $\mathcal{A} := \{A_1, A_2, \ldots, A_{\alpha\omega+1}\}$, where A_1 is the stable set occurring in the theorem; fixing an ω-coloration of each of the α graphs $G\backslash s$ ($s \in A_1$), $A_2, \ldots, A_{\alpha\omega+1}$ denote the stable sets occurring as a color-class in one of these colorations. Define $\mathcal{B} := \{B_1, B_2, \ldots, B_{\alpha\omega+1}\}$, where B_i is an ω-clique of $G\backslash A_i$. Now it is straightforward to check that $A^T B = J - I$ (see [4]). Since $G\backslash s$ ($s \in A_1$) has a partition into ω stable sets, $n \leq \alpha\omega + 1$ is obvious. On the other hand, A has full column rank, and $n \geq \alpha\omega + 1$ follows. Thus, $n = \alpha\omega + 1$ and $(\mathcal{A}, \mathcal{B})$ is an (α, ω)-design. The fact that $\mathcal{B}$ is the set of all ω-cliques of G follows now by noticing that an arbitrary ω-clique Q is disjoint from exactly one element $A_i \in \mathcal{A}$: its incidence vector is the unique solution of the equation $A^T\mathbf{x} = \mathbf{1} - \mathbf{e}_i$. On the other hand, one of the columns of B also satisfies this equation. In the same way $\mathcal{A}$ is the set of all α-stable sets.

Notice that Theorem 11 immediately implies Theorem 2 and all the properties of minimal imperfect graphs shown by Padberg [11].

From the proof of Theorem 11 we get:

Corollary 6 ([16]). *Let G be a graph and S_0 be an α-stable set in G. If $n = \alpha\omega+1$ and for each vertex $s \in S_0$ and α-stable set $S \subset V$; $\chi(G\backslash s) = \omega = \omega(G\backslash S)$, then G is an (α, ω)-graph.*

Here is another interesting consequence of Theorem 9:

Corollary 7 ([16]). *G is partitionable iff for some $p, q \geq 2$ such that $n \leq pq+1$, G has a family of n stable sets, $\mathcal{A}$, such that each vertex is in at least p of the sets $\mathcal{A}$; and $\mathcal{A}$ has no sets intersecting every q-clique.*

Proof. Let B be the matrix the ith column of which is the incidence vector of a q-clique disjoint from the stable set corresponding to the ith column of A. Then $\mathbf{1}^T A^T B \mathbf{1} \geq pqn$, hence $n = pq + 1$ and $A^T B = J - I$.

9   Design Vertices and Facets in Polyhedra

The following two theorems contain both Padberg's theorem on minimally imperfect polytopes [11] and Lehman's theorem on minimally non-ideal polyhedra [8], and the second one also contains the part of Sebő [14], which corresponds to nonsingular matrix equations. In the proofs we mainly use the ideas of Lehman [8], Padberg [10], several results on **d**-designs of the present work and the following simple but surprising fact communicated by Sebő [14]: *if P is a facet $\{0, 1\}$ polyhedron such that, for each $i \leq n$, both $P\backslash i$ and P/i are $\{0, 1\}$-polyhedra then P is full dimensional.*

Theorem 12. *Suppose P is a full dimensional, vertex $\{0, 1\}$ polyhedron such that, for each $1 \leq i \leq n$, both $P\backslash i$ and P/i are $\{0, 1\}$-polyhedra, and F is a non-$\{0, 1\}$ facet of P. Then F is a **d**-design facet. Moreover, if we denote by*

(A, B) *the* **d**-*design corresponding to* F, *then the following cases are possible:*
1. *Either* (A, B) *is a DPP-design or*

$$(A, B) \cong \left(\begin{bmatrix} 0 & 1 \\ 1 & I \end{bmatrix}, \begin{bmatrix} 1 & 1 \\ 0 & I \end{bmatrix} \right);$$

and then F *is the unique non-*$\{0, 1\}$ *facet of* P.
2. A *is totally regular and* P *has at most two non-*$\{0, 1\}$ *facets. Moreover, if* B *is nonsingular, then* F *either is an* (r, s)-*design facet or a* C-*design facet.*

Proof. Let $F = \{\mathbf{x} \in P : \mathbf{a} \cdot \mathbf{x} = 1\}$. As for each i, P/i is a $\{0, 1\}$-polyhedron, it follows that $\mathbf{a}$ has full support. Denote by A the matrix, the ith column of which is the ith vertex of F. Since $\mathbf{a}$ has full support, it follows that F is bounded, A has full row rank and has no all one column. On the other hand, as $P \backslash i$ is a $\{0, 1\}$-polyhedron, it follows that for each i, *affine hull* (A/i) can be given with the help of $\{0, 1\}$-constraints. Hence by Theorem 4, A is a DE-matrix. Now, it follows from Lemma 4 that all the neighboring facets of F are $\{0, 1\}$. F cannot have neighboring facets of type $x_i = 0$, for otherwise $\mathbf{a}_{1.} = \mathbf{e}_1^T$ and $A = I$, which is impossible. Hence F is a **d**-design facet.

Thus, by Theorem 8, either $A \cong$ DPP or A is totally regular. If $A \cong$ DPP then it is not difficult to proof that we have case 1. As P can have at most two parallel facets, the proof is finished by Theorem 10.

Theorem 13. *Let* P *be a facet* $\{0, 1\}$ *polyhedron. If for each* $i \leq n$, *both* $P \backslash i$ *and* P/i *are* $\{0, 1\}$-*polyhedra then either* $P = P_{\geq}(A)$, *where* $A \cong$ DPP, *or* P *has at most two fractional vertices, both of which are* (r, s)-*design vertices.*

Acknowledgments. I am very grateful to András Sebő for several helpful communications, which have been essential in preparing this article, and to Hasmik Lazaryan for detecting some errors in the preliminary versions.

References

1. R. S. Bose. A note on Fisher's inequality for balanced incomplete block design. *Ann. Math. Stat.*, 20:619–620, 1949.
2. N. G. de Bruijn and P. Erdős. On a combinatorial problem. *Indag. Math.*, 10:421–423, 1948.
3. V. Chvátal, R. L. Graham, A. F. Perold, and S. H. Whitesides. Combinatorial designs related to the strong perfect graph conjecture. *Discrete Math.*, 26:83–92, 1979.
4. G. S. Gasparyan. Minimal imperfect graphs: A simple approach. *Combinatorica*, 16(2):209–212, 1996.
5. G. S. Gasparyan and A. Sebő. Matrix equations in polyhedral combinatorics. In preparation.
6. S. Hougardy and V. Gurvich. Partitionable Graphs. Working paper.
7. A. Lehman. No the width-length inequality. *Math. Programming*, 17:403–413, 1979.

8. A. Lehman. The width-lenght inequality and degenerated projective planes. In W. Cook and P. D. Seymour, editors, *Polyhedral Combinatorics, DIMACS, Vol. 1*, pages 101–105, 1990.

9. L. Lovász, A characterization of perfect graphs. *J. of Combin. Theory*, 13:95–98, 1972.

10. M. Padberg. Lehman's forbidden minor characterization of ideal 0–1 matrices. *Discrete Math.*, 111:409–420, 1993.

11. M. Padberg. Perfect zero-one matrices. *Math. Programming*, 6:180–196, 1974.

12. H. Ryser. An extension of a theorem of de Bruijn and Erdős on combinatorial designs. *J. Algebra*, 10:246–261, 1968.

13. A. Schrijver. *Theory of Linear and Integer Programming*. Wiley, New York, 1986.

14. A. Sebő. Characterizing noninteger polyhedra with 0–1 constraints. In R. E. Bixby, E. A. Boyd, and R. Z. Ríos-Mercado, editors, *Proceedings of the 6th International IPCO Conference, LNCS, Vol. 1412*, pages 36–51. Springer, 1998. This volume.

15. P. D. Seymour. On Lehman's width-length characterization. *DIMACS*, 1:107–117, 1990.

16. F. B. Shepherd. Nearly-perfect matrices. *Math. Programming*, 64:295–323, 1994.

Characterizing Noninteger Polyhedra with 0–1 Constraints

András Sebő [*]

CNRS, Laboratoire Leibniz-IMAG, Grenoble, France
http://www-leibniz.imag.fr/DMD/OPTICOMB/Membres/sebo/sebo.html

Abstract. We characterize when the intersection of a set-packing and a set-covering polyhedron or of their corresponding minors has a noninteger vertex. Our result is a common generalization of Lovász's characterization of 'imperfect' and Lehman's characterization of 'nonideal' systems of inequalities, furthermore, it includes new cases in which both types of inequalities occur and interact in an essential way. The proof specializes to a conceptually simple and short common proof for the classical cases, moreover, a typical corollary extracting a new case is the following: *if the intersection of a perfect and an ideal polyhedron has a noninteger vertex, then they have minors whose intersection's coefficient matrix is the incidentce matrix of an odd circuit graph.*

1 Introduction

Let $A^{\leq}$ and $A^{\geq}$ be *0–1-matrices* (meaning that each entry is 0 or 1) with n columns. We will study the integrality of the intersection $P(A^{\leq}, A^{\geq}) := P^{\leq}(A^{\leq}) \cap P^{\geq}(A^{\geq})$ of the *set-packing polytope* $P^{\leq}(A^{\leq}) = \{x \in \mathbb{R}^n : A^{\leq} x \leq 1, x \geq 0\}$ and the *set-covering polyhedron* $P^{\geq}(A^{\geq}) = \{x \in \mathbb{R}^n : A^{\geq} x \geq 1, x \geq 0\}$. We will speak about $(A^{\leq}, A^{\geq})$ as a system of inequalities, or simply *system.*

Obviously, one can suppose that both the rows of $A^{\leq}$ and those of $A^{\geq}$ are incidence ('characteristic') vectors of a *clutter*, that is of a family of sets none of which contains the other. The sets in the clutters and their 0–1 incidence vectors will be confused, and with the same abuse of terminology, clutters and their matrix representations (where the rows are the members of the clutter) will not be distinguished. If $A^{\leq}$ and $A^{\leq}$ do not have equal rows, that is (explicit) *equalities*, we will say that $(A^{\leq}, A^{\geq})$ is *simple.*

The constraints defining $P^{\leq}(A^{\leq})$ will be called of *packing* type, and those defining $P^{\geq}(A^{\geq})$ of *covering* type. A vertex of $P(A^{\leq}, A^{\geq})$ can also be classified to be of *packing* type, of *covering* type, or of *mixed* type, depending on whether all nonequality constraints containing the vertex are of packing type, of covering type, or both types occur.

---

[*] Visiting the Research Institute for Mathematical Sciences, Kyoto University.

R. E. Bixby, E. A. Boyd, and R. Z. Ríos-Mercado (Eds.): IPCO VI
LNCS 1412, pp. 37–52, 1998. © Springer–Verlag Berlin Heidelberg 1998

The blocker of a clutter $A^{\geq}$ (or the antiblocker of $A^{\leq}$) is the set of inclusionwise minimal (resp. maximal) integer vectors in $P^{\geq}(A^{\geq})$ (resp. $P^{\leq}(A^{\leq})$). These are 0–1-vectors and define another clutter.

A *polyhedron* in this paper is the set of all (real) solutions of a system of linear inequalities with integer coefficients. A *polytope* is a bounded polyhedron. For basic definitions and statements about polyhedra we refer to Schrijver [11], and we only repeat now shortly the definition of the terms we are using directly. A *face* of a polyhedron is a set we get if we replace certain defining inequalities with the equality so that the resulting polyhedron is nonempty. A polyhedron is *integer* if each of its faces contains an integer point, otherwise it is *noninteger*.

If $X \subseteq \mathbb{R}^n$, we will denote by $r(X)$ the (linear) *rank* of X, and by $\dim(X)$ the *dimension* of X, meaning the rank of the differences of pairs of vectors in X, that is, $\dim(X) := r(\{x - y : x, y \in X\})$.

If P is a polyhedron, then its faces of dimension $\dim(P) - 1$ are called *facets*, and its faces of dimension 0, *vertices*. All (inclusionwise) minimal faces of P have the same dimension. We say that P *has vertices*, if this dimension is 0.

It is easy to see that $P(A^{\leq}, A^{\geq})$ has vertices for all $A^{\leq}, A^{\geq}$. (If a minimal face is not of dimension 0, it contains an entire line, contradicting some non-negativity constraint.) So $P(A^{\leq}, A^{\geq})$ is integer if and only if it has integer vertices.

A vertex of a full dimensional polyhedron is *simplicial*, if it is contained in exactly n facets. A simplicial vertex has n neighbouring vertices. Neighbours share $n - 1$ facets.

If $A^{\leq}$ is empty, a combinatorial coNP characterization of the integrality of $P(A^{\leq}, A^{\geq})$ is well-known (Lovász [8], Padberg [9]). If $A^{\geq}$ is empty, a recent result of Lehman solves the problem (Lehman [6], Seymour [12]). A common generalization of these could be a too modest goal: if for every $i \in \{1, \ldots, n\}$ either the i-th column of $A^{\leq}$ or that of $A^{\geq}$ is 0, then the nonintegrality of $P(A^{\leq}, A^{\geq})$ can be separated to the two 'classical' special cases. Such systems $(A^{\leq}, A^{\geq})$ contain both special cases, but nothing more. There are less trivial examples where $P^{\leq}(A^{\leq})$ and $P^{\geq}(A^{\geq})$ do not really interact in the sense that all fractional vertices of $P^{\leq}(A^{\leq}, A^{\geq})$ are vertices of $P^{\leq}(A^{\leq})$ or of $P^{\geq}(A^{\geq})$.

In this work we characterize when the intersection of a set-packing and a set-covering polyhedron or that of any of their corresponding minors is noninteger. The results contain the characterizations of perfect and ideal polyhedra and new cases involving mixed vertices. The special cases are not used and are not treated separately by the proof: a common proof is provided for them instead.

Graphs $G = (V, E)$ are always undirected, $V = V(G)$ is the vertex-set, $E = E(G)$ the edge-set; $\underline{1}$ is the all 1 vector of appropriate dimension.

If $x \in \mathbb{R}^n$, the *projection* of x parallel to the i-th coordinate is the vector $x^i = (x_1, \ldots, x_{i-1}, x_{i+1}, \ldots, x_n)$. Let us fix the notation $V := \{1, \ldots, n\}$. If $X \subseteq \mathbb{R}^n$, the projection parallel to the i-th coordinate of the set X is $X^i := \{x^i : x \in X\}$; if $I \subseteq V$, X^I is the result of successively projecting parallel to $i \in I$ (the order does not matter).

Let $P := P^{\leq}(A^{\leq})$ or $P := P^{\geq}(A^{\geq})$, and $I, J \subseteq V$, $I \cap J = \emptyset$. A *minor* of P is a polyhedron $P \setminus I / J := (P \cap \{x : x_i = 0 \text{ if } i \in I\})^{I \cup J}$. The set I is said to be *deleted*, whereas J is *contracted*. For set-packing polyhedra the contraction of J is the same as the deletion of J.

It is easy to see that $P^{\leq} \setminus I/J = P^{\leq}(A'^{\leq})$, and $P^{\geq} \setminus I/J = P^{\geq}(A'^{\geq})$, where $A'^{\leq}$, $A'^{\geq}$ arise from $A^{\leq}$, resp. $A^{\geq}$ in a simple way: delete the columns indexed by I, and then delete those rows that are no more maximal, resp. minimal; for $A^{\leq}$ do the same with J; for $A^{\geq}$ delete the columns indexed by J and also delete all the rows having a 1 in at least one of these columns. Hence *minors of set-packing or set-covering polyhedra are of the same type.*

We do not use the terms 'contraction' or 'deletion' for matrices (or clutters), because that would be confusing here for several reasons, one of which being that these operations do not only depend on the matrix (or clutter) itself. But we define the *minors of the ordered pair* $(A^{\leq}, A^{\geq})$: $(A^{\leq}, A^{\geq}) \setminus I/J := (A'^{\leq}, A'^{\geq})$, where I, J, $A'^{\leq}$, $A'^{\geq}$ are as defined above. The polyhedra $P^{\leq}(A'^{\leq})$ and $P^{\geq}(A'^{\geq})$ will be called *corresponding minors* of the two polyhedra $P^{\leq}(A'^{\leq})$ and $P^{\geq}(A'^{\geq})$. Parallelly, for a clutter (matrix) $\mathcal{A}$ and $v \in V$ we define the clutter $\mathcal{A} - v := \{A \in \mathcal{A} : v \notin A\}$ on $V \setminus \{v\}$.

If $P = P^{\leq}(A^{\leq})$ is integer, then P and $A^{\leq}$ are called *perfect*, whereas if $P = P^{\geq}(A^{\geq})$ is integer, P and $A^{\geq}$ are called *ideal*. All minors of perfect and ideal matrices are also ideal or perfect respectively. If a matrix is not perfect (not ideal) but all its proper minors are, then it is called *minimal imperfect*, or *minimal nonideal* respectively. It is easy to see that the family $\mathcal{H}_n^{n-1}$ of the $n - 1$-tuples of an n-set is minimal imperfect, and it is also an easy and well-known exercise to show that matrices not containing such a minor (or equivalently having the 'dual Helly property') can be represented as the (inclusionwise) maximal cliques of a graph. We will call $\mathcal{H}_n^{n-1}$ $(n = 3, 4, \ldots,)$ *minimal nongraph* clutters. The *degenerate projective plane* clutters $\mathcal{F}_n = \{\{1, \ldots, n-1\}, \{1, n\}, \{2, n\}, \ldots, \{n-1, n\}\}$, $(n = 3, 4, \ldots)$ are minimal nonideal.

It is easy to show that the blocker of the blocker is the original clutter. The antiblocker of the antiblocker of $\mathcal{H}_n^{n-1}$ is not itself, and this is the only exception: it is another well-known exercise to show that the antiblocker of the antiblocker of a clutter that has no $\mathcal{H}_n^{n-1}$ minor (dual Helly property), is itself.

A graph G is called *perfect* or *minimal imperfect* if its clique-matrix is so. It is said to be *partitionable*, if it has $n = \alpha\omega + 1$ vertices $(\alpha, \omega \in \mathbb{N})$, and for all $v \in V(G)$, $G - v$ can be partitioned both into α cliques and into ω stable-sets. Lovász [8] proved that minimal imperfect graphs are partitionable and Padberg [9] proved further properties of partitionable graphs.

Analogous properties have been proved for nondegenerate minimal nonideal clutters by Lehman [6], from which we extract: a pair of clutters $(\mathcal{A}, \mathcal{B})$, where $\mathcal{B}$ is the blocker or the antiblocker of $\mathcal{A}$ will be called *partitionable*, if they are defined on $V := \{1, \ldots, n\}$, $n = rs - \mu + 1$, $(r, s \in \mathbb{N}, \mu \in \mathbb{Z}, 0 \leq \mu \leq \min\{r, s\}), \mu \neq 1$, and for all $v \in V$ there exist sets $A_1, \ldots, A_s \in \mathcal{A}$ and sets $B_1, \ldots, B_r \in \mathcal{B}$ such that $v \in A_1, \ldots, A_\mu, B_1, \ldots, B_\mu$ and both $\{A_1 \setminus v, \ldots, A_\mu \setminus v, A_{\mu+1}, \ldots, A_s\}$ and $\{B_1 \setminus v, \ldots, B_\mu \setminus v, B_{\mu+1}, \ldots, B_r\}$ are partitions of $V \setminus \{v\}$.

Remark 1. The clique matrix of a partitionable graph is a partitionable clutter with $\omega = r, \alpha = s, \mu = 0$.

Supposing that $(\mathcal{A}, \mathcal{B})$ is partitionable, it is easy to see that they are antiblockers of each other if and only if $\mu = 0$, and they are blockers of each other if and only if $\mu \geq 2$. Indeed, if $(\mathcal{A}, \mathcal{B})$ are partitionable it can be shown (Padberg [9], [10]) that $\mathcal{A}$ and $\mathcal{B}$ have exactly n members and these can be indexed $A_1, \ldots, A_n$, $B_1, \ldots, B_n$ so that $|A_i \cap B_j|$ is 1 if $i \neq j$ and is μ if $i = j$. These properties will be proved directly for general 'minimal noninteger systems'.

We will call $\mathcal{A}$ partitionable, if $(\mathcal{A}, \mathcal{B})$ is partitionable, where $\mathcal{B}$ is the blocker or the antiblocker of $\mathcal{A}$ – we will always make clear which of the two is meant.

Let $\mathcal{A}$ be partitionable. Clearly, $1/r\underline{1} \in P^{\leq}(\mathcal{A})$ if $\mu = 0$, and $1/r\underline{1} \in P^{\geq}(\mathcal{A})$ if $\mu \geq 2$ (it is actually the unique full support noninteger vertex of $P^{\leq}$ or $P^{\geq}$, for minimal nonideal or minimal imperfect polyhedra, it is the unique fractional vertex). Let us call this the *regular* vertex of $P^{\leq}(\mathcal{A})$, or of $P^{\geq}(\mathcal{A})$. The regular vertex of $\mathcal{F}_n$ and that of $\mathcal{H}_n^{n-1}$ is defined as their unique fractional vertex.

The idea of this work originates in the frustrating similarities between minimal imperfect and minimal nonideal matrices and the proofs of the results. This similarity becomes fascinating when comparing Seymour's proof [12] of Lehman's, and Gasparyan's direct proof [3] of Lovász's and Padberg's theorems.

Despite these similarities, the generalization has to deal with several new phenomena, for instance $P(A^{\leq}, A^{\geq})$ can be empty, and its dimension can also vary. (Antiblocking and blocking polyhedra are trivially full dimensional !) We will meet many other difficulties that oblige us to generalize the notions and arguments of the special cases – without making the solution much more complicated. The proof synthesizes polyhedral and combinatorial arguments, moreover a lemma involving the divisibility relations between the parameters will play a crucial role when mixed fractional vertices occur.

We show now an example with mixed vertices. Surprisingly, this will be the only essential ('minimal noninteger') new example where the two types of inequalities interact in a nontrivial way. In a sense, a kind of 'Strong Perfect Graph Conjecture' is true for mixed polyhedra.

If $A^{\leq} \cup A^{\geq} = E(C_{2k+1}) \subseteq 2^{V(G)}$ ($k \in \mathbb{N}$), and neither $A^{\leq}$ nor $A^{\geq}$ is empty, then $P(A^{\leq}, A^{\geq})$ will be called a *mixed odd circuit polyhedron*, and $(A^{\leq}, A^{\geq})$ will be called a *mixed odd circuit*. The unique fractional vertex of a mixed odd circuit polyhedron is $1/2\underline{1}$.

Let now $(A^{\leq}, A^{\geq})$ be a simple odd circuit. Let us define B_i to be the (unique) subset of vertices of the graph C_{2k+1} having exactly one common vertex with every edge of C_{2k+1} except with $(i, i + 1)$; the number of common vertices of B_i with the edge $(i, i + 1)$ is required to be zero or two depending on whether its incidence vector is in $A^{\leq}$ or $A^{\geq}$ respectively ($i = 1, \ldots, 2k + 1$, $i + 1$ is understood mod $n = 2k + 1$). The neighbors of the vertex $1/2\underline{1}$ on $P(A^{\leq}, A^{\geq})$ are the characteristic vectors of the B_i, ($i = 1, \ldots, n = 2k + 1$). Follow these and other remarks on C_7:

Example 1. (an odd circuit polyhedron) Let us define $P(A^{\leq}, A^{\geq}) \subseteq \mathbb{R}^7$ with:
$x_i + x_{i+1} \leq 1$ ($i = 1, 2, 3, 4$), $x_i + x_{i+1} \geq 1$ ($i = 5, 6, 7$; for $i = 7$, $i + 1 := 1$).

This polyhedron *remains noninteger* after projecting 1: indeed, the inequality $x_7 - x_2 \geq 0$ is a sixth facet-inducing inequality (containing the vertex $1/2\,\underline{1}$) besides the five remaining edge-inequalities. These six inequalities are linearly independent ! (The projection of a vertex is still a vertex if and only if the projection is parallel to a coordinate which is nonzero both in some set-packing and some set-covering facet containing the vertex.) But the new inequality is not 0–1 ! However, a study of nonintegrality should certainly include this example.

The vertices of $P(A^{\leq}, A^{\geq})$ are, besides $(1/2)\underline{1}$, the sets B_i, $(i = 1, \ldots, 7)$. These are the shifts of $B_2 := \{1, 4, 6\}$ by ± 1 and $0, 2$, and of $B_6 := \{2, 4, 6, 7\}$ by $0, \pm 1$. Note that the vector $(0, 1, 1, 1, 0, -1, -1)$ is orthogonal to all the B_i $(i = 1, \ldots, 7)$, whence the 7×7 matrix $\mathcal{B}$ whose rows are these, *is singular!*

In general, if $(A^{\leq}, A^{\geq})$ is a simple mixed odd circuit, and $A^{\leq}$ has one more row than $A^{\geq}$, then $\underline{1}^T A^{\leq} - \underline{1}^T A^{\geq}$ (defines a Chvátal-Gomory cut and) is orthogonal to all the B_i-s $(i = 1, \ldots, 2k + 1)$, so *they are linearly dependent !*

Linear independence of the neighbors of fractional vertices play a fundamental role in the special case of Padberg [9],[10], Lehman[6], and also in Gasparyan [3],[4]. Mixed odd circuits show that *we have to work here without this condition.* As a consequence we will not be able to stay within matrix terms, but will have to mix combinatorial and polyhedral arguments: Lemma 8 is mostly a self-contained lemma on matrices, where the polyhedral context, through Lemma 7 brings in a stronger combinatorial structure: '$r = 2$'. The matricial part of Lemma 8 reoccurs in papers [4] and [5], studying the arising matrix equations. The latter avoids the 'nonsingularity assumption' replacing Lemma 7 by combinatorial (algebraic) considerations.

This paper is organized as follows: Section 2 states the main result, its corollaries, and reformulations. The proof of the main result is provided in sections 3 and 4. Section 5 is devoted to some more examples and other comments.

2   Results

When this does not cause missunderstanding, we will occasionnally use the shorter notations $P^{\leq} := P^{\leq}(A^{\leq})$, $P^{\geq} := P^{\geq}(A^{\geq})$, $P := P(A^{\leq}, A^{\geq}) = P^{\leq} \cap P^{\geq}$. Recall that the polyhedra $P^{\leq}(A'^{\leq}) := P^{\leq} \setminus I/J$ and $P^{\geq}(A'^{\geq}) := P^{\geq} \setminus I/J$, $(I, J \subseteq V := \{1, \ldots, n\}, \ I \cap J = \emptyset)$ are called corresponding minors, and $(A'^{\leq}, A'^{\geq}) =: (A^{\leq}, A^{\geq}) \setminus I/J$ is a minor of $(A^{\leq}, A^{\geq})$. (Note that two minors are corresponding if and only if the two $I \cup J$ are the same, since for set-packing polyhedra deletion is the same as contraction.) Furthermore, if for all such I, J the polyhedron $(P^{\leq}(A^{\leq}) \setminus I/J) \cap (P^{\geq}(A^{\geq}) \setminus I/J)$ is integer, then the system $(A^{\leq}, A^{\geq})$ will be called *fully integer.*

Theorem 1. *Let $A^{\leq}$ and $A^{\geq}$ be 0–1-matrices with n columns. Then $(A^{\leq}, A^{\geq})$ is not fully integer if and only if it has a minor $(A'^{\leq}, A'^{\geq})$ for which at least one of the following three statements holds:*

– $A'^\leq$ is a minimal nongraph clutter, or it is partitionable with $\mu = 0$, moreover
   in either case the regular vertex of $P^\leq(A'^\leq)$ is in $P^\geq(A'^\geq)$, and it is the
   unique packing type fractional vertex of $P^\leq(A'^\leq) \cap P^\geq(A'^\geq)$.
– $A'^\geq$ is a degenerate projective plane, or it is partitionable with $\mu \geq 2$, moreover in either case the regular vertex of $P^\geq(A'^\geq)$ is in $P^\leq(A'^\leq)$, and it is
   the unique covering type fractional vertex of $P^\leq(A'^\leq) \cap P^\geq(A'^\geq)$.
– $(A'^\leq, A'^\geq)$ is a mixed odd circuit.

Lovász's NP-characterization of imperfect graphs [8] (with the additional
properties proved by Padberg[10]), follow:

Corollary 1. *Let $A^\leq$ be a 0–1-matrix with n columns. Then $A^\leq$ is imperfect
if and only if it has either a minimal nongraph or a partitionable minor $A'^\leq$,
moreover $P(A'^\leq)$ has a unique fractional vertex.*

Specializing Theorem 1 to set-covering polyhedra one gets Lehman's celebrated
result [6], see also Seymour [12]:

Corollary 2. *Let $A^\geq$ be a 0–1-matrix with n columns. Then $A^\geq$ is nonideal if
and only if it has either a degenerate projective plane or a partitionable minor
$A'^\geq$, moreover $P(A'^\geq)$ has a unique fractional vertex.*

The following two consequences are stated in a form helpful for coNP characterization theorems (see Section 5):

Corollary 3. *Let $A^\leq$ and $A^\geq$ be 0–1-matrices with n columns. Then $(A^\leq, A^\geq)$
is not fully integer if and only if at least one of the following statements holds:*

– $A^\leq$ has a minimal nongraph or a partitionable, furthermore minimal imperfect minor with its regular vertex in the corresponding minor of $P^\geq(A^\geq)$,
– $A^\geq$ has a degenerate projective plane or a partitionable minor with its regular
   vertex in the corresponding minor $P^\leq(A'^\leq)$ of $P^\leq(A^\leq)$, where $A'^\leq$ is perfect.
– $(A^\leq, A^\geq)$ has a mixed odd circuit minor.

If we concentrate on the structural properties of the matrices $A^\leq$ and $A^\geq$ implied
by the existence of a fractional vertex we get the following. This statement is not
reversible: if $A^\leq$ consists of the maximal stable-sets of an odd antihole, and $A^\geq$
of one maximal but not maximum stable-set, then $(A^\leq, A^\geq)$ is fully integer,
although $A^\leq$ is minimal imperfect !

Corollary 4. *Let $A^\leq$ and $A^\geq$ be 0–1-matrices with n columns and assume that
$P^\leq(A^\leq) \cap P^\geq(A^\geq)$ is a noninteger polyhedron. Then*

– either $A^\leq$ has a minimal imperfect minor,
– or $A^\geq$ has a degenerate projective plane, or a partitionable minor,
– or $(A^\leq, A^\geq)$ has a mixed odd circuit minor.

Note the asymmetry between 'minimal imperfect' in the first, and 'partitionable'
in the second case (for an explanation see 5.2).
   The results certainly provide a coNP characterization in the following case:

Corollary 5. *Let $A^{\leq}$ be a perfect, and $A^{\geq}$ an ideal 0–1-matrix with the same number of columns. Then $(A^{\leq}, A^{\geq})$ is fully integer if and only if it has no mixed odd circuit minor.*

These results provide a certificate for the intersection of a set-covering polyhedron and a set-packing polytope or of their corresponding minors to be noninteger. This certificate can be checked in polynomial time in the most interesting cases (see Section 5). We will however prove Theorem 1 in the following, slightly sharper form which leaves the possibility to other applications open – and corresponds better to our proof method:

We call $(A^{\leq}, A^{\geq})$ *combinatorially minimal noninteger*, if $P := P(A^{\leq}, A^{\geq})$ is noninteger, but $(P^{\leq} \setminus i) \cap (P^{\geq} \setminus i)$ and $(P^{\leq}/i) \cap (P^{\geq}/i)$ are fully integer for all $i = 1, \ldots, n$. Clearly, mixed odd circuits have this property.

Note the difference with the following definition which takes us out of 0–1 constraints: P is *polyhedrally minimal noninteger*, if it is noninteger, but $P \cap \{x \in \mathbb{R}^n : x_i = 0\}$ and P^i are integer for all $i \in V$.

Both the combinatorial and the polyhedral definitions require that the intersection of P with each hyperplane $x_i = 0$ ($i \in V$) is integer.

The two definitions are different only in what they require from projections, and this is what we are going to generalize now. When we are contracting an element, combinatorially minimal noninteger systems require the integrality of $P^{\leq}(A^{\leq})^i \cap P^{\geq}(A^{\geq})^i$ instead of the integrality of $\left[P^{\leq}(A^{\leq}) \cap P^{\geq}(A^{\geq})\right]^i$ in the polyhedral definition, and this is the only difference between the two. It is easy to see that $P^{\leq}(A^{\leq})^i \cap P^{\geq}(A^{\geq})^i \supseteq \left[P^{\leq}(A^{\leq}) \cap P^{\geq}(A^{\geq})\right]^i$, and we saw (see Example 1) that the equality does not hold in general, so the integrality of $P^{\leq}(A^{\leq})^i \cap P^{\geq}(A^{\geq})^i$ and that of $\left[P^{\leq}(A^{\leq}) \cap P^{\geq}(A^{\geq})\right]^i$ are seemingly independent of each other. The combinatorial definition looks actually rather restrictive, since it also requires that fixing a variable to 1 in $P^{\geq}(A^{\geq})$, and fixing the same variable to 0 in $P^{\leq}(A^{\leq})$ the intersection of the two polyhedra we get is integer.

Note however, that surprisingly, the results confirm the opposite: the combinatorial definition is *less* restrictive, since besides partitionable, minimal nongraph and degenerate projective clutters, it also includes mixed odd circuit polyhedra, which are not polyhedrally minimal noninteger !

Our proofs will actually not use more about the projections than the following simple *sandwich property* of P which is clearly implied by both combinatorial and polyhedral minimal nonintegrality (Q_i can be chosen to be the polyhedron on the left hand side or the one on the right hand side respectively):

for all $i = 1, \ldots, n$, there exists an integer polyhedron Q_i such that
$$\left[P^{\leq}(A^{\leq}) \cap P^{\geq}(A^{\geq})\right]^i \subseteq Q_i \subseteq P^{\leq}(A^{\leq})^i \cap P^{\geq}(A^{\geq})^i.$$
Let us call the system $(A^{\leq}, A^{\geq})$ *minimal noninteger*, if

- P is noninteger, and
- $P \cap \{x \in \mathbb{R}^n : x_i = 0\} (= P^{\leq} \cap \{x \in \mathbb{R}^n : x_i = 0\} \cap P^{\geq} \cap \{x \in \mathbb{R}^n : x_i = 0\})$ is an integer polyhedron for all $i \in V$, and
- P has the sandwich property.

Theorem 2. *If* $(A^\le, A^\ge)$ *is minimal noninteger, simple, and* $w \in P$ *is a fractional vertex, then* P *is full dimensional,* w *is simplicial, and at least one of the following statements hold:*

- w *is of packing type, and then* $A^\le$ *is either a minimal nongraph clutter, or the clique-matrix of a partitionable graph,*
- w *is of covering type, and then* $A^\ge$ *is either a degenerate projective plane or a partitionable clutter,* $\mu \ge 2$,
- w *is a mixed vertex, and then* $(A^\le, A^\ge)$ *is a mixed odd circuit.*

Moreover, P *has at most one fractional vertex of covering type, at most one of packing type, and if it has a vertex of mixed type, then that is the unique fractional vertex of* P.

Note that Theorem 2 sharpens Theorem 1 in two directions: first, the constraint of Theorem 2 does not speak about all minors, but only about the deletion and contraction of elements; second, the integrality after the contraction of elements is replaced by the sandwich property.

The corollaries about combinatorial and polyhedral minimal nonintegrality satisfy the condition of Theorem 2 for two distinct reasons. In the combinatorial case *simplicity does not necessarily hold,* but deleting the certain equalities from $A^\ge$, the system remains combinatorially minimal noninteger (see 5.2).

Corollary 6. *If* $(A^\le, A^\ge)$ *is combinatorially minimal noninteger, then at least one of the following statements holds:*

- $A^\le$ *is a minimal nongraph or a partitionable clutter with* $\mu = 0$, *furthermore it is minimal imperfect, and the regular vertex of* $P^\le(A^\le)$ *is the unique packing type fractional vertex of* $P^\le(A^\le) \cap P^\ge(A^\ge)$.
- $A^\ge$ *is a degenerate projective plane, or a partitionable clutter with* $\mu \ge 2$, *while* $A^\le$ *is perfect, and the regular vertex of* $P^\ge(A^\ge)$ *is in* $P^\le(A^\le)$.
- $(A^\le, A^\ge)$ *is a mixed odd circuit, and* $1/2\underline{1}$ *is its unique fractional vertex.*

This easily implies Theorem 1 and its corollaries using the following remark. (it is particularly close to Corollary 3), while the next corollary does not have similar consequences. This relies on the following:

- If P is noninteger, $(A^\le, A^\ge)$ *does contain* a combinatorially minimal noninteger minor. (Proof: In both $P^\le$ and $P^\ge$ delete and contract elements so that the intersection is still noninteger. Since the result *has still 0–1 constraints this can be applied successively* until arriving at a combinatorially minimal noninteger system.)
- If P is noninteger, one *does not necessarily arrive* at a polyhedrally minimal noninteger polyhedron with deletions and restrictions of variables. (Counterexample: Example 1.)

Corollary 7. *If* $P^\le(A^\le) \cap P^\ge(A^\ge)$ *is polyhedrally minimal noninteger, then at least one of the following statements holds:*

- *either $A^{\leq}$ is a minimal nongraph or a partitionable clutter with $\mu = 0$, and the regular vertex of $P^{\leq}(A^{\leq})$ is the unique packing type fractional vertex of $P^{\leq}(A^{\leq}) \cap P^{\geq}(A^{\geq})$.*
- *or $A^{\geq}$ is a degenerate projective plane, or a partitionable clutter with $\mu \geq 2$, and the regular vertex of $P^{\geq}(A^{\geq})$ is in $P^{\leq}(A^{\leq})$.*

Proof. Express w^i as a convex combination of vertices of P^i. Replacing the vectors in this combination by their lift, we get a vector which differs from w exactly in the i-th coordinate $(i = 1, \ldots, n)$ – if it did not differ, w would be the convex combination of integer vertices of P. So the i-th unit vector is in the linear space generated by P for all $i = 1, \ldots, n$, proving that P is full dimensional, in particular, simple. So Theorem 2 can be applied, and its third alternative cannot hold (see Example 1). $\qquad\square$

Gasparyan [4] has deduced this statement by proving that in the polyhedral minimal case the matrices involved in the matrix equations are nonsingular (see comments concerning nonsingularity in Example 1).

The main frame of the present paper tries to mix (the polar of) Lehman's polyhedral and Padberg's matricial approaches so as to arrive at the simplest possible proof. Lemmas 1–4 and Lemma 7 are more polyhedral, Lemma 5, Lemma 6 and Lemma 8 are matricial and combinatorial. When specializing these to ideal clutters, their most difficult parts fall out and quite short variants of proofs of Lehman's or Padberg's theorem are at hand.

3 From Polyhedra to Combinatorics

The notation $\mathcal{A}$, $\mathcal{B}$ will be used for families of sets. (We will also use the notation $\mathcal{A}$ for the matrices whose rows are the members of $\mathcal{A}$.) The *degree* $d_{\mathcal{A}}(v)$ of v in $\mathcal{A}$ is the number of $A \in \mathcal{A}$ containing v.

Given $w \in P$, let $\mathcal{A}_w$ be the set of those rows A of $A^{\geq}$ or of $A^{\leq}$ for which $w(A) = 1$. (We do not give multiplicites to the members of $\mathcal{A}_w$, regardless of whether some of its elements are contained in both $A^{\geq}$ and $A^{\leq}$!) We also define these if the polyhedron also has non-0–1-constraints. Then $A^{\geq}$ and $A^{\leq}$ denote the set-covering and set-packing inequalities in the defining system.

If P is integer, we define $\mathcal{B}_w$ as the family of those 0–1 vectors (vertices of P) which are on the minimal face of P containing w. (Equivalently, $\mathcal{B}_w$ is the set of vertices having a nonzero coefficient in some convex combination expressing w.) If $A \in \mathcal{A}_w$, and $B \in \mathcal{B}_w$, then $|A \cap B| = 1$. Clearly, $r(\mathcal{A}_w) + r(\mathcal{B}_w) = \dim P + 1$. If it is necessary in order to avoid misunderstanding, we will write $\mathcal{A}_w(P)$, $\mathcal{B}_w(P)$.

The following lemma is based on the polar (in the sense of interchanging vertices and facets) of a statement implicit in arguments of Lehman's and Seymour's work (see Seymour [12]).

Lemma 1. *If Q is a polyhedron with 0–1 vertices (and not necessarily 0–1-constraints) and $w \in Q$, $w > 0$, then $\bigcup_{B \in \mathcal{B}_w} B = V$, and*
$$r(\mathcal{B}_w) \geq \max \{|A| : A \in \mathcal{A}_w\}, \quad r(\mathcal{A}_w) \leq n - \max \{|A| : A \in \mathcal{A}_w\} + 1.$$

Proof. Indeed, since Q is integer, w is the convex combination of 0–1 vertices in $\mathcal{B}_w$, whence $\bigcup_{B \in \mathcal{B}_w} B = V$. In particular, for $A \in \mathcal{A}_w$ and all $a \in A$ there exists $B_a \in \mathcal{B}_w$, such that $a \in B_a$.

Since $A \in \mathcal{A}_w$, and $B_a \in \mathcal{B}_w$, we have $|A \cap B_a| = 1$, and consequently $A \cap B_a = \{a\}$. Thus $\{B_a : a \in A\}$ consists of $|A|$ linearly independent sets of $\mathcal{B}_w$, whence $r(\mathcal{B}_w) \geq |A|$. $\qquad\square$

Remark 2. Compare Lemma 1 with Fonlupt, Sebő [2]: a graph is perfect if and only if the linear rank of the maximum cliques (as vertex-sets) in every induced subgraph is at most $n - \omega + 1$ where ω is the size of the maximum clique in the subgraph; the equality holds if and only if the subgraph is uniquely colorable.

We note and use in the sequel without reference that if P is minimal non-integer, then $w > 0$ for all fractional vertices w of P ($w_i = 0$ would imply that $(P^{\leq} \setminus i) \cap (P^{\geq} \setminus i)$ is also noninteger).

In sections 3 and 4 I will denote the identity matrix, J the all 1 matrix of appropriate dimensions; A is called r-*regular*, if $\underline{1}A = r\underline{1}$, and r-*uniform* if $A\underline{1} = r\underline{1}$; $\mathcal{A}^c := \{V \setminus A : A \in \mathcal{A}\}$. $\mathcal{A}$ is said to be *connected* if V cannot be partitioned into two nonempty classes so that every $A \in \mathcal{A}$ is a subset of one of the two classes. There is a unique way of partitioning $\mathcal{A}$ and V into the *connected components* of $\mathcal{A}$.

Lemma 2. *If $(A^{\leq}, A^{\geq})$ is minimal noninteger, w is a fractional vertex of $P := P(A^{\leq}, A^{\geq})$, and $\mathcal{A} \subseteq \mathcal{A}_w$ is a set of n linearly independent members of $\mathcal{A}_w$, then every connected component K of $\mathcal{A}^c$ is $n - r_K$-regular and $n - r_K$-uniform ($r_K \in \mathbb{N}$), and $r(\mathcal{A} - v) = n - d_{\mathcal{A}}(v)$.*

Proof. Recall that $w > 0$. If P is minimal noninteger, then for arbitrary $i \in V$ the sandwich property provides us $Q_i \subseteq \mathbb{R}^{V \setminus \{v\}}$, $w^i \in \left[P^{\leq}(A^{\leq}) \cap P^{\geq}(A^{\geq})\right]^i \subseteq Q_i \subseteq P^{\leq}(A^{\leq})^i \cap P^{\geq}(A^{\geq})^i$, that is, $w^i \in Q_i$ and $w^i > 0$. Applying the inequality in Lemma 1 to Q_i and w^i, and using the *trivial but crucial fact that* $\mathcal{A}_{w^i}(Q_i) \supseteq \mathcal{A} - i$, we get the inequality $r(\mathcal{A} - i) \leq n - \max\{|A| : A \in \mathcal{A} - i\}$.

On the other hand, $r(\mathcal{A}) = n$ by assumption. One can now finish in a few lines like Conway proves de Bruijn and Erdős's theorem [7], which is actually the same as Seymour [12, Lemma 3.2]:

Let $\mathcal{H} := \mathcal{A}^c$ for the simplicity of the notation. What we have proved so far translates as $d_{\mathcal{H}}(v) \leq |H|$ for all $v \in H \in \mathcal{H}$. But then,

$$n = \sum_{H \in \mathcal{H}} 1 = \sum_{H \in \mathcal{H}} \sum_{v \in H} 1/|H| = \sum_{v \in V} \sum_{H \in \mathcal{H}, v \in H} 1/|H| = \sum_{v \in V} d_{\mathcal{H}}(v)/|H| \leq \sum_{v \in V} 1,$$

and the equality follows. $\qquad\square$

Remark 3. The situation of the above proof will be still repeated several times: when applying Lemma 1, the 0–1 vectors that have an important auxiliary role for bounding the rank of some sets are in $\mathcal{B}_w(Q_i)$, and are not necessarily vertices *of* P. The reader can check on mixed odd circuits that the neighbors $\mathcal{B} = \{B_1, \ldots, B_n\}$ of $1/2\underline{1}$ are not suitable for the same task (unlike in the special cases): the combinatorial ways that use $\mathcal{B}$ had to be replaced by this more general polyhedral argument. Watch for the same technique in Lemma 7 !

The next lemma synthesizes two similar proofs occurring in the special cases:

Lemma 3. *If* $(A^{\leq}, A^{\geq})$ *is minimal noninteger, w and w' are fractional vertices of P, then defining $\mathcal{A}$ and $\mathcal{A}'$ to be a set of n linearly independent vectors from $\mathcal{A}_w$, $\mathcal{A}_{w'}$ respectively, $\mathcal{A}$ and $\mathcal{A}'$ cannot have exactly $n - 1$ common elements.*

Proof. (Sketch) Apply Lemma 6 to both w and w'. With the exception of some degenerate cases easy to handle, any member of an r-regular clutter can be uniquely reconstructed from the others. □

Lemma 4. *If* $(A^{\leq}, A^{\geq})$ *is minimal noninteger and simple, and w is a fractional vertex of P, then $P := P(A^{\leq}, A^{\geq})$ is full dimensional, w is simplicial, and the vertices neighbouring w on P are integer.*

The proof can be summarized with the sentence: a minimal noninteger system cannot contain implicit equalities (only 'explicit' equalities).

Proof. Let us first prove that P is full dimensional. By Lemma 3 $\mathcal{A}_w$ is linearly independent (recall that every member was included only once). Suppose $0 \in \mathbb{R}^n$ can be written as a nontrivial nonnegative linear combination of valid inequalities. Clearly, all of these are implicit equalities (see [11]) of P. In particular their coefficient vectors are in $\mathcal{A}_w$. In this nontrivial nonnegative combination there is no nonnegativity constraint $x_i \geq 0$, because otherwise $P \subseteq \{x : x_i = 0\}$, contradicting $w > 0$. So everything participating in it is in $\mathcal{A}_w$ contradicting its linear independence.

Since $\mathcal{A}_w$ is linearly independent, w is simplicial. If a neighbour w' of w is noninteger, we arrive at a contradiction with Lemma 3. □

We will say that a polyhedron P is *minimal noninteger* if $P = P(A^{\leq}, A^{\geq})$ for some simple, minimal noninteger system. (Since P is full dimensional by Lemma 4, it determines $(A^{\leq}, A^{\geq})$ uniquely.)

Given a minimal noninteger polyhedron P and a fractional vertex w of P, fix $\mathcal{A} := \mathcal{A}(w) := \mathcal{A}_w$ and let $\mathcal{B} := \mathcal{B}(w)$ denote the set of vertices neighboring w in P.

Note that $\cup_{i=1}^n \mathcal{B}_{w^i}(P^i) = \mathcal{B}(w)$ holds in the polyhedrally minimal noninteger case, but does not necessarily hold otherwise, and therefore we need essential generalizations. Do not confuse $\mathcal{B}_w$ (which is just $\{w\}$) with $\mathcal{B}(w)$.

We will say that a vertex $B \in \mathcal{B}$ and a facet $A \in \mathcal{A}$ not containing it are *associates*. By Lemma 4 w is simplicial, whence this relation perfectly matches $\mathcal{A}$ and $\mathcal{B}$. We will suppose that the associate of the i-th row A_i of A is the associate B_i of A_i; $\mu_i := |A_i \cap B_i|$. Clearly, $\mu_i \neq 1$ $(i = 1, \ldots, n)$. Denoting by $\mathrm{diag}(d_1, \ldots, d_n)$ the $n \times n$ diagonal matrix whose diagonal entries are $d_1, d_2, \ldots, d_n$, we have proved:

Lemma 5. $\mathcal{A}\mathcal{B}^T = J + \text{diag}(\mu_1 - 1, \ldots, \mu_n - 1)$, where $\mu_i \neq 1$, $(i = 1, \ldots, n)$.

If μ_i does not depend on i, we will simply denote it by μ. (This notation is not a coincidence: in this case $(\mathcal{A}, \mathcal{B})$ turns out to be partitionable where μ is the identically denoted parameter.) By Lemma 5, $\mu \neq 1$.

The main content of Lemma 3, 5, some aspects of Lemma 4 and most of Lemma 6 are already implicitly present already in Padberg[9].

4   Associates and the Divisibility Lemma

The following lemma extracts and adapts to our needs well-known statements from Lehman's, Seymour's and Padberg's works, and reorganizes these into one statement. It can also be deduced by combining results of Gasparyan [4], which investigate combinatorial properties implied by matrix equations. For instance the connectivity property of Lemma 6 below is stated in [4] in a general self-contained combinatorial setting.

Lemma 6. If P is minimal noninteger, and $w \in P$ is a fractional vertex of P, then $\mathcal{A} = \mathcal{A}(w)$ is nonsingular and connected, moreover,

- if the clutter $\mathcal{A}^c$ is connected, then $\underline{1}\mathcal{A} = \mathcal{A}\underline{1} = r\underline{1}$, $r \geq 2$.
- if the clutter $\mathcal{A}^c$ has two components, then $\mathcal{A}$ is a degenerate projective plane.
- if the clutter $\mathcal{A}^c$ has at least three components, then $\mathcal{A} = \mathcal{H}_n^{n-1}$.

Proof. (Sketch) If $\mathcal{A}^c$ has at least two components, then any two sets whose complements are in different components cover V. This, and the matrix equation of Lemma 5 determine a degenerate combinatorial structure. (For instance one can immediately see that the associate of a third set has cardinality at most two, and it follows that all but at most one members of $\mathcal{B}$ have at most two elements.)

If $\mathcal{A}^c$ has one component, then the uniformity and regularity of $\mathcal{A}^c$ claimed by Lemma 2 implies that of $\mathcal{A}$.   □

Recall that the nonsingularity of $\mathcal{B}$ cannot be added to Lemma 6 !

It is well-known that both for minimal imperfect and minimal nonideal matrices the associates of intersecting sets are (almost) disjoint. In our case they can also *contain one another*, and the proof does not fit into the combinatorial properties we have established (namely Lemma 5). We have to go back to our polyhedral context (established in the proof of Lemma 2, see also Remark 3):

Let us say that $\mathcal{A}$ with $\mathcal{A}\mathcal{B}^T = J + \text{diag}(\mu_1 - 1, \ldots, \mu_n - 1)$, where $\mu_i \neq 1$ $(i = 1, \ldots, n)$ is *nice*, if for $A_1, A_2 \in \mathcal{A}$, $v \in A_1 \cap A_2$ the associates $B_1, B_2 \in \mathcal{B}$ of A_1 and A_2 respectively, either satisfy $B_1 \cap B_2 \setminus \{v\} = \emptyset$ or $B_1 \setminus \{v\} = B_2 \setminus \{v\}$. (In the latter case, since B_1 and B_2 cannot be equal, one of the two contains v.)

Lemma 7. Let P be minimal noninteger, and $\mathcal{A} = \mathcal{A}(w)$, $\mathcal{B} = \mathcal{B}(w)$ for some noninteger vertex $w \in P$. Then $\mathcal{A}$ is nice.

Check the statement for the mixed C_7 of Example 1 ! (It can also be instructive to follow the proof on this example.)

Proof. Let $v \in A_1$, $A_2 \in \mathcal{A}$, and let $B_1, B_2 \in \mathcal{B}$ be their associates. Moreover assume $u \in B_1 \cap B_2 \setminus \{v\}$. Let $A_0 \in \mathcal{A}$, $u \in A$, $v \notin A_0$. (There exists such an $A_0 \in \mathcal{A}$ since for instance Lemma 6 implies that a column of $\mathcal{A}$ cannot dominate another.) Since P is minimal noninteger, there exists an integer polyhedron Q_v such that $[P^\leq(A^\leq) \cap P^\geq(A^\geq)]^v \subseteq Q_v \subseteq P^\leq(A^\leq)^v \cap P^\geq(A^\geq)^v$. Now because of $\mathcal{A}_{w^v}(Q_v) \supseteq \mathcal{A} - i$, the scalar product of the vertices of $\mathcal{B}_{w^v}(Q_v)$ with all vectors in $\mathcal{A} - i$ is 1, and the proof method of Lemma 1 can be applied:

For every $a \in A_0 \setminus u$ fix some $B_a \in \mathcal{B}_{w^v}(Q_v)$ so that $a \in B_a$. Now $\{B_a : a \in A_0 \setminus u\} \cup \{B_1 \setminus v, B_2 \setminus v\}$ are $r + 1$ vectors in $\mathbb{R}^{V \setminus v}$ all of which have exactly one common element with each $A \in \mathcal{A} - v$. On the other hand, by Lemma 2 $r(\mathcal{A} - v) = n - r = (n - 1) - (r - 1)$, so there can be at most r linearly independent sets with this property. Hence there exists a nontrivial linear combination $\lambda_1(B_1 \setminus v) + \lambda_2(B_2 \setminus v) + \sum_{a \in A \setminus u} \lambda_a B_a = 0$. Since for $a \in A_0 \setminus u$ the unique vector in this linear combination which contains a is B_a, one gets that $\lambda_a = 0$ for all $a \in A \setminus u$. It follows that $B_1 \setminus v = B_2 \setminus v$, and $\lambda_1 = \lambda_2$. □

Although the following statement is the heart of our proof, it is independent of the other results. The very root of the statement is the simple observation that $n + d_j$ is a multiple of r. Note that in order to deduce $r = 2$ we need more than just the matrix equation !

Lemma 8. *Assume that $\mathcal{A}$, $\mathcal{B}$ are 0–1 matrices, $\mathbf{1}\mathcal{A} = \mathcal{A}\mathbf{1} = r\mathbf{1}$, and $\mathcal{A}\mathcal{B}^T = J + \mathrm{diag}(\mu_1 - 1, \ldots, \mu_n - 1)$, $\mu_i \neq 1$. Then*

 - *either $\mu_1 = \ldots = \mu_n =: \mu$, and then $\mathcal{A}\mathcal{B}^T = \mathcal{B}^T\mathcal{A} = J + (\mu - 1)I$, $\mathcal{B}J = JB = sJ$, $(s = (n + \mu - 1)/r)$,*
 - *or $\{\mu_1, \ldots, \mu_n\} = \{0, r\}$, and if $\mathcal{A}$ is connected and nice, then $r = 2$.*

Proof. If $\mu_1 = \mu_2 = \ldots = \mu_n =: \mu$, then we finish easily, like [3]: since $\mu \neq 0$, $\mathcal{A}$ is invertible; since $\mathcal{A}$ commutes with I, and by assumption with J too, so does its inverse; now expressing $\mathcal{B}^T$ from $\mathcal{A}\mathcal{B}^T = J + (\mu - 1)I$ we get that it is the product of two matrices which commute with both A and J. So $\mathcal{B}^T$ also commutes with these matrices, proving the statement concerning this case. (The matrices X and Y are said to commute, if $XY = YX$.)

So suppose that there exist $i, j \in V$ such that $\mu_i \neq \mu_j$.

Claim (1). $r|B_j| = n + \mu_j - 1$, and $0 \leq \mu_j \leq r$, $(j = 1, \ldots, n)$.

Indeed, $r\mathbf{1}\mathcal{B}^T = (\mathbf{1}\mathcal{A})\mathcal{B}^T = \mathbf{1}(\mathcal{A}\mathcal{B}^T) = \mathbf{1}(J + \mathrm{diag}(\mu_1 - 1, \ldots, \mu_n - 1)) = (n + \mu_1 - 1, \ldots, n + \mu_n - 1)$.
The inequality is obvious: $0 \leq \mu_j = |A_j \cap B_j| \leq |A_j| = r, (j = 1, \ldots, n)$.

Claim (2). If there exist $i, j \in V$, $\mu_i \neq \mu_j$, then $\mu_j \in \{0, r\}$ for all $j \in V$.

Indeed, according to Claim (1) we have $n + \mu_j - 1 \equiv 0 \bmod r$, where μ_j is in an interval of $r + 1$ consecutive integers representing every residue class mod r exactly once, except 0, which is represented twice, by 0 and r. Hence if $\{\mu_1, \ldots, \mu_n\}$ contains two different values, then these values can only be 0 and r as claimed.

Claim (3). If $v \in A_1 \cap A_2$, $\mu_1 = 0$, $\mu_2 = r$, then $B_1 = B_2 \setminus \{v\}$.

Indeed, let $u \in A_2 \cap B_1$. (Because of the matrix equation in the constraint, we also know $|A_2 \cap B_1| = 1$.) We have $|A_1 \cap B_1| = \mu_1 = 0$, and since $|A_2 \cap B_2| = \mu_2 = r = |A_2|$, we also have $A_2 \subseteq B_2$.

Since $v \in A_1$ and $A_1 \cap B_1 = \emptyset$: $u \neq v$. Because of $A_2 \subseteq B_2$ we have $u \in (B_1 \cap B_2) \setminus \{v\}$. So we must have $B_1 \setminus \{v\} = B_2 \setminus \{v\}$ by the condition, and since $v \notin B_1$, $v \in B_2$: $B_1 = B_2 \setminus \{v\}$. The claim is proved.

Now we finish the proof. Since there exist $i, j \in V$ such that $\mu_i \neq \mu_j$, by Claim (2) $\mu_j \in \{0, r\}$ for all $j \in V$. Since $\mathcal{A}$ is a connected clutter, there exists $v \in V$ so that $v \in A_i \cap A_j$ and $\mu_i = 0$, $\mu_j = r$. After possible renumbering, we can assume $i = 1$, $j = 2$.

So let $A_1, A_2 \in \mathcal{A} = \mathcal{A}(w)$, $v \in A_1 \cap A_2$, $\mu_1 = 0$, $\mu_2 = r$ and denote the associates of A_1, A_2 by B_1, B_2 respectively.

By Claim (3), $1 = |A_2 \cap B_1| = |A_2 \cap B_2| - 1 = r - 1$, so $r = 2$. □

Proof of Theorem 2. (Sketch) Let $(A^\leq, A^\geq)$ be minimal noninteger. Furthermore, let $w \in P$ a fractional vertex of P. Let $\mathcal{A} := \mathcal{A}(w)$ and $\mathcal{B} := \mathcal{B}(w)$. Then we have the matrix equation of Lemma 5.

Case 1. $\mathcal{A}^c$ is connected: according to Lemma 6 and Lemma 7 the conditions of Lemma 8 are satisfied, and using Lemma 8 it is straightforward to finish.

Case 2. $\mathcal{A}^c$ has two components: by Lemma 6 $\mathcal{A}$ is a degenerate projective plane. It can be checked then that either $\mathcal{A} = A^\geq$ or $\mathcal{A}$ is not minimal noninteger. We prove this with the following technique (and use similar arguments repeatedly in the sequel): we prove first that there exist an $i \in V$ so that $(P^\geq/i) \cap (P^\leq/i)$ is noninteger. It turns out then that the maximum p of the sum of the coordinates of a vector on $(P^\geq/i) \cap (P^\leq/i)$ and the maximum q of the same objective function on P^i are close to each other: $[p, q]$ does not contain any integer (we omit the details). So for all Q_i such that $P^i \subseteq Q_i \subseteq (P^\geq/i) \cap (P^\leq/i)$ the maximum of the sum of coordinates on Q_i must lie in the interval $[p, q]$. Thus Q_i cannot be chosen to be integer, whence P does not have the sandwich property.

Case 3. $\mathcal{A}^c$ has at least three components: by Lemma 6 $\mathcal{A}$ is the set of $n - 1$-tuples of an n-set. If $\mathcal{A} = A^\leq$, then we are done (again the first statement holds in the theorem). In all the other cases P turns out not to be minimal noninteger (with the above-described technique). □

5   Comments

5.1   Further Examples

A system $(A^{\leq}, A^{\geq})$ for which a $P(A^{\leq}, A^{\geq}) \subseteq \mathbb{R}^5$ is integer, but $(A^{\leq}, A^{\geq})$ is not fully integer: the rows of $A^{\leq}$ are $(1,1,0,0,0)$, $(0,1,1,0,0)$, $(1,0,1,0,0)$ and $(0,0,1,1,1)$; $A^{\geq}$ consists of only one row, $(0,0,0,1,1)$.

We mention that a class of minimal noninteger simple systems $(A^{\leq}, A^{\geq})$ with the property that $(A^{\leq}, A^{\geq}) \setminus i$ $(i \in V)$ defines an integer, but not always fully integer polyhedron, can be defined with the help of 'circular' minimal imperfect and minimal nonideal systems (see Cornuéjols and Novick [1]): define $A^{\leq} := C_n^r$, $A^{\geq} := C_n^s$, where $r \leq s$ and $A^{\leq}$ is minimal imperfect, $A^{\geq}$ is minimal nonideal.

Such examples do not have mixed vertices, so they also show that the first two cases of our results can both occur in the same polyhedron.

5.2   A Polynomial Certificate

We sketch why Corollary 6 follows from Theorem 2. Note that Corollary 6 immediately implies Corollary 3.

In a combinatorially minimal noninteger system $(A^{\leq}, A^{\geq})$, $A^{\leq}$ is in fact minimal imperfect or perfect. This is a simple consequence of the following:

Claim. If $P^{\leq} := P^{\leq}(A^{\leq})$ or $P^{\geq} := P^{\geq}(A^{\geq})$ is partitionable with a regular vertex $w \in P := P^{\leq} \cap P^{\geq}$, and $P^{\leq}/I$ $(I \subseteq V)$ is partitionable with regular vertex w', then $w' \in P^{\geq}/I$.

Indeed, suppose that w is the regular vertex of a polyhedron whose defining clutter has parameters (r, s), and let the parameters of w' be (r', s'). So $w := 1/r\underline{1}$ and $w' := 1/r'\underline{1}$.

Now $r' \leq r$, because the row-sums of the defining matrix of $P^{\leq}/I$ (which is a submatrix of $A^{\leq}$) do not exceed the row-sums of $A^{\leq}$. Since $w \in P^{\leq}(A^{\leq})$, the row-sums of $A^{\leq}$ are at most r.

But then, if we replace in $1/r\underline{1}$ some coordinates by $1/r'$ some others by 1 the vector w'' we get majorates $1/r\underline{1} \in P^{\geq}(A^{\geq})$ whence it is also in $P^{\geq}(A^{\geq})$. Since $w' \in P^{\geq}/I$ is equivalent to the belonging to $P^{\geq}$ of such a vector w'', the claim is proved.

To finish the proof of Corollary 6 one can show that after deleting from $A^{\geq}$ an equality from '$\mathcal{A}_w$', the system remains minimal noninteger.

Using appropriate oracles, Corollary 3 provides a polynomial certificate. (For the right assumptions about providing the data and certifying the parameters of a partitionable clutter we refere to Seymour [12]. We need an additional oracle for the set-covering part.)

The polynomial certificates can be proved from the Claim using the fact that for partitionable clutters and perfect graphs the parameters can be certified in polynomial time.

For the non-full-integrality of the intersection of perfect and ideal polyhedra a simple polynomial certificate is provided by Corollary 5.

Acknowledgments

I am thankful to Grigor Gasparyan and Myriam Preissmann for many valuable comments. Furthermore, I feel lucky to have learnt Lehman's results and especially to have heard the main ideas of Padberg's work from Grigor Gasparyan.

I would like to thank András Frank for comparing a lemma in [12] concerning ideal matrices to Erdős and de Bruijn's theorem: this helped me getting closer to ideal matrices and strengthened my belief in a common generalization (a particular case of Fisher's inequality is implicit in proofs for minimal imperfect graphs as well, see [3]).

Last but not least I am indebted to Kazuo and Tomo Murota for their miraculous help of various nature: due to them, it was possible to convert an extended abstract to a paper during five jet-lag-days.

References

1. G. Cornuéjols and B. Novick. Ideal 0–1 matrices. *J. of Comb. Theory B*, 60(1):145–157, 1994.
2. J. Fonlupt and A. Sebő. The clique rank and the coloration of perfect graphs. In R. Kannan and W. Pulleyblank, editors, *Integer Programming and Combinatorial Optimization I*. University of Waterloo Press, 1990.
3. G. Gasparyan. Minimal imperfect graphs: A simple approach. *Combinatorica*, 16(2):209–212, 1996.
4. G. Gasparyan. Bipartite designs. In R. E. Bixby, E. A. Boyd, and R. Z. Ríos-Mercado, editors, *Integer Programming and Combinatorial Optimization: Proceedings of the 6th International Conference on Integer Programming and Combinatorial Optimization, LNCS, Vol. 1412*, pages 23–35. Springer, 1998. This volume.
5. G. Gasparyan and A. Sebő. Matrix equations in polyhedral combinatorics. 1998. In preparation.
6. A. Lehman. The width-length inequality and degenerate projective planes. In W. Cook and P. D. Seymour, editors, *Polyhedral Combinatorics, DIMACS, Vol. 1*, pages 101–105, 1990.
7. J. H. van Lint and R. M. Wilson. *A Course in Combinatorics*. Cambridge University Press, 1992.
8. L. Lovász. A characterization of perfect graphs. *J. of Comb. Theory B*, 13:95–98, 1972.
9. M. Padberg. Perfect zero-one matrices. *Math. Programming*, 6:180–196, 1974.
10. M. Padberg. Lehman's forbidden minor characterization of ideal 0–1 matrices. *Discrete Mathematics*, 111:409–420, 1993.
11. A. Schrijver. *Theory of Linear and Integer Programming*. Wiley, 1986.
12. P. D. Seymour. On Lehman's width-length characterization. In *Polyhedral Combinatorics, DIMACS, Vol. 1*, pages 107–117, 1990.

A Theorem of Truemper

Michele Conforti and Ajai Kapoor *

Dipartimento di Matematica Pura ed Applicata, Università di Padova
Via Belzoni 7, 35131 Padova, Italy

Abstract. An important theorem due to Truemper characterizes the graphs whose edges can be labelled so that all chordless cycles have prescribed parities. This theorem has since proved an essential tool in the study of balanced matrices, graphs with no even length chordless cycle and graphs with no odd length chordless cycle of length greater than 3. In this paper we prove this theorem in a novel and elementary way and we derive some of its consequences. In particular, we show how to obtain Tutte's characterization of regular matrices.

1 Truemper's Theorem

Let β be a 0,1 vector indexed by the chordless cycles of an undirected graph $G = (V, E)$. G is β-*balanceable* if its edges can be labelled with labels 0 and 1 such that $l(C) \equiv \beta_C \bmod 2$ for every chordless cycle C of G, where $l(e)$ is the label of edge e and $l(C) = \sum_{e \in E(C)} l(e)$.

We denote by β^H the restriction of the vector β to the chordless cycles of an induced subgraph H of G.

In [14] Truemper showed the following theorem:

Theorem 1. *A graph G is β-balanceable if and only if every induced subgraph H of type (a), (b), (c) and (d) (Figure 1) is β^H-balanceable.*

Graphs of type (a), (b) or (c) are referred to as *3-path configurations* (3PC's). A graph of type (a) is called a $3PC(x, y)$ where node x and node y are connected by three internally disjoint paths P_1, P_2 and P_3. A graph of type (b) is called a $3PC(xyz, u)$, where xyz is a triangle and P_1, P_2 and P_3 are three internally disjoint paths with endnodes x, y and z respectively and a common endnode u. A graph of type (c) is called a $3PC(xyz, uvw)$, consists of two node disjoint triangles xyz and uvw and disjoint paths P_1, P_2 and P_3 with endnodes x and u, y and v and z and w respectively. In all three cases the nodes of $P_i \cup P_j$ $i \neq j$ induce a chordless cycle. This implies that all paths P_1, P_2, P_3 of (a) have length greater than one. Graphs of type (d) are *wheels* (H, x). These consist of a chordless cycle H called the *rim* together with a node x called the *center*, that has at least three neighbors on the cycle. Note that a graph of type (b) may also be a wheel.

---

* Supported in part by a grant from Gruppo Nazionale Delle Ricerche-CNR.

R. E. Bixby, E. A. Boyd, and R. Z. Ríos-Mercado (Eds.): IPCO VI
LNCS 1412, pp. 53–68, 1998. © Springer–Verlag Berlin Heidelberg 1998

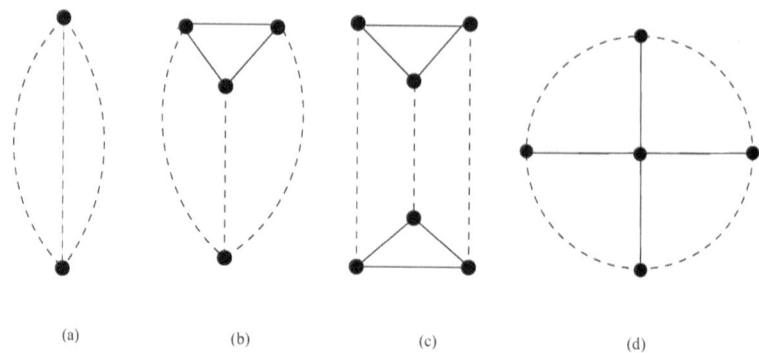

(a)   (b)   (c)   (d)

Fig. 1. 3-path configurations and wheel

In this paper, we give an alternative simple proof of Theorem 1 and we highlight its importance by deriving some well known theorems, such as the Tutte's characterization of regular matrices, the characterization of balanceable matrices and of even and odd signable graphs. Finally we show how to use Theorem 1 to obtain decomposition theorems.

Truemper [14] derived the excluded minor characterization of matroids representable over GF3 as a consequence of Theorem 1. From known results, see [13], this implies Tutte's theorem. Here we offer a more direct derivation.

First some definitions. $N(v)$ is the set of neighbors of node v. A *signed* graph G is a graph whose edges are labeled with 0 or 1. Given a $0, 1$ vector β indexed by the chordless cycles of G, if a chordless cycle C of G satisfies $l(C) \equiv \beta_C$ mod 2, then C is *signed correctly*, otherwise C is *signed incorrectly*. A *β-balancing* of G is one for which each of its chordless cycles is correctly signed. The operation of *scaling* on a cut $\delta(S)$ of G consists of switching (from 0 to 1 and viceversa) the labels on all edges in $\delta(S)$. Since cuts and cycles of G have even intersections, we have the following:

Remark 2. *Let G' be a signed graph obtained from G by scaling on a cut. A chordless cycle C is correctly signed in G' if and only if C is correctly signed in G.*

Assume G is connected and contains a clique cutset K_l with l nodes and let $G'_1, G'_2, \ldots, G'_n$ be the components of the subgraph induced by $V(G) \setminus K_l$. The *blocks* of G are the subgraphs G_i induced by $V(G'_i) \cup K_l$, $i = 1, \ldots, n$.

Remark 3. *If G contains a K_1 or K_2 cutset, then G is β-balanceable if and only if all of its blocks are β^{G_i}-balanceable.*

Proof: The "only if" part is obvious. We prove the "if" statement when G has a K_2 cutset $\{u, v\}$, since the other case is again immediate. All blocks have a β-balancing in which edge uv has the same label, since all blocks G_i are β^{G_i}-balanceable and we can always scale on a cut of G_i separating u and v. The signings of the blocks induce a β-balancing of G, since every chordless cycle of G belongs to one of the blocks G_i.   □

The following observation appears in [14].

Remark 4. *A graph G which is a wheel or a 3-path configuration is β-balanceable if and only if any signing of G produces an even number of incorrectly signed chordless cycles.*

Proof: If G is a wheel or a 3-path configuration, every edge of G belongs to exactly two chordless cycles. Therefore switching the label of edge uv changes the parities of the labels of the two chordless cycles containing uv and of no other chordless cycle. Now if G is a wheel or a 3-path configuration and has at least two chordless cycles that are signed incorrectly, then either G contains two chordless cycles that are signed incorrectly and have a common edge or G contains three chordless cycles C_1, C_2 and C_3 such that C_1 and C_3 are signed incorrectly, C_2 is signed correctly and C_2 has common edges with both C_1 and C_3, but C_1 and C_3 do not have a common edge.

Therefore by switching the label of at most two edges, the number of cycles that are incorrectly signed decreases by two.   □

An ordering $e_1, \ldots, e_n$ of the edges of a connected graph G is *consistent* if the first edges in the sequence belong to a spanning tree T of G and all other edges e_j have the property that e_j, together with some edges having smaller indices, closes a chordless cycle C_j of G. Note that for any spanning tree T of G, G admits a consistent ordering where the edges of T appear first.

Assume a connected graph G, a vector β are given and consider the following **signing algorithm**:

Let $e_1, \ldots, e_n$ be a consistent ordering of the edges of G. Label the edges of T arbitrarily and label the remaining edges e_j so that the chordless cycles C_j are signed in accordance with the components β_{C_j} of β.

Since every edge of T belongs to a cut of G, containing no other edge of T, then Remark 2 shows that if G is β-balanceable, an arbitrary labeling of the edges of T can always be extended to a β-balancing of G. Therefore the above signing algorithm will produce a β-balancing of G, whenever G is β-balanceable. Conversely, given a consistent ordering where a tree T appears first and a β-balancing of G, this same signing of G is produced by the algorithm when T is signed as in the signing of G.

Remark 5. *Let G be a β-balanceable graph and let G_v be the subgraph of G, obtained by removing node v. Then every β-balancing of G_v with respect to β^{G_v} can be extended to a β-balancing of G.*

Proof: We assume that v is not a cutnode of G, else by Remark 3, we can argue on the blocks of G. Now G contains a spanning tree T where v is a leaf of T. Order the neighbors of v in G_v as $v_0, v_1, \ldots, v_k$, where v_0 is the neighbor of v

in T and v_i is chosen so that amongst all nodes in $N(v) \setminus \{v_0, \ldots, v_{i-1}\}$ the path between nodes v and v_i is shortest in the subgraph of G with edge set $E(G_v) \cup \{vv_0, \ldots, vv_{i-1}\}$. Now place first the edges of T, then the other edges of G_v in a consistent ordering with respect to $T \setminus \{v\}$, then $vv_1, \ldots, vv_k$. This ordering is a consistent ordering for G and the signing algorithm can be applied to produce from the β-balancing of G_v, a β-balancing of G. $\qquad\square$

Lemma 6. *Let G' be an induced subgraph of G, containing a given chordless cycle C and satisfying the following properties:*

1) G' is connected and contains no K_1 or K_2 cutset.
2) C belongs to G' and $G' \setminus C$ is nonempty.
3) $V(G')$ is minimal with respect to 1) and 2).

Then G' is a 3-path configuration or a wheel containing C.

Proof: Let G'' be the subgraph of G', induced by $V(G') \setminus V(C)$. If G'' is a single node, say u, and u has only two neighbors c_i and c_j in C, then c_i and c_j are nonadjacent and G' is a $3PC(c_i, c_j)$. Otherwise G' is a wheel with u as center.

If G'' contains more than one node, by 3) we have that G'' is connected and that:

4) Every node of G'' has at most two neighbors in C and these two neighbors are adjacent.
5) G'' contains at most one pair of nodes, say x_1 and x_n such that both x_1 and x_n have neighbors in C and $(N(x_1) \cup N(x_n)) \cap V(C)$ either contains at least three nodes or two nonadjacent nodes.

(Indeed by 3), we have that G'' is connected. So if G'' contains more that one such pair, let x_1, x_n be chosen satisfying 5) and closest in G''. Let $P = x_1, \ldots, x_n$ be a shortest path in G'' connecting them. The subgraph G^* of G, induced by $V(C) \cup V(P)$ satisfies 1) and 2). Then if more that one such pair exists, G^* is a proper subgraph of G' and this contradicts 3).)

Let $C = c_1, \ldots, c_m$ and assume first that G'' contains one pair of nodes, x_1, x_n satisfying 5). Then by 3), G'' is a path $P = x_1, \ldots, x_n$. If a node of C, say c_i, is adjacent to a node x_i, $1 < i < n$ of P, then by 3) and 4), x_1 is adjacent to c_{i-1}, possibly to c_i and no other node of C. Node x_n is adjacent to c_{i+1}, possibly to c_i (indices $\mathrm{mod}\, m$) and no other node of C. Therefore no other node of C is adjacent to an intermediate node of P. In this case, G' is a wheel with center c_i.

If no node of C is adjacent to an intermediate node of P, then by 4) we can assume w.l.o.g. that x_1 is adjacent to c_{i-1} and possibly c_i and x_n is adjacent to c_{j+1} and possibly c_j. If x_1 or x_n has two neighbors in C and $i = j$, then G' is a wheel with center c_i. In the remaining cases G' is a 3-path configuration.

If G'' contains no pair of nodes satisfying 5), by 1) and 4) we have that C is a triangle c_1, c_2, c_3, all three nodes of C have neighbors in G'' and no node of G'' has more than one neighbor in C. If G'' is a chordless path $P = x_1, \ldots, x_n$ with

x_1 adjacent to, say c_1 and x_n adjacent to c_2, then c_3 has some neighbor in P and G' is a wheel with center c_3. Otherwise let P_{12} be a shortest path connecting c_1 and c_2 and whose intermediate nodes are in G''. If c_3 has a neighbor in P_{12} we have a wheel with center c_3. Otherwise let P_3 be a shortest path connecting c_3 and $V(P_{12}) \setminus \{c_1, c_2\}$ and whose intermediate nodes are in G''. By 3), G' is made up by C, together with P_{12} and P_3, furthermore P_3 meets P_{12} either in a node x or in two adjacent nodes t_1, t_2. In the first case, we have a $3PC(c_1 c_2 c_3, x)$, otherwise we have a $3PC(c_1 c_2 c_3, t_1 t_2 t_3)$. □

For $e \in E(G)$, G^e denotes the graph whose node set represents the chordless cycles of G containing e and whose edges are the pairs C_1, C_2 in $V(G^e)$, such that C_1 and C_2 belong to a 3-path configuration or a wheel.

Lemma 7. *If $e = \{u, v\}$ is not a K_2 cutset of G, G^e is connected.*

Proof: Assume not. Let G_1^e and G_2^e be two components of G^e. Let G_i be the subgraph of G induced by the node set $\cup_{C \in G_i^e} V(C)$, for $i = 1, 2$.

Assume first that $\{u, v\}$ is a K_2 cutset separating G_1 from G_2 in the graph induced by $V(G_1) \cup V(G_2)$. Pick $C_1 \in G_1^e$ and $C_2 \in G_2^e$ and a path P in G such that in the subgraph G' of G induced by $V(C_1) \cup V(C_2) \cup V(P)$, $\{u, v\}$ is not a K_2 cutset and C_1, C_2 and P are chosen so that $|P|$ is minimized. (Note that P exists since $\{u, v\}$ is not a K_2 cutset of G). Then by the minimality of P, no node of P is contained in a chordless cycle containing edge e. By Lemma 6, C_1 is a chordless cycle in a 3-path configuration or wheel H, contained in G'. Since any edge in a 3-path configuration or wheel is contained in two chordless cycles, $V(C_1) \cup V(C_2) \subseteq V(H)$. But then $C_1 C_2$ is an edge of G^e, a contradiction.

So $\{u, v\}$ is not a K_2 cutset in the graph induced by $V(G_1) \cup V(G_2)$. Let $C_2 \in G_2^e$, such that for some $C \in G_1^e$, $\{u, v\}$ is not a K_2 cutset in the graph induced by $V(C) \cup V(C_2)$. Let C_2 be $u = v_1, \ldots, v_m = v$. Let v^C be the node of lowest index in $V(C_2) \setminus V(C)$ and let S_C be the component of the graph induced by $V(C_2) \setminus V(C)$ containing node v^C. Amongst all $C \in G_1^e$ such that $\{u, v\}$ is not a K_2 cutset in the graph induced by $V(C) \cup V(C_2)$, let C_1 be the chordless cycle for which the node v^{C_1} has the highest index and with respect to that $|S_{C_1}|$ is smallest possible. By Lemma 6, C_1 is a chordless cycle of a 3-path configuration or wheel H contained in $V(C_1) \cup V(S_{C_1})$. Let C_3 be the chordless cycle of H distinct from C_1 containing edge e. We show that C_3 contradicts the choice of C_1. Since H contains C_1 and C_3, $C_3 \in G_1^e$. Also C_2 and C_3 have a common node which is distinct from u or v and so uv is not a K_2 cutset in the subgraph of G, induced by $V(C_3) \cup V(C_2)$. If v^{C_1} is contained in $V(C_3)$ then v^{C_3} has an index higher than i, a contradiction, otherwise since $S_{C_3} \subseteq S_{C_1}$ and some node of S_{C_1} belongs to C_3, $|S_{C_3}| < |S_{C_1}|$, a contradiction. □

Proof of Theorem 1: The necessity of the condition is obvious. We prove the sufficiency by contradiction. Assume that G and β are chosen so that G is a counterexample to the theorem with respect to β and $V(G)$ is minimal. Then G is connected and by Remark 5 G contains no K_1 or K_2 cutset.

Let $e = uv$ be any edge of G and let $G_u = G \setminus u$, $G_v = G \setminus v$ and $G_{uv} = G \setminus \{u, v\}$. By the minimality of graph G, G_u, G_v and G_{uv} are respectively β^{G_u}-, β^{G_v}- and $\beta^{G_{uv}}$- balanceable and Remark 3 shows that a β-balancing of G_{uv} can be extended to a β-balancing of G_u and to a β-balancing of G_v. To complete the signing of G, label uv arbitrarily. Now we have signed G so that:

Every chordless cycle of G which is incorrectly signed contains edge $e = uv$.

Let the chordless cycles of G containing edge uv be partitioned into the incorrectly signed $\mathcal{B}$ and the correctly signed $\mathcal{C}$. Both sets are nonempty, else, by possibly switching the label of uv, we have a β-balancing of G. Furthermore $\{u, v\}$ is not a K_2 cutset of G.

Hence by connectivity of G^e (Lemma 7), there exists an edge C_1C_2 in G^e, where $C_1 \in \mathcal{B}$ and $C_2 \in \mathcal{C}$. Any edge in a 3-path configuration or a wheel is contained in exactly two chordless cycles, thus G contains a 3-path configuration or a wheel with exactly one chordless cycle in $\mathcal{B}$ and by Remark 4, we are done. $\square$

2 Even and Odd-Signable Graphs

A *hole* in an undirected graph G is a chordless cycle of length greater than three. Signed graphs provide a useful setting for studying graphs without even or odd holes.

A graph G is *even-signable* if G is β-balanceable for the vector $\beta_C = 1$ if C is a triangle of G and $\beta_C = 0$ if C is a hole of G. A graph G is *odd-signable* if G is β-balanceable for the vector β of all ones. Even-signable graphs were introduced in [6] and odd-signable in [4].

Note that G contains no odd hole if and only if G is even-signable with all labels equal to one and G contains no even hole if and only if G is odd-signable with all labels equal to one.

A graph of type (b) may also be a wheel of type (d), when at least one of the paths P_1, P_2, P_3 is of length one. To separate these cases, from now on, we impose that all three paths in graph of type (b) have length greater than one. With this assumption, all chordless cycles of graphs of type (a), (b) and (c) are holes except the triangle of (b) and the two triangles of (c). Furthermore the rim of a wheel is a hole unless the wheel is K_4.

We now derive from Theorem 1 co-NP characterizations of even-signable and odd-signable graphs.

For graphs of type (d) (the wheels), when the center together with the nodes of the hole induces an odd number of triangles the wheel is called an *odd wheel*. When the center has an even number of neighbors on the hole the wheel is called an *even wheel*. (Notice that a wheel may be both odd and even and K_4 is a wheel that is neither even nor odd).

In a signed graph G the *weight* of a subgraph H is the sum of the labels of the edges contained in H.

Theorem 8. *A graph is even-signable if and only if it contains no $3PC(xyz, u)$ and no odd wheel.*

Theorem 9. *A graph is odd-signable if and only if it contains no $3PC(x, y)$, no $3PC(xyz, uvw)$ and no even wheel.*

Proof of Theorem 8: In a $3PC(xyz, u)$ the sum of the weights of the three holes modulo 2 is equivalent to the weight of the edges of the triangle, since all other edges are counted precisely twice. So if a $3PC(xyz, u)$ is signed so that its holes have even weight, then the triangle also has even weight. Thus a $3PC(xyz, u)$ is not odd-signable.

Similarly in a wheel the sum of the weights of the chordless cycles containing the center is equivalent modulo 2 to the weight of the rim. In an odd wheel the number of triangles containing the center is odd and so if the wheel is signed so that the weights of the chordless cycles containing the center are correct then the weight of the rim is odd.

Consider graphs (a) and (c). By labeling 1 all edges in the triangles and 0 all other edges, we obtain a correct labeling of these graphs. In wheel (H, x), that is not odd label 1 all edges of H that belong to a triangle of (H, x) and 0 all other edges. □

Proof of Theorem 9: In a $3PC(x, y)$ the sum of the weights of two of the holes is equivalent modulo 2 to the weight of the third, since the edges in the intersection of the two are counted twice and the remainder induce the third. Thus if the graph is signed so that these two holes have odd weight then the weight of the third is even. So a $3PC(x, y)$ is not odd-signable.

Similarly in a $3PC(xyz, uvw)$ the sum of the weights of all three holes is equivalent modulo 2 to the sum of the weights of the two triangles. If the graph is signed so that the weight of the three holes is odd then at least one of the triangles must have even weight.

In an even wheel the weight of the rim is equivalent modulo 2 to the sum of the weights of the other chordless cycles. Since there are an even number of these, each with odd weight, the rim has even weight.

Consider a graph of type (b). By labeling 1 all edges in triangles and 0 all other edges, we obtain an odd signing of these graphs. To label a wheel (H, x) that is not even, on every subpath of H with endnodes adjacent to x and no intermediate node adjacent to x, label 1 one edge and label 0 all other edges of (H, x). This gives an odd signing of the wheel. □

The recognition problem for both even-signable and odd-signable graphs is still open. In [4] both problems are solved for graphs that do not contain a cap as induced subgraph. (A cap is a hole H plus a node that has two neighbors in H and these neighbors are adjacent).

In [3] a polynomial time recognition algorithm is given, to test if a graph G contains no even hole (i.e. G is odd-signable with all labels equal to one).

3   Universally Signable Graphs

Let G be a graph which is β-balanced for all $0, 1$ vectors β that have an entry of 1 corresponding to the triangles of G. Such a graph we call *universally signable*.

Clearly triangulated graphs i.e. graphs that do not contain a hole are universally signable. In [5] these graphs are shown to generalize many of the structural properties of triangulated graphs. Here we show a decomposition theorem that follows easily from the co-NP characterization of these graphs as given by Theorem 1.

Theorem 10. *A graph G is universally signable if and only if G contains no graph of type (a), (b), (c) or (d) which is distinct from K_4.*

In view of the previous remark, the above condition is equivalent to the condition: "no hole of G belongs to a graph of type (a), (b), (c) or (d)". Now the proof of the above theorem follows from Theorem 1.

As a consequence of Theorem 10 and Lemma 6 we have the following decomposition theorem.

Theorem 11. *A connected universally signable graph that is not a hole and is not a triangulated graph contains a K_1 or K_2 cutset.*

It was the above decomposition theorem that prompted us to look for a new proof for Theorem 1.

Now Theorem 11 and the following result of Hajnal and Suranyi [11] can be used to decompose with clique cutsets a universally signable graph into holes and cliques.

Theorem 12. *A triangulated graph that is not a clique contains a clique cutset.*

4 α-Balanced Graphs, Regular and Balanceable Matrices

Actually, Truemper proved the following theorem that he also showed to be equivalent to Theorem 1.

Let α be a vector with entries in $\{0, 1, 2, 3\}$ indexed by the chordless cycles of a graph G. A graph $G = (V, E)$ is α-*balanceable* if its edges can be labeled with labels of -1 and $+1$ so that for every chordless cycle C of G, $l(C) \equiv \alpha_C \bmod 4$. Such a signing is an α-*balancing* of G.

Theorem 13. *A graph is α-balanceable if and only if α_C is even for all even length chordless cycles C and odd otherwise and every subgraph H of G of type (a), (b), (c) or (d) is α^H-balanceable.*

To see that the two theorems are equivalent note that an α-balancing of G with labels of 1 and -1, is implied by a β-balancing with $\beta = (\frac{\alpha_C - |E(C)|}{2}) \bmod 2$, by replacing the 0's by -1's. Similarly the β-balancing of G with labels of 0 and 1 is implied by an α-balancing with $\alpha_C = (2\beta_C + |E(C)|) \bmod 4$, by replacing the -1's by 0's.

Balanceable and Balanced Matrices

The *bipartite graph* $G(A)$ of a matrix A has the row and column sets of A as color classes and for all entries $a_{ij} \neq 0$, $G(A)$ has an edge ij of label a_{ij}.

A $0, \pm 1$ matrix A is *balanced* if $G(A)$ is α-balanced for the vector α of all zeroes. A $0, 1$ matrix A is *balanceable* if $G(A)$ is α-balanceable for the vector α of all zeroes. (From now on, *signing* consists of replacing some of the $1's$ with $-1's$).

Note that the same signing algorithm of Section 1, applied to $G(A)$, can be used to obtain a balanced matrix from A, when A is balanceable. Here signing the edges of $G(A)$ means assigning labels ± 1.

We can now derive from Theorem 13 a co-NP characterization of balanceable matrices:

Theorem 14. *A $0, 1$ matrix A is balanceable if $G(A)$ contains no $3PC(x, y)$ where x and y belong to opposite sides of the bipartition and no wheel, where the center node has odd degree.*

Proof: By Theorem 13 we only need to find in $G(A)$ the subgraphs of type (a), (b), (c) or (d) that are not balanceable. Since $G(A)$ is bipartite it cannot contain graphs of type (b) or (c). Graphs of type (a) with both endnodes in the same side of the bipartition are seen to be balanceable by signing the edges so that the three paths have the same length mod 4. When the two nodes of degree 3 belong to opposite sides of the bipartition then since two of the paths have the same length mod4, either 1 mod 4 or 3 mod 4 there exists a chordless cycle signed incorrectly with respect to α of all zeroes.

For a wheel (H, x), let $C_1, \ldots, C_k$ be the chordless cycles of (H, x) containing x. Obtain a signing of the graph so that $C_1, \ldots, C_k$ are signed correctly. For $F \subseteq E$, let $l(F) = \sum_{e \in F} l(e)$. Then $\sum_{i=1}^{k} l(C_i) \equiv 0 \bmod 4$. But $l(H) = \sum_{i=1}^{k} l(C_i) - 2l(S)$ where S consists of all edges with one endpoint the center node of the wheel. Since $2l(S) \equiv 2|S| \bmod 4$, clearly $l(H) \equiv 0 \bmod 4$ if and only if $k = |S|$ is even. □

In [8], [2], a polynomial algorithm is given, to recognize if a matrix is balanceable or balanced. Balanced $0, \pm 1$ matrices have interesting polyhedral properties and have been recently the subject of several investigations, see [7] for a survey.

Totally Unimodular and Regular Matrices: A Theorem of Tutte

A matrix $\tilde{A}$ is *totally unimodular* (TU, for short) if every square submatrix of $\tilde{A}$ has determinant $0, \pm 1$. Consequently a TU matrix is a $0, \pm 1$ matrix. If $\tilde{A}$ is a $0, \pm 1$ matrix such that $G(\tilde{A})$ is a chordless cycle C, then $det(\tilde{A}) = 0$ if $l(C) \equiv 0 \bmod 4$ and $det(\tilde{A}) = \pm 2$ if $l(C) \equiv 2 \bmod 4$. So if $\tilde{A}$ is TU, then $\tilde{A}$ is balanced.

A $0, 1$ matrix A is *regular* if A can be signed to be TU. An example of a $0, 1$ matrix that is not regular is one whose bipartite graph is a wheel with a rim of length 6 (and the center node has obviously three neighbors in the rim). We will see this later in this section.

To state the theorem of Tutte characterizing regular matrices, we need to introduce the notion of *pivoting* in a matrix. Pivoting on an entry $\epsilon \neq 0$ of a matrix $A = \begin{bmatrix} \epsilon & y^T \\ x & D \end{bmatrix}$, we obtain the matrix $B = \begin{bmatrix} -\epsilon & y^T \\ x & D - \epsilon x y^T \end{bmatrix}$.

Remark 15. *Let B be obtained from A by pivoting on the nonzero entry a_{ij}. Then:*

- *A can be obtained from B by pivoting on the same entry.*
- *Let a_{ij} be the pivot element. Then $b_{ij} = -a_{ij}$. For $l \neq j$, $b_{il} = a_{il}$ and for $k \neq i$, $b_{kj} = a_{kj}$. For $l \neq j$ and $k \neq i$, $b_{kl} = a_{kl} - a_{ij}a_{il}a_{kj}$*
- *$det(A) = \pm det(D - \epsilon x y^T)$ and $det(B) = \pm det(D)$.*

We are interested in performing the pivot operations on A both over the reals (R-pivoting) and over $GF2$ ($GF2$-pivoting). Let B be a matrix obtained from A by performing a $GF2$-pivot or an R-pivot. We next show how to obtain $G(B)$ from $G(A)$.

Remark 16. *Let B be the $0,1$ matrix obtained from a $0,1$ matrix A by $GF2$-pivoting on $a_{ij} = 1$. Then $G(B)$ is obtained from $G(A)$ as follows:*

1) For every 4-cycle $C = u_1, i, j, v_1$ of $G(A)$ remove edge $u_1 v_1$.
2) For every induced chordless path $P = u_1, i, j, v_1$ of $G(A)$, add edge $u_1 v_1$.

Proof: Follows from Remark 15.   □

It is easy to check that a 2×2, ± 1 matrix is singular if and only if the sum of its entries is equivalent to 0 mod 4. A $0, \pm 1$ matrix A is *weakly balanced* if every 4-cycle C of $G(A)$ satisfies $l(C) \equiv 0$ mod 4. Equivalently, A is weakly balanced if every 2×2 submatrix of A has determinant $0, \pm 1$.

Remark 17. *Let $\tilde{B}$ be the matrix obtained from a weakly balanced $0, \pm 1$ matrix $\tilde{A}$ by R-pivoting on a non-zero entry $a_{ij} = \epsilon$. Then $\tilde{B}$ is a $0, \pm 1$ matrix and $G(\tilde{B})$ is obtained from $G(\tilde{A})$ as follows:*

1) Edge ij has label $-\epsilon$.
2) For every 4-cycle u_1, i, j, v_1 in $G(\tilde{A})$ remove edge $u_1 v_1$.
3) For every induced chordless path $P = u_1, i, j, v_1$ in $G(\tilde{A})$ add edge $u_1 v_1$ and label it so that, for the resulting cycle $C = u_1, i, j, v_1$ in $G(\tilde{B})$, $l(C) \equiv 0$ mod 4.

Proof: 1) is trivial. By Remark 15, for $k \neq i$ and $l \neq j$, $b_{kl} = a_{kl} - \epsilon a_{kj} a_{il}$. Note that $\epsilon^2 = 1$. So ϵb_{kl} is the value of the determinant of the 2×2 submatrix of $\tilde{A}$ with rows i, k and columns j, l. Since $\tilde{A}$ is weakly balanced, ϵb_{kl} and b_{kl}, have values in $0, \pm 1$. For 2), note that all 2×2 submatrices of $\tilde{A}$ with all four entries non-zero have determinant 0. Finally since the 2×2 submatrix of $\tilde{B}$ has determinant 0, part 3) follows.   □

Corollary 18. *Let $\tilde{A}$ be a weakly balanced $0, \pm 1$ matrix and A be the $0, 1$ matrix with the same support. Let $\tilde{B}$ and B be the matrices obtained by R-pivoting $\tilde{A}$ and GF2-pivoting A on the same entry. Then $G(\tilde{B})$ and $G(B)$ have the same edge set. (Equivalently, $\tilde{B}$ and B have the same support).*

Tutte [16], [17] proves the following:

Theorem 19. *A $0, 1$ matrix A is regular if and only if for no matrix B, obtained from A by GF2-pivoting, $G(B)$ contains a wheel whose rim has length 6.*

To prove the above result, we need the following three lemmas (the first is well known):

Lemma 20. *A $0, 1$ matrix A is regular if and only if any matrix obtained from A by GF2-pivoting is regular.*

Proof: Follows by Remark 15 and Corollary 18.   □

Lemma 21. *Let $\tilde{A}$ be a balanced $0, \pm 1$ matrix, $\tilde{B}$ be obtained by R-pivoting $\tilde{A}$ on $\tilde{a}_{ij}$ and B be the $0, 1$ matrix with the same support as $\tilde{B}$. If $\tilde{B}$ is not balanced, then B is not balanceable.*

Proof: We show that $G(\tilde{B})$ can be obtained from $G(B)$ by applying the signing algorithm. Let T be any tree in $G(\tilde{B})$, chosen to contain all edges in $\{ij\} \cup \{ix : x \in N(i)\} \cup \{jy : y \in N(j)\}$. Then T is also a tree of $G(\tilde{A})$. Let $S = t_1, \ldots, t_{|T|-1}, e_1, \ldots, e_l$ be a consistent ordering of the edges of $G(\tilde{B})$, where t_i are edges in T. We show that the signing of $G(\tilde{B})$ can be obtained by the signing algorithm with sequence S, where the edges of T are labeled as in $G(\tilde{B})$. Let e_k be an edge of S and C_{e_k} be a chordless cycle of $G(B)$ containing e_k and edges in $S \setminus e_{k+1}, \ldots, e_m$, such that C_{e_k} has the largest possible intersection with $\{i, j\}$ and, subject to this, C_{e_k} is shortest. We show that C_{e_k} forces e_k to be signed as in $G(\tilde{B})$.

Remark 17 shows that if C_{e_k} contains both nodes i and j and has length 4, then C_{e_k} forces e_k to be signed as in $G(\tilde{B})$.

All other edges e_k are labeled the same in $G(\tilde{B})$ and $G(\tilde{A})$. We show that C_{e_k} forces this signing of edge e_k.

If C_{e_k} contains both nodes i and j and has length bigger than 4, then in $G(\tilde{A})$ the nodes of C_{e_k} induce a cycle with unique chord $i_1 j_1$, where i_1 and j_1 are the neighbors of i and j in C_{e_k}. By Remark 17, the sum of the labels on the edges $i_1 i, ij, jj_1$ in $G(\tilde{B})$ is equivalent modulo 4 to the label of edge $i_1 j_1$, in $G(\tilde{A})$. Thus the cycle C'_{e_k} of $G(\tilde{A})$ induced by $V(C_{e_k}) \setminus \{i, j\}$ and the cycle C_{e_k} of $G(\tilde{B})$ force e_k to be signed the same.

If C_{e_k} contains one of $\{i, j\}$, say i, then by choice of C_{e_k}, node j has i as unique neighbor in C_{e_k}. For, if not, e_k either belongs to a chordless cycle of $G(\tilde{B})$ of to a chordless cycle that is shorter that C_{e_k} and contains node j (this happens when (C_{e_k}, j) is the rim of a wheel with center j and no hole of (C_{e_k}, j) contains i, j and e_k), a contradiction to our assumption.

But then C_{e_k} is also a chordless cycle of $G(\tilde{A})$ and forces e_k to be signed as in $G(\tilde{B})$.

If C_{e_k} contains neither i nor j and at most one neighbor of i or j then C_{e_k} is also a chordless cycle of $G(\tilde{A})$ and forces e_k to be signed as in $G(\tilde{B})$. Otherwise by the choice of C_{e_k}, node i has a unique neighbor i' in C_{e_k}, node j has a unique neighbor j' in C_{e_k} and i', j' are adjacent. So, by Remark 17, $G(\tilde{A})$ contains a hole C'_{e_k}, whose node set is $V(C_{e_k}) \cup \{i, j\}$. This hole C'_{e_k} and the hole C_{e_k} of $G(\tilde{B})$ force e_k to be signed the same.   □

Lemma 22. *From every $0, 1$ matrix A that is not regular, we can obtain a $0, 1$ matrix that is not balanceable by a sequence of GF2-pivots.*

Proof: Let A be the smallest $0, 1$ matrix (in terms of the sum of the number of rows and columns) that is not regular but cannot be pivoted to a matrix that is not balanceable. Since A is obviously balanceable, let $\tilde{A}$ be a corresponding balanced $0, \pm 1$ matrix. By minimality, we can assume that $\tilde{A}$ is square and $|det(\tilde{A})| \geq 2$. By Remark 15, we can R-pivot on any nonzero entry of $\tilde{A}$ to obtain a $0, \pm 1$ matrix $\tilde{B}$ which contains a proper submatrix $\tilde{C}$ with the same determinant value as $\tilde{A}$. Since $\tilde{A}$ is weakly balanced, by Remark 17, $\tilde{B}$, and hence $\tilde{C}$, is a $0, \pm 1$ matrix. Let B be the $0, 1$ matrix with the same support as $\tilde{B}$ and C the submatrix of B corresponding to $\tilde{C}$. By Corollary 18, B is obtained from A with a $GF2$-pivot on the same element. Assume $\tilde{B}$ is balanced. Then $\tilde{C}$ would be a balanced matrix which is not TU. However, this implies that C is not regular (this was already known to Camion [1]): Indeed, $\tilde{C}$ is a signing of C which is balanced but not TU: So $\tilde{C}$ can be obtained by applying the signing algorithm on $G(C)$, starting with a tree T of $G(C)$. Assume C has a TU signing $\tilde{C}'$. Since $\tilde{C}'$ is also a balanced matrix, then $G(\tilde{C}')$ can be obtained through the signing algorithm by signing T as in $G(\tilde{C}')$. So $G(\tilde{C})$ and $G(\tilde{C}')$ differ on some fundamental cuts of T. So $\tilde{C}$ can be transformed in $\tilde{C}'$ by multiplying by -1 the rows and columns corresponding to the nodes in on shore of this cut. However this operation preserves the TU property.

So $\tilde{B}$ is not balanced and by Lemma 21, B is not balanceable.   □

Proof of Theorem 19: By Lemma 20, regular matrices are closed under $GF2$-pivoting and if A is a $0, 1$ matrix such that $G(A)$ contains a wheel $G(W)$ whose rim has length 6, then W (hence A) is obviously not regular.

For the sufficiency part, if A is a $0, 1$ matrix which is not regular, then by Lemma 22, we can obtain by $GF2$-pivots a $0, 1$ matrix B which is not balanceable. By Theorem 14, $G(B)$ contains a $3PC(x, y)$ where x and y belong to distinct color classes, or a wheel with rim H and center v and v has an odd number, greater than one, of neighbors in H.

If $G(B)$ contains a $3PC(x, y)$, Remark 16 shows that we can $GF2$-pivot on B so that all of its paths have length three and by doing a last $GF2$-pivot on an entry corresponding to an edge incident to x, we obtain a wheel whose rim has length 6.

If $G(B)$ contains a wheel (H, x) and x has an odd number of neighbors in the rim H, Remark 16 shows that we can $GF2$-pivot on an entry corresponding to an edge of H, incident with a neighbor of x, to obtain a wheel (H', x), where x has two less neighbors in H' than in H. When x has only three neighbors in H', to obtain a wheel whose outer cycle has length 6, $GF2$-pivot so that all the subpaths of H', between two consecutive neighbors of x have length two.   □

Tutte's original proof of the above theorem is quite difficult. A short, self-contained proof can be found in [10]. In [12], a decomposition theorem for regular matrices in given, together with a polynomial algorithm to test if a matrix is regular or TU. A faster algorithm is given in [15].

5   Decomposition

The co-NP characterizations obtained in Theorems 8, 9 and 14 are used in [3], [4], [8], [2] to obtain the decomposition results for graphs without even holes, cap-free graphs and balanceable matrices. However the proofs of these theorems are long and technical. We have seen how Theorem 1 can be used to decompose universally signable graphs with K_1 and K_2 cutsets into holes and triangulated graphs. Here we further illustrate in two easy cases the use of a co-NP characterization to obtain decomposition results and polynomial recognition algorithms.

Restricted Unimodular and Totally Odd Matrices

A $0, \pm 1$ matrix A is *restricted unimodular* (RU, for short) if every cycle C (possibly with chords) of $G(A)$ satisfies $l(C) \equiv 0 \bmod 4$. A $0, 1$ matrix A is *signable to be RU* if there exists a RU $0, \pm 1$ matrix that has the same support. RU matrices are a known subclass of TU matrices, see e.g. [18].

A $0, \pm 1$ matrix A is *totally odd* (TO, for short) if every cycle C of $G(A)$ satisfies $l(C) \equiv 2 \bmod 4$. A $0, 1$ matrix A is *signable to be TO* if there exists a TO $0, \pm 1$ matrix that has the same support. TO matrices are studied in [9].

A *weak 3-path configuration* between nodes x and y $(W3PC(x, y))$ is made up by three paths P_1, P_2, P_3 connecting x and y such that $P_i \cup P_j$, $i \neq j$, $i, j = 1, 2, 3$ induces a cycle (possibly with chords). So P_i may be a single edge or may contain chords and edges may have endnodes in distinct paths P_i and P_j. If G is a bipartite graph, a $W3PC(x, y)$ is *homogeneous* is x and y belong to the same color class of G and is *heterogeneous* otherwise.

Theorem 23. *A $0, 1$ matrix A is signable to be RU if and only if $G(A)$ contains no weak 3-path configuration which is heterogeneous and A is signable to be TO if and only if $G(A)$ contains no weak 3-path configuration which is homogeneous.*

Proof: Let G' be the bipartite graph obtained from $G(A)$ by replacing each edge with a path of length 3 and A' be the $0, 1$ matrix such that $G' = G(A')$. Then there is a correspondence between the cycles of $G(A)$ and the holes of G'. So A is signable to be RU if and only if G' is α-balanceable for the vector α of all zeroes (i.e. A' is a balanceable matrix) and A is signable to be TO if and only if G' is α-balanceable for the vector α of all twos. In the first case, the theorem follows from Theorem 14. The proof for the second case in analogous and is left as an exercise. $\qquad\square$

A *bridge* of a cycle C is either a chord of C or a subgraph of G, whose node set contains all nodes of a connected component of $G \setminus V(C)$, say G', together with the nodes of C, adjacent to at least one node in G' and whose edges are the edges of G with at least one endnode in G'. The *attachments* of a bridge B are the nodes of $V(B) \cap V(C)$. A bridge is *homogeneous* if all of its attachments belong to the same color class of G and is *heterogeneous* if no two of its attachments belong to the same color class of G. Obviously, if B is a heterogeneous bridge, then B has at most two attachments.

Lemma 24. *Let A be $0,1$ matrix that is signable to be RU and C any cycle of $G(A)$. Then every bridge of C is homogeneous.*
Let A be $0,1$ matrix that is signable to be TO and C any cycle of $G(A)$. Then every bridge of C is heterogeneous.

Proof: We prove the first statement. Let x and y be two attachments of a bridge B of a cycle C. If x and y belong to distinct color classes of $G(A)$, then $G(A)$ contains a heterogeneous $W3PC(x,y)$ where P_1, P_2 are the two xy-subpaths of C and P_3 is any xy-path in B. The proof of the second statement is similar. $\square$

Bridges B_1 and B_2 of C *cross* if there exist attachments x_1, y_1 of B_1 and x_2, y_2 of B_2 that are distinct and appear in the order x_1, x_2, y_1, y_2 when traversing C in one direction.

Lemma 25. *Let A be $0,1$ matrix that is signable to be RU and C any cycle of $G(A)$. Then no pair of homogeneous bridges of C, having attachments in distinct color classes of $G(A)$, cross.*
Let A be $0,1$ matrix that is signable to be TO, C any cycle of $G(A)$. Then no pair of heterogeneous bridges of C cross.

Proof: To prove the first statement, assume B_1 and B_2 are homogeneous bridges of C having attachments x_1, y_1 of B_1 and x_2, y_2 of B_2, appearing in the order x_1, x_2, y_1, y_2 when traversing C. If x_1, y_1 and x_2, y_2 are in distinct color classes of $G(A)$, we have a heterogeneous $W3PC(x_1, x_2)$, where P_1 is the subpath of C, connecting x_1, x_2 and not containing y_1. P_2 and P_3 contain respectively a x_1, y_1-path in B_1 and a x_2, y_2-path in B_2. The proof of the second part is similar. $\square$

Theorem 26. *Let A be $0,1$ matrix that is signable to be TO, C any cycle of $G(A)$ and B be a heterogeneous bridge of C with two attachments x and y. Then $G(A) \setminus \{x, y\}$ is disconnected.*

Proof: By Lemma 24, x and y are the only two attachments of B. Let P_1, P_2 be the two subpaths of C, connecting x and y. By Lemma 25, no bridge of C has an attachments in both $P_1 \setminus \{x, y\}$ and $P_2 \setminus \{x, y\}$. So no two of B, P_1 and P_2 are in the same component of $G(A) \setminus \{x, y\}$ and, since at least two of them are not edges, $G(A) \setminus \{x, y\}$ contains at least two components.   □

Theorem 27 ([18]). *Let A be $0, 1$ matrix that is signable to be RU and C any cycle of $G(A)$ containing homogeneous bridges B_1 and B_2 with attachments in distinct color classes of $G(A)$. Then C contains two edges whose removal disconnects $G(A)$ and separates B_1 and B_2.*

Proof: Assume that the attachments of B_1 and B_2 belong to the "red" and "blue" sides of the bipartition of $G(A)$. Let P_1 be be minimal subpath of C with the following property:

P_1 *contains all the attachments of B_1 and no bridge of C with red attachments has all its attachments either in P_1 or outside P_1*

The subpath P_2 is similarly defined, with respect to B_2 and the bridges with blue attachments. By Lemma 25 P_1 and P_2 can be chosen to be nonoverlapping. Furthermore by minimality of P_1 and P_2, the endnodes a_1, b_1 of P_1 are red nodes and the endnodes a_2, b_2 of P_2 are blue nodes. Let $C = a_1, P_1, b_1, P_{b_1 b_2}, b_2, P_2, a_2,$ $P_{a_2 a_1}$, let b be any edge of $P_{b_1 b_2}$ and a any edge of $P_{a_1 a_2}$. By Lemma 25 and the construction of P_1 and P_2, $P_1 \cup B_1$ and $P_2 \cup B_2$ belong to distinct components of $G \setminus \{a, b\}$.   □

Clearly to test if A is signable to be RU, we can assume that $G(A)$ is biconnected, otherwise we work on the biconnected components.

If $G(A)$ is biconnected and contains no cycle with homogeneous bridges with attachments in distinct color classes of $G(A)$, then A has two ones per row or per column. (This is easy from network flows). In this case A is obviously RU: Sign A so that each row or column contains a 1 and a -1 to obtain a network matrix (or its transpose). From this fact and the above theorem yield in a straightforward way a polytime algorithm to test if a $0, 1$ is signable to be RU. This algorithm, combined with the signing algorithm of Section 1, gives a procedure to test if a $0, \pm 1$ matrix is RU.

In a similar manner, see [9], Theorem 26 and the signing algorithm give procedures to test if a $0, 1$ matrix is signable to be TO and to test if a $0, \pm 1$ matrix is TO.

References

1. P. Camion. Caractérisation des matrices totalement unimodulaires. *Cahiers Centre Études Rech. Op.*, 5:181–190, 1963.
2. M. Conforti, G. Cornuéjols, A. Kapoor, and K. Vušković. Balanced $0, \pm 1$ matrices, Parts I–II. 1994. Submitted for publication.
3. M. Conforti, G. Cornuéjols, A. Kapoor, and K. Vušković. Even-hole-free graphs, Parts I–II. Preprints, Carnegie Mellon University, 1997.

4. M. Conforti, G. Cornuéjols, A. Kapoor, and K. Vušković. Even and odd holes in cap-free graphs. 1996. Submitted for publication.
5. M. Conforti, G. Cornuéjols, A. Kapoor, and K. Vušković. Universally signable graphs. *Combinatorica*, 17(1):67–77, 1997.
6. M. Conforti, G. Cornuéjols, A. Kapoor, and K. Vušković. A Mickey-Mouse decomposition theorem. In Balas and Clausen, editors, *Proceedings of 4th IPCO Conference*, Springer Verlag, 1995.
7. M. Conforti, G. Cornuéjols, A. Kapoor, M. R. Rao, and K. Vušković. Balanced matrices. *Proceedings of the XV International Symposium on Mathematical Programming*, University of Michigan Press, 1994.
8. M. Conforti, G. Cornuéjols,and M. R. Rao. Decomposition of balanced 0,1 matrices, Parts I–VII. 1991. Submitted for publication.
9. M. Conforti, G. Cornuéjols, and K. Vušković. Balanced cycles and holes in bipartite graphs. 1993. Submitted for publication.
10. A. M. H. Gerards. A short proof of Tutte's characterization of totally unimodular matrices. *Linear Algebra and its Applications*, 14:207–212, 1989.
11. A. Hajnal and T. Suryani. Uber die auflosung von graphen vollstandiger teilgraphen. *Ann. Univ. Sc. Budapest. Eotvos Sect. Math.*, 1, 1958.
12. P. Seymour. Decomposition of regular matroids. *Journal of Combinatorial Theory B*, 28:305–359, 1980.
13. K. Truemper. On balanced matrices and Tutte's characterization of regular matroids. Working paper, University of Texas at Dallas, 1978.
14. K. Truemper. Alpha-balanced graphs and matrices and GF(3)-representability of matroids. *Journal of Combinatorial Theory B*, 32:112–139, 1982.
15. K. Truemper. A decomposition theory of matroids V. Testing of matrix total unimodularity. *Journal of Combinatorial Theory B*, 49:241–281, 1990.
16. W. T. Tutte. A homotopy theorem for matroids I, II. *Trans. Amer. Math. Soc.*, 88:144–174, 1958.
17. W. T. Tutte. Lectures on matroids. *J. Nat. Bur. Standards B*, 69:1–47, 1965.
18. M. Yannakakis. On a class of totally unimodular matrices. *Mathematics of Operations Research*, 10:280–304, 1985.

The Generalized Stable Set Problem for Claw-Free Bidirected Graphs

Daishin Nakamura and Akihisa Tamura

Department of Computer Science and Information Mathematics
University of Electro-Communications
1-5-1 Chofugaoka, Chofu, Tokyo 182-8585, Japan
{daishin, tamura}@@im.uec.ac.jp

Abstract. Bidirected graphs are a generalization of undirected graphs. The generalized stable set problem is an extension of the maximum weight stable set problem for undirected graphs to bidirected graphs. It is known that the latter problem is polynomially solvable for claw-free undirected graphs. In this paper, we define claw-free bidirected graphs and show that the generalized stable set problem is also polynomially solvable for claw-free bidirected graphs.

1 Introduction

Let $G = (V, E)$ be an undirected graph. A subset S of V is called a *stable set* if any two elements of S are nonadjacent. Given a weight vector $w \in \Re^V$, a *maximum weight stable set* is a stable set S maximizing $w(S) = \sum_{i \in S} w_i$. The problem of finding a maximum weight stable set is called the *maximum weight stable set problem* (MWSSP). It is well known that the problem can be formulated as the following integer programming problem:

$$[\text{MWSSP}] \quad \text{maximize} \quad w \cdot x \quad \text{subject to } x_i + x_j \leq 1 \text{ for } (i,j) \in E,$$
$$x_i \in \{0,1\} \text{ for } i \in V.$$

In this paper, we consider the problem generalized as follows: for a given finite set V and for given $P, N, I \subseteq V \times V$,

$$[\text{GSSP}] \quad \text{maximize} \quad w \cdot x \quad \text{subject to} \quad x_i + x_j \leq 1 \quad \text{for } (i,j) \in P,$$
$$-x_i - x_j \leq -1 \text{ for } (i,j) \in N,$$
$$x_i - x_j \leq 0 \quad \text{for } (i,j) \in I,$$
$$x_i \in \{0,1\} \quad \text{for } i \in V.$$

Here we call this problem the *generalized stable set problem* (GSSP). We note that the GSSP is equivalent to the generalized set packing problem discussed in [1,2]. To deal with the GSSP, a 'bidirected' graph is useful. A *bidirected graph*

R. E. Bixby, E. A. Boyd, and R. Z. Ríos-Mercado (Eds.): IPCO VI
LNCS 1412, pp. 69–83, 1998. © Springer–Verlag Berlin Heidelberg 1998

$G = (V, E)$ has a set of vertices V and a set of edges E, in which each edge $e \in E$ has two vertices $i, j \in V$ as its endpoints and two associated signs (plus or minus) at i and j. The edges are classified into three types: the $(+, +)$-edges with two plus signs at their endpoints, the $(-, -)$-edges with two minus signs, and the $(+, -)$-edges (and the $(-, +)$-edges) with one plus and one minus sign. Given an instance of the GSSP, we obtain a bidirected graph by making $(+, +)$-edges, $(-, -)$-edges and $(+, -)$-edges for vertex-pairs of P, N and I respectively. Conversely, for a given bidirected graph with a weight vector on the vertices, by associating a variable x_i with each vertex, we may consider the GSSP. We call a $0-1$-vector satisfying the inequality system arising from a bidirected graph G a *solution* of G. We also call a subset of vertices a solution of G if its incidence vector is a solution of G. The GSSP is an optimization problem over the solutions of a bidirected graph.

Since several distinct bidirected graphs may have the same set of solutions, we deal with some kind of 'standard' bidirected graphs. A bidirected graph is said to be *transitive*, if whenever there are edges $e_1 = (i, j)$ and $e_2 = (j, k)$ with opposite signs at j, then there is also an edge $e_3 = (i, k)$ whose signs at i and k agree with those of e_1 and e_2. Obviously, any bidirected graph and its transitive closure have the same solutions. A bidirected graph is said to be *simple* if it has no loop and if it has at most one edge for each pair of distinct vertices. Johnson and Padberg [3] showed that any transitive bidirected graph can be reduced to simple one without essentially changing the set of solutions, or determined to have no solution. We note that a transitive bidirected graph has no solution if and only if it has a vertex with both a $(+, +)$-loop and a $(-, -)$-loop. For any bidirected graph, the associated simple and transitive bidirected graph can be constructed in time polynomial in the number of vertices.

Given a bidirected graph G, its *underlying graph*, denoted by $\underline{G}$, is defined as the undirected graph obtained from G by changing all the edges to $(+, +)$-edges. A bidirected graph is said to be *claw-free* if it is simple and transitive and if its underlying graph is claw-free (i.e., does not contain a vertex-induced subgraph which is isomorphic to the complete bipartite graph $K_{1,3}$).

It is well known that the MWSSP is NP-hard for general undirected graphs (and hence, the GSSP is also NP-hard). However, for several classes of undirected graphs, the MWSSP is polynomially solvable. For example, Minty [4] proposed a polynomial time algorithm for the MWSSP for claw-free undirected graphs. On the other hand, there are several polynomial transformations from the GSSP to the MWSSP (see [5,6]). Unfortunately, we cannot easily derive the polynomial solvability of the GSSP for claw-free bidirected graphs by using these transformations, because these do not preserve claw-freeness. Our aim in this paper is to verify that the GSSP for claw-free bidirected graphs is polynomially solvable.

2   Canonical Bidirected Graphs and Their Solutions

In this section, we will give several definitions and discuss basic properties of solutions of bidirected graphs. Let $G = (V, E)$ be a simple and transitive bidirected graph and w be a weight vector on V. For any subset $U \subseteq V$, we call the transformation which reverse the signs of the u side of all edges incident to each $u \in U$ the *reflection* of G at U, and we denote it by $G{:}U$. Obviously, reflection preserves simpleness and transitivity. Let $w{:}U$ denote the vector defined by $(w{:}U)_i = -w_i$ if $i \in U$; otherwise $(w{:}U)_i = w_i$. For two subsets X and Y of V, let $X \triangle Y$ denote the symmetric difference of X and Y.

Lemma 1. *Let X be any solution of G. Then, $X \triangle U$ is a solution of $G{:}U$. The GSSP for (G, w) is equivalent to the GSSP for $(G{:}U, w{:}U)$.*

Proof. The first assertion is trivial from the definition of $G{:}U$. The second assertion follows from the equation $w{:}U(X \triangle U) = \sum_{i \in X \setminus U} w_i + \sum_{i \in U \setminus X} (-x_i) = w(X) - \sum_{i \in U} w_i$, (the last term is a constant). $\qquad\square$

We say that a vertex is *positive* (or *negative*) if all edges incident have plus (or minus) signs at it, and that a vertex is *mixed* if it is neither positive nor negative. If a bidirected graph has no $(-, -)$-edge, it is said to be *pure*. We say that a bidirected graph is *canonical* if it is simple, transitive and pure and it has no negative vertex. For any instance (G, w) of the GSSP, we can transform it to equivalent one whose bidirected graph is canonical as follows. From the previous section, we can assume that G is simple and transitive. Johnson and Padberg [3] proved that G has at least one solution $U \subseteq V$. From Lemma 1, $G{:}U$ has the solution $U \triangle U = \emptyset$, that is, $G{:}U$ must be pure. Let W be the set of negative vertices of $G{:}U$. Then $G{:}U{:}W$ has no negative vertex, and furthermore, it is pure because any edge (v, w) of $G{:}U$ with $w \in W$ must be a $(+, -)$-edge. Since this transformation is done in polynomial time, we assume that a given bidirected graph of the GSSP is canonical in the sequel.

For any solution X of a canonical bidirected graph G, we partition X into two parts:

$$X_B = \{i \in X \mid N_G^{-+}(i) \cap X = \emptyset\} \quad \text{and} \quad X_I = \{i \in X \mid N_G^{-+}(i) \cap X \neq \emptyset\},$$

where $N_G^{-+}(i)$ denotes the set of vertices adjacent to i by a $(-, +)$-edge incident to i with a minus sign, $N_G^{+-}(i)$ is defined analogously. Here we call X_B a *base* of X. Let

$$\mathrm{ex}(X_B) = X_B \cup \{i \in V \mid i \in N_G^{+-}(x) \text{ for some } x \in X_B\}.$$

If $S \subseteq V$ is a stable set of $\underline{G}$, we say that S is a stable set of G. It is not difficult to show the following lemmas.

Lemma 2. *For any solution X of a canonical bidirected graph G, $X = \mathrm{ex}(X_B)$, and hence, $(\mathrm{ex}(X_B))_B = X_B$.*

Lemma 3. *For any solution X of a canonical bidirected graph G, its base X_B is a stable set of G.*

Lemma 4. *For any stable set S of a canonical bidirected graph G, $\mathrm{ex}(S)$ is a solution of G.*

Thus there is a one-to-one correspondence between the solutions and the stable sets of G.

For any subset U of V, let $G[U]$ denote the subgraph induced by U. We call $H \subseteq V$ a connected component of G if H induces a connected component of $\underline{G}$.

Lemma 5. *Let X and Y be solutions of a canonical bidirected graph G. For any connected component H of $G[X_B \triangle Y_B]$, $X_B \triangle H$ and $Y_B \triangle H$ are bases of certain solutions of G.*

Proof. From Lemma 3, X_B and Y_B are stable sets of G. Thus $X_B \triangle H$ and $Y_B \triangle H$ are also stable sets of G. Hence Lemma 4 implies the assertion. □

Let X be a specified solution of G. For any solution Y of G, let $H_1, \ldots, H_\ell$ be the connected components of $G[X_B \triangle Y_B]$. We define the weight of H_i, denoted by $\delta^X(H_i)$ or simply $\delta(H_i)$, by

$$\delta^X(H_i) = w(\mathrm{ex}(X_B \triangle H_i)) - w(X).$$

We remark that the equation $w(Y) - w(X) = \sum \delta(H_i)$ may not hold because there may exist a vertex v such that $N_G^{-+}(v)$ contains several vertices of distinct connected components, that is, w_v may be doubly counted. In order to avoid this obstacle, we require some additional conditions.

Lemma 6. *For any solution X of G, there exists $U \subseteq V$ such that $G' = G{:}U$ and $X' = X \triangle U$ satisfy*

(a) *G' is canonical,*
(b) *X' is a stable set of G', i.e., $X' = (X')_B$,*
(c) *for each mixed vertex $v \notin X'$, there is a vertex $u \in X'$ adjacent to v.*

Proof. Let M be the set of all mixed vertices v such that $v \notin X$, v is adjacent to no vertex of X_B and $N_G^{+-}(v) = \emptyset$. For any (inclusion-wise) maximal stable set S of $G[M]$, $U = X_I \cup S$ satisfies the assertion. □

We note that a subset U having the conditions of Lemma 6 can be found in polynomial time. The conditions of Lemma 6 overcome the above obstacle.

Lemma 7. *Let G be a canonical bidirected graph and X be a solution of G satisfying the conditions of Lemma 6. For any solution Y, let $H_1, \ldots, H_\ell$ be the connected components of $G[X_B \triangle Y_B]$. Then,*

$$w(Y) - w(X) = \sum_{i=1}^{\ell} \delta^X(H_i) = \sum_{i=1}^{\ell} \{w(\mathrm{ex}(X_B \triangle H_i)) - w(X)\}.$$

Proof. Suppose to the contrary that there exists a mixed vertex v such that $N_G^{-+}(v)$ contains two vertices u and w of distinct connected components H_i and H_j. Since X is a stable set, $u, w \in Y_B$. Let x be a vertex of X adjacent to v. From the transitivity, x must be adjacent to both u and w. This contradicts the fact that H_i and H_j are distinct connected components of $G[X_B \triangle Y_B]$. □

3 A Basic Idea for Finding an Optimal Solution of the GSSP

Given an instance (G, w) of the GSSP, for each $i = 0, 1, \ldots, |V|$, let

$$\mathcal{S}_i = \{X \subseteq V \mid X \text{ is a solution of } G \text{ and has exactly } i \text{ positive vertices }\},$$
$$w^i = \max_{X \in \mathcal{S}_i} w(X),$$
$$\mathcal{S}_i^* = \{X \in \mathcal{S}_i \mid w(X) = w^i\}.$$

Suppose that N denotes the smallest number j with $w^j = \max_i w^i$. Minty [4] showed that if a given undirected graph is claw-free, then $w^0 < \cdots < w^N$. More precisely, $(0, w^0), \ldots, (N, w^N)$ lie on an increasing concave curve. Minty's algorithm for solving the MWSSP for claw-free undirected graphs finds an optimal solution by tracing (i, w^i) one by one. However, even if a given bidirected graph is claw-free, this fact does not hold as an example in Figure 1 where $(+, +)$-edges are drawn by lines and $(+, -)$-edges by arrows whose heads mean minus signs. Thus, it seems to be difficult to trace (i, w^i) one by one for the GSSP.

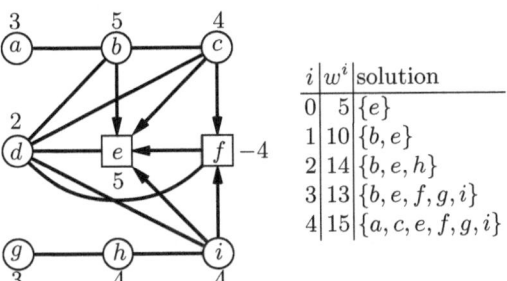

i	w^i	solution
0	5	$\{e\}$
1	10	$\{b, e\}$
2	14	$\{b, e, h\}$
3	13	$\{b, e, f, g, i\}$
4	15	$\{a, c, e, f, g, i\}$

Fig. 1.

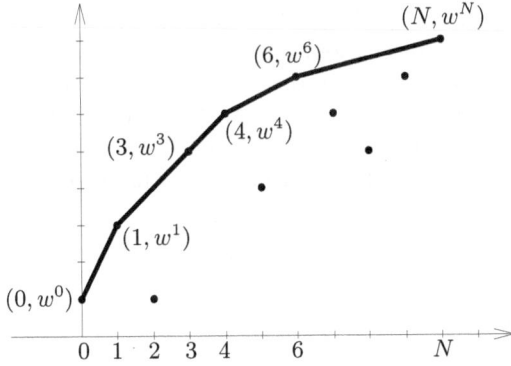

Fig. 2.

We will use a technique of the fractional programming. Let us consider the upper envelope of the convex hull of the set of pairs $(0, w^0), (1, w^1), \ldots, (N, w^N)$ as in Figure 2. We call (i, w^i) a *Pareto-optimal pair* if it lies on the envelope, and their solutions *Pareto-optimal solutions*. Obviously, $(0, w^0)$ and (N, w^N) are always Pareto-optimal. In Figure 2, $(0, w^0)$, $(1, w^1)$, $(3, w^3)$, $(4, w^4)$, $(6, w^6)$ and (N, w^N) are Pareto-optimal.

Let X^i be a Pareto-optimal solution with $X^i \in \mathcal{S}_i$. Suppose that $\mathcal{F}$ is a subset of all the solutions of G such that $X^i \in \mathcal{F}$ and $\mathcal{F}$ is defined independently to the weight vector w. Let us also consider the Pareto-optimal solutions for the restriction on $\mathcal{F}$. Obviously, X^i is also Pareto-optimal in $\mathcal{F}$. We consider the following two problems

$$[\text{MAX}\delta] \quad \max_{Y \in \mathcal{F}} \left\{ \delta(Y) = w(Y) - w(X^i) \right\}, \quad \text{and}$$

$$[\text{MAX}\rho] \quad \max_{Y \in \mathcal{F}} \left\{ \rho(Y) = \frac{\delta(Y)}{\nu(Y)} \mid \delta(Y) > 0 \right\},$$

where $\nu(Y)$ denotes the difference of the numbers of all the positive vertices of Y and X^i. We denote $\rho(\cdot)$ and $\delta(\cdot)$ for a weight vector $\bar{w}$ by $\rho_{\bar{w}}(\cdot)$ and $\delta_{\bar{w}}(\cdot)$ explicitly. Suppose that X^i is not optimal in $\mathcal{F}$. Let Y^1 be an optimal solution of the MAXδ for $\bar{w}^0 = w$. We set $r = \rho_{\bar{w}^0}(Y^1)$ and consider the new weight vector $\bar{w}^1$ defined by

$$\bar{w}_i^1 = \begin{cases} \bar{w}_i^0 - r & \text{if } i \text{ is a positive vertex,} \\ \bar{w}_i^0 & \text{otherwise.} \end{cases} \quad (1)$$

Then, $\delta_{\bar{w}^1}(Y^1) = 0$. For any solution $Y \in \mathcal{F}$,

$$\rho_{\bar{w}^1}(Y) = \frac{\delta_{\bar{w}^0}(Y) - r \cdot \nu(Y)}{\nu(Y)} = \rho_{\bar{w}^0}(Y) - r.$$

Thus, X^i is Pareto-optimal in $\mathcal{F}$ for $\bar{w}^1$. We now assume that there is a solution Y^* with $\rho_{\bar{w}^0}(Y^*) > \rho_{\bar{w}^0}(Y^1)$ and $\delta_{\bar{w}^0}(Y^*) > 0$. Then, evidently, $0 < \nu(Y^*) < \nu(Y^1)$. We also have $\delta_{\bar{w}^1}(Y^*) = [\delta_{\bar{w}^0}(Y^*) - r \cdot \nu(Y^*)] = \nu(Y^*)[\rho_{\bar{w}^0}(Y^*) - \rho_{\bar{w}^0}(Y^1)] > 0$. Conversely, if $\delta_{\bar{w}^1}(Y^*) > 0$ then $\rho_{\bar{w}^0}(Y^*) > \rho_{\bar{w}^0}(Y^1)$ and $\delta_{\bar{w}^0}(Y^*) > 0$. Summing up the above discussion, for an optimal solution Y^2 of the MAXδ for $\bar{w}^1$, if $\delta_{\bar{w}^1}(Y^2) = 0$ then Y^1 is an optimal solution of the MAXρ for w; otherwise, by repeating the above process at most $|V|$ times, the MAXρ for w can be solved, because of the fact that $\nu(Y^1) > \nu(Y^2) > \cdots > 0$.

From the above discussion, for each Pareto-optimal solution $X^i \in \mathcal{S}_i^*$, if we can easily define a subset $\mathcal{F}$ such that

(A1) $X^i \in \mathcal{F}$ and $\mathcal{S}_j^* \cap \mathcal{F} \neq \emptyset$ where (j, w^j) is the next Pareto-optimal pair, and

(A2) the MAXδ for $\mathcal{F}$ and for any w can be solved in time polynomial in the number of vertices of G,

then we can either determine X^i is optimal or find a Pareto-optimal solution $X^k \in \mathcal{S}_k^*$ with $i < k \leq N$ in polynomial time. (We may find $(4, w^4)$ from $(1, w^1)$

in Figure 2.) In addition, if $X^0 \in \mathcal{S}_0^*$ can be found in polynomial time, the GSSP for (G, w) can be solved in polynomial time. In fact, this initialization is not so difficult if we can apply the above technique for any vertex-induced subgraph of G, because it is sufficient to solve the GSSP for the bidirected graph obtained from the current one by deleting all the positive vertices, recursively.

Finally we introduce a tool in order to trace Pareto-optimal pairs. Let X^i be a Pareto-optimal solution with $i < N$. Without loss of generality, we assume that X^i and G satisfy the conditions of Lemma 6. We say that $H \subseteq V$ is an *alternating set* for X^i if H is connected in G and if $X^i \triangle H$ is a stable set of G. We define the weight $\delta(H)$ of an alternating set H with respect to w by $w(\mathrm{ex}(X^i \triangle H)) - w(X^i)$.

Lemma 8. *Let (j, w^j) be the next Pareto-optimal pair of (i, w^i). Then, for any $X^j \in \mathcal{S}_j^*$, there exists a connected component H of $G[X_B^i \triangle X_B^j]$ such that $\mathrm{ex}(X_B^i \triangle H)$ is a Pareto-optimal solution with more positive vertices than X^i.*

Proof. For each connected component H, we denote by $\nu(H)$ the difference of numbers of positive vertices of $X^j \cap H$ and $X^i \cap H$. It is not difficult to show that either $(\delta(H) > 0$ and $\nu(H) > 0)$ or $(\delta(H) = 0$ and $\nu(H) = 0)$ or $(\delta(H) < 0$ and $\nu(H) < 0)$. Here we ignore the second case. Since $i < j$, $G[(X_B^i \triangle X_B^j)]$ must have at least one connected component of the first case. For each connected component H of the first or third case, let $\rho(H) = \delta(H)/\nu(H)$. From the minimality of j and the Pareto-optimality of (j, w^j), if H is of the first case, then $\rho(H) \leq (w^j - w^i)/(j - i)$. Similarly, if H is of the third case, then $\rho(H) \geq (w^j - w^i)/(j - i)$. By combining the above inequalities and Lemma 7, one can obtain $\rho(H) = \frac{w^j - w^i}{j - i}$ for any H of the first or third case. Hence, any connected component H of the first case satisfy the assertion. □

Lemma 8 says that we can trance Pareto-optimal solutions by using alternating sets.

4   Finding a Next Pareto-Optimal Solution

Let G, w and X be a given claw-free bidirected graph, a given weight vector on the vertices and a Pareto-optimal solution with respect to w. Without loss of generality, we assume that G and X satisfy the conditions of Lemma 6. In this section, we explain how to find a next Pareto-optimal solution.

We first give several definitions. We call the vertices of X *black* and the other vertices *white*. Any white vertex is adjacent to at most two black vertices, since otherwise G must have a claw. A white vertex is said to be *bounded* if it is adjacent to two black vertices, *free* if it is adjacent to exactly one black vertex and otherwise *super free*. A cycle (or path) is called an *alternating cycle* (or *path*) if white and black vertices appear alternately, and its white vertices form a stable set. An alternating path is called *free* if its endpoints are either black or free or super free. Alternating cycles and free alternating paths are alternating sets, and vice versa in claw-free cases. Thus, Lemma 8 guarantees that we deal with only alternating cycles and free alternating paths in order to find a next Pareto-optimal solution. An alternating cycle or a free alternating path is called an *augmenting cycle* or an *augmenting path* respectively if it has a positive weight. For two distinct black vertices x and y, let W denote the set of all the bounded vertices adjacent to both x and y. If W is not empty, W is called a *wing* adjacent to x (and y). A black vertex is called *regular* if it is adjacent to three or more wings, *irregular* if it is adjacent to exactly two wings, and otherwise *useless*. An alternating cycle is said to be *small* if it has at most two regular vertices; otherwise *large*. Here we call $C_1, \ldots, C_k$ a *large augmenting cycle family* if each C_i is a large augmenting cycle and each vertex in C_i is adjacent to no vertex in C_j for $1 \leq i < j \leq k$. From Lemma 7, $\delta(C_1 \cup \cdots \cup C_k) = \delta(C_1) + \cdots + \delta(C_k)$ holds.

Our algorithm for finding a next Pareto-optimal solution is described by using the technique discussed in the previous section:

(0) $w^0 \leftarrow w$ and $i \leftarrow 0$;
(1) Find a small augmenting cycle A_{i+1} of the maximum weight for w^i if it exists, otherwise go to (2) ;
     Construct the new weight w^{i+1} by applying (1), $i \leftarrow i+1$ and repeat (1) ;
(2) Find a large augmenting cycle family A_{i+1} of the maximum weight for w^i if it exists, otherwise go to (3) ;
     Construct the new weight w^{i+1} by applying (1), $i \leftarrow i+1$ and repeat (2) ;
(3) Find an augmenting path A_{i+1} of the maximum weight for w^i if it exists, otherwise go to (4) ;
     Construct the new weight w^{i+1} by applying (1), $i \leftarrow i+1$ and repeat (3) ;
(4) If $i = 0$ then X is optimal, otherwise $\mathrm{ex}(X \triangle A_i)$ is a next Pareto-optimal solution.

Note that in (2) there is no small augmenting cycle since these are eliminated in (1), and that in (3) there is no augmenting cycle since these are eliminated in (1) and (2). These facts are important in the following sense.

Theorem 9. *For any weight vector,*

1. *a maximum weight small augmenting cycle can be found in polynomial time,*

2. *a maximum weight large augmenting cycle family can be found in polynomial time if no small augmenting cycle exists,*

3. *a maximum weight augmenting path can be found in polynomial time if no augmenting cycle exists.*

By Lemma 8 and Theorem 9, our algorithm find a next Pareto-optimal solution in polynomial time. Summing up the above discussions, we obtain our main theorem.

Theorem 10. *The GSSP for claw-free bidirected graphs is polynomially solvable.*

In the rest of the section, we briefly explain a proof of Theorem 9. Our approach is an extension of Minty's algorithm for undirected claw-free graphs. This, however, does not seem a straightforward extension because we must overcome several problems. A significant problem is how to deal with 'induced weights'. Let A be an alternating cycle or a free alternating path. Then its weight is expressed as

$$\delta_X(A) = w(A - X) - w(X \cap A) + \sum \{w(v) \mid v \text{ is mixed}, N_G^{-+}(v) \cap (A - X) \neq \emptyset\}.$$

We call the $\sum$ term the *induced weight*, which appears in the bidirected case but not in the undirected case.

We first consider cycles. Let $x_1, \ldots, x_k$ with $k \geq 3$ be distinct black vertices and $W_1, \ldots, W_k, W_{k+1} = W_1$ be wings such that x_i is adjacent to W_i and W_{i+1} for $i = 1, \ldots, k$. Then $(W_1, x_1, W_2, \ldots, W_k, x_k, W_1)$ is called a *cycle of wings*. It is easy to show the following:

Lemma 11 ([4]). *Let $(W_1, x_1, W_2, \ldots, W_k, x_k, W_1)$ with $k \geq 3$ be a cycle of wings and $y_i \in W_i$ for $i = 1, \ldots, k$. Then $(y_1, x_1, y_2, \ldots, y_k, x_k, y_{k+1} = y_1)$ is an alternating cycle if and only if y_i is not adjacent to y_{i+1} for $i = 1, \ldots, k$.*

Lemma 12. *Let v be a mixed vertex such that $N_G^{-+}(v)$ has a bounded vertex but is not included in a wing. Then there uniquely exists a black vertex x such that $[x = v$ or x is adjacent to $v]$ and all the vertices in $N_G^{-+}(v)$ are adjacent to x.*

Proof. It is trivial if v is black. Suppose that v is white. Let y be a bounded vertex in $N_G^{-+}(v)$, and let x_1 and x_2 be the black vertices adjacent to y. Since they are in X which is a stable set, x_1 is not adjacent to x_2. Thus, without loss of generality, we can assume that v is adjacent to x_1 since otherwise $\{y, v, x_1, x_2\}$ induces a claw. The edge (v, x_1) is not a $(-, +)$-edge because v is white and x_1 is black. That is, the sign of this edge at v is $+$. Let y' be any vertex in $N_G^{-+}(v) - \{y\}$. Then y' must be adjacent to x_1 from the transitivity. Finally note that v is not adjacent to x_2, since otherwise from the same discussion any vertex in $N_G^{-+}(v) - \{y\}$ must be adjacent to x_2 and $N_G^{-+}(v)$ is included in the wing adjacent to x_1 and x_2, a contradiction. □

Lemma 13. *Let $\mathcal{C} = (W_1, x_1, W_2, \ldots, W_k, x_k, W_1)$ be a cycle of wings ($k \geq 3$). Then a maximum weight alternating cycle included in $\mathcal{C}$ can be found in polynomial time.*

Proof. Let $W_{k+1} = W_1$ and $W_0 = W_k$. For $i = 1, \ldots, k$ and for each pair $y \in W_i$ and $z \in W_{i+1}$ such that y is not adjacent to z, draw a directed 'red' edge from y to z with weight $w(y) - w(x_i) + \sum \{w(v) \mid v$ is mixed, $N_G^{-+}(v) \cap W_{i-1} = \emptyset, N_G^{-+}(v) \cap W_i \neq \emptyset$ and $[y \in N_G^{-+}(v)$ or $z \in N_G^{-+}(v)]\}$. From Lemma 11, there is a one-to-one mapping between all the directed cycles of red edges and all the alternating cycles in $\mathcal{C}$.

Let v be a mixed vertex such that $N_G^{-+}(v) \cap (W_1 \cup \cdots \cup W_k) \neq \emptyset$. From Lemma 12, there uniquely exists $i \in \{1, \ldots, k\}$ such that $N_G^{-+}(v) \cap W_{i-1} = \emptyset$ and $N_G^{-+}(v) \cap W_i \neq \emptyset$. Moreover from Lemma 12 again, for such i, $N_G^{-+}(v) \cap ((W_1 \cup \cdots \cup W_k) - (W_i \cup W_{i+1})) = \emptyset$. Hence the mapping conserves weights. A maximum weight directed cycle of red edges can be found in polynomial time by the breadth first search.   □

Lemma 14. *A maximum weight small augmenting cycle can be found in polynomial time.*

Proof. The number of alternating cycles of length 4 is polynomially bounded. Thus we can easily find one having the maximum weight. On the other hand, each alternating cycle of length at least 6 is included in a certain cycle of wings. The number of all cycles of wings containing at most two regular vertices is polynomially bounded. We can also enumerate these in polynomial time. By Lemma 13, we can find a maximum weight small augmenting cycle in polynomial time.   □

Unfortunately, a maximum weight large augmenting cycle cannot be found in polynomial time in the same way because the number of the cycles of wings having three or more regular vertices cannot be polynomially bounded. Before considering the step (2) in our algorithm, we introduce a useful property relative to wings around regular vertices. For convenience, we will use some notations as below:

- $v_1 \sim v_2$ means that v_1 and v_2 are adjacent, and $v_1 \not\sim v_2$ means v_1 and v_2 are not adjacent.
- $v_1 \overset{+-}{\sim} v_2$ says there is an edge having plus and minus sings at v_1 and v_2 respectively, and $v_1 \overset{+-}{\not\sim} v_2$ is its negation.
- $v_1 \overset{+}{\sim} v_2$ denotes either $v_1 \overset{++}{\sim} v_2$ or $v_1 \overset{+-}{\sim} v_2$, and $v_1 \overset{+}{\not\sim} v_2$ is the negation of $v_1 \overset{+}{\sim} v_2$.
- $v_1 \diamond v_2$ says that v_1 and v_2 are contained in the same wing, and $v_1 \not\diamond v_2$ is its negation.

Lemma 15 ([4]). *Given a regular vertex x, let $B(x) = \{v \mid v \sim x$ and v is bounded$\}$. Then there exists a partition of $B(x)$, namely $[N^1(x), N^2(x)]$, such that for any $v_1, v_2 \in B(x)$ with $v_1 \not\diamond v_2$,*

$$v_1 \sim v_2 \quad \Longleftrightarrow \quad [v_1, v_2 \in N^1(v) \text{ or } v_1, v_2 \in N^2(v)].$$

Moreover this partition is uniquely determined, and hence, it can be found in polynomial time.

This is the key lemma of Minty's algorithm. If a large alternating cycle or a free alternating path passes through $v_1 \in N^1(v)$ and a regular vertex v, then it must pass through a vertex v_2 such that $v_2 \in N^2(v)$ and $v_2 \not\sim v_1$. From this property Minty showed that by constructing a graph called the "Edmonds' graph" and by finding a maximum weight perfect matching of it, a maximum weight augmenting path for any Pareto-optimal stable set can be found in polynomial time. To deal with induced weights, we require an additional property of the partition of vertices adjacent to a regular vertex.

Lemma 16. *For a regular vertex x and a vertex v such that $v = x$ or $v \sim x$, we define*

$$N^1(x) \succ_v N^2(x) \overset{\text{def}}{\Longleftrightarrow} \exists a \in N^1(x), \exists b \in N^2(x) \text{ such that } a \not\sim b, a \overset{+-}{\sim} v \text{ and } b \overset{+-}{\not\sim} v,$$

$$N^2(x) \succ_v N^1(x) \overset{\text{def}}{\Longleftrightarrow} \exists c \in N^2(x), \exists d \in N^1(x) \text{ such that } c \not\sim d, c \overset{+-}{\sim} v \text{ and } d \overset{+-}{\not\sim} v.$$

Then at most one of $N^1(x) \succ_v N^2(x)$ and $N^2(x) \succ_v N^1(x)$ holds.

Proof. Let us consider the case $v = x$. If $b \overset{+-}{\not\sim} x$, then $b \overset{-}{\sim} x$ because $b \sim x$. In addition, if $a \overset{+-}{\sim} x$, then $a \overset{+}{\sim} b$. Hence neither $N^1(x) \succ_x N^2(x)$ nor $N^2(x) \succ_x N^1(x)$ holds.

Suppose to the contrary that $v \sim x$, $N^1(x) \succ_v N^2(x)$ and $N^2(x) \succ_v N^1(x)$. There exist $a, d \in N^1(x)$ and $b, c \in N^2(x)$ such that $a \not\sim b$, $c \not\sim d$, $a \overset{+-}{\sim} v$, $b \overset{+-}{\not\sim} v$, $c \overset{+-}{\sim} v$ and $d \overset{+-}{\not\sim} v$. Note that b, d and v are mutually distinct. Assume to the contrary that $b \sim v$. Then $b \overset{+}{\sim} v$ because $b \overset{+-}{\not\sim} v$. But $a \overset{+-}{\sim} v$ and $v \overset{+}{\sim} b$ induce $a \overset{+}{\sim} b$, contradicting $a \not\sim b$. Hence $b \not\sim v$ and similarly $d \not\sim v$. Now $b \sim d$ since otherwise $\{x, b, d, v\}$ induces a claw. Thus $b \diamond d$ from Lemma 15.

Suppose that $a \diamond c$. Because x is regular, i.e., x is adjacent to at least three wings, there exists $e \in N(x)$ such that $e \not\diamond a \diamond c$ and $e \not\diamond b \diamond d$. Suppose that $e \in N^1(x)$. Then $e \not\sim b$ and $e \not\sim c$ from Lemma 15. If $e \overset{+-}{\not\sim} v$, then replace d by e, and from the above discussion, $b \diamond e$, a contradiction. Hence $e \overset{+-}{\sim} v$, and we can replace a by e. Similarly if $e \notin N^1(x)$, i.e., $e \in N^2(x)$, then we can replace c by e. Henceforth we assume that $a \not\diamond c$.

Suppose to the contrary that $a \not\diamond d$. From Lemma 15, $a \sim d$. Since a is bounded, a is adjacent to two black vertices: x and namely y. Then $d \not\sim y$, since otherwise $d \sim x$ and $d \sim y$ imply $a \diamond d$, a contradiction. Now $v \sim y$ since otherwise $\{a, d, v, y\}$ induces a claw. Note that $v \overset{+}{\sim} y$ since otherwise $v \overset{-+}{\sim} y$, contradicting the fact that y is black and v is white. Thus $c \overset{+-}{\sim} v$ and $v \overset{+}{\sim} y$ induce $c \overset{+}{\sim} y$. However, $c \sim x$ and $c \sim y$ imply $a \diamond c$, a contradiction. Hence $a \diamond d$ and similarly $c \diamond b$. Since $a \diamond d$, $d \diamond b$ and $b \diamond c$, $a \diamond c$ holds. However, this contradicts the assumption $a \not\diamond c$. □

We add the induced weight of an alternating cycle or a free alternating path to weights of appropriate vertices in it. We define $\tilde{w} : (V \cup (V \times V)) \to \Re$ by the following procedure: let $\tilde{w} \leftarrow 0$ and for each mixed vertex v,

- if $B^{-+}(v) = \{u \mid u \text{ is bounded}, v \overset{-+}{\sim} u\}$ is empty or included in a wing, $\tilde{w}(u) \leftarrow \tilde{w}(u) + w(v)$ for each $u \in B^{-+}(v)$,

- otherwise there uniquely exists a black vertex x such that $x = v$ or $x \sim v$, from Lemma 12,
    * if x is regular, then
        - if $N^2(x) \succ_v N^1(x)$, then $\tilde{w}(u) \leftarrow \tilde{w}(u) + w(v)$ for each $u \in B^{-+}(v) \cap N^2(x)$,
        - otherwise $\tilde{w}(u) \leftarrow \tilde{w}(u) + w(v)$ for each $u \in B^{-+}(v) \cap N^1(x)$,
    * otherwise x must be irregular, and $\tilde{w}(t, u) \leftarrow \tilde{w}(t, u) + w(v)$ for each pair of vertices $t, u \in B(x)$ such that $t \not\sim u$, $t \not\sim u$ and $[t \overset{+-}{\sim} v$ or $u \overset{+-}{\sim} v]$.

By combining Lemmas 15 and 16, we can prove the next lemma.

Lemma 17. *Let* $C = (y_1, x_1, y_2, x_2, \ldots, y_k, x_k, y_{k+1} = y_1)$ *be an alternating cycle with white vertices* $y_1, \ldots, y_k$ *and black vertices* $x_1, \ldots, x_k$ ($k \geq 3$). *Then*

$$\delta_X(C) = \sum_{i=1}^{k} w(y_i) - \sum_{i=1}^{k} w(x_i) + \sum_{i=1}^{k} \tilde{w}(y_i) + \sum_{i=1}^{k} \tilde{w}(y_i, y_{i+1}).$$

If there is no small augmenting cycle, by using Lemma 17, we can construct the Edmonds' graph $\hat{G}$ such that

1. each edge of $\hat{G}$ is colored black or white, and it has a weight $\hat{w}$,
2. all the black edges form a perfect matching M of $\hat{G}$,
3. if M is a maximum weight perfect matching of $\hat{G}$ then there is no large augmenting cycle family in G and
4. if $\hat{w}(M) < \hat{w}(M^*)$ for a maximum weight perfect matching M^* of $\hat{G}$, let $\hat{C}_1, \ldots, \hat{C}_k$ be all the augmenting cycles in $M^* \triangle M$; then $\hat{C}_1, \ldots, \hat{C}_k$ correspond to a maximum weight large augmenting cycle family $C_1, \ldots, C_k$ in G.

In the next section, we show that the Edmonds' graph can be constructed in polynomial time. Hence the step (2) in our algorithm can be done in polynomial time. Analogously, if there is no augmenting cycle, for any pair of vertices a and b, we can find a maximum weight augmenting path whose endpoints are a and b, if it exists, by constructing the Edmonds' graph and by finding a maximum weight perfect matching in it. Now we can find a maximum weight augmenting path by trying all the pairs of vertices a and b.

5   Constructing and Modifying the Edmonds' Graph

We now describe how to construct the Edmonds' graph to find a maximum weight large alternating cycle family. We note that Edmonds' graphs for finding a maximum weight augmenting path can be obtained by modifying the construction.

A white alternating path P is called an *irregular white alternating path* *(IWAP)* if all black vertices of P are irregular and no wing contains two white vertices of P. For an IWAP $P = (y_1, z_1, y_2, z_2, \ldots, z_{k-1}, y_k)$, we define its *weight*, denoted by $\tilde{\delta}_X(P)$ as

$$\tilde{\delta}_X(P) = \sum_{i=1}^{k} w(y_i) - \sum_{i=1}^{k-1} w(z_i) + \sum_{i=1}^{k} \tilde{w}(y_i) + \sum_{i=1}^{k-1} \tilde{w}(y_i, y_{i+1}).$$

Then Lemma 17 can be described in terms of IWAP:

Lemma 18. *Let $C = (P_1, x_1, P_2, x_2, \ldots, P_k, x_k, P_{k+1} = P_1)$ be an alternating cycle of length at least 6 such that $k \geq 2$, $x_1, \ldots, x_k$ are distinct regular vertices and $P_1, \ldots, P_k$ are IWAPs. Then*

$$\delta_X(C) = \sum_{i=1}^k \tilde{\delta}_X(P_i) - \sum_{i=1}^k w(x_i).$$

Lemma 19. *Let A and B be subsets of bounded vertices. Then a maximum weight IWAP whose endpoints are in A and B respectively can be found in polynomial time.*

Proof. We can reduce this problem to find maximum weight directed paths in directed acyclic graphs. □

Now we make the *Edmonds' graph* G_{Ed}. Let $x_1, \ldots, x_r$ be all the regular vertices. G_{Ed} has $2r$ vertices, namely x_i^1, x_i^2 ($i = 1, \ldots, r$). Join x_i^1 and x_i^2 by a black edge with weight $\hat{w}(x_i^1, x_i^2) = w(x_i)$ ($i = 1, \ldots, r$). For each pair of regular vertices x_i and x_j and for $p, q \in \{1, 2\}$, if there exists an IWAP whose endpoints are in $N^p(x_i)$ and $N^q(x_j)$, join x_i^p and x_j^q by a white edge whose weight $\hat{w}(x_i^p, x_j^q)$ is the maximum weight among such IWAPs. Now we finish constructing the Edmonds' graph.

Let M be the set of all the black edges. Note that M is a perfect matching. An alternating cycle $\hat{C}$ of length $2k \geq 6$ in G_{Ed} corresponds to a large alternating cycle C in G where C has k regular vertices and $\delta_M(\hat{C}) = \delta_X(C)$.

But this is not true for $k = 2$. Let $\hat{C} = (x_i^1, x_j^1, x_j^2, x_i^2, x_i^1)$ be an alternating cycle of G_{Ed}. Here x_i and x_j are distinct regular vertices. We denote P_{pq} as the maximum weight IWAP corresponding to the edge (x_i^p, x_j^q) in the Edmonds' graph for $p, q \in \{1, 2\}$. If $\delta_M(\hat{C})$ is not positive, then there is no problem in our purpose. So suppose that its weight is positive, i.e. $\delta_M(\hat{C}) = \tilde{\delta}_X(P_{11}) + \tilde{\delta}_X(P_{22}) - w(x_i) - w(x_j) > 0$.

If P_{11} and P_{22} have no vertex in common, then $C = (x_i, P_{11}, x_j, P_{22}, x_i)$ is a small augmenting cycle, contradicting to that we have already eliminated all the small augmenting cycles. Hence we can denote

$$P_{11} = (y_1^1, z_1, y_2^1, z_2, \ldots, y_{\ell-1}^1, z_{\ell-1}, y_\ell^1) \text{ and}$$
$$P_{22} = (y_1^2, z_1, y_2^2, z_2, \ldots, y_{\ell-1}^2, z_{\ell-1}, y_\ell^2).$$

Here $z_1, \ldots, z_{\ell-1}$ are irregular vertices, both y_k^1 and y_k^2 are in a common wing W_k for $k = 1, \ldots, \ell$, $y_1^1 \in N^1(x_i)$, $y_1^2 \in N^2(x_i)$, $y_\ell^1 \in N^1(x_j)$ and $y_\ell^2 \in N^2(x_j)$.

We first discuss an easy situation. A wing W is said to be *irregular reachable* to a regular vertex x if there exist an integer $m \geq 1$, distinct irregular vertices $z_1, \ldots, z_{m-1}$ and distinct wings $W_1(= W), W_2, \ldots, W_m$ such that W_1 is adjacent to z_1 and W_k is adjacent to z_{k-1} and z_k for $k = 2, \ldots, m$, where $z_m = x$. Let $W(x_i, x_j)$ denote the union of all the wings that are irregular reachable to both x_i and x_j.

Lemma 20. If $N^1(x_j) \subseteq W(x_i, x_j)$, then any large alternating cycle in G passes through neither P_{12} nor P_{22}. That is, we can delete the edges (x_i^1, x_j^2) and (x_i^2, x_j^2) from G_{Ed}. Similarly if $N^2(x_j) \subseteq W(x_i, x_j)$, we can delete (x_i^1, x_j^1) and (x_i^2, x_j^1). If $N^1(x_i) \subseteq W(x_i, x_j)$, we can delete (x_i^2, x_j^1) and (x_i^2, x_j^2). If $N^2(x_i) \subseteq W(x_i, x_j)$, we can delete (x_i^1, x_j^1) and (x_i^1, x_j^2).

Proof. Suppose that a large alternating cycle C passes x_i, P_{12} (or P_{22}) and x_j. Before x_j, it passes a vertex in $N^2(x_j)$. Hence after x_j, it must pass a vertex $v \in N^1(x_j) \subseteq W(x_i, x_j)$. Hence C contains exactly two regular vertices x_i and x_j, contradicting to that C is large. □

In the sequel, we suppose that none of $N^1(x_i)$, $N^2(x_i)$, $N^1(x_j)$ nor $N^2(x_j)$ is contained in $W(x_i, x_j)$.

Lemma 21. There exists k such that $y_k^1 = y_k^2$ and $2 \le k \le \ell - 1$, or there exists k such that $y_k^1 \sim y_k^2$ and $1 \le k \le \ell$.

Proof. Suppose that this lemma does not hold, i.e. $y_k^1 \ne y_k^2$ and $y_k^1 \nsim y_k^2$ for all $k = 1, \ldots, \ell$ (Note that $y_1^1 \ne y_1^2$ and $y_\ell^1 \ne y_\ell^2$ since $B^1(x_i) \cap B^2(x_i) = B^1(x_j) \cap B^2(x_j) = \emptyset$). Let $z_0 = x_i$, $z_\ell = x_j$ and $C_k = (y_k^1, z_k, y_k^2, z_{k-1}, y_k^1)$ $(k = 1, \ldots, \ell)$. Then C_k is a small alternating cycle for all $k = 1, \ldots, \ell$. We can show that $\sum_{k=1}^{\ell} \delta_X(C_k) = \tilde{\delta}_X(P_{11}) + \tilde{\delta}_X(P_{22}) - w(x_i) - w(x_j)(> 0)$. (The proof is slightly complicated because we must consider about the induced weight $\tilde{w}$.) Hence at least one C_k is a small augmenting cycle, a contradiction. □

Now we can show the next two lemmas, but proofs are omitted.

Lemma 22. If $\ell = 1$, any large alternating cycle passes through neither P_{11} nor P_{22}. Hence we can delete the edges (x_i^1, x_j^1) and (x_i^2, x_j^2) from G_{Ed}.

Lemma 23. If $\ell \ge 2$, the followings hold.

1. There exists k such that $2 \le k \le \ell - 1$ and $y_k^1 = y_k^2$, or there exists k such that $1 \le k \le \ell - 1$, $y_k^1 \ne y_k^2$, $y_{k+1}^1 \ne y_{k+1}^2$, $y_k^1 \nsim y_{k+1}^2$ and $y_k^2 \nsim y_{k+1}^1$.
2. For such k, let

$$P_{11i} = (y_1^1, z_1, y_2^1, \ldots, z_{k-1}, y_k^1), \qquad P_{11j} = (y_{k+1}^1, z_{k+1}, \ldots, y_{\ell-1}^1, z_{\ell-1}, y_\ell^1),$$
$$P_{22i} = (y_1^2, z_1, y_2^2, \ldots, z_{k-1}, y_k^2) \text{ and } P_{22j} = (y_{k+1}^2, z_{k+1}, \ldots, y_{\ell-1}^2, z_{\ell-1}, y_\ell^2),$$

and let $P'_{12} = (P_{11i}, z_k, P_{22j})$ and $P'_{21} = (P_{22i}, z_k, P_{11j})$.
Then $\tilde{\delta}_X(P'_{12}) + \tilde{\delta}_X(P'_{21}) = \tilde{\delta}_X(P_{11}) + \tilde{\delta}_X(P_{22})$, P'_{12} is an IWAP between $B^1(x_i)$ and $B^2(x_j)$, and P'_{21} is an IWAP between $B^2(x_i)$ and $B^1(x_j)$.
3. $\tilde{\delta}_X(P_{11}) + \tilde{\delta}_X(P_{22}) = \tilde{\delta}_X(P_{12}) + \tilde{\delta}_X(P_{21})$.
4. $\tilde{\delta}_X(P'_{12}) = \tilde{\delta}_X(P_{12})$ and $\tilde{\delta}_X(P'_{21}) = \tilde{\delta}_X(P_{21})$.

Summing up the above discussion, dealing with three cases, i.e. Lemmas 20, 22 and 23, we modify the Edmonds' graph. In the first two cases, elimination of augmenting cycles of a form $(x_i^1, x_j^1, x_j^2, x_i^2, x_i^1)$ or $(x_i^1, x_j^2, x_j^1, x_i^2, x_i^1)$ can be easily done by deleting edges. In the last case, we modify G_{Ed} as below:

1. Delete four edges (x_i^1, x_j^1), (x_i^2, x_j^2), (x_i^1, x_j^2) and (x_i^2, x_j^1) (Lemma 23 guarantees the existence of these four edges),
2. Add two new vertices z_k^i and z_k^j, join z_k^i and z_k^j by a black edge and assign its weight $\hat{w}((z_k^i, z_k^j))$ to be 0, where k satisfies the conditions of Lemma 23,
3. Add four white edges (x_i^1, z_k^i), (x_i^2, z_k^i), (x_j^1, z_k^j) and (x_j^2, z_k^j), and assign their weights to be $\hat{w}((x_i^1, z_k^i)) = \tilde{\delta}_X(P_{11})$, $\hat{w}((x_i^2, z_k^i)) = \tilde{\delta}_X(P_{22})$, $\hat{w}((x_j^1, z_k^j)) = 0$ and $\hat{w}((x_j^2, z_k^j)) = \tilde{\delta}_X(P_{12}) - \tilde{\delta}_X(P_{11})(= \tilde{\delta}_X(P_{22}) - \tilde{\delta}_X(P_{21}))$.

All large alternating cycles through black edges (x_i^1, x_i^2) and (x_j^1, x_j^2) can be preserved by our revision, because (x_i^p, x_j^q) in the original Edmonds' graph $(p, q \in \{1, 2\})$ is interpreted by the path $(x_i^p, z_k^i, z_k^j, x_j^q)$ in the revised Edmonds' graph. Furthermore, Lemma 23 guarantees that weights of these four edges are equal to those of such four paths, respectively.

Lemma 24. *A maximum weight large alternating cycle family can be found in polynomial time if there is no small augmenting cycle.*

Proof. Make the Edmonds' graph. Then eliminate all the augmenting cycles of a form $(x_i^1, x_j^1, x_j^2, x_i^2, x_i^1)$ or $(x_i^1, x_j^2, x_j^1, x_i^2, x_i^1)$. Let G'_{Ed} be the modified graph and M' be the set of its black edges. Note that M' is perfect. Let M^* be a maximum weight perfect matching and $\hat{C}_1, \ldots, \hat{C}_k$ be all the augmenting cycle in $M' \triangle M^*$ (k may be zero). Note that $(\bigcup_{i=1}^{k} \hat{C}_i)$ is a maximum weight alternating cycle family of G'_{Ed}. Then each $\hat{C}_i$ has length at least 6 because we eliminate all augmenting cycles of length 4, and hence $\hat{C}_i$ corresponds to a large augmenting cycle C_i of X such that $\delta_X(C_i) = \delta_{M'}(\hat{C}_i)$. Moreover $C_1, \ldots, C_k$ are disjoint because $\hat{C}_1, \ldots, \hat{C}_k$ are vertex-disjoint. Now from construction and modification of the Edmonds' graph, we can conclude that $(\bigcup_{i=1}^{k} C_i)$ is a maximum weight large alternating cycle family of X. □

References

1. E. Boros and O. Čepek, O. On perfect $0, \pm 1$ matrices. *Discrete Math.*, 165/166:81–100, 1997.
2. M. Conforti, G. Cornuéjols, and C. De Francesco. Perfect $0, \pm 1$ matrices. *Linear Algebra Appl.*, 253:299–309, 1997.
3. E. L. Johnson and M. W. Padberg. Degree-two inequalities, clique facets, and biperfect graphs. *Ann. Discrete Math.*, 16:169–187, 1982.
4. G. J. Minty. On maximal independent sets of vertices in claw-free graphs. *J. Combin. Theory Ser. B*, 28:284–304, 1980.
5. E. C. Sewell. Binary integer programs with two variables per inequality. *Math. Programming*, 75:467–476, 1996.
6. A. Tamura. The generalized stable set problem for perfect bidirected graphs. *J. Oper. Res. Soc. Japan*, 40:401–414, 1997.

On a Min-max Theorem of Cacti

Zoltán Szigeti *

Equipe Combinatoire, Université Paris 6
75252 Paris, Cedex 05, France
Zoltan.Szigeti@@ecp6.jussieu.fr

Abstract. A simple proof is presented for the min-max theorem of Lovász on cacti. Instead of using the result of Lovász on matroid parity, we shall apply twice the (conceptionally simpler) matroid intersection theorem.

1   Introduction

The graph matching problem and the matroid intersection problem are two well-solved problems in Combinatorial Theory in the sense of min-max theorems and polynomial algorithms for finding an optimal solution. The matroid parity problem, a common generalization of them, turned out to be much more difficult. For the general problem there does not exist polynomial algorithm [2], [3]. Moreover, it contains NP-hard problems. On the other hand, for linear matroids Lovász [3] provided a min-max formula and a polynomial algorithm. There are several earlier results which can be derived from Lovász' theorem, e.g. Tutte's result on f-factors [9], a result of Mader on openly disjoint A-paths [5], a result of Nebesky concerning maximum genus of graphs [6]. Another application which can be found in the book of Lovász and Plummer [4] is the problem of cacti. It is mentioned there that *"a direct proof would be desirable."* Our aim is to fill in this gap, that is to provide a simpler proof for this problem. We remark here that we shall apply the matroid intersection theorem twice. We refer the reader to [7] for basic concepts of matroids.

A graph K is called *cactus* if each block (maximal 2-connected subgraph) of K is a triangle (cycle of length three). The size of a cactus K is the number of its blocks. Lovász derived a min-max theorem for the maximum size of a cactus contained in a given graph G from his general min-max theorem on linear matroid parity problem. Here we shall give a simple proof for this result on cacti. The proof follows the line of Gallai's (independently Anderson's [1]) proof for Tutte's theorem on the existence of perfect matchings.

In fact, we shall solve the graphic matroid parity problem in the special case when for each pair the two edges have exactly one vertex in common. The graphic

---

* This work was done while the author visited Laboratoire LEIBNIZ, Institut IMAG, Grenoble.

R. E. Bixby, E. A. Boyd, and R. Z. Ríos-Mercado (Eds.): IPCO VI
LNCS 1412, pp. 84–95, 1998. © Springer–Verlag Berlin Heidelberg 1998

matroid parity problem is the following. Given a graph G and a partition of its edge set into pairs, what is the maximum size of a forest which consists of pairs, in other words, what is the maximum number of pairs whose union is a forest. A pair of edges is called *v-pair* if these two edges have exactly one vertex in common and they are not loops. If G is an arbitrary graph and $\mathcal{V}$ is a partition of the edge set of G into v-pairs then $(G, \mathcal{V})$ is called *v-graph*. From now on a *cactus* of $(G, \mathcal{V})$ is a forest of G consisting of v-pairs in $\mathcal{V}$. The size of a cactus is the number of v-pairs contained in it. The *v-graphic matroid parity problem* consists of finding the maximum size $\beta(G, \mathcal{V})$ of a cactus in a v-graph $(G, \mathcal{V})$.

The original cactus problem can be formulated as a v-graphic matroid parity problem as follows. Let $(G', \mathcal{V})$ be the following v-graph. The vertex set of G' is the same as of G. We define the edge set of G' and the partition $\mathcal{V}$ of the edge set into v-pairs as follows. For each triangle T of G we introduce a v-pair in $(G', \mathcal{V})$: choose any two edges of T, add them to the edge set of G' and add this v-pair to $\mathcal{V}$. (G' will contain lots of parallel edges. In fact, G' is obtained from G by multiplying edges.) Obviously, there is a one to one correspondence between the cacti of G and the forests of G' being the union of v-pairs. Thus the problem is indeed a v-graphic matroid parity problem.

To state the theorem on cacti we need some definitions. Let $(G, \mathcal{V})$ be a v-graph. Let $\mathcal{P} := \{V_1, V_2, ..., V_l\}$ be a partition of the vertex set $V(G)$. Let $\mathcal{V_P} \subseteq \mathcal{V}$ ($\mathcal{S_P} \subseteq \mathcal{V}$) be the set of those v-pairs whose end vertices belong to three (two) different members of $\mathcal{P}$. Let $\mathcal{Q} := \{H_1, H_2, ..., H_k\}$ be a partition of $\mathcal{V_P} \cup \mathcal{S_P}$. Let us denote by $p(H_i)$ the number of V_j's for which there exists at least one v-pair in H_i with a vertex in V_j. We say that $(\mathcal{P}, \mathcal{Q})$ is a *cover* of $(G, \mathcal{V})$. The value $val(\mathcal{P}, \mathcal{Q})$ of a cover is defined as follows.

$$val(\mathcal{P}, \mathcal{Q}) := n - l + \sum_{H_i \in \mathcal{Q}} \lfloor \frac{p(H_i) - 1}{2} \rfloor,$$

where $n = |V(G)|$, $l = |\mathcal{P}|$ and $k = |\mathcal{Q}|$.

Now, we are able to present the min-max result of Lovász [4] on cacti in our terminology.

Theorem 1. *Let $(G, \mathcal{V})$ be a v-graph. Then $\beta(G, \mathcal{V}) = \min\{val(\mathcal{P}, \mathcal{Q})\}$, where the minimum is taken over all covers $(\mathcal{P}, \mathcal{Q})$ of $(G, \mathcal{V})$.*

Remark 1. In the definition of a cover, $\mathcal{Q}$ could be defined as the partition of $\mathcal{V}$ and not of $\mathcal{V_P} \cup \mathcal{S_P}$. Indeed, if $(\mathcal{P}, \mathcal{Q})$ is a cover, then the pairs in $\mathcal{V} - \mathcal{V_P} \cup \mathcal{S_P}$ can be added to $\mathcal{Q}$ as new members of the partition without changing the value of the cover since $p(T) = 1$ for all $T \in \mathcal{V} - \mathcal{V_P} \cup \mathcal{S_P}$. We mention that if $T \in \mathcal{V} - \mathcal{V_P} \cup \mathcal{S_P}$, then T can not be in a maximal cactus.

A cactus of $(G, \mathcal{V})$ is called *perfect* if it is a spanning tree of G. If G contains only one vertex v and no edge, then the vertex v is considered as a perfect cactus of $(G, \mathcal{V})$. Since a spanning tree contains $n - 1$ edges, $\beta(G, \mathcal{V}) \leq \lfloor \frac{n-1}{2} \rfloor$, for any v-graph $(G, \mathcal{V})$.

Let $(\mathcal{P}, \mathcal{Q})$ be a cover of a v-graph $(G, \mathcal{V})$. The elements $H_i \in \mathcal{Q}$ are called *components* of the cover. $G/\mathcal{P}$ will denote the graph obtained from G by contracting each set V_i in $\mathcal{P}$ into one vertex. We identify the edge sets of G and $G/\mathcal{P}$. $(G_{\mathcal{P}}, \mathcal{V}_{\mathcal{P}})$ is the v-graph, where $\mathcal{V}_{\mathcal{P}}$ is defined as above, it contains those v-pairs of $\mathcal{V}$ which remain v-pairs after the contraction, the vertex set of $G_{\mathcal{P}}$ is the same as of $G/\mathcal{P}$ and the edge set of $G_{\mathcal{P}}$ is the set of edges of the v-pairs in $\mathcal{V}_{\mathcal{P}}$, that is, $G_{\mathcal{P}}$ is obtained from $G/\mathcal{P}$ by deleting the edges which do not belong to any v-pair in $\mathcal{V}_{\mathcal{P}}$. For $H_i \in \mathcal{Q}$, $(G_{\mathcal{P}}[H_i], H_i)$ will denote the v-graph for which the edge set of $G_{\mathcal{P}}[H_i]$ is the set of edges of the v-pairs in H_i and the vertex set of $G_{\mathcal{P}}[H_i]$ contains those vertices of $G_{\mathcal{P}}$ for which at least one v-pair of H_i is incident. Then $p(H_i)$ is the number of vertices of $G_{\mathcal{P}}[H_i]$. (Note that if $H_i \in \mathcal{S}_{\mathcal{P}}$, then $(G_{\mathcal{P}}[H_i], H_i)$ contains two edges which are parallel or one of them is a loop, that is it is not really a v-graph. However, we shall need this type of "v-graphs" in the proof.) If F is a subset of edges of a v-graph $(G, \mathcal{V})$ then the number of v-pairs of $\mathcal{V}$ contained in F is denoted by $v_{\mathcal{V}}(F)$.

For a graph G on n vertices and with c connected components, a forest of G containing $n - c$ edges is called *spanning*. For a connected graph G, a forest F of G containing $n - 2$ edges (that is, F has exactly two connected components) is called *almost spanning*. A v-graph will be called *(cactus)-critical* if by identifying any two vertices the v-graph obtained has a perfect cactus. Especially, this means that in a critical v-graph there exists a cactus which is *almost perfect*, that is, it is an almost spanning tree consisting of v-pairs. Critical v-graphs will play an important role in the proof, like factor-critical graphs play the key role in the proof of Tutte's theorem. A component $H_i \in \mathcal{Q}$ is said to be *critical* in $(G, \mathcal{V})$ if the v-graph $(G_{\mathcal{P}}[H_i], H_i)$ is critical. If $H_i \in \mathcal{S}_{\mathcal{P}}$, then $(G_{\mathcal{P}}[H_i], H_i)$ is considered to be critical.

We say that the partition $\mathcal{P}$ of V is the *trivial partition* if $l := |\mathcal{P}| = n := |V|$. The cover $(\mathcal{P}, \mathcal{Q})$ is the *trivial cover* if $l = n$ and $k := |\mathcal{Q}| = 1$. Let $\mathcal{P}' = \{V_1^1, ..., V_1^{r_1}, V_2^1, ..., V_2^{r_2}, ..., V_l^1, ..., V_l^{r_l}\}$, where $\cup_j V_i^j = V_i$ for all i, then the partition $\mathcal{P}'$ is called a *refinement* of the partition $\mathcal{P}$. If $\mathcal{P}'$ is a refinement of $\mathcal{P}$ so that $|\mathcal{P}'| = |\mathcal{P}| + 1$, then we say it is an *elementary refinement*. If $V_i \in \mathcal{P}$ then the partition obtained from $\mathcal{P}$ by replacing V_i by its singletons will be denoted by $\mathcal{P} \div \{V_i\}$. If $\mathcal{P}'$ is a refinement of $\mathcal{P}$, then we shall use $p'(H_i)$ instead of $p(H_i)$.

We shall need later two auxiliary graphs B and D. These graphs will depend on a v-graph $(G, \mathcal{V})$ and a cover $(\mathcal{P}, \mathcal{Q})$ of this v-graph. We suppose that for each component H_i, $p(H_i)$ is even. First we define the graph $B = (V(G), E(B))$. $e = uv$ will be an edge of B if and only if there exist $u, v \in V_j \in \mathcal{P}$, $H_i \in \mathcal{Q}$ and a cactus K in $(G_{\mathcal{P} \div \{V_j\}}[H_i], H_i)$ consisting of $p(H_i)/2$ v-pairs so that exactly two vertices u and v of V_j are connected in K, not necessarily by an edge but by a path in K, that is u and v are in the same connected component of K. (Note that K contains a cactus of size $(p(H_i) - 2)/2$ in $(G_{\mathcal{P}}[H_i], H_i)$. We mention that (by Lemma 2, see later) $(G_{\mathcal{P}}[H_i], H_i)$ will always contain a cactus consisting of $(p(H_i) - 2)/2$ v-pairs of $\mathcal{V}$.) We call this edge e an *augmenting edge* for H_i. In other words, the *trace* of the cactus K in $\mathcal{P}$ is the edge e. We will call the edges of B as *augmenting edges*. Note that an edge of B may be augmenting for more

$H_i \in \mathcal{Q}$. Let $\mathcal{P}'$ be a refinement of $\mathcal{P}$. Then the set $A_{\mathcal{P}'}$ of augmenting edges connecting vertices in different sets of $\mathcal{P}'$ will be called the augmenting edges *with respect to the refinement* $\mathcal{P}'$.

The second auxiliary graph D will be a bipartite graph with colour classes $E(B)$ (the edge set of B) and $\mathcal{Q}$. Two vertices $e \in E(B)$ and $H_i \in \mathcal{Q}$ are connected if and only if e is an augmenting edge for H_i. As usually, the set of neighbours of a vertex set X of one of the colour classes of D will be denoted by $\Gamma_D(X)$.

Finally, some words about the ideas of the proof. As it was mentioned earlier we shall follow the proof of Tutte's theorem. Let us briefly summarize the steps of this proof. We suppose that the Tutte condition is satisfied for a given graph G and we have to construct a perfect matching of G. Let X be a maximal set satisfying the condition with equality. The maximality of X implies that all the components of $G - X$ are factor-critical, thus it is enought to find a perfect matching in an auxiliary bipartite graph D, where one of the color classes corresponds to X while the other to the (critical) components. Hall's theorem (or the matroid intersection theorem) provides easily the existence of a perfect matching M in D. The desired perfect matching of G can be obtained from M and from the almost perfect matchings of the critical components. We mention that this is a lucky case because the union of these almost perfect matchings will be automatically a matching in G.

In the case of cacti it is not easier to prove that version where we have to find a perfect cactus, so we shall prove directly the min-max theorem. We shall choose a minimal cover $(\mathcal{P}, \mathcal{Q})$ of $(G, \mathcal{V})$ which is maximal in some certain sense. This will imply that the minimal cover of $(G_{\mathcal{P}}[H_i], H_i)$ is unique for each component H_i. This fact has two consequences, namely (i) each component H_i is critical (hence $p(H_i)$ is even) and (ii) for any component H_i and for any refinement $\mathcal{P}'$ of $\mathcal{P}$, either there exists an augmenting edge for H_i with respect to $\mathcal{P}'$ or its cover rests minimal in $(G_{\mathcal{P}'}[H_i], H_i)$.

We shall construct the cactus of size $val(\mathcal{P}, \mathcal{Q})$ in $(G, \mathcal{V})$ as follows. (1) For $n - l$ components H_i, we shall find a cactus K_i in $(G_{\mathcal{P} \div \{V_j\}}[H_i], H_i)$ of size $p(H_i)/2$ so that the trace of K_i in $\mathcal{P}$ is an edge and the corresponding augmenting edges form a spanning forest of the auxiliary graph B. (We shall see that the size of a spanning forest of B is indeed $n-l$.) (2) For the other components H_j we shall need a cactus in $(G_{\mathcal{P}}[H_j], H_j)$ of size $p(H_j)/2-1$, and (3) the union of all of these forests will be a forest, that is a cactus of size $val(\mathcal{P}, \mathcal{Q})$. Using (i), for the latter components H_j it is enough to find an arbitrary almost spanning tree in $G_{\mathcal{P}}[H_j]$ (and then using that H_j is critical, this forest can be replaced by a convenient cactus containing the same number of edges, that is of size $p(H_i)/2 - 1$). By the definition of augmenting edge, for the former components H_i it is enough to consider an arbitrary spanning tree in $G_{\mathcal{P}}[H_i]$ so that $(*)$ there exist augmenting edges for these components whose union will be a spanning forest of B. Thus we have to find a forest F in G so that (a) $E(F) \cap E(G_{\mathcal{P}}[H_j])$ is either a spanning tree or an almost spanning tree in $G_{\mathcal{P}}[H_j]$, (b) for $n - l$ components H_i we have spanning tree, (c) for these components $(*)$ is satisfied.

The existence of a forest with (a) and (b) can be proved, using (ii), by a matroid partition theorem (for a graphic matroid and a truncated partitional matroid). We shall see in Lemma 5 that if for all such forests we consider the components where the corresponding forest is a spanning tree then we get the set of basis of a matroid on the set of indices of the components.

Two matroids will be defined on the edge set of the auxiliary graph D, one of them will be defined by the above introduced matroid, and the other one will be defined by the cycle matroid of B. The matroid intersection theorem will provide a forest of G with (a), (b) and (c). As we mentioned earlier, each part of the forest, which corresponds to a component, can be replaced by a convenient cactus, and thus the desired cactus has been found.

2   The Proof

Proof. (max $\leq$ min) Let F be an arbitrary cactus in $(G, \mathcal{V})$ and let $(\mathcal{P}, \mathcal{Q})$ be any cover of $(G, \mathcal{V})$. Contract each $V_i \in \mathcal{P}$ $i = 1, 2, \ldots, l$ into a vertex and let F' be a subset of F of maximum size so that F' is a forest in the contracted graph $G/\mathcal{P}$. For the number c (c') of connected components of F in G (of F' in $G/\mathcal{P}$) we have obviously, $c' \leq c$. Thus $|F| = n - c \leq n - c' = (l - c') + (n - l) = |F'| + (n - l)$. It follows that $v_{\mathcal{V}}(F) \leq v_{\mathcal{V}}(F') + n - l$. Let F'' be the maximum subforest of F' in $G_{\mathcal{P}}$ consisting of v-pairs in $\mathcal{V}$. Obviously, F'' forms a cactus in each $(G_{\mathcal{P}}[H_i], H_i)$ $H_i \in \mathcal{Q}$. By definition, $\bigcup_{H_i \in \mathcal{Q}} H_i$ covers all the v-pairs contained in F''. Thus $v_{\mathcal{V}}(F') = v_{\mathcal{V}}(F'') = v_{\mathcal{V}_{\mathcal{P}}}(F'') = \sum_{H_i \in \mathcal{Q}} v_{H_i}(F'') \leq \sum_{H_i \in \mathcal{Q}} \lfloor \frac{p(H_i) - 1}{2} \rfloor$, and the desired inequality follows.   □

Proof. (max $\geq$ min) We prove the theorem by induction on the number n of vertices of G. For $n = 3$ the result is trivially true.

Let $(\mathcal{P}, \mathcal{Q})$ be a minimum cover of $(G, \mathcal{V})$ for which l is as small as possible and subject to this k is as large as possible. Note that by the maximality of k, each pair in $\mathcal{S}_{\mathcal{P}}$ will form a component because for each $H_i \in \mathcal{S}_{\mathcal{P}}$, $\lfloor \frac{p(H_i) - 1}{2} \rfloor = 0$.

Lemma 1. *For each $H_i \in \mathcal{Q}$, the unique minimum cover of $(G_{\mathcal{P}}[H_i], H_i)$ is the trivial one.*

Proof. Let $(\mathcal{P}', \mathcal{Q}')$ be a minimum cover of $(G_{\mathcal{P}}[H_i], H_i)$. Clearly, $val(\mathcal{P}', \mathcal{Q}') \leq \lfloor \frac{p(H_i) - 1}{2} \rfloor$. Using this cover, a new cover $(\mathcal{P}^*, \mathcal{Q}^*)$ of $(G, \mathcal{V})$ can be defined as follows. Let the partition $\mathcal{P}^*$ of $V(G)$ be obtained from $\mathcal{P}$ by taking the union of all those V_r and V_s whose corresponding vertices in $G_{\mathcal{P}}$ are in the same set of $\mathcal{P}'$. Then $l^* = l - p(H_i) + l'$, where $l' = |\mathcal{P}'|$. Let $\mathcal{Q}^*$ be obtained from $\mathcal{Q}$ by deleting H_i and by adding $\mathcal{Q}'$. We claim that the new cover is also a minimum cover.

$$val(\mathcal{P}^*, \mathcal{Q}^*) \leq n - l^* + \sum_{H_j \in \mathcal{Q} - \{H_i\}} \left\lfloor \frac{p(H_j) - 1}{2} \right\rfloor + \sum_{H'_j \in \mathcal{Q}'} \left\lfloor \frac{p'(H'_j) - 1}{2} \right\rfloor$$

$$= n - l^* + \sum_{H_j \in \mathcal{Q} - \{H_i\}} \left\lfloor \frac{p(H_j) - 1}{2} \right\rfloor$$

$$+ (val(\mathcal{P}', \mathcal{Q}') - (p(H_i) - l'))$$

$$\leq n - (l - p(H_i) + l') + \sum_{H_j \in \mathcal{Q} - \{H_i\}} \left\lfloor \frac{p(H_j) - 1}{2} \right\rfloor$$

$$+ \left\lfloor \frac{p(H_i) - 1}{2} \right\rfloor - p(H_i) + l' = val(\mathcal{P}, \mathcal{Q}).$$

It follows that equality holds everywhere, so $val(\mathcal{P}', \mathcal{Q}') = \lfloor \frac{p(H_i)-1}{2} \rfloor$, thus the trivial cover of $(G_\mathcal{P}[H_i], H_i)$ is minimal. Furthermore, by the minimality of l, $\mathcal{P}'$ is the trivial partition of $V(G_\mathcal{P}[H_i])$ and by the maximality of k, $\mathcal{Q}'$ may contain only one set and we are done.   □

Lemma 2. *Each component $H_i \in \mathcal{Q}$ is critical.*

Proof. Suppose that there exists a component $H_i \in \mathcal{Q}$ for which $(G_\mathcal{P}[H_i], H_i)$ is not critical, that is there are two vertices a and b in $G_\mathcal{P}[H_i]$ so that after identifying a and b the new v-graph (G', V') has no perfect cactus. By the hypothesis of the induction, it follows that there is a cover $(\mathcal{P}', \mathcal{Q}')$ of (G', V') so that in G' $val(\mathcal{P}', \mathcal{Q}') < \frac{(p(H_i)-1)-1}{2} \leq \lfloor \frac{p(H_i)-1}{2} \rfloor$. This cover can be considered as a cover $(\mathcal{P}'', \mathcal{Q}'')$ of $(G_\mathcal{P}[H_i], H_i)$ and $val(\mathcal{P}'', \mathcal{Q}'') = val(\mathcal{P}', \mathcal{Q}') + 1$. Thus $val(\mathcal{P}'', \mathcal{Q}'') \leq \lfloor \frac{p(H_i)-1}{2} \rfloor$, that is $(\mathcal{P}'', \mathcal{Q}'')$ is a minimal cover of $(G_\mathcal{P}[H_i], H_i)$ but not the trivial one (a and b are in the same member of $\mathcal{P}''$), which contradicts Lemma 1.   □

Corollary 1. *If $H_i \in \mathcal{Q}$ and a, b are two vertices of $G_\mathcal{P}[H_i]$, then there exists an almost perfect cactus K in $(G_\mathcal{P}[H_i], H_i)$ so that a and b belong to different connected components of K.*

Proof. By Lemma 2, H_i is critical, so by identifying a and b in $G_\mathcal{P}[H_i]$, the v-graph obtained has a perfect cactus K. Clearly, K has the desired properties in $(G_\mathcal{P}[H_i], H_i)$.   □

Remark 2. By Corollary 1, for any component H_i the v-graph $(G_\mathcal{P}[H_i], H_i)$ (and consequently (G, V)) contains a cactus containing $\lfloor \frac{p(H_i)-1}{2} \rfloor$ v-pairs. However, at this moment we can not see whether we can choose a cactus containing $\lfloor \frac{p(H_i)-1}{2} \rfloor$ v-pairs for all H_i so that their union is a cactus as well. Note that by Corollary 1, $p(H_i)$ is even for each component $H_i \in \mathcal{Q}$, that is $\lfloor \frac{p(H_i)-1}{2} \rfloor = \frac{p(H_i)-2}{2}$.

Proposition 1. *If $l = n-1, k = 1$ and $val(\mathcal{P}, \mathcal{Q}) = 1 + \lfloor \frac{n-2}{2} \rfloor$, then (G, V) has a perfect cactus.*

Proof. Let (uv, vw) be one of the v-pairs in $\mathcal{V}$. Let us consider the following cover $(\mathcal{P}', \mathcal{Q}')$ of $(G, \mathcal{V})$. Each set of $\mathcal{P}'$ contains exactly one vertex of G except one which contains u and v, and $\mathcal{Q}'$ contains exactly one set H (containing all v-pairs in $\mathcal{V}$). Then, clearly, this is a minimum cover. By the assumptions for l and k this cover also minimizes l and maximizes k, thus by Lemma 2, its unique component H is critical, that is the v-graph $(G_{\mathcal{P}'}, \mathcal{V}_{\mathcal{P}'})$ is critical. Let F be a perfect cactus of the v-graph obtained from $(G_{\mathcal{P}'}, \mathcal{V}_{\mathcal{P}'})$ by identifying v and w. Obviously, $F \cup uv \cup vw$ is a perfect cactus of $(G, \mathcal{V})$ and we are done. □

Lemma 3. *Let $\mathcal{P}'$ be a refinement of $\mathcal{P}$ and let $H_i \in \mathcal{Q}$ for which $H_i \notin \Gamma_D(A_{\mathcal{P}'})$. Then the trivial cover is a minimal cover of $(G_{\mathcal{P}'}[H_i], H_i)$ with value $\lfloor \frac{p(H_i)-1}{2} \rfloor$.*

Proof. By Corollary 1, $(G_{\mathcal{P}}[H_i], H_i)$ (and consequently $(G_{\mathcal{P}'}[H_i], H_i)$) contains a cactus of size $\lfloor \frac{p(H_i)-1}{2} \rfloor$, hence the value of a minimum cover of $(G_{\mathcal{P}'}[H_i], H_i)$ is at least $\lfloor \frac{p(H_i)-1}{2} \rfloor$. Thus what we have to show is that the trivial cover has value $\lfloor \frac{p(H_i)-1}{2} \rfloor$.

First, we prove this when $\mathcal{P}'$ is an arbitrary elementary refinement of $\mathcal{P}$, say $V_j^1 \cup V_j^2 = V_j$. We shall denote the vertices of $G_{\mathcal{P}'}[H_i]$ corresponding to V_j^1 and V_j^2 by v_1 and v_2. In this case we have to prove the following.

Proposition 2. *v_1 or v_2 does not belong to $G_{\mathcal{P}'}[H_i]$.*

Proof. $H_i \notin \Gamma_D(A_{\mathcal{P}'})$ implies that there exists no augmenting edge for H_i with respect to $\mathcal{P}'$ that is $(G_{\mathcal{P}'}[H_i], H_i)$ has no perfect cactus. By Proposition 1, we can use the induction hypothesis (of Theorem 1), that is there exists a cover $(\mathcal{P}'', \mathcal{Q}'')$ of $(G_{\mathcal{P}'}[H_i], H_i)$ so that $val(\mathcal{P}'', \mathcal{Q}'') \leq \frac{(p(H_i)+1)-1}{2} - 1 = \frac{p(H_i)-2}{2}$. This cover gives a cover $(\mathcal{P}^*, \mathcal{Q}^*)$ of $(G_{\mathcal{P}}[H_i], H_i)$ with $val(\mathcal{P}^*, \mathcal{Q}^*) \leq \frac{p(H_i)-2}{2}$. So $(\mathcal{P}^*, \mathcal{Q}^*)$ is a minimum cover of $(G_{\mathcal{P}}[H_i], H_i)$ and by Lemma 1, it is the trivial cover. Moreover, v_1 and v_2 are in different sets of $\mathcal{P}''$ (otherwise, $val(\mathcal{P}^*, \mathcal{Q}^*) < \frac{p(H_i)-2}{2}$, a contradiction), hence $(\mathcal{P}^*, \mathcal{Q}^*)$ is the trivial cover of $(G_{\mathcal{P}'}[H_i], H_i)$ and its value is $\frac{p(H_i)-2}{2}$. It follows that v_1 or v_2 is not a vertex in $G_{\mathcal{P}'}[H_i]$ and the proposition is proved. □

Let $\mathcal{P}' = \{V_1^1, ..., V_1^{r_1}, V_2^1, ..., V_2^{r_2}, ..., V_l^1, ..., V_l^{r_l}\}$ where $\cup_j V_i^j = V_i$ for all i, be a refinement of $\mathcal{P}$. It is enough to prove that for all i where $r_i \geq 2$ there exists an elementary refinement $\mathcal{P}^*$ of $\mathcal{P}$ with $V_i = V_i^j \cup (V_i - V_i^j)$ for some $1 \leq j \leq r_i$ so that in $G_{\mathcal{P}^*}[H_i]$ the vertex corresponding to $V_i - V_i^j$ is isolated. Applying Proposition 2, at most r_i times, we see that such an elementary refinement exists indeed. □

Corollary 2. *The vertex sets of the connected components of the graph B (defined by the augmenting edges) are exactly the sets in $\mathcal{P}$.*

Proof. By definition, there is no edge of B between two different sets of $\mathcal{P}$. Let us consider an elementary refinement $\mathcal{P}'$ of $\mathcal{P}$. If there was no augmenting edge with respect to this refinement, then by Lemma 3, the value of the cover $(\mathcal{P}', \mathcal{Q}')$ would be $val(\mathcal{P}, \mathcal{Q}) - 1$, where $\mathcal{Q}'$ is obtained from $\mathcal{Q}$ by adding the elements of $S_{\mathcal{P}'} - S_{\mathcal{P}}$ as new members of the partition, contradicting the minimality of the cover $(\mathcal{P}, \mathcal{Q})$. This implies that the subgraphs of B spanned on the sets V_i in $\mathcal{P}$ are connected. □

Let F_i be an arbitrary spanning tree of $G_\mathcal{P}[H_i]$ for all $H_i \in \mathcal{Q}$ (by Corollary 1, $G_\mathcal{P}[H_i]$ is connected). Then $E(F_i) \cap E(F_j) = \emptyset$ if $i \neq j$ because the components of $\mathcal{Q}$ are disjoint. Let $W = (V(G_\mathcal{P}), E(W))$ where $E(W) := \bigcup_{H_i \in \mathcal{Q}} E(F_i)$. Let $\mathcal{P}'$ be a refinement of $\mathcal{P}$ with $|\mathcal{P}'| = l'$. Let $\mathcal{Q}_1 := \Gamma_D(A_{\mathcal{P}'})$ and $\mathcal{Q}_2 := \mathcal{Q} - \mathcal{Q}_1$. We define two matroids on $E(W)$. Let $\mathcal{G}$ be the cycle matroid of W with rank function $r_\mathcal{G}$, that is the edge sets of the forests are the independent sets. Let $\mathcal{F}_{\mathcal{P}'} := \mathcal{F}_1 + \mathcal{F}_2$ (direct sum), where $\mathcal{F}_j$ will be the following (truncated) partitional matroid (with rank function r_j) on $E_j := \bigcup_{H_i \in \mathcal{Q}_j} E(F_i)$ $j = 1, 2$. Let $\mathcal{F}_1$ contain those sets $F \subseteq E_1$ for which $|F \cap E(F_i)| \leq 1$ for all i and the intersection can be 1 at most $t_0 := |\mathcal{Q}_1| - (l' - l)$ times. Let $\mathcal{F}_2$ contain those sets $F \subseteq E_2$ for which $|F \cap E(F_i)| \leq 1$ for all i. For the rank function r' of $\mathcal{F}_{\mathcal{P}'}$ $r'(X) = r_1(X \cap E_1) + r_2(X \cap E_2)$.

Lemma 4. *For any refinement $\mathcal{P}'$ of $\mathcal{P}$, $E(W)$ can be written as the union of an independent set in $\mathcal{G}$ and an independent set in $\mathcal{F}_{\mathcal{P}'}$.*

Proof. This is a matroid partition problem. It is well-known (see for example [7]) that the lemma is true if and only if for any $Y \subseteq E(W)$, $|Y| \leq r_\mathcal{G}(Y) + r'(Y)$. Suppose that this is not true, and let Y be a maximum cardinality set violating the above inequality. Then, clearly, Y is closed in $\mathcal{F}_{\mathcal{P}'}$. Thus Y can be written in the form $Y = \bigcup_{H_i \in \mathcal{Q}^*} E(F_i)$, for some $\mathcal{Q}^* \subseteq \mathcal{Q}$. Let $K_1, \ldots, K_c$ be the connected components of the graph K^* on the vertex set $V(G_\mathcal{P})$ with edge set Y. Then $r_\mathcal{G}(Y) = l - c = \sum_1^c (p(K_j) - 1)$. Let $t := |\mathcal{Q}^* \cap \mathcal{Q}_1|$.

Let $\mathcal{Q}''$ be obtained from $\mathcal{Q}$ by taking the unions of all those H_m and $H_{m'}$ in $\mathcal{Q}^*$ for which F_m and $F_{m'}$ are in the same connected component of K^*, that is each member $H_j'' \in \mathcal{Q}'' - \mathcal{Q}$ corresponds to some K_j, so $p(H_j'') = p(K_j)$.

Case 1. $t \leq t_0$. Then $r'(Y) = |\mathcal{Q}^*|$. Let us consider the cover $(\mathcal{P}, \mathcal{Q}'')$ of $(G, \mathcal{V})$. Since $0 \leq val(\mathcal{P}, \mathcal{Q}'') - val(\mathcal{P}, \mathcal{Q}) = \sum_1^c \lfloor \frac{p(H_j'') - 1}{2} \rfloor - \sum_{H_i \in \mathcal{Q}^*} \frac{p(H_i) - 2}{2}$,

$$|Y| = \sum_{H_i \in \mathcal{Q}^*} (p(H_i) - 1) = 2 \sum_{H_i \in \mathcal{Q}^*} \frac{p(H_i) - 2}{2} + |\mathcal{Q}^*|$$

$$\leq 2 \sum_1^c \frac{p(H_j'') - 1}{2} + |\mathcal{Q}^*| = r_\mathcal{G}(Y) + r'(Y)$$

contradicting the assumption for Y.

Case 2. $t > t_0$. Now, by the closedness of Y in $\mathcal{F}_{\mathcal{P}'}$, Y contains all the trees F_i for which $H_i \in \mathcal{Q}_1$. Thus $r'(Y) = r_1(Y \cap E_1) + r_2(Y \cap E_2) = t_0 + (|\mathcal{Q}^*| - |\mathcal{Q}_1|) =$

$|Q^*| - (l' - l)$. Let us consider the following cover $(\mathcal{P}', \mathcal{Q}^3)$ of $(G, \mathcal{V})$, where $\mathcal{Q}^3$ is obtained from the above defined $\mathcal{Q}''$ by adding each element in $\mathcal{S}_{\mathcal{P}'} - \mathcal{S}_{\mathcal{P}}$ as a component and adding the v-pairs in $\mathcal{V}_{\mathcal{P}'} - \mathcal{V}_{\mathcal{P}}$ to appropriate members of $\mathcal{Q}''$. If $L \in \mathcal{V}_{\mathcal{P}'} - \mathcal{V}_{\mathcal{P}}$, then L corresponds in W to a vertex or an edge and in the latter case $L \in \mathcal{Q}_1$ so L corresponds to an edge of a connected component K_j of K^*. We add L to the member H_j'' of $\mathcal{Q}''$ corresponding to K_j. (If K_j is an isolated vertex, then the corresponding H_j'' of $\mathcal{Q}''$ was empty earlier.) Now, $(\mathcal{P}', \mathcal{Q}^3)$ is a cover of $(G, \mathcal{V})$ indeed. We shall denote the members of $\mathcal{Q}^3 - \mathcal{Q}$ by H_j^3 $1 \le j \le c$. Clearly, $\sum_1^c p'(H_j^3) \le l'$. By Lemma 3, the value of the new cover is the following.

$$val(\mathcal{P}', \mathcal{Q}^3) = n - l' + \sum_1^c \left\lceil \frac{p'(H_j^3) - 1}{2} \right\rceil + \sum_{H_i \in \mathcal{Q} - \mathcal{Q}^*} \frac{p(H_i) - 2}{2}$$

$$\le n - l' + \frac{l' - c}{2} + \sum_{H_i \in \mathcal{Q} - \mathcal{Q}^*} \frac{p(H_i) - 2}{2}.$$

Using that $val(\mathcal{P}, \mathcal{Q}) \le val(\mathcal{P}', \mathcal{Q}^3)$ we have the following inequality.

$$|Y| = \sum_{H_i \in \mathcal{Q}^*} (p(H_i) - 1) = 2 \sum_{H_i \in \mathcal{Q}^*} \frac{p(H_i) - 2}{2} + |Q^*|$$

$$\le (l - c) + (|Q^*| - (l' - l)) = r_{\mathcal{G}}(Y) + r'(Y),$$

contradicting the assumption for Y. The proof of Lemma 4 is complete.   □

By Lemma 4, for the trivial partition $\mathcal{P}'$ of $V(G)$, the following fact is immediate.

Corollary 3. *There exists a forest F in the graph W so that for $n - l$ indices i $E(F_i) \subseteq E(F)$ and $E(F) \cap E(F_i)$ is an almost spanning tree of $G_{\mathcal{P}}[H_i]$ for the other indices.*   □

We shall need the following claim whose proof is trivial.

Proposition 3. *Let F be a forest on a vertex set S. Let F' be a subgraph of F with two connected components F_1' and F_2'. If F_1' and F_2' belong to the same connected component of F then let us denote by a and b the two end vertices of the shortest path in F connecting F_1' and F_2', otherwise let $a \in V(F_1')$ and $b \in V(F_2')$ be two arbitrary vertices. Let F'' be any forest on $V(F')$ with two connected components so that a and b are in different connected components of F''. Then $(F - E(F')) \cup E(F'')$ is a forest on S.*   □

Remark 3. By Corollary 3, there exists a forest F in W and consequently in $G_{\mathcal{P}}$ so that $E(F) \cap E(F_i)$ is a forest with two connected components on $V(G_{\mathcal{P}}[H_i])$ for all components H_i. Let H_i be an arbitrary component of $\mathcal{Q}$. By Corollary 1, for the two vertices a and b defined in Proposition 3 ($F' = E(F) \cap E(F_i)$), there exists an almost perfect cactus K in $(G_\mathcal{P}[H_i], H_i)$ so that a and b belong to different components of K. Then, by Proposition 3, $F - (E(F) \cap E(F_i)) \cup E(K)$ is a forest. We can do this for all components, so the v-graph $(G, \mathcal{V})$ contains a cactus containing $\sum_{H_i \in \mathcal{Q}} \lfloor \frac{p(H_i)-1}{2} \rfloor$ v-pairs.

Now we define a matroid $(\mathcal{Q}, \mathcal{M})$ on the sets of $\mathcal{Q}$. Let $\mathcal{Q}' \subseteq \mathcal{Q}$ be in $\mathcal{M}$ if and only if there is $f_i \in E(F_i)$ for each $H_i \in \mathcal{Q} - \mathcal{Q}'$ so that $E(W) - \cup f_i$ is a forest in W.

Lemma 5. $(\mathcal{Q}, \mathcal{M})$ *is a matroid.*

Proof. We show that $\mathcal{M}$ satisfies the three properties of independent sets of matroids.
(1) By Lemma 4, for $\mathcal{P}' = \mathcal{P}$, $\emptyset \in \mathcal{M}$.
(2) If $\mathcal{Q}'' \subseteq \mathcal{Q}' \in \mathcal{M}$, then $\mathcal{Q}'' \in \mathcal{M}$ because any subgraph of a forest is a forest.
(3) Let $\mathcal{Q}', \mathcal{Q}'' \in \mathcal{M}$ so that $|\mathcal{Q}''| < |\mathcal{Q}'|$. By definition, there are $f_i' \in E(F_i)$ for $H_i \in \mathcal{Q} - \mathcal{Q}'$ and $f_i'' \in E(F_i)$ for $H_i \in \mathcal{Q} - \mathcal{Q}''$ so that $T' := E(W) - \cup f_i'$ and $T'' := E(W) - \cup f_i''$ are forests in W. Choose these two forests T' and T'' so that they have edges in common as many as possible. $|\mathcal{Q}''| < |\mathcal{Q}'|$ implies that T' has more edges than T'' has. T' and T'' are two independent sets in the matroid $\mathcal{G}$ thus there is an edge $e \in T' - T''$ so that $T'' \cup e$ is also a forest in W. Then, clearly, $e = f_i''$ for some i. If $e \in E(F_i)$ with $H_i \notin \mathcal{Q}'$ then replace f_i'' by f_i' and the new forest T^* with T' contradicts the assumption on T' and T''. Thus $e \in E(F_i)$ so that $H_i \in \mathcal{Q}'$ and then obviously $\mathcal{Q}'' \cup \{H_i\} \in \mathcal{M}$ and we are done. $\qquad\square$

We shall apply the matroid intersection theorem for the following two matroids on the edge set of the graph D. For a set $Z \subseteq E(D)$, let us denote the end vertices of Z in the colour class $E(B)$ $(\mathcal{Q})$ by Z_1 $(Z_2$, respectively). The rank of Z in the first matroid will be $r_B(Z_1)$ and $r_\mathcal{M}(Z_2)$ in the second matroid, where r_B is the rank function of the cycle matroid of the graph B and $r_\mathcal{M}$ is the rank function of the above defined matroid $\mathcal{M}$. Note that if a vertex x of D is in the colour class $E(B)$ $(\mathcal{Q})$ then the edges incident to x correspond to parallel elements of the first (second) matroid.

Remark 4. By Corollary 2, $r_B(E(B)) = n-l$ and by Corollary 3, $r_\mathcal{M}(\mathcal{Q}) \geq n-l$. Moreover, if $A_{\mathcal{P}'}$ is the set of augmenting edges of some refinement $\mathcal{P}'$ of $\mathcal{P}$, then by Lemma 4,

$$l' - l \leq r_\mathcal{M}(\Gamma_D(A_{\mathcal{P}'})). \tag{1}$$

Lemma 6. *There exists a common independent set of size* $n - l$ *of these two matroids.*

Proof. By the matroid intersection theorem (see for example [7]) we have to prove that for any set $Z \subseteq E(D)$ $(+)$ $n - l \leq r_B(E(D) - Z) + r_\mathcal{M}(Z)$.

Suppose that there is a set Z violating $(+)$. Clearly, we may assume that $E(D) - Z$ is closed in the first matroid. This implies that there is a set $J \subseteq E(B)$ so that $E(D) - Z$ is the set of all edges of D incident to J and J is closed in the cycle matroid of B. Let us denote by $V_1', V_2', ..., V_{l'}'$ the vertex sets of the connected components of the graph on vertex set $V(B)$ with edge set J. Then by the closedness of J, $E(B) - J$ is the set of augmenting edges of the refinement $\mathcal{P}' := \{V_1', V_2', ..., V_{l'}'\}$ of $\mathcal{P}$, that is, $A_{\mathcal{P}'} = E(B) - J$. (Obviously, Z is the set of all edges incident to $E(B) - J$ in D.) Then $r_{\mathcal{M}}(Z) = r_{\mathcal{M}}(\Gamma_D(A_{\mathcal{P}'}))$ and $r_B(E(D) - Z) = r_B(J) = n - l'$. By (1), $l' - l \leq r_{\mathcal{M}}(\Gamma_D(A_{\mathcal{P}'}))$ and thus $n - l = (l' - l) + (n - l') \leq r_{\mathcal{M}}(Z) + r_B(E(D) - Z)$, contradicting the fact that Z violates $(+)$. $\qquad \square$

The Construction of the Desired Cactus. Let $N \subseteq E(D)$ be a common independent set of size $n - l$. (By Lemma 6, such a set exists.) It follows that N is a matching in D so that it covers a basis E' in the cycle matroid of B and an independent set Q' in $\mathcal{M}$. Thus there exists a forest F' on $V(G_{\mathcal{P}})$ so that it spans the spanning trees F_i in $G_{\mathcal{P}}[H_i]$ for $H_i \in Q'$ and almost spanning trees $F_i - f_i$ in $G_{\mathcal{P}}[H_i]$ (for appropriate f_i) for $H_i \in Q - Q'$. Let us denote by c the number of connected components of F'. Clearly, $E' \cup E(F')$ is a forest on $V(G)$ containing $2(n - l + \sum_{H_i \in Q} \frac{p(H_i) - 2}{2})$ edges and it has c connected components. ($|E' \cup E(F')| = |E'| + |E(F')| = n - l + \sum_{H_i \in Q'}(p(H_i) - 1) + \sum_{H_i \in Q - Q'}(p(H_i) - 2) = 2(n - l) + \sum_{H_i \in Q}(p(H_i) - 2)$.)

We shall change the trees and forests by appropriate ones obtaining a cactus of the desired size. As in Remark 3, for each $H_i \in Q - Q'$ we may replace in F' $F_i - f_i$ by an almost perfect cactus in $G_{\mathcal{P}}[H_i]$ obtaining a forest F'' on $V(G_{\mathcal{P}})$ with the same number of edges. As above, $E' \cup E(F'')$ is a forest on $V(G)$. For all $e \in E'$ e is an augmenting edge for $H_e \in Q'$, where H_e is the pair of e in the matching N. Thus there exists a cactus K_e in $(G_{\mathcal{P} \div V_i}[H_e], H_e)$ of size $p(H_i)/2$ so that the trace of K_e in V_i is the edge e, where $V_i \in \mathcal{P}$ contains the edge e. (Note that each K_e corresponds to a connected graph F_e' in $G_{\mathcal{P}}[H_e]$.) Replace $E' \bigcup_{H_i \in Q'} E(F_i)$ by $\bigcup_{e \in E'} E(K_e)$. We obtain again a forest of G with the same number of edges. (Indeed, first in F'' we replace $\bigcup_{H_i \in Q'} E(F_i)$ by $\bigcup_{e \in E'} E(F_e')$ and obviously we obtained a graph with c connected components, and, clearly, the edge set of this graph corresponds to a subgraph of G with c connected components. Since the number of edges in this subgraph is the same as in $E' \cup E(F')$ it is a forest of the same size.) The forest obtained consists of v-pairs, that is it is a cactus with size $n - l + \sum_{H_i \in Q} \lfloor \frac{p(H_i) - 1}{2} \rfloor$. $\qquad \square$

Remark 5. While I was writing the final version of this paper I realized that the same proof (after the natural changes) works for the general graphic matroid parity problem. The details will be given in a forthcoming paper [8].

Acknowledgement. I am very grateful to Gábor Bacsó for the fruitful discussions on the topic.

References

1. I. Anderson. Perfect matchings of a graph. *Journal of Combinatorial Theory, Series B,* 10:183–186, 1971.
2. P. Jensen and B. Korte, Complexity of matroid property algorithms. *SIAM J. Comput.,* 11:184–190, 1982.
3. L. Lovász. Matroid matching problem. In *Algebraic Methods in Graph Theory.* Colloquia Mathematica Societatis J. Bolyai 25, Szeged, 1978.
4. L. Lovász and M. D. Plummer. *Matching Theory.* North Holland, Amsterdam, 1986.
5. W. Mader. Über die maximalzahl kreuzungsfreier H-wege. *Archiv der Mathematik,* 31, 1978.
6. L. Nebesky. A new characterization of the maximum genus of a graph. *Czechoslovak Mathematical Journal,* 31, 1981.
7. A. Recski. *Matroid Theory and its Applications in Electric Network Theory and in Statics.* Akadémiai Kiadó, Budapest, 1989.
8. Z. Szigeti. On the graphic matroid parity problem. In preparation.
9. W. T. Tutte. Graph Factors. *Combinatorica,* 1:79-97, 1981.

Edge-Splitting and Edge-Connectivity Augmentation in Planar Graphs

Hiroshi Nagamochi[1] and Peter Eades[2]

[1] Kyoto University
naga@@kuamp.kyoto-u.ac.jp
[2] University of Newcastle
eades@@cs.newcastle.edu.au

Abstract. Let $G = (V, E)$ be a k-edge-connected multigraph with a designated vertex $s \in V$ which has even degree. A *splitting operation* at s replaces two edges (s, u) and (s, v) incident to s with a single edge (u, v). A set of splitting operations at s is called *complete* if there is no edge incident to s in the resulting graph. It is known by Lovász (1979) that there always exists a complete splitting at s such that the resulting graph G' (neglecting the isolated vertex s) remains k-edge-connected. In this paper, we prove that, in the case where G is planar and k is an even integer or $k = 3$, there exists a complete splitting at s such that the resulting graph G' remains k-edge-connected and planar, and present an $O(|V|^3 \log |V|)$ time algorithm for finding such a splitting. However, for every odd $k \geq 5$, there is a planar graph G with a vertex s which has no complete splitting at s which preserves both k-edge-connectivity and planarity. As an application of this result, we show that the problem of augmenting the edge-connectivity of a given outerplanar graph to an even integer k or to $k = 3$ can be solved in polynomial time.

1   Introduction

Let $G = (V, E)$ stand for an undirected multigraph, where an edge with end vertices u and v is denoted by (u, v). For a subset[1] $S \subseteq V$ in G, $G[S]$ denotes the subgraph induced by S. For two disjoint subsets $X, Y \subset V$, we denote by $E_G(X, Y)$ the set of edges (u, v) with $u \in X$ and $v \in Y$, and by $c_G(X, Y)$ the number of edges in $E_G(X, Y)$. The set of edges $E_G(u, v)$ may alternatively be represented by a single *link* (u, v) with multiplicity $c_G(u, v)$. In this way, we also represent a multigraph $G = (V, E)$ by an edge-weighted simple graph $N = (V, L_G, c_G)$ (called a *network*) with a set V of vertices and a set L_G of links weighted by $c_G : L_G \to Z^+$, where Z^+ is the set of non-negative integers. We denote $|V|$ by n, $|E|$ by e and $|L_G|$ by m. A *cut* is defined as a subset X of V

---

[1] A singleton set $\{x\}$ may be simply written as x, and " $\subset$ " implies proper inclusion while " $\subseteq$ " means " $\subset$ " or " $=$ ".

R. E. Bixby, E. A. Boyd, and R. Z. Ríos-Mercado (Eds.): IPCO VI
LNCS 1412, pp. 96–111, 1998. © Springer–Verlag Berlin Heidelberg 1998

with $\emptyset \neq X \neq V$, and the *size* of the cut X is defined by $c_G(X, V - X)$, which may also be written as $c_G(X)$. If $X = \{x\}$, $c_G(x)$ denotes the degree of vertex x. For a subset $X \subseteq V$, define its *inner-connectivity* by $\lambda_G(X) = \min\{c_G(X') \mid \emptyset \neq X' \subset X\}$. In particular, $\lambda_G(V)$ (i.e., the size of a minimum cut in G) is called the *edge-connectivity* of G. For a vertex $v \in V$, a vertex u adjacent to v is called a *neighbor* of v in G. Let $\Gamma_G(v)$ denote the set of neighbors of v in G.

Let $s \in V$ be a *designated vertex* in V. A cut X is called s-*proper* if $\emptyset \neq X \subset V - s$. The size $\lambda_G(V - s)$ of a minimum s-proper cut is called the s-*based-connectivity* of G. Hence $\lambda_G(V) = \min\{\lambda_G(V - s), c_G(s)\}$. A splitting at s is (k, s)-*feasible* if $\lambda_{G'}(V - s) \geq k$ holds for the resulting graph G'. Lovász [6] showed the following important property:

Theorem 1 ([2,6]). *Let a multigraph $G = (V, E)$ have a designated vertex $s \in V$ with even $c_G(s)$, and k be an integer with $2 \leq k \leq \lambda_G(V - s)$. Then there is a complete (k, s)-feasible splitting.*   □

Since a complete (k, s)-feasible splitting effectively reduces the number of vertices in a graph while preserving its s-based-connectivity, it plays an important role in solving many graph connectivity problems (e.g., see [1,2,9]).

In this paper, we prove an extension of Lovász's edge-splitting theorem, aiming to solve the edge-connectivity augmentation problem with an additional constraint that preserves the planarity of a given planar graph. Firstly, we consider the following type of splitting; for a multigraph $G = (V, E)$ with a designated vertex s, let $\Gamma_G(s) = \{w_0, w_1, \ldots, w_{p-1}\}$ ($p = |\Gamma_G(s)|$) of neighbors of s, and assume that a cyclic order $\pi = (w_0, w_1, \ldots, w_{p-1})$ of $\Gamma_G(s)$ is given. We say that two edges $e_1 = (w_h, w_i)$ and $e_2 = (w_j, w_\ell)$ are *crossing* (with respect to π) if e_1 and e_2 are not adjacent and the four end vertices appear in the order of w_h, w_j, w_i, w_ℓ along π (i.e., $h + a = j + b = i + c = \ell \pmod{p}$ holds for some $1 \leq c < b < a \leq p - 1$). A sequence of splittings at s is called *noncrossing* if no two split edges resulting from the sequence are crossing. We prove that there always exists a complete and noncrossing (k, s)-feasible splitting for even integers k, and such a splitting can be found in $O(n^2(m + n \log n))$ time.

Next we consider a planar multigraph $G = (V, E)$ with a vertex $s \in V$ of even degree. A complete splitting at s is called *planarity-preserving* if the resulting graph from the splitting remains planar. Based on the result of noncrossing splitting, we prove that, if k is an even integer with $k \leq \lambda_G(V - s)$, then there always exists a complete (k, s)-feasible and planarity-preserving splitting, and the splitting can be found in $O(n^3 \log n)$ time. For $k = 3$, we prove by a separate argument that there exists a complete (k, s)-feasible and planarity-preserving splitting if the resulting graph is allowed to be re-embedded in the plane.

Example 1. (a) Fig. 1(a) shows a graph $G_1 = (V, E)$ with $c_{G_1}(s, w_i) = 1$ and $c_{G_1}(w_i, w_{i+1}) = a$, $0 \leq i \leq 3$ for a given integer $a \geq 1$. Clearly, $\lambda_{G_1}(V - s) = k$ for $k = 2a + 1$. For a cyclic order $\pi = (w_0, w_1, w_2, w_3)$, G_1 has a unique complete (k, s)-feasible splitting (i.e., splitting pair of $(s, w_0), (s, w_2)$ and a pair of $(s, w_1), (s, w_3)$), which is crossing with respect to π. This implies that, for every odd $k \geq 3$, there is a graph G with a designated vertex s and a cyclic order of

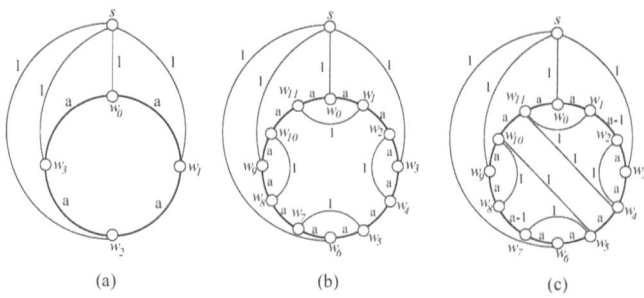

Fig. 1. Example of three planar graphs (a) G_1, (b) G_2, and (c) G_3.

$\Gamma_G(s)$ which has no complete and noncrossing (k, s)-feasible splitting. Note that the planar G_1 has a complete and planarity-preserving (k, s)-feasible splitting (by putting one of the split edges in the inner area of cycle $C_1 = \{w_0, w_1, w_2, w_3\}$).

(b) Fig. 1(b) shows a planar graph $G_2 = (V, E)$ with $c_{G_2}(w_i, w_{i+1}) = a$ (mod 12) for $0 \le i \le 11$ and $c_{G_2}(e) = 1$ otherwise for an integer $a \ge 1$, which satisfies $\lambda_{G_2}(V - s) = k$ for $k = 2a + 1$. The G_2 has a unique complete (k, s)-splitting, which is not planarity-preserving unless the embedding of subgraph $G_2[V - s]$ is not changed; if $G_2[V - s]$ is re-embedded in the plane so that block components $\{w_2, w_3, w_4\}$ and $\{w_8, w_9, w_{10}\}$ of $G_2[V - s]$ are flipped and two vertices w_3 and w_9 share the same inner face, then the complete (k, s)-splitting is now planarity-preserving. From this, we see that for every odd $k \ge 3$, there is a planar graph G with a designated vertex s which has no complete and planarity-preserving (k, s)-feasible splitting (unless the embedding of G is re-embedded).

(c) Let $a \ge 2$ be an integer, and consider the graph $G_3 = (V, E)$ in Fig. 1(c), where $c_{G_3}(w_i, w_{i+1}) = a$ for $i \in \{1, 7\}$, $c_{G_3}(w_i, w_{i+1}) = a$ (mod 12) for $i \in \{0, 1, \dots, 11\} - \{1, 7\}$, and $c_{G_3}(e) = 1$ otherwise. Clearly, $\lambda_{G_3}(V - s) = k$ for $k = 2a + 1 \ (\ge 5)$. It is easily observed that the unique complete (k, s)-feasible splitting is not planarity-preserving for any choice of re-embedding of G_3 in the plane. This implies that for every odd $k \ge 5$, there exists a graph which has no complete and planarity-preserving (k, s)-feasible splitting even if re-embedding after splitting is allowed.   □

2   Preliminaries

2.1   Computing s-Based Connectivity

The vertex set V of a multigraph $G = (V, E)$ are denoted by $V(G)$. We say that a cut X *separates* two disjoint subsets Y and Y' of V if $Y \subseteq X \subseteq V - Y'$ (or $Y' \subseteq X \subseteq V - Y$). The *local edge-connectivity* $\lambda_G(x, y)$ for two vertices $x, y \in V$ is defined to be the minimum size of a cut in G that separates x and y. A cut X *crosses* another cut Y if none of subsets $X \cap Y$, $X - Y$, $Y - X$ and $V - (X \cup Y)$ is empty.

An ordering $v_1, v_2, \ldots, v_n$ of all vertices in V is called a *maximum adjacency* (MA) *ordering* in G if it satisfies $c_G(\{v_1, v_2, \ldots, v_i\}, v_{i+1}) \geq c_G(\{v_1, v_2, \ldots, v_i\}, v_j)$, $1 \leq i < j \leq n$.

Lemma 1. [8] *Let $G = (V, E)$ be a multigraph, and v_1 be a vertex in V.*
(i) *An MA ordering $v_1, v_2, \ldots, v_n$ of vertices in G can be found in $O(m + n \log n)$ time.*
(ii) *The last two vertices v_{n-1} and v_n satisfy $\lambda_G(v_{n-1}, v_n) = c_G(v_n)$.*   □

Using this lemma repeatedly, we can compute $\lambda_G(V - s)$ by the next algorithm.

Algorithm CONTRACT
Input: A multigraph $G = (V, E)$ with $|V| \geq 3$ and a designated vertex $s \in V$.
Output: an s-proper cut X^* with $c_G(X^*) = \lambda_G(V - s) < \lambda_G(X^*)$.
  1 $\alpha := \min\{c_G(v) \mid v \in V - s\}$;
  2 Let $X := \{v\}$ for a vertex $v \in V - s$ with $c_G(v) = \alpha$;
  3 $H := G$;
  4 **while** $|V(H)| \geq 4$ **do**   { $c_H(u) \geq \alpha$ holds for all $u \in V(H) - s$ }
  5   Find an MA-ordering in H starting from $v_1 = s$, and let $v, w \ (\neq s)$ be the
          last two vertices in this ordering;   { $\lambda_H(v, w) = c_H(w)$ by Lemma 1(ii) }
  6   Contract v and w into a vertex, say z, and let H be the resulting graph;
  7   **if** $c_H(z) < \alpha$ **then**
  8   Let X^* be the set of all vertices in $V - s$ contracted into z so far;
              { $c_H(z) = c_G(X^*)$ }
  9   **end** { if }
10 **end**. { while }

It should be noted that for each $u \in V(H) - s$ $c_H(u) \geq \alpha$ holds before every iteration of the while-loop. The last two vertices v, w in an MA ordering in line 5, which are clearly distinct from s, satisfy $\lambda_H(v, w) = c_H(w)$ by Lemma 1(ii). Let X^* be the cut output by CONTRACT, and α^* be the final value of α (i.e., $\alpha^* = c_G(X^*)$). Note that any two vertices v and w in line 5 have been contracted into a single vertex only when $\lambda_H(v, w) \geq \alpha^*$ holds. We prove that $\alpha^* = \lambda_G(V - s)$. For a vertex $u \in V(H) - s$, let X_u denote the set of all vertices in $V - s$ contracted so far. Assume that there is an s-proper cut Y with $c_G(Y) < \alpha^*$. Clearly, the final graph H has three vertices z_1, z_2 and s, and satisfies $\alpha^* \geq \min\{c_H(z_1), c_H(z_2)\}$, and thus $Y \neq X_{z_1}, X_{z_2}$. Hence there is a vertex pair $v, w \in V(H)$ chosen in line 5 at some iteration of the while-loop such that $X_v \subseteq Y$ and $X_w \subseteq (V - s) - Y$ (or $X_w \subseteq Y$ and $X_v \subseteq (V - s) - Y$). Assume that v and w are the vertices in the earliest iteration of the while-loops among such pairs of vertices. This implies that when v and w are contracted into a vertex, the current graph H has a subset $Y' \subset V(H) - s$ such that $\cup_{y \in Y'} X_y = Y$. However, $\lambda_H(v, w) \leq c_H(Y') = c_G(Y) < \alpha^*$, contradicting $\lambda_H(v, w) \geq \alpha^*$. Therefore, $\alpha^* = \lambda_G(V - s)$.

Now we show $\lambda_G(X^*) > \alpha^*$. Assume that the output cut X^* is found in line 8 in the i-th iteration of the while-loop (the case where the X^* is found at line 3 is trivial), and let α' be the value of α before the i-th iteration. Then $\alpha' > \alpha^*$, and any two vertices v and w in line 5 in any earlier iteration have been contracted into a single vertex only when $\lambda_H(v, w) \geq \alpha'$. Analogously with the above argument, we see that when CONTRACT finds the final cut X^* in line 8, $\lambda_G(u, u') \geq \alpha'$ holds for $u, u' \in X^*$ (hence $\lambda_G(X^*) > \alpha^*$). This leads to the next lemma, where the running time clearly follows from Lemma 1(i).

Lemma 2. *For a multigraph $G = (V, E)$ with a designated vertex $s \in V$, CON-TRACT computes a cut X^* such that $c_G(X^*) = \lambda_G(V - s) < \lambda_G(X^*)$ in $O(n(m + n \log n))$ time.* □

2.2 Splitting Edges for a Pair of Neighbors

Given a multigraph $G = (V, E)$, a designated vertex $s \in V$, vertices $u, v \in \Gamma_G(s)$ (possibly $u = v$) and a non-negative integer $\delta \leq \min\{c_G(s, u), c_G(s, v)\}$, we construct graph $G' = (V, E')$ from G by deleting δ edges from $E_G(s, u)$ and $E_G(s, u)$, respectively, and adding new δ edges to $E_G(u, v)$. We say that G' is obtained from G by *splitting* δ pairs of edges (s, u) and (s, v) by *size* δ, and denote the resulting graph G' by $G/(u, v, \delta)$. Clearly, for any s-proper cut X, we see that

$$c_{G/(u,v,\delta)}(X) = \begin{cases} c_G(X) - 2\delta & \text{if } u, v \in X \\ c_G(X) & \text{otherwise.} \end{cases} \tag{1}$$

Given an integer k satisfying $0 \leq k \leq \lambda_G(V - s)$, we say that splitting δ pairs of edges (s, u) and (s, v) is (k, s)-*feasible* if $\lambda_{G/(u,v,\delta)}(V - s) \geq k$.

For an integer k, let $\Delta_G(u, v, k)$ be the maximum δ such that splitting edges (s, u) and (s, v) with size δ is (k, s)-feasible in G. In this subsection, we show how to compute $\Delta_G(u, v, k)$. An s-proper cut X is called (k, s)-*semi-critical* in G if it satisfies $c_G(s, X) > 0$, $k \leq c_G(X) \leq k + 1$ and $\lambda_G(X) \geq k$.

An algorithm, called MAXSPLIT(u, v, k), for computing $\Delta_G(u, v, k)$ is described as follows.

1. Let $\delta_{max} = \min\{c_G(s, u), c_G(s, v)\}$ if $u \neq v$, and $\delta_{max} = \lfloor c_G(s, u)/2 \rfloor$ if $u = v$, and let $G_{max} = G/(u, v, \delta_{max})$.
2. Compute $\lambda_{G_{max}}(V - s)$ and an s-proper cut X with $c_{G_{max}}(X) = \lambda_{G_{max}}(V - s) < \lambda_{G_{max}}(X)$ (such X exists by Lemma 2). If $\lambda_{G_{max}}(V - s) \geq k$, then $\Delta_G(u, v, k) = \delta_{max}$, where at least one of u and v is no longer a neighbor of s in G_{max} in the case $u \neq v$, or $c_{G_{max}}(s, u) \leq 1$ in the case $u = v$.
3. If $k - \lambda_{G_{max}}(V - s) \geq 1$, then $u, v \in X$ (for otherwise $c_G(X) = c_{G_{max}}(X) < k$ would hold). Output $\Delta_G(u, v, k) = \delta_{max} - \lceil \frac{1}{2}(k - \lambda_{G_{max}}(V - s)) \rceil$ and the s-proper cut X as such a (k, s)-semi-critical cut with $u, v \in X$.

The correctness of step 2 is clear. In step 3, we see from (1) that $G' = G/(u, v, \delta)$ with $\delta = \delta_{max} - \lceil \frac{1}{2}(k - \lambda_{G_{max}}(V - s)) \rceil$ satisfies $\lambda_{G'}(V - s) = k$ or

$k + 1$. This implies that $\Delta_G(u, v, k) = \delta$. We show that the X has a property that

$$c_{G'}(Z) > c_{G'}(X) \text{ for any } Z \text{ with } u, v \in Z \subset X,$$

where we call such a (k, s)-semi-critical cut X with $u, v \in X$ admissible (with respect to u, v) in G'. For any Z with $u, v \in Z \subset X$, we have $c_{G'}(Z) = c_{G_{max}}(Z) + 2\lceil \frac{1}{2}(k - \lambda_{G_{max}}(V - s)) \rceil > c_{G'}(X)$, since $\lambda_{G_{max}}(X) > \lambda_{G_{max}}(V - s)$ implies $c_{G_{max}}(Z) > c_{G_{max}}(X)$. By summarizing this, we have the next result.

Lemma 3. *For a multigraph $G = (V, E)$ with a designated vertex $s \in V$, and vertices $u, v \in \Gamma_G(s)$ (possibly $u = v$), let k be a nonnegative integer with $k \le \lambda_G(V - s)$, and let $G' = G/(u, v, \delta)$ for $\delta = \Delta_G(u, v, k)$. Then:*
(i) If $c_{G'}(s, u) > 0$ and $c_{G'}(s, v) > 0$ in the case $u \ne v$ or if $c_{G'}(s, u) \ge 2$ in the case $u = v$, then G' has an admissible cut X.
(ii) The cut X in (i) (if any) and $\Delta_G(u, v, k)$ can be computed in $O(mn + n^2 \log n)$ time. □

3   Noncrossing Edge Splitting

For a cyclic order $\pi = (w_0, w_1, \ldots, w_{p-1})$ of $\Gamma_G(s)$, a sequence of splittings at s is called *noncrossing* (with respect to π) if no two split edges (w_h, w_i) and (w_j, w_ℓ) are crossing with respect to π (see Section 1 for the definition). In this section, we show that for any even $k \le \lambda_G(V - s)$, there always exists a complete and noncrossing (k, s)-splitting. However, as observed in Example 1(a), for every odd $k \ge 3$, there is a graph that has no such splitting.

3.1 (k, s)-Semi-critical Collections

Before computing a complete (k, s)-feasible splitting, we first find a family $\mathcal{X}$ of subsets of $V - s$ (by performing some noncrossing edge splittings at s) as follows. For a multigraph $G = (V, E)$ and $s \in V$, a family $\mathcal{X} = \{X_1, X_2, \ldots, X_r\}$ of disjoint subsets $X_i \subset V - s$ is called a *collection* in $V - s$. A collection $\mathcal{X}$ may be empty. A collection $\mathcal{X}$ is called *covering* if $\sum_{i=1}^r c_G(s, X_i) = c_G(s)$. A collection $\mathcal{X}$ in $V - s$ is called (k, s)-*semi-critical* in G either if $\mathcal{X} = \emptyset$ or if all $X_i \in \mathcal{X}$ are (k, s)-semi-critical. We can easily see that a (k, s)-semi-critical covering collection in G with $c_G(s) > 0$ satisfies $|\mathcal{X}| \ge 2$ [9].

Let X be an s-proper cut with $c_G(X) \le k + 1$. Clearly, splitting two edges (s, u) and (s, v) such that $u, v \in X$ is not (k, s)-feasible. Then the size of any cut $Z \subseteq X$ remains unchanged after any (k, s)-feasible splitting in G. We say that two s-proper cuts X and Y s-*cross* each other if X and Y cross each other and $c_G(s, X \cap Y) > 0$. It is not difficult to prove the following properties by using submodularity of cut function c_G (the detail is omitted).

Lemma 4. *Let $G = (V, E)$ be a multigraph with a designated vertex s, and k be an integer with $k \le \lambda_G(V - s)$. Then:*

(i) *If two (k, s)-semi-critical cuts X and Y s-cross each other, then $c_G(X) = c_G(Y) = k + 1$, $c_G(X - Y) = c_G(Y - X) = k$ and $c_G(X \cap Y, V - (X \cup Y)) = 1$.*
(ii) *Let X be an admissible cut with respect to $u, u' \in V - s$ (possibly $u = u'$), and Y be a (k, s)-semi-critical cut. If X and Y cross each other, then $c_G(X) = c_G(Y) = k + 1$ and $c_G(Y - X) = k$.*
(iii) *Let X_i (resp., X_j) be admissible cuts with respect to u_i, u'_i (resp., with respect to u_j, u'_j), where possibly $u_i = u'_i$ or $u_j = u'_j$ holds. If $\{u_i, u'_i\} \cap \{u_j, u'_j\} = \emptyset$ or $c_G(s, u) \geq 2$ for some $u \in \{u_i, u'_i\} \cap \{u_j, u'_j\}$, then two cuts X_i and X_j do not cross each other.* □

We now describe an algorithm, called COLLECTION, which computes a (k, s)-semi-critical covering collection $\mathcal{X}$ in a graph G^* obtained from G by a noncrossing sequence of (k, s)-feasible splittings. Let $\pi = (w_0, w_1, \ldots, w_{p-1})$ be a cyclic order of $\Gamma_G(s)$. In a graph G' obtained from G by a noncrossing sequence of (k, s)-feasible splittings, a vertex w_j is called a *successor* of a vertex w_i if $c_{G'}(s, w_j) \geq 1$ and $h > 0$ with $j = i + h \pmod{p}$ is minimum.
0. Initialize $\mathcal{X}$ to be $\emptyset$.
1. for each w_i, $i := 0, \ldots, p - 1$ **do**
if w_i is not in any cut $X \in \mathcal{X}$ **then** execute MAXSPLIT(w_i, w_i, k) to compute $G' = G/(w_i, w_i, \delta)$ with $\delta = \Delta_G(w_i, w_i, k)$ and an admissible cut X_{w_i} in G' (if $c_{G'}(s, w_i) \geq 2$); let $G := G'$;
if $c_G(s, w_i) \geq 2$ **then** $\mathcal{X} := \mathcal{X} \cup \{X_{w_i}\}$, discarding all $X \in \mathcal{X}$ with $X \subset X_{w_i}$ from $\mathcal{X}$.
end { for }
2. for each w_i such that $c_G(s, w_i) = 1$, $i := 0, \ldots, p - 1$ **do**
if w_i is not in any cut $X \in \mathcal{X}$ **then** execute MAXSPLIT(w_i, w_j, k) for w_i and the successor w_j of s in the current G to compute $G' = G/(w_i, w_j, \delta)$ with $\delta = \Delta_G(w_i, w_j, k)$ and an admissible cut X_{w_i} in G' (if $c_{G'}(s, w_i) = 1$); let $G := G'$;
if $c_G(s, w_i) = 1$ **then** $\mathcal{X} := \{X - X_{w_i} \mid c_G(s, X - X_{w_i}) > 0, \ X \in \mathcal{X}\} \cup \{X_{w_i}\}$.
else (if $c_G(s, w_i) = 0$) remove any cut X with $c_G(s, X) = 0$ from $\mathcal{X}$.
end { for }
Output $G^* := G$. □
Clearly, the resulting sequence of splitting is (k, s)-feasible and noncrossing.

Lemma 5. *Algorithm COLLECTION correctly computes a (k, s)-semi-critical covering collection $\mathcal{X}$ in the output graph G^*.*

Proof: Let $\mathcal{X}$ be the set of cuts obtained after step 1. If two cuts $X_{w_i}, X_{w_j} \in \mathcal{X}$ $(0 \leq i < j \leq p - 1)$ has a common vertex v, then $w_j \notin X_{w_i}$ and $X_{w_i} - X_{w_j} \neq \emptyset$ (otherwise, X_{w_i} must have been discarded). However, this implies that X_{w_i} and X_{w_j} cross each other, contradicting Lemma 4(iii). Thus, the $\mathcal{X}$ is a (k, s)-semi-critical collection.
Now we prove by induction that $\mathcal{X}$ is a (k, s)-semi-critical collection during step 2. Assume that MAXSPLIT(w_i, w_j, k) is executed to compute $G' = G/(w_i, w_j, \delta)$ with $\delta = \Delta_G(w_i, w_j, k)$. If $c_{G'}(s, w_i) = 0$, then a cut $X \in \mathcal{X}$ with $w_j \in X$ may satisfy $c_{G'}(s, X) = 0$ after the splitting. However, such a cut will

be removed from $\mathcal{X}$. If $c_{G'}(s, w_i) = 1$, then an admissible cut X_{w_i} in G' is found. Clearly, any $X \in \mathcal{X}$ satisfies one of the followings: (i) $X \cap X_{w_i} = \emptyset$, (ii) $X \subset X_{w_i}$, and (iii) $X \cap X_{w_i} \neq \emptyset \neq X - X_{w_i}$. Since $\mathcal{X}$ is updated to $\{X - X_{w_i} \mid c_G(s, X - X_{w_i}) > 0, \ X \in \mathcal{X}\} \cup \{X_{w_i}\}$, it is sufficient to show that $c_{G'}(X - X_{w_i}) = k$ holds in the case (iii) (note that $\lambda_{G'}(X - X_{w_i}) \geq k$ follows from $\lambda_{G'}(X) \geq k$). Since two cuts X and X_{w_i} cross each other in the case (iii), $c_{G'}(X - X_{w_i}) = k$ follows from Lemma 4(ii). This proves that $\mathcal{X}$ remains to be a (k, s)-semi-critical collection, which becomes covering after step 2.   □

3.2   Algorithm for Noncrossing Edge-Splitting

In this subsection, k is assumed to be a positive even integer. We can prove the next property by Lemma 4(i) and the evenness of k (the detail is omitted).

Lemma 6. *Let $G = (V, E)$ be a multigraph with a designated vertex s, and k be an even integer with $k \leq \lambda_G(V - s)$. Further, let X be a (k, s)-semi-critical cut, and Y and Y' be (k, s)-semi-critical cuts with $Y \cap Y' = \emptyset$. Then X can s-cross at most one of Y and Y'.*   □

Using the lemma, we now describe an algorithm that constructs a complete and noncrossing (k, s)-feasible splitting from a given (k, s)-semi-critical covering collection $\mathcal{X}$ in a graph G with a designated vertex s of even $c_G(s)$.

If s has at most three neighbors, then any complete (k, s)-feasible splitting is noncrossing (with respect to any cyclic order of $\Gamma_G(s)$) and such a splitting can be found by applying MAXSPLIT at most three times. In what follows, we assume that $|\Gamma_G(s)| \geq 4$.

First, we define a notion of segment. For a given covering collection $\mathcal{X}$ with $|\mathcal{X}| \geq 2$ in a multigraph G with a designated vertex s and a cyclic order $\pi = (w_0, w_1, \ldots, w_{p-1})$ of $\Gamma_G(s)$, a subset $P \subset \Gamma_G(s)$ of neighbors of s which are consecutive in the cyclic order is called *segment* if there is a cut $X \in \mathcal{X}$ with $P \subseteq X$ such that P is maximal subject to this property. Note that there may be two segments P and P' with $P \cup P' \subseteq X$ for the same cut $X \in \mathcal{X}$. A segment P with $|P| = 1$ is called *trivial*. We now describe the two cases.

Case 1: There is a nontrivial segment $P = \{w_i, w_{i+1}, \ldots, w_j\}$ (with respect to $\mathcal{X}$). Let X_1, X_2 and X_3 be the cuts in $\mathcal{X}$ such that $w_{i-1} \in X_1$, $\{w_i, \ldots, w_j\} \subseteq X_2$ and $w_{j+1} \in X_3$ (possibly $X_1 = X_3$). We execute MAXSPLIT(w_{i-1}, w_i, k) and then MAXSPLIT(w_j, w_{j+1}, k). Let G'' be the resulting graph. We first remove all cuts $X \in \mathcal{X}$ with $c_{G''}(s, X) = 0$ from $\mathcal{X}$. If one of $w_{i-1}, w_i, w_j, w_{j+1}$ is no longer a neighbor of s in G'', then the number of neighbors of s decreases at least by one. Let us consider the case where all of $w_{i-1}, w_i, w_j, w_{j+1}$ remain neighbors of s in G''. Thus, G'' has admissible cuts Y_i and Y_j (with respect to w_{i-1}, w_i and w_j, w_{j+1}, respectively). In the case of $w_{i-1} = w_{j+1}$, it holds $c_{G''}(s, w_{i-1}) \geq 2$, since otherwise $c_{G''}(s, X_2) \geq 3$ (by evenness of $c_{G''}(s)$) would imply $c_{G''}(V - X_2) = c_{G''}(X_2) - c_{G''}(s, X_2) + c_{G''}(s, X_1) \leq (k+1) - 3 + 1 < k$, contradicting $\lambda_{G''}(V - s) \geq k$. Thus, by Lemma 4(iii), two cuts Y_i and Y_j do not cross each other. There are two subcases (a) $Y_i \cap Y_j = \emptyset$ and (b) $Y_i \subseteq Y_j$ or $Y_j \subseteq Y_i$.

(a) $Y_i \cap Y_j = \emptyset$. We prove that $X_1 \neq X_3$ and $X_1 \cup X_2 \cup X_3 \subseteq Y_i \cup Y_j$. Then two (k, s)-semi-critical cuts Y_i and X_2 s-cross each other, and by Lemma 4(i) $c_{G''}(X_2 - Y_i) = c_{G''}(Y_i - X_2) = k$. Note that $X_2 - Y_i$ is a (k, s)-semi-critical cut. Thus $X_2 - Y_i$ cannot cross another admissible cut Y_j (otherwise $c_{G''}(X_2 - Y_i) = k$ would contradict Lemma 4(ii)), and hence $X_2 \subset Y_i \cup Y_j$. By Lemma 6, Y_i which already crosses X_2 cannot s-cross X_1, and thus $X_1 \subset Y_i$. Similarly, we have $X_3 \subset Y_j$. Therefore, $X_1 \neq X_3$ and $X_1 \cup X_2 \cup X_3 \subseteq Y_i \cup Y_j$. There may be some cut $X \in \mathcal{X} - \{X_1, X_2, X_3\}$ which crosses one of Y_i or Y_j. If the X crosses Y_i, then $c_{G''}(X - Y_i) = k$ by Lemma 4(ii). We see that $c_{G''}(s, X - Y_i) = c_{G''}(s, X) \geq 1$, because if $c_{G''}(s, X \cap Y_i) \geq 1$ (i.e., X and Y_i s-cross each other) then $c_{G''}(Y_i - X) = k < k + 1 = c_{G''}(Y_i)$ by Lemma 4(i), contradicting the admissibility of Y_i (note $\{w_{i-1}, w_i\} \subseteq Y_i - X$). Thus $X - Y_i$ is a (k, s)-semi-critical cut in G''. Similarly, If the X crosses Y_j, then $X - Y_j$ is a (k, s)-semi-critical cut in G''. Hence if $X \cap Y_i \neq \emptyset \neq X \cap Y_j$, then $X - Y_i - Y_j$ is also a (k, s)-semi-critical cut in G''. Therefore, we can update $\mathcal{X}$ by $\mathcal{X} := \{X - Y_i - Y_j \mid X \in \mathcal{X}\} \cup \{Y_i, Y_j\}$.

(b) $Y_i \subseteq Y_j$ or $Y_j \subseteq Y_i$. Without loss of generality, assume $Y_j \subseteq Y_i$. Since $c_{G''}(s, X_2 \cap Y_i) \geq 2$ holds by $w_i, w_j \in X_2 \cap Y_i$, we see by Lemma 4(i) that X_2 cannot cross Y_i (hence, $X_2 \subset Y_i$). If $X_1 = X_3$, then $c_{G''}(s, X_1 \cap Y_i) \geq 2$ (including the case $w_{i-1} = w_{j+1}$) and by Lemma 4(i) $X_1 = X_3$ cannot cross Y_i (hence, $X_1 = X_3 \subset Y_i$). In the case $X_1 \neq X_3$, at most one of X_1 and X_3 can s-cross Y_i by Lemma 6. Thus $X_1 \subset Y_i$ or $X_3 \subset Y_i$. For any cut $X \in \mathcal{X} - \{X_3\}$ which crosses Y_i, we can show that $X - Y_i$ is a (k, s)-semi-critical cut in G'' using similar reasoning as for Case 1(a). Therefore, we can update $\mathcal{X}$ by $\mathcal{X} := \{X - Y_i \mid c_{G''}(s, X - Y_i) > 0, \ X \in \mathcal{X}\} \cup \{Y_i\}$ (note that $c_{G''}(s, X_1 - Y_i)$ or $c_{G''}(s, X_3 - Y_i)$ may be 0).

Notice that the number of cuts in $\mathcal{X}$ in cases (a) and (b) decreases at least by one after updating.

Case 2: All segments are trivial. We choose a neighbor w_i of s and its successor w_j, and execute MAXSPLIT(w_i, w_j, k). Assume that MAXSPLIT(w_i, w_j, k) finds an admissible cut Y (otherwise, the number of neighbors of s decreases at least by one). Let X_1 and X_2 be the cuts in $\mathcal{X}$ which contain w_i and w_j, respectively. We see that $X_1 \subset Y$ or $X_2 \subset Y$, because otherwise Y would s-cross both X_1 and X_2 (contradicting Lemma 6). If X_1 s-crosses Y, then we see by Lemma 4(ii) that $X_1 - Y$ is a (k, s)-semi-critical cut if $c_{G''}(s, X_1 - Y) \geq 1$. The case where X_2 s-crosses Y is similar. For any cut $X \in \mathcal{X} - \{X_1, X_2\}$ which crosses Y, we can show that $X - Y$ is a (k, s)-semi-critical cut in G'' using similar reasoning as for Case 1(a). We update $\mathcal{X}$ by $\mathcal{X} := \{X - Y \mid c_G(s, X - Y) \geq 1, \ X \in \mathcal{X}\} \cup \{Y\}$. In this case, the number of cuts in $\mathcal{X}$ never increases, but it may not decrease, either. However, Y contains a nontrivial segment in the new $\mathcal{X}$, and we can apply the above argument of Case 1.

By applying the above argument to Case 1, at least one vertex is no longer a neighbor of s or the number of cuts in a collection $\mathcal{X}$ is decreased at least by one. After applying the argument of Case 2, at least one vertex is no longer a neighbor of s or a nontrivial segment is created. Therefore, by executing MAXSPLIT at most $4(|\Gamma_G(s)|+|\mathcal{X}|) = O(|\Gamma_G(s)|)$ times, we can obtain a complete (k, s)-feasible splitting of a given graph G, which is obviously noncrossing.

Theorem 2. *Given a multigraph $G = (V, E)$ with a designated vertex $s \in V$ of even degree, a positive even integer $k \leq \lambda_G(V - s)$, and a cyclic order π of neighbors of s, a complete and noncrossing (k, s)-feasible splitting can be found in $O(|\Gamma_G(s)|n(m + n \log n))$ time.* ☐

4 Planarity-Preserving Splitting

In this section, we assume that a given graph G with a designated vertex s of even degree and an integer $k \leq \lambda_G(V - s)$ is *planar*, and consider whether there is a complete and planarity-preserving (k, s)-feasible splitting. We prove that such splitting always exists if k is even or $k = 3$, but may not exist if k is odd and $k \geq 5$, as observed in Example 1(c). We initially fix an embedding ψ of G in the plane, and let π_ψ be the order of neighbors of s that appear around s in the embedding χ of G.

4.1 The Case of Even k

Clearly, a complete splitting at s is planarity-preserving if it is noncrossing with respect to π_ψ. Therefore, in the case of even integers k, the next theorem is immediate from Theorem 2 and the fact that m is $O(n)$ in a planar graph G.

Theorem 3. *Given a planar multigraph $G = (V, E)$ with a designated vertex $s \in V$ of even degree, and a positive even integer $k \leq \lambda_G(V - s)$, there exists a complete and planarity-preserving (k, s)-feasible splitting (which also preserves the embedding of $G[V - s]$ in the plane), and such splitting can be found in $O(|\Gamma_G(s)|n^2 \log n)$ time.* ☐

4.2 The Case of $k = 3$

For $k = 3 \leq \lambda_G(V - s)$, we can prove that there is a complete and planarity-preserving (k, s)-feasible splitting. However, as observed in Example 1(b), in this case we may need to re-embed the subgraph $G[V - s]$ in the plane to obtain such a splitting. We will show how to obtain a complete $(3, s)$-feasible and planarity-preserving splitting. Firstly, however, we describe a preprocessing algorithm based on the following Lemma 7 (the proof is omitted).

A set Y of vertices in G (or the induced graph $G[Y]$) is called a *k-component* if $\lambda_G(u, v) \geq k$ for all $u, v \in Y$ and $|Y|$ is maximal subject to this property. It is easy to see that the set of k-components in a graph is unique, and forms a partition of the vertex set. Such a partition can be found in linear time for $k = 1, 2$ [10] and for $k = 3$ [7,12]. A k-component Y (or the induced graph $G[Y]$) is called a *leaf k-component* if $c_G(Y) < k$. Note that $\lambda_{G[V-s]}(V - s)$ means the edge-connectivity of the subgraph $G[V - s]$.

Lemma 7. *Let $G = (V, E)$ be a multigraph with a designated vertex s of even degree such that $\lambda_G(V - s) \geq 3$.*

(i) *Any s-proper cut X with $c_G(X) \leq 4$ induces a connected subgraph $G[X]$.*

(ii) *Assume $\lambda_{G[V-s]}(V - s) = 0$, and let u and v be two neighbors of s such that they belong to different 1-components in $G[V - s]$. Then $\Delta_G(u, v, 3) \geq 1$.*

(iii) *Assume $\lambda_{G[V-s]}(V - s) = 1$. Let Y be a leaf 2-component in $G[V - s]$. If $\Delta_G(u, v, 3) = 0$ for some neighbors $u \in \Gamma_G(s) \cap Y$ and $v \in \Gamma_G(s) - Y$, then $\Gamma_G(s) - Y - v \neq \emptyset$ and $\Delta_G(u', v', 3) \geq 1$ for any neighbors $v' \in \Gamma_G(s) - Y - v$ and $u' \in \Gamma_G(s) \cap Y$.*

(iv) *Assume $\lambda_{G[V-s]}(V - s) = 2$. Let Y be a 3-component with $c_G(s, Y) \geq 2$ in $G[V - s]$. Then $\Delta_G(u, v, 3) \geq 1$ for any neighbors $u \in \Gamma_G(s) \cap Y$ and $v \in \Gamma_G(s) - Y$.*

(v) *Assume $\lambda_{G[V-s]}(V - s) = 2$. Let Y be a non-leaf 3-component with $c_G(s, Y) = 1$ in $G[V - s]$. If $\Delta_G(u, v, 3) = 0$ for some neighbors $u \in \Gamma_G(s) \cap Y$ and $v \in \Gamma_G(s) - Y$, then $\Gamma_G(s) - Y - v \neq \emptyset$ and $\Delta_G(u, v', 3) \geq 1$ for any neighbor $v' \in \Gamma_G(s) - Y - v$.* □

Based on this lemma, for a given cyclic order π of $\Gamma_G(s)$, we can find a noncrossing sequence of (k, s)-feasible splittings (which may not be complete) such that the resulting graph G^* satisfies either $c_{G^*}(s) = 0$ (i.e., the obtained splitting is complete) or the following condition:

$$\lambda_{G^*[V-s]}(V - s) = 2, \text{ and } c_{G^*}(s, Y) = 1 \text{ for all leaf 3-components } Y, \text{ and } c_{G^*}(s, Y') = 0 \text{ for all non-leaf 3-components } Y' \text{ in } G^*[V - s]. \tag{2}$$

The entire preprocessing is described as follows.

Algorithm PREPROCESS

Input: A multigraph $G = (V, E)$ (which is not necessarily planar), a designated vertex $s \in V$ with even degree and $\lambda_G(V - s) \geq 3$, and a cyclic order π of $\Gamma_G(s)$.
Output: A multigraph G^* obtained from G by a noncrossing $(3, s)$-feasible splitting such that G^* satisfies either $c_{G^*}(s) = 0$ or (2).

1    $G' := G$;
2   **while** $G'[V - s]$ is not connected **do**
3   Find a neighbor $w \in \Gamma_{G'}(s)$ and its successor $w' \in \Gamma_{G'}(s)$ such that w and w' belong to different 1-components in $G'[V - s]$;
4        $G' := G'/(w, w', 1)$
5   **end**; { while }
6   **while** $G'[V - s]$ is not 2-edge-connected (i.e., $\lambda_{G'[V-s]}(V - s) = 1$) **do**
7   Choose a leaf 2-component Y in $G'[V - s]$;
8   Find a neighbor $w \in \Gamma_{G'}(s) \cap Y$ and its successor $w' \in \Gamma_{G'}(s) - Y$;
9   **if** $\Delta_G(w, w', 3) \geq 1$ **then** $G' := G'/(w, w', 1)$
10   **else**
11   Find a neighbor $w'' \in \Gamma_{G'}(s) - Y$ and its successor $w''' \in \Gamma_{G'}(s) \cap Y$, and $G' := G'/(w'', w''', 1)$
12   **end** { if }

13   **end**; { while }
14   **while** $G'[V - s]$ has a 3-component Y with $c_{G'}(s, Y) \geq 2$ or a non-leaf
        3-component Y with $c_{G'}(s, Y) = 1$ **do**
15   Find neighbors $w \in \Gamma_{G'}(s) \cap Y$ and $w' \in \Gamma_{G'}(s) - Y$ such that w' is
          the successor of w;
16   **if** $\Delta_G(w, w', 3) \geq 1$ **then** $G' := G'/(w, w', 1)$
17   **else** { Y is a non-leaf 3-component with $c_{G'}(s, Y) = 1$ }
18   Find a neighbor $w'' \in \Gamma_{G'}(s) - Y$ such that w is the successor of w'',
              and $G' := G'/(w'', w, 1)$
19   **end** { if }
20   **end**; { while }
21   Output $G^* := G'$.

Correctness of PREPROCESS easily follows from Lemma 7. Clearly, the
number of splitting operations carried out in PREPROCESS is $O(n)$, and each
splitting (including testing if $\Delta_G(w, w', 3) \geq 1$ in lines 9 and 16) can be performed in linear time. Therefore, PREPROCESS runs in $O(n^2)$ time.

Lemma 8. *For a multigraph $G = (V, E)$, a designated vertex $s \in V$ with an even
degree and $\lambda_G(V - s) \geq 3$, and a cyclic order π of $\Gamma_G(s)$, there is a noncrossing
sequence of $(3, s)$-feasible splitting such that the resulting graph G^* satisfies either
$c_{G^*}(s) = 0$ or (2), and such a splitting can be found in $O(n^2)$ time.*   □

Now we describe how to obtain a complete $(3, s)$-feasible and planarity-
preserving splitting given (2).

Let $G^* = (V, E^*)$ be a multigraph satisfying (2), and let $G'' = (V'', E'')$
denote the graph $G^*[V - s]$. Then $\lambda_{G''}(V'') = 2$. A cut in a graph is a 2-cut
if its cut size is 2. Now we consider a representation of all 2-cuts in G''. Let
$\{Y_1, Y_2, \ldots, Y_r\}$ be the set of 3-components in G'', and $\mathcal{G}_{G''} = (\mathcal{V}, \mathcal{E})$ denote
the graph obtained from G'' by contracting each 3-component Y_i into a single
vertex y_i. For a vertex $v \in V$, $\varphi_{G''}(v)$ denotes the vertex $y_i \in \mathcal{V}$ into which
v (possibly together with other vertices) is contracted. We can easily observe
that $\mathcal{G}_{G''}$ is a *cactus*, i.e., a connected graph such that any edge is contained in
exactly one cycle, where a cycle may be of length two. Clearly a cactus is an
outerplanar graph, and any two cycles in a cactus have at most one common
vertex. A vertex with degree 2 in a cactus is called a *leaf vertex*. Since any non-
leaf vertex in a cactus is a cut-vertex, there may be many ways of embedding a
cactus in the plane. For a vertex $z \in \mathcal{V}$ in the cactus $\mathcal{G}_{G''}$, we denote by $\varphi_{G''}^{-1}(z)$
the 3-component Y_i such that Y_i is contracted into z. It is easy to see that a
subset $Y \subset V$ in G'' is a leaf 3-component if and only if there is a leaf vertex
$z \in \mathcal{V}$ with $\varphi_{G''}^{-1}(z) = Y$ in $\mathcal{G}_{G''}$. From this, we have the next lemma.

Lemma 9. *Let $G^* = (V, E^*)$ be a multigraph, and s be a designated vertex with
even degree. Assume that $\lambda_{G^*}(V - s) \geq 3$ and (2) holds. Let $L(\mathcal{G}_{G''})$ be the set
of leaf vertices in cactus $\mathcal{G}_{G''}$ of $G'' = G^*[V - s]$. Then $\varphi_{G''}$ defines a bijection
between $\Gamma_G(s)$ and $L(\mathcal{G}_{G''})$.*   □

Now assume that a given graph $G = (V, E)$ with $\lambda_G(V - s) \geq 3$ is planar. Fix a planar embedding ψ of G, and let π_ψ be the cyclic order of $\Gamma_G(s)$ in which neighbors in $\Gamma_G(s)$ appear along the outer face of $G[V - s]$. By applying PRE-PROCESS to G and π_ψ, a multigraph G^* is obtained from G by a noncrossing (hence planarity-preserving) $(3, s)$-feasible splitting satisfying either $c_{G^*}(s) = 0$ or (2). If $c_{G^*}(s) = 0$ (i.e., the splitting is complete) then we are done. Thus assume that G^* satisfies (2). It is not difficult to see that $G'' = G^*[V - s]$ and cactus $\mathcal{G}_{G''}$ have the following properties:

(i) For any 2-cut X in G'', $Z = \cup_{x \in X} \varphi_{G''}(x)$ is a 2-cut in $\mathcal{G}_{G''}$. (i.e., the two edges in $E_{G''}(X)$ exist in the same cycle in $\mathcal{G}_{G''}$).

(ii) For any 2-cut Z in $\mathcal{G}_{G''}$, $X = \cup_{z \in Z} \varphi_{G''}^{-1}(z)$ is a 2-cut in G''.

In other words, cactus $\mathcal{G}_{G''}$ represents all 2-cuts in $G'' = G^*[V - s]$. By Lemma 9, there is a bijection between $E_{G^*}(s)$ and $L(\mathcal{G}_{G''})$, and thus $\mathcal{G}_{G''}$ has an even number of leaf vertices. A set σ of new edges which pairwise connect all leaf vertices in a cactus is called a *leaf-matching*. Hence finding a complete (k, s)-feasible splitting in G^* is to obtain a leaf-matching σ in $\mathcal{G}_{G''}$ such that adding the leaf-matching destroys all 2-cuts in the cactus $\mathcal{G}_{G''}$.

However, to ensure that the complete splitting corresponding to a leaf-matching preserves the planarity of G^*, we need to choose a leaf-matching σ of $\mathcal{G}_{G''}$ carefully. An embedding χ of a cactus in the plane is called *standard* if all leaf vertices are located on the outer face of χ. In particular, for a cyclic order π of leaf vertices, a standard embedding χ of a cactus is called a standard π-*embedding* if the leaf vertices appear in the order of π along the outer face of χ. Note that a standard π-embedding of a cactus is unique (unless we distinguish one edge from the other in a cycle of length two). We define a flipping operation in an embedding χ of cactus $\mathcal{G} = (\mathcal{V}, \mathcal{E})$ as follows. Choose a cycle C in $\mathcal{G}$ and a vertex z in C. We see that removal of the two edges in C incident to z creates two connected components, say $\mathcal{G}'$ and $\mathcal{G}''$; we assume that $z \in V(\mathcal{G}')$. Let $\mathcal{G}[C, z]$ denote the subgraph $\mathcal{G}'$ of $\mathcal{G}$. We say that the embedding χ of $\mathcal{G}$ is *flipped* by (C, z) if we fold the subgraph $\mathcal{G}[C, z]$ back into the other side of area surrounded by C while fixing the other part of the embedding χ. An embedding obtained from a standard π-embedding of a cactus by a sequence of flipping operations is called a π'-*embedding*.

Recall that neighbors in $\Gamma_{G^*}(s)$ appear in cyclic order $\pi'_\psi = (w_1, \ldots, w_r)$ in an embedding χ_ψ of G^*. We also use π'_ψ to represent the cyclic order of leaf vertices $z_1 = \varphi(w_1), z_2 = \varphi(w_2), \ldots, z_r = \varphi(w_r)$ in cactus $\mathcal{G}_{G''}$. Then the standard π'_ψ-embedding χ_ψ of $\mathcal{G}_{G''}$ has the following property:

each vertex $z \in \mathcal{V}$ in $\mathcal{G}_{G''}$ can be replaced with the subgraph $G^*[\varphi^{-1}(z)]$ without creating crossing edges in χ.   (3)

Observe that a flipping operation preserves property (3), and hence any π-embedding χ of $\mathcal{G}_{G''}$ also satisfies the property (3).

A pair (σ, χ) of a leaf-matching σ on leaf vertices in a cactus $\mathcal{G}$ and a π-embedding χ of $\mathcal{G}$ is called a π-*configuration*. A π-configuration (σ, χ) of $\mathcal{G}$ is called *good* if it satisfies the following conditions:

(a) all 2-cuts in cactus $\mathcal{G} = (\mathcal{V}, \mathcal{E})$ are destroyed by adding σ (i.e., for any 2-cut X, σ contains an edge (z, z') with $z \in X$ and $z' \in V - X$), and

(b) all edges in σ can be drawn in π-embedding χ of $\mathcal{G}$ without creating crossing edges.

Now the problem of computing a complete and planarity-preserving $(3, s)$-feasible splitting in G^* is to find a good π'_ψ-configuration (σ, χ) of $\mathcal{G}_{G''}$. To show that such a good π'_ψ-configuration always exists in $\mathcal{G}_{G''}$, it suffices to prove the next lemma (the proof is omitted).

Lemma 10. *For a cactus $\mathcal{G} = (\mathcal{V}, \mathcal{E})$ and a cyclic order π of leaf vertices, assume that there is a standard π-embedding of $\mathcal{G}$. Then there always exists a good π-configuration (σ, χ) of $\mathcal{G}$, and such a configuration can be found in $O(|\mathcal{V}|^2)$ time.*   □

By this lemma, a complete and planarity-preserving $(3, s)$-feasible splitting in a graph G^* which satisfies (2) can be computed in $O(n^2)$ time. This and Lemma 8 establish the following theorem.

Theorem 4. *Given a planar multigraph $G = (V, E)$ with a designated vertex $s \in V$ of even degree, and $\lambda_G(V - s) \geq 3$, there exists a complete and planarity-preserving $(3, s)$-feasible splitting, and such a splitting can be found in $O(n^2)$ time.*   □

5   Augmenting Edge-Connectivity of Outerplanar Graphs

Given a multigraph $G = (V, E)$ and a positive integer k, the k-edge-connectivity (resp., k-vertex-connectivity) augmentation problem asks to find a minimum number of new edges to be added to G such that the augmented graph becomes k-edge-connected (resp., k-vertex-connected). Watanabe and Nakamura [11] proved that the k-edge-connectivity augmentation problem for general k is polynomially solvable. In such applications as graph drawing (see [3]), a planar graph G is given, and we may want to augment its edge- (or vertex-) connectivity optimally while preserving its planarity. Kant and Boldlaender [5] proved that the planarity-preserving version of 2-vertex-connectivity augmentation problem is NP-hard. Kant [4] also showed that, if a given graph G is outerplanar, then the planarity-preserving versions of both the 2-edge-connectivity and 2-vertex-connectivity can be solved in linear time. For a planar graph G, let $\gamma_k(G)$ (resp., $\tilde{\gamma}_k(G)$) denote the minimum number of new edges to be added to G so that the resulting graph G' becomes k-edge-connected (resp., so that the resulting graph G'' becomes k-edge-connected and remains planar). Clearly, $\gamma_k(G) \leq \tilde{\gamma}_k(G)$ for any planar graph G and $k \geq 1$. From the results in the preceding sections, we can show the next result.

Theorem 5. *Let $G = (V, E)$ be an outerplanar graph. If $k \geq 0$ is an even integer or $k = 3$, then $\gamma_k(G) = \tilde{\gamma}_k(G)$ and the planarity-preserving version of the k-edge-connectivity augmentation problem can be solved in $O(n^2(m + n \log n))$ time.*

Proof: (Sketch) Based on Theorems 3 and 4, we can apply the approach of Cai and Sun [1] (also see [2]) for solving the k-edge-connectivity augmentation problem by using the splitting algorithm. The detail is omitted.   □

Furthermore, for every odd integer $k \geq 5$, there is an outerplanar graph G such that $\gamma_k(G) < \tilde{\gamma}_k(G)$. Consider the graph G_3' obtained from the graph G_3 in Example 1(c) by deleting s and the edges in $E_{G_3}(s)$. It is easy to see that $\gamma_k(G_3') = 2 < 3 = \tilde{\gamma}_k(G_3')$.

Remark: Given an undirected outerplanar network $N = (V, L, c)$ and a real $k > 0$, we consider the k-edge-connectivity augmentation problem which asks how to augment N by increasing link weights and by adding new links so that the resulting network $N' = (V, L \cup L', c')$ is k-edge-connected and remains planar while minimizing $\sum_{e \in L}(c'(e) - c(e)) + \sum_{e \in L'} c'(e)$, where c and c' are allowed to be nonnegative reals. It is not difficult to observe that this problem can be solved in $O(n^2(m + n \log n))$ time by the argument given so far in this paper. (It would be interesting to see whether the problem can be formulated as a linear programming or not; if the planarity is not necessarily preserved then the problem is written as a linear programming.)

Acknowledgments

This research was partially supported by the Scientific Grant-in-Aid from Ministry of Education, Science, Sports and Culture of Japan, the grant from the Inamori Foundation and Kyoto University Foundation, and the Research Management Committee from The University of Newcastle.

References

1. G.-R. Cai and Y.-G. Sun. The minimum augmentation of any graph to k-edge-connected graph. *Networks*, 19:151–172, 1989.
2. A. Frank. Augmenting graphs to meet edge-connectivity requirements. *SIAM J. Disc. Math.*, 5:25–53, 1992.
3. G. Kant. *Algorithms for Drawing Planar Graphs.* PhD thesis, Dept. of Computer Science, Utrecht University, 1993.
4. G. Kant. Augmenting outerplanar graphs. *J. Algorithms*, 21:1–25, 1996.
5. G. Kant and H. L. Boldlaender. Planar graph augmentation problems. *LNCS, Vol. 621*, pages 258–271. Springer-Verlag, 1992.
6. L. Lovász. *Combinatorial Problems and Exercises.* North-Holland, 1979.
7. H. Nagamochi and T. Ibaraki. A linear time algorithm for computing 3-edge-connected components in multigraphs. *J. of Japan Society for Industrial and Applied Mathematics*, 9:163–180, 1992.
8. H. Nagamochi and T. Ibaraki. Computing edge-connectivity of multigraphs and capacitated graphs. *SIAM J. Disc. Math.*, 5:54–66, 1992.
9. H. Nagamochi and T. Ibaraki. Deterministic $\tilde{O}(nm)$ time edge-splitting in undirected graphs. *J. Combinatorial Optimization*, 1:5–46, 1997.
10. R. E. Tarjan. Depth-first search and linear graph algorithms. *SIAM J. Comput.*, 1:146–160, 1972.

11. T. Watanabe and A. Nakamura. Edge-connectivity augmentation problems. *J. Comp. System Sci.*, 35:96–144, 1987.

12. T. Watanabe, S. Taoka and K. Onaga. A linear-time algorithm for computing all 3-edge-components of a multigraph. *Trans. Inst. Electron. Inform. Comm. Eng. Jap.*, E75-A:410–424, 1992.

A New Bound for the 2-Edge Connected Subgraph Problem

Robert Carr[1]* and R. Ravi[2]**

[1] Sandia National Laboratories
Albuquerque, NM, USA
bobcarr@@cs.sandia.gov
[2] GSIA, Carnegie Mellon University
Pittsburgh, PA, USA
ravi@@cmu.edu

Abstract. Given a complete undirected graph with non-negative costs on the edges, the *2-Edge Connected Subgraph Problem* consists in finding the minimum cost spanning 2-edge connected subgraph (where multi-edges are allowed in the solution). A lower bound for the minimum cost 2-edge connected subgraph is obtained by solving the *linear programming relaxation* for this problem, which coincides with the *subtour relaxation* of the traveling salesman problem when the costs satisfy the triangle inequality.

The simplest fractional solutions to the subtour relaxation are the $\frac{1}{2}$-*integral solutions* in which every edge variable has a value which is a multiple of $\frac{1}{2}$. We show that the minimum cost of a 2-edge connected subgraph is at most four-thirds the cost of the minimum cost $\frac{1}{2}$-integral solution of the subtour relaxation. This supports the long-standing $\frac{4}{3}$ *Conjecture* for the TSP, which states that there is a *Hamilton cycle* which is within $\frac{4}{3}$ times the cost of the optimal subtour relaxation solution when the costs satisfy the triangle inequality.

1 Introduction

The 2-Edge Connected Subgraph Problem is a fundamental problem in Survivable Network Design. This problem arises in the design of communication networks that are resilient to single-link failures and is an important special case in the design of survivable networks [11,12,14].

1.1 Formulation

An integer programming formulation for the 2-Edge Connected Subgraph Problem is as follows. Let $K_n = (V, E)$ be the complete graph of feasible links on

* Supported by NSF grant DMS9509581 and DOE contract AC04-94AL85000.
** Research supported in part by an NSF CAREER grant CCR-9625297.

R. E. Bixby, E. A. Boyd, and R. Z. Ríos-Mercado (Eds.): IPCO VI
LNCS 1412, pp. 112–125, 1998. © Springer–Verlag Berlin Heidelberg 1998

which the 2-Edge Connected Subgraph Problem is formulated. We denote an edge of this graph whose endpoints are $i \in V$ and $j \in V$ by ij. For each vertex $v \in V$, let $\delta(v) \subset E$ denote the set of edges incident to v. For each subset of vertices $S \subset V$, let $\delta(S) \subset E$ denote the set of edges in the *cut* which has S as one of the *shores*, i.e. the set of edges having exactly one endpoint in S. Denote the edge variable for $e \in E$ by x_e, which is 0,1, or 2 depending on whether e is absent, occurs singly or doubly in the 2-edge connected subgraph. For $A \subset E$, let $x(A)$ denote the sum $\sum_{e \in A} x_e$. Let c_e denote the cost of edge e. We have the following integer programming formulation.

$$\text{minimize} \quad c \cdot x$$
$$\text{subject to}$$
$$\begin{aligned} x(\delta(v)) &\geq 2 & \text{for all } v \in V, \\ x(\delta(S)) &\geq 2 & \text{for all } S \subset V, \\ x_e &\geq 0 & \text{for all } e \in E, \\ x_e & \text{ integral.} \end{aligned} \tag{1}$$

The *LP relaxation* is obtained by dropping the integrality constraint in this formulation. This LP relaxation is almost the same as the *subtour relaxation* for the Traveling Salesman Problem (TSP). The Traveling Salesman Problem consists in finding the minimum cost *Hamilton cycle* in a graph (a Hamilton cycle is a cycle which goes through all the vertices). The subtour relaxation for the TSP is as follows.

$$\text{minimize} \quad c \cdot x$$
$$\text{subject to}$$
$$\begin{aligned} x(\delta(v)) &= 2 & \text{for all } v \in V, \\ x(\delta(S)) &\geq 2 & \text{for all } S \subset V, \\ x_e &\geq 0 & \text{for all } e \in E. \end{aligned} \tag{2}$$

The constraints of the subtour relaxation are called the *degree constraints*, the *subtour elimination constraints*, and the *non-negativity constraints* respectively.

If one has the relationship $c_{ij} \leq c_{ik} + c_{jk}$ for all distinct $i, j, k \in V$, then c is said to satisfy the *triangle inequality*. An interesting known result is that if the costs satisfy the triangle inequality, then there is an optimal solution to (1) which is also feasible and hence optimal for (2). This follows from a result of Cunningham [11] (A more general result called the *Parsimonious Property* is shown by Goemans and Bertsimas in [7]). We can show that this equivalence holds even when the costs do not satisfy the triangle inequality. In the latter case, we replace the given graph by its *metric completion*, namely, for every edge ij such that c_{ij} is greater than the cost of the shortest path between i and j in the given graph, we reset the cost to that of this shortest path. The intent is that if this edge is chosen in the solution of (1), we may replace it by the shortest cost path connecting i and j. Since multiedges are allowed in the 2-edge connected graph this transformation is valid. Hence without loss of generality, we can assume that the costs satisfy the triangle inequality.

1.2   Our Result and its Significance

Our main result is the following.

Theorem 1. *The minimum cost of a 2-edge connected subgraph is within $\frac{4}{3}$ times the cost of the optimal half-integral subtour solution for the TSP.*

This result is a first step towards proving the following conjecture we offer.

Conjecture 1. The minimum cost of a 2-edge connected subgraph is within $\frac{4}{3}$ times the cost of the optimal subtour solution for the TSP.

By our remarks in the end of Section 1.1, it would follow from Conjecture 1 that the minimum cost of a 2-edge connected subgraph is also within $\frac{4}{3}$ times the cost of an optimal solution to the linear programming relaxation (1).

We formulated Conjecture 1 as an intermediate step in proving the following stronger "four-thirds conjecture" on the subtour relaxation for the TSP, which would directly imply Conjecture 1.

Conjecture 2. If the costs satisfy the triangle inequality, then the minimum cost of a Hamilton cycle is within $\frac{4}{3}$ times the cost of the optimal subtour solution for the TSP.

Note that Theorem 1 and Conjecture 1 imply similar relations between the fractional optimum of the subtour relaxation and a minimum-cost 2-vertex connected subgraph when the costs obey the triangle inequality. In particular, Theorem 1 implies that when the costs satisfy the triangle inequality, the minimum cost 2-vertex connected spanning subgraph is within $\frac{4}{3}$ times the cost of the optimal half-integral subtour solution for the TSP. This follows from the simple observation that from the minimum-cost 2-edge connected graph, we can short-cut "over" any cut vertices without increasing the cost by using the triangle inequality [5,11].

1.3   Related Work

A heuristic for finding a low cost Hamilton cycle was developed by Christofides in 1976 [4]. An analysis of this heuristic shows that the ratio is no worse than $\frac{3}{2}$ in both Conjecture 1 and Conjecture 2. This analysis was done by Wolsey in [16] and by Shmoys and Williamson in [15]. A modification of the Christofides heuristic to find a low cost 2-vertex connected subgraph when the costs obey the triangle inequality was done by Fredrickson and Ja Ja in [5]. The performance guarantee for this heuristic to find a 2-vertex connected subgraph is $\frac{3}{2}$. There has also been a spate of work on approximation algorithms for survivable network design problems generalizing the 2-edge connected subgraph problem [7,8,9,10,13,17]; however, the performance guarantee for the 2-edge connected subgraph problem from these methods is at best $\frac{3}{2}$ when the costs obey the triangle inequality (shown in [5,7]) and at best 2 when they do not (shown in [9]).

Both Conjecture 2 and Conjecture 1 have remained open since Christofides developed his heuristic. In this paper, we suggest a line of attack for proving Conjecture 1.

2   Motivation

In this section we discuss two distinct motivations that led us to focus on half-integral extreme points and prove a version of Conjecture 1 for this special case. One follows from a particular strategy to prove Conjecture 1 and the other from examining subclasses of subtour extreme points that are sufficient to prove Conjectures 1 and 2.

2.1   A Strategy for Proving Conjecture 1

Let an arbitrary point x^* of the subtour polytope for K_n be given. Multiply this by $\frac{4}{3}$ to obtain the vector $\frac{4}{3}x^*$. Denote the edge incidence vector for a given 2-edge connected subgraph H in K_n by χ^H. Note that edge variables could be 0,1, or 2 in this incidence vector. Suppose we could express $\frac{4}{3}x^*$ as a convex combination of incidence vectors of 2-edge connected subgraphs H_i for $i = 1, 2, \ldots, k$. That is, suppose that

$$\frac{4}{3}x^* = \sum_{i=1}^{k} \lambda_i \chi^{H_i}, \tag{3}$$

where $\lambda_i \geq 0$ for $i = 1, 2, \ldots, k$ and

$$\sum_{i=1}^{k} \lambda_i = 1.$$

Then, taking dot products on both sides of (3) with the cost vector c yields

$$\frac{4}{3}c \cdot x^* = \sum_{i=1}^{k} \lambda_i c \cdot \chi^{H_i}. \tag{4}$$

Since the right hand side of (4) is a weighted average of the numbers $c \cdot \chi^{H_i}$, it follows that there exists a $j \in \{1, 2, \ldots, k\}$ such that

$$c \cdot \chi^{H_j} \leq \frac{4}{3}c \cdot x^*. \tag{5}$$

If we could establish (5) for any subtour point x^*, then it would in particular be valid for the optimal subtour point, which would prove Conjecture 1.

In an attempt at proving Conjecture 1, we aim at contradicting the idea of a minimal counterexample, that is, a subtour point x^* having the fewest number of vertices n' such that (3) can not hold for any set of 2-edge connected subgraphs. First we have the following observation.

Theorem 2. *At least one of the minimal counterexamples x^* to (3) holding (for some set of 2-edge connected subgraphs) is an extreme point of the subtour polytope.*

Proof. Suppose $x^* = \sum_{l} \mu_l x^l$, where each x^l is an extreme point which is not a minimal counterexample, and the μ_l's satisfy the usual constraints for a set of convex multipliers. Thus, for each l, we can find a set of 2-edge connected subgraphs H_i^l such that

$$\frac{4}{3}x^l = \sum_i \lambda_i^l \chi^{H_i^l},$$

where the λ_i^l's satisfy the usual constraints for a set of convex multipliers. Then

$$\frac{4}{3}x^* = \sum_l \frac{4}{3}\mu_l x^l = \sum_l \mu_l \sum_i \lambda_i^l \chi^{H_i^l}. \tag{6}$$

Since we have that

$$\sum_l \mu_l \cdot \left(\sum_i \lambda_i^l\right) = \sum_l \mu_l \cdot (1) = 1,$$

Equation (6) shows that $\frac{4}{3}x^*$ can be expressed as a convex combination of 2-edge connected subgraphs as well, from which this theorem follows.

Thus we need to focus only on minimal counterexamples x^* in $K_{n'}$ which are extreme points. To carry out the proof, we wish to find a *substantial tight cut* $\delta(H)$ for x^*, i.e. an $H \subset V$ such that $3 \le |H| \le n' - 3$ and

$$x^*(\delta(H)) = 2.$$

We can then split x^* into 2 smaller subtour solutions x^1 and x^2 in the following way. Take the vertices of $V \setminus H$ in x^* and contract them to a single vertex to obtain x^1. Likewise, take the vertices of H in x^* and contract them to a single vertex to obtain x^2. An example of this is shown in Figure 1.

Since x^1 and x^2 are not counterexamples to our conjecture, we would be able to decompose $\frac{4}{3}x^1$ and $\frac{4}{3}x^2$ into combinations of 2-edge connected subgraphs, which we may then attempt to glue together to form a similar combination for $\frac{4}{3}x^*$, thereby showing that x^* is not a counterexample (We show how this can be accomplished for the case of half-integral extreme points in Case 1 in the Proof of Theorem 6).

What if there are no tight substantial cuts however? The following proposition which is shown in [1] shows us what we need to do.

Proposition 1. *If x^* is an extreme point of the subtour polytope and has no substantial tight cuts, then x^* is a $1/2$-integer solution.*

This led us to focus on $1/2$-integral solutions x^*, and we were able to complete the proof for this special case. In the next section, we show our main result that if x^* is a $1/2$-integer subtour solution, then (3) can always be satisfied.

2.2   The Important Extreme Points

Consider any extreme point x^*. We wish to express $\frac{4}{3}x^*$ as a convex combination of 2-edge connected graphs for Conjecture 1 or a convex combination of Eulerian graphs for Conjecture 2. An important question is what features of x^* make it

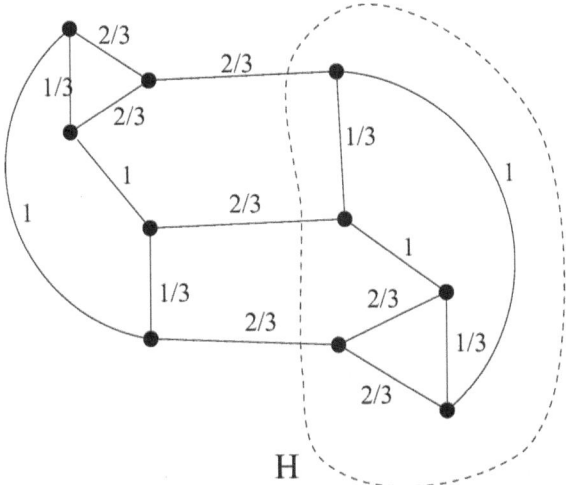

Fig. 1. An idea for splitting a minimal counterexample into two smaller instances. Note that H defines a substantial tight cut, i.e., both H and $V \setminus H$ have at least three vertices and $x(\delta(H)) = 2$.

difficult to do this? In an effort to answer this question, we try to transform x^* into another extreme point $\overline{x}^*$ on a larger graph so that $\overline{x}^*$ belongs to a subclass of the extreme points, but $\frac{4}{3}\overline{x}^*$ is at least as hard to express as a convex combination of 2-edge connected graphs (or Eulerian graphs) as $\frac{4}{3}x^*$ is. The idea then is that we only have to be able to express $\frac{4}{3}\overline{x}$ as a convex combination of 2-edge connected graphs (or Eulerian graphs) for all extreme points $\overline{x}$ belonging to this particular subclass in order to prove Conjecture 1 (or Conjecture 2). If we have a subclass S of extreme points $\overline{x}$ such that being able to express $\frac{4}{3}\overline{x}$ as a convex combination of 2-edge connected graphs for all extreme points $\overline{x}$ belonging to this particular subclass is sufficient to prove Conjecture 1, then we say that S is *sufficient* to prove Conjecture 1. Likewise, a subclass S can be sufficient to prove Conjecture 2.

We have found two different subclasses of extreme points which are sufficient to prove both Conjecture 1 and Conjecture 2. In some sense, the extreme points in such a subclass are the hardest extreme points to deal with when proving Conjecture 1 or Conjecture 2. One class, termed *fundamental extreme points*, can be found in [2].

Definition 1. *A fundamental extreme point is an extreme point for the subtour relaxation satisfying the following conditions.*

   (i) The support graph is 3-regular,
   (ii) There is a 1-edge incident to each vertex,
   (iii) The fractional edges form disjoint cycles of length 4.

A second class of such sufficient extreme points is described below. We will restrict our attention to showing that the subclass described below is sufficient to prove Conjecture 1, although showing that it is also sufficient to prove Conjecture 2 requires only minor modifications in our arguments.

Consider any extreme point x^*. Pick the smallest integer k such that x_e^* is a multiple of $\frac{1}{k}$ for every edge $e \in E$. Then form a $2k$-regular $2k$-edge connected multigraph $G_k = (V, E_k)$ as follows. For every edge $e = uv \in E$, put l edges between u and v, where $l := kx_e^*$. Then showing that $\frac{4}{3}E_k$ can be expressed as a convex combination of 2-edge connected graphs is equivalent to showing that $\frac{4}{3}x^*$ can be so expressed. But suppose every vertex in G_k is replaced by a circle of $2k$ nodes, each node with one edge from E_k, and $2k-1$ new edges linking this node to its two neighboring nodes in the circle, all in such a way that the resulting graph $\overline{G}_k = (\overline{V}, \overline{E}_k)$ is still $2k$-regular and $2k$-edge connected. Note that loosely speaking, we have $E_k \subset \overline{E}_k$. We seek to then show that if we can express $\frac{4}{3}\overline{E}_k$ as a convex combination of 2-edge connected graphs, then we can do so for $\frac{4}{3}E_k$ as well. The graph $\overline{G}_k$ will turn out to corresponds to a subtour extreme point $\overline{x}^*$ (in the same way that G_k corresponds to x^*). It is more convenient to define this subtour extreme point $\overline{x}^*$ than to define $\overline{G}_k$.

Let us now define $\overline{x}^*$.

Definition 2. *Expand each vertex in V into a circle of $2k$ nodes, with an edge of E_k leaving each such node, as described in the previous paragraph. Take the equivalent of an Eulerian tour through all the edges of E_k by alternately traversing these edges and jumping from one node to another node in the same circle until you have traversed all of the edges in E_k and have come back to the edge in E_k you started with. When you jump from node u to node v in the same circle in this Eulerian tour, define $\overline{x}_{uv}^* := \frac{k-1}{k}$. For every edge $e \in E_k$, we naturally define $\overline{x}_e^* := \frac{1}{k}$. For each circle C_v of nodes corresponding to the vertex $v \in V$, we pick an arbitrary perfect matching M_v on the nodes in C_v, including in M_v only edges e which have not yet been used in the definition of $\overline{x}^*$. We then define $\overline{x}_e^* := 1$ for all $e \in M_v$.*

We have the following:

Theorem 3. $\overline{x}^*$ *in Definition 2 is a subtour extreme point.*

Proof. The support graph of $\overline{x}^*$ is 3-regular, with the fractional edges in $\overline{x}^*$ forming a Hamilton cycle on the vertices $\overline{V}$. Call the edges in $\overline{x}^*$'s support graph $\overline{E}_k$.

We first show that $\overline{x}^*$ is a feasible subtour point. If it were not, there would have to be a cut in the graph $\overline{G}_k = (\overline{V}, \overline{E}_k)$ of value less than 2. Clearly, such a cut $\mathcal{C}$ would have to go through some circle C_v of nodes since G_k is $2k$-edge connected. But the contribution of the edges from the circle C_v to any cut crossing it is at least 1 since the edges in the circle C_v each have a value greater than or equal to $1/2$. Hence, the contribution from the non-circle edges in the cut $\mathcal{C}$ is less than 1. But this is not possible because when v is ripped out of x^*, the minimum cut in the remaining solution is greater than or equal to 1. Hence, $\overline{x}^*$ is a feasible subtour point.

We show that $\bar{x}^*$ is an extreme point by showing that it can not be expressed as $\frac{1}{2}x^1 + \frac{1}{2}x^2$, where x^1 and x^2 are distinct subtour points. Suppose $\bar{x}^*$ could be so expressed. Then the support graphs of x^1 and x^2 would coincide with or be subgraphs of the support graph $\bar{E}_k$ of $\bar{x}^*$. Because of the structure of the support graph, setting the value of just one fractional edge determines the entire solution due to the degree constraints. Hence, all the edges $e \in \bar{E}_k$ such that $x_e = \frac{1}{k}$ would have to say be smaller than $\frac{1}{k}$ in x^1 and larger than $\frac{1}{k}$ in x^2. But, then a cut separating any circle of nodes C_v from the rest of the vertices in x^1 would have a value less than 2, which contradicts x^1 being a subtour point.

We now have the following:

Theorem 4. *If $\frac{4}{3}\bar{x}^*$ can be expressed as a convex combination of 2-edge connected graphs spanning $\bar{V}$, then $\frac{4}{3}x^*$ can be expressed as a convex combination of 2-edge connected graphs spanning V.*

Proof. Suppose $\frac{4}{3}\bar{x}^*$ can be expressed as a convex combination

$$\frac{4}{3}\bar{x}^* = \sum_i \lambda_i \chi^{\bar{H}_i}, \tag{7}$$

where the $\bar{H}_i$'s are 2-edge connected graphs spanning $\bar{V}$. For each i, contract each circle of nodes C_v back to the vertex $v \in V$ in $\bar{H}_i$. Call the resulting graph H_i. Since contraction preserves edge connectivity, H_i is a 2-edge connected graph spanning V. When one performs this contraction on $\bar{x}^*$, one gets x^*. As a result, we obtain that

$$\frac{4}{3}x^* = \sum_i \lambda_i \chi^{H_i}, \tag{8}$$

which proves our theorem.

We can now define the subclass of *important extreme points*.

Definition 3. *An important extreme point is an extreme point for the subtour relaxation satisfying the following conditions.*

   (i) The support graph is 3-regular,
   (ii) There is a 1-edge incident to each vertex,
   (iii) The fractional edges form a Hamilton cycle.

We are now ready for the culminating theorem of this section.

Theorem 5. *The subclass of important extreme points is sufficient to prove Conjecture 1.*

Proof. If there is an extreme point x^* such that $\frac{4}{3}x^*$ cannot be expressed as a convex combination of 2-edge connected graphs, then by Theorem 4, the important extreme point $\frac{4}{3}\bar{x}^*$ cannot be expressed as a convex combination of 2-edge connected graphs either. Hence, our theorem follows.

The analogous theorem for the class of fundamental extreme points can be found in [2].

3   The Proof of Theorem 1

Let x^* be a 1/2-integer subtour solution on $K_n = (V, E)$. Denote the edges of the *support graph* of x^* (the set of edges $e \in E$ such that $x_e^* > 0$) by $\hat{E}(x^*)$. Construct the multigraph $G(x^*) = (V, E(x^*))$, where $E(x^*) \supset \hat{E}(x^*)$ and differs from $\hat{E}(x^*)$ only in that there are two copies in $E(x^*)$ of every edge $e \in \hat{E}(x^*)$ for which $x_e^* = 1$. Note that the parsimonious property [7] implies that there are no edges e with $x_e > 1$ in the optimal fractional solution.

Because of the constraints of the subtour relaxation, it follows that $G(x^*)$ is a 4-regular 4-edge connected multigraph. Similarly, corresponding to every 4-regular 4-edge connected multigraph is a 1/2-integer subtour solution, although this solution may not be an extreme point.

Showing (3) for some choice of 2-edge connected subgraphs H_i for every 1/2-integer subtour solution x^* would prove Conjecture 1 whenever the optimal subtour solution was 1/2-integer, as was discussed in the last section. So, equivalently to showing (3) for some choice of 2-edge connected subgraphs H_i for every 1/2-integer subtour solution x^*, we could show

$$\tfrac{2}{3}\chi^{E(G)} = \sum_i \lambda_i \chi^{H_i}, \tag{9}$$

where this expression is a convex combination of some chosen set of 2-edge connected subgraphs H_i, for every 4-regular 4-edge connected multigraph $G = (V, E(G))$. These are equivalent because of the remarks in the previous paragraph and the observation that $G(x^*)$ behaves like $2x^*$.

It turns out that (9) is very difficult to show directly, but the following slight strengthening of it makes the task easier. Consider any 4-regular 4-edge connected multigraph $G = (V, E(G))$ and any edge $e \in E(G)$. Then, we prove instead that

$$\tfrac{2}{3}\chi^{E(G)\backslash\{e\}} = \sum_i \lambda_i \chi^{H_i} \tag{10}$$

where this expression is a convex combination of some chosen set of 2-edge connected subgraphs H_i.

For technical reasons, we will prove (10) with the additional restriction that none of the H_i's may use more than one copy of any edge in $E(G)$. Note however that G may itself have multiedges so H may also have multiedges. In the latter case, we think of two parallel multiedges in H as being copies of two distinct multiedges in G.

For any 4-regular 4-edge connected graph G and any edge $e \in E(G)$, we define $P(G, e)$ to be the following statement.

Statement 1. $P(G, e) \Leftrightarrow$ For some finite set of 2-edge connected subgraphs H_i, we have (10), where $\lambda_i \geq 0$ for all i and $\sum_i \lambda_i = 1$, and none of the H_i's may use more than one copy of any edge in $E(G)$.

As noted above, Statement 1 does not rule out the possibility of doubled edges in the H_i's because there may be doubled edges in G.

We define a *tight cut* for a 4-edge connected graph G to be a cut which has exactly 4-edges in it. We define a *non-trivial cut* for such a graph to be a cut where both shores have at least 2 vertices each. We have the following lemma.

Lemma 1. *Let $G = (V, E)$ be a 4-regular 4-edge connected graph which has no tight non-trivial cut which includes an edge $e = uv \in E$. Let the other 3 (not necessarily distinct) neighbors of v be x, y, and z. Then either ux or yz is a loop or $G' = G - v + ux + yz$ is 4-regular and 4-edge connected, and likewise for the other combinations.*

Proof. Let $G = (V, E)$ and $e = uv \in E$ be given, where the neighbors of v are as stated. First, note that any cut in G containing all four edges incident on v has size at least 8, since the cut formed by moving v to the opposite side of the cut must have size at least 4 since G is 4-edge connected.

Suppose neither ux or yz is a loop. Then clearly, G' is a 4-regular connected graph. Since it is 4-regular, every cut has an even number of edges in it. By our earlier observation, there can be no cuts $\delta(H)$ in G' of cardinality zero. Suppose G' has a non-trivial cut $\delta(H)$ with only 2 edges in it. Consider $\hat{G} = G + ux + yz$ with vertex v back in. The two non-trivial cuts $\delta(H \cup \{v\})$ and $\delta((V \setminus H) \cup \{v\})$ can each have at most 3 more edges each (for a total of 5 edges each) since as observed earlier, these cuts could not have all 4 edges incident to v in them. But, $G = \hat{G} - ux - yz$ has only cuts with an even number of edges in them since it is 4-regular. Hence the cuts $\delta(H \cup \{v\})$ and $\delta((V \setminus H) \cup \{v\})$ in G have at most 4 edges in them. One of these two cuts is a tight non-trivial cut which contains e, which yields the lemma.

We are now ready for our main theorem.

Theorem 6. *Let x^* be a minimum cost 1/2-integer subtour solution. Then there exists a 2-edge connected subgraph H such that $c \cdot \chi^H \le \frac{4}{3} c \cdot x^*$.*

Proof. As remarked in the discussion before this theorem, it is sufficient to prove $P(G, e)$ for all 4-regular 4-edge connected multigraphs G and for all $e \in E(G)$. To prove this, we show that a minimal counterexample to $P(G, e)$ can not happen.

Let $G = (V, E(G))$ be a 4-regular 4-edge connected multigraph and $e \in E(G)$ which has the minimum number of vertices such that $P(G, e)$ does not hold. Since by inspection, we can verify that $P(G, e)$ holds when G has 3 vertices, we can assume that $|V| > 3$. We now consider the cases where G has a tight non-trivial cut which includes edge e and where G has no tight non-trivial cut which includes e.

Case 1: G has a tight non-trivial cut which includes edge e.

Choose such a tight non-trivial cut and denote the edges other than e in this cut by a, b, and c. As before, consider contracting one of the shores of this cut to a single vertex v_1. Denote the edges incident to v_1, which corresponded to e, a, b, and c, by e_1, a_1, b_1, and c_1 respectively. This resulting graph $G_1 = (V_1, E_1)$ can be seen to be 4-regular and 4-edge connected. (To see this, suppose there was a cut of cardinality less than four in G_1 and let H_1 be the shore of this cut not containing v_1. Then the cut $\delta(H_1)$ in G shows that G is not 4-edge-connected, a contradiction.) Since (G, e) was a minimal counterexample to $P(G, e)$, we have $P(G_1, e_1)$. By contracting the other shore, we can get a 4-regular 4-edge connected graph G_2, and we know that $P(G_2, e_2)$ also holds.

By $P(G_1, e_1)$ we have

$$\tfrac{2}{3}\chi^{E(G_1)\setminus\{e_1\}} = \sum_i \lambda_i \chi^{H_i^1}, \tag{11}$$

and by $P(G_2, e_2)$ we have

$$\tfrac{2}{3}\chi^{E(G_2)\setminus\{e_2\}} = \sum_i \mu_i \chi^{H_i^2}. \tag{12}$$

In (11), consider the edges incident to v_1 in each of the H_i^1's. There are clearly at least 2 such edges for every H_i^1. The values of edges a_1, b_1, c_1, and e_1 in $\tfrac{2}{3}\chi^{E(G_1)\setminus\{e_1\}}$ are $\tfrac{2}{3}, \tfrac{2}{3}, \tfrac{2}{3}$, and 0 respectively. This adds up to 2. Hence, since we are dealing with convex combinations, which are weighted averages, when the weights are taken into account, the H_i^1's have on average 2 edges incident to v_1 each. But since every H_i^1 has at least 2 such edges, it follows that every H_i^1 has exactly 2 edges incident to v_1 in it.

For each 2-edge connected subgraph H_i^1 which has edges a_1 and b_1, denote the corresponding convex multiplier by λ_i^{ab}. Define λ_i^{ac} and λ_i^{bc} similarly. One can see that the only way for the variable values of edges a_1, b_1, and c_1 to end up all being $\tfrac{2}{3}$ in $\tfrac{2}{3}\chi^{E(G_1)\setminus\{e_1\}}$ is for the following to hold:

$$\sum_i \lambda_i^{ab} = \sum_i \lambda_i^{ac} = \sum_i \lambda_i^{bc} = \tfrac{1}{3}. \tag{13}$$

Similarly, we must have

$$\sum_i \mu_i^{ab} = \sum_i \mu_i^{ac} = \sum_i \mu_i^{bc} = \tfrac{1}{3}. \tag{14}$$

Call the three types of 2-edge connected graphs H_i^j as ab-graphs, ac-graphs, and bc-graphs. Our strategy is to combine say each ab-graph H_i^1 of G_1 with an ab-graph H_j^2 of G_2 to form an ab-graph H_{ij}^{ab} of G which is also 2-edge connected. So, we define

$$H_{ij}^{ab} := (H_i^1 - v_1) + (H_j^2 - v_2) + a + b, \tag{15}$$

where H_i^1 and H_j^2 are ab-graphs. Since $H_i^1 - v_1$ and $H_j^2 - v_2$ are both connected, it follows that H_{ij}^{ab} is 2-edge connected. Similarly define H_{ij}^{ac} and H_{ij}^{bc}.

Now consider the following expression:

$$\sum_{i,j} 3\lambda_i^{ab}\mu_j^{ab}H_{ij}^{ab} + \sum_{i,j} 3\lambda_i^{ac}\mu_j^{ac}H_{ij}^{ac} + \sum_{i,j} 3\lambda_i^{bc}\mu_j^{bc}H_{ij}^{bc}. \tag{16}$$

One can verify that this is in fact a convex combination. Any edge f in say $G_1 - v_1$ occurs in (16) with a weight of

$$\sum_{\{i \,|\, f \in H_i^1\}} (\lambda_i^{ab} \cdot (3 \cdot \sum_j \mu_j^{ab}) + \lambda_i^{ac} \cdot (3 \cdot \sum_j \mu_j^{ac}) + \lambda_i^{bc} \cdot (3 \cdot \sum_j \mu_j^{bc})). \tag{17}$$

In light of (14) we have that (17) evaluates to

$$\sum_{\{i \mid f \in H_i^1\}} \lambda_i = \frac{2}{3}. \tag{18}$$

We have a similar identity when f is in $G_2 - v_2$ and we also have that edges $a, b,$ and c each occur in (16) with a weight of $\frac{2}{3}$ as well. Therefore we have

$$\sum_{i,j} 3\lambda_i^{ab} \mu_j^{ab} H_{ij}^{ab} + \sum_{i,j} 3\lambda_i^{ac} \mu_j^{ac} H_{ij}^{ac} + \sum_{i,j} 3\lambda_i^{bc} \mu_j^{bc} H_{ij}^{bc} = \frac{2}{3} \chi^{E(G)\setminus\{e\}}, \tag{19}$$

which contradicts (G, e) being a minimal counterexample.

Case 2: G has no tight non-trivial cut which includes edge e.

Denote the endpoints of e by $u \in V$ and $v \in V$, and denote the other 3 not necessarily distinct neighbors of v in G by $x, y, z \in V$. Because e is in no tight non-trivial cut, we have that $x \neq y \neq z$. (If any two of the neighbors x, y and z are the same, say $x = y$, then the cut $\delta(\{v, x\})$ will be a tight non-trivial cut). Thus, without loss of generality, if any two neighbors are the same vertex, we can assume that they are u and z. Hence, $u \neq x$ and $u \neq y$.

Define the graph $G_1 = (V_1, E_1)$ by

$$G_1 = G - v + ux + yz, \tag{20}$$

and define $e_1 = ux$. We know by Lemma 1 that G_1 is 4-regular and 4-edge connected. Since (G, e) is a minimal counterexample, we therefore know that $P(G_1, e_1)$ holds. Similarly, define the graph $G_2 = (V_2, E_2)$ by

$$G_2 = G - v + uy + xz, \tag{21}$$

and define $e_2 = uy$. As before, we know that $P(G_2, e_2)$ holds as well.

So, we can form the following convex combinations of 2-edge connected graphs:

$$\frac{2}{3} \chi^{E_1 \setminus \{e_1\}} = \sum_i \lambda_i \chi^{H_i^1}, \tag{22}$$

and

$$\frac{2}{3} \chi^{E_2 \setminus \{e_2\}} = \sum_i \mu_i \chi^{H_i^2}. \tag{23}$$

Define $\hat{H}_i^1$ by

$$\hat{H}_i^1 = \begin{cases} H_i^1 - yz + yv + zv & \text{for } yz \in H_i^1, \\ H_i^1 + yv + xv & \text{for } yz \notin H_i^1. \end{cases} \tag{24}$$

Likewise, define $\hat{H}_i^2$ by

$$\hat{H}_i^2 = \begin{cases} H_i^2 - xz + xv + zv & \text{for } xz \in H_i^2, \\ H_i^2 + yv + xv & \text{for } xz \notin H_i^2. \end{cases} \tag{25}$$

Consider the convex combination of 2-edge connected subgraphs

$$\frac{1}{2}\sum_i \lambda_i \chi^{\hat{H}_i^1} + \frac{1}{2}\sum_i \mu_i \chi^{\hat{H}_i^2}. \tag{26}$$

Every edge in $f \in E \setminus \delta(v)$ occurs with a total weight of $\frac{2}{3}$ in (26) since f occured with that weight in both (22) and (23). Since yz occurs with a total weight of $\frac{2}{3}$ in (22) and xz occurs with a total weight of $\frac{2}{3}$ in (23), one can verify that xv, yv, and zv each occur with a total weight of $\frac{2}{3}$ in (26) as well. Therefore, we have

$$\frac{2}{3}\chi^{E \setminus \{e\}} = \frac{1}{2}\sum_i \lambda_i \chi^{\hat{H}_i^1} + \frac{1}{2}\sum_i \mu_i \chi^{\hat{H}_i^2}, \tag{27}$$

which contradicts G, e being a minimal counterexample.

4   Concluding Remarks

An obvious open problem arising from our work is to extend our strategy and settle Conjecture 1. In another direction, it would be interesting to apply our ideas to design a $\frac{4}{3}$-approximation algorithm for the minimum cost 2-edge- and 2-vertex-connected subgraph problems.

Another interesting question is the tightness of the bound proven in Theorem 1. The examples we have been able to construct seem to demonstrate an asymptotic ratio of $\frac{6}{5}$ between the cost of a minimum cost 2-edge connected subgraph and that of an optimal half-integral subtour solution. Finding instances with a worse ratio or improving our bound in Theorem 1 are open problems.

References

1. M. Balinski. On recent developments in integer programming. In H. W. Kuhn, editor, *Proceedings of the Princeton Symposium on Mathematical Programming*, pages 267–302. Princeton University Press, NJ, 1970.
2. S. Boyd and R. Carr. Finding low cost TSP and 2-matching solutions using certain half-integer subtour vertices. Manuscript, March 1998.
3. S. Boyd and R. Carr. A new bound for the 2-matching problem. Report TR-96-07, Department of Computer Science, University of Ottawa, Ottawa, 1996.
4. N. Christofides. Worst case analysis of a new heuristic for the traveling salesman problem. Report 388, Graduate School of Industrial Administration, Carnegie Mellon University, Pittsburgh, 1976.
5. G. N. Fredrickson and J. Ja Ja. On the relationship between the biconnectivity augmentation and traveling salesman problems. *Theoretical Computer Science*, 19:189–201, 1982.
6. M. X. Goemans. Worst-case comparison of valid inequalities for the TSP. *Math. Programming*, 69:335–349, 1995.
7. M. X. Goemans and D. J. Bertsimas. Survivable networks, linear programming relaxations and the parsimonious property. *Math. Programming*, 60:145–166, 1993.
8. M. X. Goemans, A. Goldberg, S. Plotkin, D. Shmoys, É. Tardos, and D. P. Willamson. Approximation algorithms for network design problems. *Proceedings of the Fifth Annual ACM-SIAM Symposium on Discrete Algorithms (SODA '94)*, pages 223–232, 1994.

9. S. Khuller and U. Vishkin. Biconnectivity approximations and graph carvings. *J. Assoc. Comput. Mach.*, 41(2):214–235, 1994.

10. P. Klein and R. Ravi. When cycles collapse: A general approximation technique for constrained two-connectivity problems. *Proceedings of the Conference on Integer Programming and Combinatorial Optimization (IPCO '93)*, pages 39–56, 1993.

11. C. L. Monma, B. S. Munson, and W. R. Pulleyblank. Minimum-weight two-connected spanning networks. *Math. Programming*, 46:153–171, 1990.

12. C. L. Monma and D. F. Shallcross. Methods for designing communication networks with certain two-connectivity survivability constraints. *Oper. Res.*, 37:531–541, 1989.

13. H. Saran, V. Vazirani, and N. Young. A primal-dual approach to approximation algorithms for network Steiner problems. *Proc. of the Indo-US workshop on Cooperative research in Computer Science*, pages 166–168. Bangalore, India, 1992.

14. K. Steiglitz, P. Weiner, and D. J. Kleitman. The design of minimum-cost survivable networks. *IEEE Trans. on Circuit Theory*, CT-16, 4:455–460, 1969.

15. D. B. Shmoys and D. P. Williamson. Analyzing the Held-Karp TSP bound: A monotonicity property with application, *Inf. Process. Lett.*, 35:281–285, 1990.

16. L. A. Wolsey. Heuristic analysis, linear programming and branch and bound. *Math. Program. Study*, 13:121–134, 1980.

17. D. P. Williamson, M. X. Goemans, M. Mihail, and V. Vazirani. A primal-dual approximation algorithm for generalized Steiner network problems. *Combinatorica* 15:435–454, 1995.

An Improved Approximation Algorithm for Minimum Size 2-Edge Connected Spanning Subgraphs

Joseph Cheriyan[1], András Sebő[2], and Zoltán Szigeti[3]

[1] Department of Combinatorics and Optimization
University of Waterloo, Waterloo, Ontario, Canada N2L 3G1
jcheriyan@@dragon.uwaterloo.ca
[2] Departement de Mathematiques Discretes
CNRS, Laboratoire LEIBNIZ (CNRS,INPG,UJF)-IMAG
46 Avenue Felix Viallet, 38000 Grenoble Cedex, France
Andras.Sebo@@imag.fr
[3] Equipe Combinatoire, Université Paris VI
4 place Jussieu, Couloir 45-46 3e, 75252 Paris, France
Zoltan.Szigeti@@ecp6.jussieu.fr

Abstract. We give a $\frac{17}{12}$-approximation algorithm for the following NP-hard problem:
Given a simple undirected graph, find a 2-edge connected spanning subgraph that has the minimum number of edges.
The best previous approximation guarantee was $\frac{3}{2}$. If the well known TSP $\frac{4}{3}$ conjecture holds, then there is a $\frac{4}{3}$-approximation algorithm. Thus our main result gets half-way to this target.

1 Introduction

Given a simple undirected graph, consider the problem of finding a 2-edge connected spanning subgraph that has the minimum number of edges. The problem is NP-hard, since the Hamiltonian cycle problem reduces to it. A number of recent papers have focused on approximation algorithms [1] for this and other related problems, [2]. We use the abbreviation 2-ECSS for 2-edge connected spanning subgraph.

Here is an easy 2-approximation algorithm for the problem:

Take an ear decomposition of the given graph (see Section 2 for definitions), and discard all 1-ears (ears that consist of one edge). Then the resulting graph is 2-edge connected and has at most $2n - 3$ edges, while the optimal subgraph has $\geq n$ edges, where n is the number of nodes.

---

[1] An *α-approximation algorithm* for a combinatorial optimization problem runs in polynomial time and delivers a solution whose value is always within the factor α of the optimum value. The quantity α is called the *approximation guarantee* of the algorithm.

R. E. Bixby, E. A. Boyd, and R. Z. Ríos-Mercado (Eds.): IPCO VI
LNCS 1412, pp. 126–136, 1998. © Springer–Verlag Berlin Heidelberg 1998

Khuller & Vishkin [8] were the first to improve on the approximation guarantee of 2. They gave a simple and elegant algorithm based on depth-first search that achieves an approximation guarantee of 1.5. In an extended abstract, Garg, Santosh & Singla [6] claimed to have a 1.25-approximation algorithm for the problem. No proof of this claim is available; on the other hand, there is no evidence indicating that achieving an approximation guarantee of 1.25 in polynomial time is impossible.

We improve Khuller & Vishkin's $\frac{18}{12}$-approximation guarantee to $\frac{17}{12}$. If the well known TSP $\frac{4}{3}$ conjecture holds, then there is a $\frac{4}{3}$-approximation algorithm, see Section 5. Thus our main result gets half-way to this target.

Let $G = (V, E)$ be the given simple undirected graph, and let n and m denote $|V|$ and $|E|$. Assume that G is 2-edge connected.

Our method is based on a matching-theory result of András Frank, namely, there is a good characterization for the minimum number of even-length ears over all possible ear decompositions of a graph, and moreover, an ear decomposition achieving this minimum can be computed efficiently, [4]. Recall that the 2-approximation heuristic starts with an arbitrary ear decomposition of G. Instead, if we start with an ear decomposition that maximizes the number of 1-ears, and if we discard all the 1-ears, then we will obtain the optimal solution. In fact, we start with an ear decomposition that maximizes the number of odd-length ears. Now, discarding all the 1-ears gives an approximation guarantee of 1.5 (see Proposition 8 below). To do better, we repeatedly apply "ear-splicing" steps to the starting ear decomposition to obtain a final ear decomposition such that the number of odd-length ears is the same, and moreover, the internal nodes of distinct 3-ears are nonadjacent. We employ two lower bounds to show that discarding all the 1-ears from the final ear decomposition gives an approximation guarantee of $\frac{17}{12}$. The first lower bound is the "component lower bound" due to Garg et al [6, Lemma 4.1], see Proposition 4 below. The second lower bound comes from the minimum number of even-length ears in an ear decomposition of G, see Proposition 7 below.

After developing some preliminaries in Sections 2 and 3, we present our heuristic in Section 4. Section 5 shows that the well known $\frac{4}{3}$ conjecture for the metric TSP implies that there is a $\frac{4}{3}$-approximation algorithm for a minimum-size 2-ECSS, see Theorem 18. Almost all of the results in Section 5 are well known, but we include the details to make the paper self-contained. Section 6 has two examples showing that our analysis of the heuristic is tight. Section 6 also compares the two lower bounds with the optimal value.

A Useful Assumption

For our heuristic to work, it is essential that the given graph be 2-node connected. Hence, in Section 4 of the paper where our heuristic is presented, we will assume that the given graph G is 2-node connected. Otherwise, if G is not 2-node connected, we compute the blocks (i.e., the maximal 2-node connected subgraphs) of G, and apply the algorithm separately to each block. We compute a 2-ECSS for each block, and output the union of the edge sets as the edge set of a 2-ECSS of G. The resulting graph has no cut edges since the subgraph found for each block has no cut edge, and moreover, the approximation guarantee for G is at most the maximum of the approximation guarantees for the blocks.

2   Preliminaries

Except in Section 5, all graphs are simple, that is, there are no loops nor multi-edges. A closed path means a cycle, and an open path means that all the nodes are distinct.

An *ear decomposition* of the graph G is a partition of the edge set into open or closed paths, $P_0 + P_1 + \ldots + P_k$, such that P_0 is the trivial path with one node, and each P_i ($1 \leq i \leq k$) is a path that has both end nodes in $V_{i-1} = V(P_0) \cup V(P_1) \cup \ldots \cup V(P_{i-1})$ but has no internal nodes in V_{i-1}. A (closed or open) *ear* means one of the (closed or open) paths $P_0, P_1, \ldots, P_k$ in the ear decomposition, and for a nonnegative integer ℓ, an ℓ-*ear* means an ear that has ℓ edges. An ℓ-ear is called *even* if ℓ is an even number, otherwise, the ℓ-ear is called *odd*. (The ear P_0 is always even.) An *open* ear decomposition $P_0 + P_1 + \ldots + P_k$ is one such that all the ears $P_2, \ldots, P_k$ are open.

Proposition 1 (Whitney [12]).

(i) A graph is 2-edge connected if and only if it has an ear decomposition.
(ii) A graph is 2-node connected if and only if it has an open ear decomposition.

An *odd* ear decomposition is one such that every ear (except the trivial path P_0) has an odd number of edges. A graph is called *factor-critical* if for every node $v \in V(G)$, there is a perfect matching in $G - v$. The next result gives another characterization of factor-critical graphs.

Theorem 2 (Lovász [9], Theorem 5.5.1 in [10]). *A graph is factor-critical if and only if it has an odd ear decomposition.*

It follows that a factor-critical graph is necessarily 2-edge connected. An *open odd* ear decomposition $P_0 + P_1 + \ldots + P_k$ is an odd ear decomposition such that all the ears $P_2, \ldots, P_k$ are open.

Theorem 3 (Lovász & Plummer, Theorem 5.5.2 in [10]). *A 2-node connected factor-critical graph has an open odd ear decomposition.*

Let $\varepsilon(G)$ denote the minimum number of edges in a 2-ECSS of G. For a graph H, let $c(H)$ denote the number of (connected) components of H. Garg et al [6, Lemma 4.1] use the following lower bound on $\varepsilon(G)$.

Proposition 4. *Let $G = (V, E)$ be a 2-edge connected graph, and let S be a nonempty set of nodes such that the deletion of S results in a graph with $c = c(G - S) \geq 2$ components. Then $\varepsilon(G) \geq |V| + c - |S|$.*

Proof. Focus on an arbitrary component D of $G - S$ and note that it contributes $\geq |V(D)| + 1$ edges to an optimal 2-ECSS, because every node in D contributes ≥ 2 edges, and at least two of these edges have exactly one end node in D. Summing over all components of $G - S$ gives the result.   □

For a set of nodes $S \subseteq V$ of a graph $G = (V, E)$, $\delta(S)$ denotes the set of edges that have one end node in S and one end node in $V - S$. For the singleton node set $\{v\}$, we use the notation $\delta(v)$. For a vector $x : E \to \mathbb{R}$, $x(\delta(S))$ denotes $\sum_{e \in \delta(S)} x_e$.

3   Frank's Theorem and a New Lower Bound for ε

For a 2-edge connected graph G, let $\varphi(G)$ (or φ) denote the minimum number of even ears of length ≥ 2, over all possible ear decompositions. For example: $\varphi(G) = 0$ if G is a factor-critical graph (e.g., G is an odd clique $K_{2\ell+1}$ or an odd cycle $C_{2\ell+1}$), $\varphi(G) = 1$ if G is an even clique $K_{2\ell}$ or an even cycle $C_{2\ell}$, and $\varphi(G) = \ell - 1$ if G is the complete bipartite graph $K_{2,\ell}$ ($\ell \geq 2$). The proof of the next result appears in [4], see Theorem 4.5 and Section 2 of [4].

Theorem 5 (A. Frank [4]). *Let $G = (V, E)$ be a 2-edge connected graph. An ear decomposition $P_0 + P_1 + \ldots + P_k$ of G having $\varphi(G)$ even ears of length ≥ 2 can be computed in time $O(|V| \cdot |E|)$.*

Proposition 6. *Let G be a 2-node connected graph. An* open *ear decomposition $P_0 + P_1 + \ldots + P_k$ of G having $\varphi(G)$ even ears of length ≥ 2 can be computed in time $O(|V| \cdot |E|)$.*

Proof. Start with an ear decomposition having $\varphi(G)$ even ears of length ≥ 2 (the ears may be open or closed). Subdivide one edge in each even ear of length ≥ 2 by adding one new node and one new edge. The resulting ear decomposition is odd. Hence, the resulting graph G' is factor critical, and also, G' is 2-node connected since G is 2-node connected. Apply Theorem 3 to construct an open odd ear decomposition of G'. Finally, in the resulting ear decomposition, "undo" the $\varphi(G)$ edge subdivisions to obtain the desired ear decomposition $P_0 + P_1 + \ldots + P_k$ of G.   □

Frank's theorem gives the following lower bound on the minimum number of edges in a 2-ECSS.

Proposition 7. *Let $G = (V, E)$ be a 2-edge connected graph. Then $\varepsilon(G) \geq |V| + \varphi(G) - 1$.*

Proof. Consider an arbitrary 2-ECSS of G. If this 2-ECSS has an ear decomposition with fewer than $\varphi(G) + 1$ even ears, then we could add the edges of G not in the 2-ECSS as 1-ears to get an ear decomposition of G with fewer than $\varphi(G) + 1$ even ears. Thus, every ear decomposition of the 2-ECSS has $\geq \varphi(G) + 1$ even ears. Let $P_0 + P_1 + \ldots + P_k$ be an ear decomposition of the 2-ECSS, where $k \geq \varphi(G)$. It is easily seen that the number of edges in the 2-ECSS is $k + |V| - 1 \geq \varphi(G) + |V| - 1$. The result follows.   □

The next result is not useful for our main result, but we include it for completeness.

Proposition 8. *Let $G = (V, E)$ be a 2-edge connected graph. Let $G' = (V, E')$ be obtained by discarding all the 1-ears from an ear decomposition $P_0 + P_1 + \ldots + P_k$ of G that has $\varphi(G)$ even ears of length ≥ 2. Then $|E'|/\varepsilon(G) \leq 1.5$.*

Proof. Let t be the number of internal nodes in the odd ears of $P_0 + P_1 + \ldots + P_k$. (Note that the node in P_0 is not counted by t.) Then, the number of edges contributed to E' by the odd ears is $\leq 3t/2$, and the number of edges contributed to E' by the even ears is $\leq \varphi + |V| - t - 1$. By applying Proposition 7 (and the fact that $\varepsilon(G) \geq |V|$) we get, $|E'|/\varepsilon(G) \leq (t/2 + \varphi + |V| - 1)/\max(|V|, \varphi + |V| - 1) \leq (t/2|V|) + (\varphi + |V| - 1)/(\varphi + |V| - 1) \leq 1.5$. □

4   Approximating ε via Frank's Theorem

For a graph H and an ear decomposition $P_0 + P_1 + \ldots + P_k$ of H, we call an ear P_i of length ≥ 2 *pendant* if none of the internal nodes of P_i is an end node of another ear P_j of length ≥ 2. In other words, if we discard all the 1-ears from the ear decomposition, then one of the remaining ears is called pendant if all its internal nodes have degree 2 in the resulting graph.

Let $G = (V, E)$ be the given graph, and let $\varphi = \varphi(G)$. Recall the assumption from Section 1 that G is 2-node connected. By an *evenmin ear decomposition* of G, we mean an ear decomposition that has $\varphi(G)$ even ears of length ≥ 2. Our method starts with an open evenmin ear decomposition $P_0 + P_1 + \ldots + P_k$ of G, see Proposition 6, i.e., for $2 \leq i \leq k$, every ear P_i has distinct end nodes, and the number of even ears is minimum possible. The method performs a sequence of "ear splicings" to obtain another (evenmin) ear decomposition $Q_0 + Q_1 + \ldots + Q_k$ (the ears Q_i may be either open or closed) such that the following holds:

Property (α)
(0) the number of even ears is the same in $P_0 + P_1 + \ldots + P_k$ and in $Q_0 + Q_1 + \ldots + Q_k$,
(1) every 3-ear Q_i is a pendant ear,
(2) for every pair of 3-ears Q_i and Q_j, there is no edge between an internal node of Q_i and an internal node of Q_j, and
(3) every 3-ear Q_i is open.

Proposition 9. *Let $G = (V, E)$ be a 2-node connected graph with $|V| \geq 4$. Let $P_0 + P_1 + \ldots + P_k$ be an open evenmin ear decomposition of G. There is a linear-time algorithm that given $P_0 + P_1 + \ldots + P_k$, finds an ear decomposition $Q_0 + Q_1 + \ldots + Q_k$ satisfying property (α).*

Proof. The proof is by induction on the number of ears. The result clearly holds for $k = 1$. Suppose that the result holds for $(j-1)$ ears $P_0 + P_1 + \ldots + P_{j-1}$. Let

$Q'_0 + Q'_1 + \ldots + Q'_{j-1}$ be the corresponding ear decomposition that satisfies property (α). Consider the open ear P_j, $j \geq 2$. Let P_j be an ℓ-ear, $v_1, v_2, \ldots, v_\ell, v_{\ell+1}$. Possibly, $\ell = 1$. (So v_1 and $v_{\ell+1}$ are the end nodes of P_j, and $v_1 \neq v_{\ell+1}$.)

Let T denote the set of internal nodes of the 3-ears of $Q'_0 + Q'_1 + \ldots + Q'_{j-1}$. Suppose P_j is an ear of length $\ell \geq 2$ with exactly one end node, say, v_1 in T. Let $Q'_i = w_1, v_1, w_3, w_4$ be the 3-ear having v_1 as an internal node. We take $Q_0 = Q'_0, \ldots, Q_{i-1} = Q'_{i-1}, Q_i = Q'_{i+1}, \ldots, Q_{j-2} = Q'_{j-1}$. Moreover, we take Q_{j-1} to be the $(\ell+2)$-ear obtained by adding the last two edges of Q'_i to P_j, i.e., $Q_{j-1} = w_4, w_3, v_1, v_2, \ldots, v_\ell, v_{\ell+1}$, and we take Q_j to be the 1-ear consisting of the first edge $w_1 v_1$ of Q'_i. Note that the parities of the lengths of the two spliced ears are preserved, that is, Q_{j-1} is even (odd) if and only if P_j is even (odd), and both Q_j and Q'_i are odd. Hence, the number of even ears is the same in $P_0 + P_1 + \ldots + P_j$ and in $Q_0 + Q_1 + \ldots + Q_j$.

Now, suppose P_j has both end nodes v_1 and $v_{\ell+1}$ in T. If there is one 3-ear Q'_i that has both v_1 and $v_{\ell+1}$ as internal nodes (so $\ell \geq 2$), then we take Q_{j-1} to be the $(\ell+2)$-ear obtained by adding the first edge and the last edge of Q'_i to P_j, and we take Q_j to be the 1-ear consisting of the middle edge $v_1 v_{\ell+1}$ of Q'_i. Also, we take $Q_0 = Q'_0, \ldots, Q_{i-1} = Q'_{i-1}, Q_i = Q'_{i+1}, \ldots, Q_{j-2} = Q'_{j-1}$. Observe that the number of even ears is the same in $P_0 + P_1 + \ldots + P_j$ and in $Q_0 + Q_1 + \ldots + Q_j$.

If there are two 3-ears Q'_i and Q'_h that contain the end nodes of P_j, then we take Q_{j-2} to be the $(\ell+4)$-ear obtained by adding the last two edges of both Q'_i and Q'_h to P_j, and we take Q_{j-1} (similarly, Q_j) to be the 1-ear consisting of the first edge of Q'_i (similarly, Q'_h). (For ease of description, assume that if a 3-ear has exactly one end node v of P_j as an internal node, then v is the second node of the 3-ear.) Also, assuming $i < h$, we take $Q_0 = Q'_0, \ldots, Q_{i-1} = Q'_{i-1}, Q_i = Q'_{i+1}, \ldots, Q_{h-2} = Q'_{h-1}, Q_{h-1} = Q'_{h+1}, \ldots, Q_{j-3} = Q'_{j-1}$. Again, observe that the number of even ears is the same in $P_0 + P_1 + \ldots + P_j$ and in $Q_0 + Q_1 + \ldots + Q_j$.

If the end nodes of P_j are disjoint from T, then the proof is easy (take $Q_j = P_j$). Also, if P_j is a 1-ear with exactly one end node in T, then the proof is easy (take $Q_j = P_j$).

The proof ensures that in the final ear decomposition $Q_0 + Q_1 + \ldots + Q_k$, every 3-ear is pendant and open, and moreover, the internal nodes of distinct 3-ears are nonadjacent. We leave the detailed verification to the reader. Therefore, the ear decomposition $Q_0 + Q_1 + \ldots + Q_k$ satisfies property (α). $\qquad\square$

Remark 10. In the induction step, which applies for $j \geq 2$ (but not for $j = 1$), it is essential that the ear P_j is open, though Q'_i (and Q'_h) may be either open or closed. Our main result (Theorem 12) does not use part (3) of property (α).

Our approximation algorithm for a minimum-size 2-ECSS computes the ear decomposition $Q_0 + Q_1 + \ldots + Q_k$ satisfying property (α), starting from an open evenmin ear decomposition $P_0 + P_1 + \ldots + P_k$. (Note that $Q_0 + Q_1 + \ldots + Q_k$ is an evenmin ear decomposition.) Then, the algorithm discards all the edges in 1-ears. Let the resulting graph be $G' = (V, E')$. G' is 2-edge connected by Proposition 1.

Let T denote the set of internal nodes of the 3-ears of $Q_0 + Q_1 + \ldots + Q_k$, and let $t = |T|$. (Note that the node in Q_0 is not counted by t.) Property (α) implies that in the subgraph of G induced by T, $G[T]$, every (connected) component has exactly two nodes. Consider the approximation guarantee for G', i.e., the quantity $|E'|/\varepsilon(G)$.

Lemma 11. $\varepsilon(G) \geq 3t/2$.

Proof. Apply Proposition 4 with $S = V - T$ (so $|S| = n - t$) and $c = c(G - S) = t/2$ to get $\varepsilon(G) \geq n - (n - t) + (t/2)$. $\square$

Theorem 12. *Given a 2-edge connected graph $G = (V, E)$, the above algorithm finds a 2-ECSS $G' = (V, E')$ such that $|E'|/\varepsilon(G) \leq \frac{17}{12}$. The algorithm runs in time $O(|V| \cdot |E|)$.*

Proof. By the previous lemma and Proposition 7,

$$\varepsilon(G) \geq \max(n + \varphi(G) - 1, \, 3t/2) \ .$$

We claim that

$$|E'| \leq \frac{t}{4} + \frac{5(n + \varphi(G) - 1)}{4} \ .$$

To see this, note that the final ear decomposition $Q_0 + Q_1 + \ldots + Q_k$ satisfies the following: (i) the number of edges contributed by the 3-ears is $3t/2$; (ii) the number of edges contributed by the odd ears of length ≥ 5 is $\leq 5q/4$, where q is the number of internal nodes in the odd ears of length ≥ 5; and (iii) the number of edges contributed by the even ears of length ≥ 2 is $\leq \varphi(G) + (n - t - q - 1)$, since there are $\varphi(G)$ such ears and they have a total of $(n - t - q - 1)$ internal nodes. (The node in Q_0 is not an internal node of an ear of length ≥ 1.)

The approximation guarantee follows since

$$\frac{|E'|}{\varepsilon(G)} \leq \frac{t/4 + 5(n + \varphi(G) - 1)/4}{\varepsilon(G)}$$

$$\leq \frac{t/4 + 5(n + \varphi(G) - 1)/4}{\max(n + \varphi(G) - 1, \, 3t/2)}$$

$$\leq \frac{t}{4}\frac{2}{3t} + \frac{5(n + \varphi(G) - 1)}{4}\frac{1}{n + \varphi(G) - 1}$$

$$= \frac{17}{12} \ .$$

$\square$

5   Relation to the TSP $\frac{4}{3}$ Conjecture

This section shows that the well known $\frac{4}{3}$ conjecture for the metric TSP (due to Cunningham (1986) and others) implies that there is a $\frac{4}{3}$-approximation algorithm for a minimum-size 2-ECSS, see Theorem 18. Almost all of the results in this section are well known, except possibly Fact 13, see [1,3,5,7,11,13]. The details are included to make the paper self-contained.

In the *metric TSP* (traveling salesman problem), we are given a complete graph $G' = K_n$ and edge costs c' that satisfy the triangle inequality ($c'_{vw} \leq c'_{vu} + c'_{uw}, \forall v, w, u \in V$). The goal is to compute c'_{TSP}, the minimum cost of a Hamiltonian cycle.

Recall our 2-ECSS problem: Given a simple graph $G = (V, E)$, compute $\varepsilon(G)$, the minimum size of a 2-edge connected spanning subgraph. Here is the multiedge (or uncapacitated) version of our problem. Given $G = (V, E)$ as above, compute $\mu(G)$, the minimum size (counting multiplicities) of a 2-edge connected spanning submultigraph $H = (V, F)$, where F is a multiset such that $e \in F \Longrightarrow e \in E$. (To give an analogy, if we take $\varepsilon(G)$ to correspond to the f-factor problem, then $\mu(G)$ corresponds to the f-matching problem.)

Fact 13. *If G is a 2-edge connected graph, then $\mu(G) = \varepsilon(G)$.*

Proof. Let $H = (V, F)$ give the optimal solution for $\mu(G)$. If H uses two copies of an edge vw, then we can replace one of the copies by some other edge of G in the cut given by $H - \{vw, vw\}$. In other words, if S is the node set of one of the two components of $H - \{vw, vw\}$, then we replace one copy of vw by some edge from $\delta_G(S) - \{vw\}$.   □

Remark 14. The above is a lucky fact. It *fails* to generalize, both for minimum-cost (rather than minimum-size) 2-ECSS, and for minimum-size k-ECSS, $k \geq 3$.

Given an n-node graph $G = (V, E)$ together with edge costs c (possibly c assigns unit costs), define its *metric completion* G', c' to be the complete graph $K_n = G'$ with c'_{vw} ($\forall v, w \in V$) equal to the minimum-cost of a v-w path in G, c.

Fact 15. *Let G be a 2-edge connected graph, and let c assign unit costs to the edges. The minimum cost of the TSP on the metric completion of G, c, satisfies $c'_{TSP} \geq \mu(G) = \varepsilon(G)$.*

Proof. Let T be an optimal solution to the TSP. We replace each edge $vw \in E(T) - E(G)$ by the edges of a minimum-cost v-w path in G, c. The resulting multigraph H is obviously 2-edge connected, and has $c'_{TSP} = c(H) \geq \mu(G)$.   □

Here is the *subtour* formulation of the TSP on G', c', where $G' = K_n$. This gives an integer programming formulation, using the subtour elimination constraints. There is one variable x_e for each edge e in G'.

$$
\begin{aligned}
c'_{TSP} = \text{minimize} \quad & c' \cdot x \\
\text{subject to} \quad & x(\delta(v)) = 2, & & \forall v \in V \\
& x(\delta(S)) \geq 2, & & \forall S \subset V, \emptyset \neq S \neq V \\
& x \geq 0, \\
& x \in \mathbb{Z} .
\end{aligned}
$$

The *subtour LP* (linear program) is obtained by removing the integrality constraints, i.e., the x-variables are nonnegative reals rather than nonnegative integers. Let z_{ST} denote the optimal value of the subtour LP. Note that z_{ST} is computable in polynomial time, e.g., via the Ellipsoid method. In practice, z_{ST} may be computed via the Held-Karp heuristic, which typically runs fast.

Theorem 16 (Wolsey [13]). *If c' is a metric, then $c'_{TSP} \leq \frac{3}{2} z_{ST}$.*

TSP $\frac{4}{3}$ Conjecture. *If c' is a metric, then $c'_{TSP} \leq \frac{4}{3} z_{ST}$.*

To derive the lower bound $z_{ST} \leq \varepsilon(G)$, we need a result of Goemans & Bertsimas on the subtour LP, [7, Theorem 1]. In fact, a special case of this result that appeared earlier in [11, Theorem 8] suffices for us.

Proposition 17 (Parsimonious property [7]). *Consider the TSP on $G' = (V, E'), c'$, where $G' = K_{|V|}$. Assume that the edge costs c' form a metric, i.e., c' satisfies the triangle inequality. Then the optimal value of the subtour LP remains the same even if the constraints $\{x(\delta(v)) = 2, \forall v \in V\}$ are omitted.*

Note that this result does not apply to the subtour integer program given above.

Let z_{2CUT} denote the optimal value of the LP obtained from the subtour LP by removing the constraints $x(\delta(v)) = 2$ for all nodes $v \in V$. The above result states that if c' is a metric, then $z_{ST} = z_{2CUT}$. Moreover, for a 2-edge connected graph G and unit edge costs $c = \mathbb{1}$, we have $z_{2CUT} \leq \mu(G) = \varepsilon(G)$, since $\mu(G)$ is the optimal value of the integer program whose LP relaxation has optimal value z_{2CUT}. (Here, z_{2CUT} is the optimal value of the LP on the metric completion of G, c.) Then, by the parsimonious property, we have $z_{ST} = z_{2CUT} \leq \varepsilon(G)$. The main result in this section follows.

Theorem 18. *Suppose that the TSP $\frac{4}{3}$ conjecture holds. Then*

$$z_{ST} \leq \varepsilon(G) \leq c'_{TSP} \leq \frac{4}{3} z_{ST} \ .$$

A $\frac{4}{3}$-approximation of the minimum-size 2-ECSS is obtained by computing $\frac{4}{3} z_{ST}$ on the metric completion of G, c, where $c = \mathbb{1}$.

The Minimum-Cost 2-ECSS Problem

Consider the weighted version of the problem, where each edge e has a nonnegative cost c_e and the goal is to find a 2-ECSS (V, E') of the given graph $G = (V, E)$ such that the cost $c(E') = \sum_{e \in E'} c_e$ is minimum. Khuller & Vishkin [8] pointed out that a 2-approximation guarantee can be obtained via the weighted matroid intersection algorithm. When the edge costs satisfy the triangle inequality (i.e., when c is a metric), Frederickson and Ja'Ja' [5] gave a 1.5-approximation algorithm, and this is still the best approximation guarantee known. In fact, they proved that the TSP tour found by the Christofides heuristic achieves an approximation guarantee of 1.5. Simpler proofs of this result based on Theorem 16 were found later by Cunningham (see [11, Theorem 8]) and by Goemans & Bertsimas [7, Theorem 4].

Consider the minimum-cost 2-ECSS problem on a 2-edge connected graph $G = (V, E)$ with nonnegative edge costs c. Let the minimum cost of a simple 2-ECSS and of a multiedge 2-ECSS be denoted by c_ε and c_μ, respectively. Clearly, $c_\varepsilon \geq c_\mu$. Even for the case of arbitrary nonnegative costs c, we know of no example where $\frac{c_\mu}{z_{ST}} > \frac{7}{6}$. There is an example G, c with $\frac{c_\mu}{z_{ST}} \geq \frac{7}{6}$. Take two copies of K_3, call them C_1, C_2, and add three disjoint length-2 paths P_1, P_2, P_3 between C_1 and C_2 such that each node of $C_1 \cup C_2$ has degree 3 in the resulting graph G. In other words, G is obtained from the triangular prism $\overline{C_6}$ by subdividing once each of the 3 "matching edges". Assign a cost of 2 to each edge in $C_1 \cup C_2$, and assign a cost of 1 to the remaining edges. Then $c_\varepsilon = c_\mu = 14$, as can be seen by taking 2 edges from each of C_1, C_2, and all 6 edges of $P_1 \cup P_2 \cup P_3$. Moreover, $z_{ST} \leq 12$, as can be seen by taking $x_e = 1/2$ for each of the 6 edges e in $C_1 \cup C_2$, and taking $x_e = 1$ for the remaining 6 edges e in $P_1 \cup P_2 \cup P_3$.

6   Conclusions

Our analysis of the heuristic is (asymptotically) tight. We give two example graphs. Each is an n-node Hamiltonian graph $G = (V, E)$, where the heuristic (in the worst case) finds a 2-ECSS $G' = (V, E')$ with $17n/12 - \Theta(1)$ edges. The first example graph, G, is constructed by "joining" many copies of the following graph H: H consists of a 5-edge path $u_0, u_1, u_2, u_3, u_4, u_5$, and 4 disjoint edges $v_1 w_1, v_2 w_2, v_3 w_3, v_4 w_4$. We take q copies of H and identify the node u_0 in all copies, and identify the node u_5 in all copies. Then we add all possible edges $u_i v_j$, and all possible edges $u_i w_j$, i.e., we add the edge set of a complete bipartite graph on all the u-nodes and all the v-nodes, and we add the edge set of another complete bipartite graph on all the u-nodes and all the w-nodes. Finally, we add 3 more nodes u_1', u_2', u_3' and 5 more edges to obtain a 5-edge cycle $u_0, u_1', u_2', u_3', u_5, u_0$. Clearly, $\varepsilon(G) = n = 12q + 5$. If the heuristic starts with the closed 5-ear $u_0, u_1', u_2', u_3', u_5, u_0$, and then finds the 5-ears $u_0, u_1, u_2, u_3, u_4, u_5$ in all the copies of H, and finally finds the 3-ears $u_0 v_j w_j u_5$ ($1 \leq j \leq 4$) in all copies of H, then we have $|E'| = 17q + 5$.

Here is the second example graph, $G = (V, E)$. The number of nodes is $n = 3 \times 5^q$, and $V = \{0, 1, 2, ..., 3 \times 5^q - 1\}$. The "first node" 0 will also be denoted 3×5^q. The edge set E consists of (the edge set of) a Hamiltonian cycle together with (the edge sets of) "shortcut cycles" of lengths $n/3, n/(3 \times 5), n/(3 \times 5^2), \ldots, 5$. In detail, $E = \{i(i+1) \mid \forall 0 \leq i \leq q-1\} \cup \{(3 \times 5^j \times i)(3 \times 5^j \times (i+1)) \mid \forall 0 \leq j \leq q-1, 0 \leq i \leq 5^{q-j} - 1\}$. Note that $|E| = 3 \times 5^q + 5^q + 5^{q-1} + ... + 5 = (17 \times 5^q - 5)/4$. In the worst case, the heuristic initially finds 5-ears, and finally finds 3-ears, and so the 2-ECSS (V, E') found by the heuristic has all the edges of G. Hence, we have $|E'|/\varepsilon(G) = |E|/n = 17/12 - 1/(12 \times 5^{q-1})$.

How do the lower bounds in Proposition 4 (call it L_c) and in Proposition 7 (call it L_φ) compare with ε? Let n denote the number of nodes in the graph. There is a 2-node connected graph such that $\varepsilon/L_\varphi \geq 1.5 - \Theta(1)/n$, i.e., the upper bound of Proposition 8 is tight. There is another 2-edge connected (but not 2-node connected) graph such that $\varepsilon/L_c \geq 1.5 - \Theta(1)/n$ and $\varepsilon/L_\varphi \geq 1.5 - \Theta(1)/n$. Among 2-node connected graphs, we have a graph with $\varepsilon/L_c \geq 4/3 - \Theta(1)/n$, but we do not know whether there exist graphs that give higher ratios. There is a 2-node connected graph such that $\varepsilon/\max(L_c, L_\varphi) \geq 5/4 - \Theta(1)/n$, but we do not know whether there exist graphs that give higher ratios.

References

1. R. Carr and R. Ravi. A new bound for the 2-edge connected subgraph problem. In R. E. Bixby, E. A. Boyd, and R. Z. Ríos-Mercado, editors, *Integer Programming and Combinatorial Optimization: Proceedings of the 6th International Conference on Integer Programming and Combinatorial Optimization, LNCS, Vol. 1412*, pages 110–123. Springer, 1998. This volume.
2. J. Cheriyan and R. Thurimella. Approximating minimum-size k-connected spanning subgraphs via matching. *Proc. 37th Annual IEEE Sympos. on Foundat. of Comput. Sci.*, pages 292–301, 1996.
3. N. Christofides. Worst-case analysis of a new heuristic for the traveling salesman problem. Technical report, G.S.I.A., Carnegie-Mellon Univ., Pittsburgh, PA, 1976.
4. A. Frank. Conservative weightings and ear-decompositions of graphs. *Combinatorica*, 13:65–81, 1993.
5. G. L. Frederickson and J. Ja'Ja'. On the relationship between the biconnectivity augmentation and traveling salesman problems. *Theor. Comp. Sci.*, 19:189–201, 1982.
6. N. Garg, V. S. Santosh, and A. Singla. Improved approximation algorithms for biconnected subgraphs via better lower bounding techniques. *Proc. 4th Annual ACM-SIAM Symposium on Discrete Algorithms*, pages 103–111, 1993.
7. M. X. Goemans and D. J. Bertsimas. Survivable networks, linear programming relaxations and the parsimonious property. *Mathematical Programming*, 60:143–166, 1993.
8. S. Khuller and U. Vishkin. Biconnectivity approximations and graph carvings. *Journal of the ACM*, 41:214–235, 1994. Preliminary version in *Proc. 24th Annual ACM STOC*, pages 759–770, 1992.
9. L. Lovász. A note on factor-critical graphs. *Studia Sci. Math. Hungar.*, 7:279–280, 1972.
10. L. Lovász and M. D. Plummer. *Matching Theory*. Akadémiai Kiadó, Budapest, 1986.
11. C. L. Monma, B. S. Munson, and W. R. Pulleyblank. Minimum-weight two-connected spanning networks. *Mathematical Programming*, 46:153–171, 1990.
12. H. Whitney. Nonseparable and planar graphs. *Trans. Amer. Math. Soc.*, 34:339–362, 1932.
13. L. A. Wolsey. Heuristic analysis, linear programming and branch and bound. *Mathematical Programming Study*, 13:121–134, 1980.

Multicuts in Unweighted Graphs with Bounded Degree and Bounded Tree-Width

Gruia Călinescu[1] *, Cristina G. Fernandes[2]**, and Bruce Reed[3]***

[1] College of Computing, Georgia Institute of Technology
Atlanta, GA 30332–0280, USA
gruia@@cc.gatech.edu
[2] Department of Computer Science, University of São Paulo
Rua do Matao, 1010   05508–900 Sao Paulo, Brazil
cris@@ime.usp.br
[3] CNRS - Paris, France, and
Department of Computer Science, University of São Paulo, Brazil
reed@@ime.usp.br

Abstract. The MULTICUT problem is defined as follows: given a graph G and a collection of pairs of distinct vertices (s_i, t_i) of G, find a smallest set of edges of G whose removal disconnects each s_i from the corresponding t_i. Our main result is a polynomial-time approximation scheme for MULTICUT in unweighted graphs with bounded degree and bounded tree-width: for any $\epsilon > 0$, we presented a polynomial-time algorithm with performance ratio at most $1 + \epsilon$. In the particular case when the input is a bounded-degree tree, we have a linear-time implementation of the algorithm. We also provided some hardness results. We proved that MULTICUT is still NP-hard for binary trees and that, unless $P = NP$, no polynomial-time approximation scheme exists if we drop any of the the three conditions: unweighted, bounded-degree, bounded-tree-width. Some of these results extend to the vertex version of MULTICUT.

1 Introduction

Multicommodity Flow problems have been intensely studied for decades [7,11,9], [13,15,17] because of their practical applications and also of the appealing hardness of several of their versions. The fractional version of a Multicut problem is the dual of a Multicommodity Flow problem and, therefore, Multicut is of similar interest [3,9,10,13,20].

* Research supported in part by NSF grant CCR-9319106.
** Research partially supported by NSF grant CCR-9319106 and by FAPESP (Proc. 96/04505–2).
*** Research supported in part by ProNEx (MCT/FINEP) (Proj. 107/97) and FAPESP (Proc. 96/12111–4).

R. E. Bixby, E. A. Boyd, and R. Z. Ríos-Mercado (Eds.): IPCO VI
LNCS 1412, pp. 137–152, 1998. © Springer–Verlag Berlin Heidelberg 1998

The WEIGHTED MULTICUT is the following problem: given an undirected graph G, a weight function w on the edges of G, and a collection of k pairs of distinct vertices (s_i, t_i) of G, find a minimum weight set of edges of G whose removal disconnects each s_i from the corresponding t_i.

The particular case in which $k = 1$ is characterized by the famous *Max-Flow Min-Cut Theorem* [6], and is solvable in strongly polynomial time [4]. For $k = 2$, a variant of the Max-Flow Min-Cut Theorem holds [11,12] and Multicut is solvable in polynomial time. For $k \geq 3$, the problem is NP-hard [3].

Since many variants of the Weighted Multicut are known to be NP-hard, we search for efficient approximation algorithms. The *performance ratio of an approximation algorithm* A for a minimization problem is the supremum, over all possible instances I, of the ratio between the weight of the output of A when running on I and the weight of an optimal solution for I. We say A is an α-*approximation algorithm* if its performance ratio is at most α. The smaller the performance ratio, the better.

The best known performance ratio for Weighted Multicut in general graphs is $O(\log k)$ [10]. Important research has been done for improving the performance ratio when the input graph G belongs to special classes of graphs. For planar graphs, Tardos and Vazirani [20], see also [13], give an approximate Max-Flow Min-Cut theorem and an algorithm with a constant performance ratio.

The case when the input graph is restricted to a tree has been studied in [9]. Unweighted Multicut problem (in which $w(e) = 1$ for all edges e of G) restricted to stars (trees of height one) is equivalent (including performance ratio) to Minimum Vertex Cover, by Proposition 1 in [9]. It follows that Unweighted Multicut restricted to stars is NP-hard and Max SNP-hard. In fact, getting a performance ratio better than two seems very hard, since getting a performance ratio better than two for Minimum Vertex Cover remains a challenging open problem [16]. Garg, Vazirani and Yannakakis give an algorithm in [9] with a performance ratio of two for the Weighted Multicut problem in trees. Note that the integral unweighted Multicommodity Flow problem in trees is solvable in polynomial time [9].

We find useful two variations of the Multicut problem. The VERTEX MULTICUT problem is: given an undirected graph G and a collection of k pairs of distinct nonadjacent vertices (s_i, t_i) of G called *terminals*, find a minimum set of nonterminal vertices whose removal disconnects each s_i from the corresponding t_i. The UNRESTRICTED VERTEX MULTICUT problem is: given an undirected graph G and a collection of k pairs of vertices (s_i, t_i) of G called *terminals*, find a minimum set of vertices whose removal disconnects each s_i from the corresponding t_i. (Note that in this variation, terminals might be removed.) Observe that Vertex Multicut is at least as hard as Unrestricted Vertex Multicut. From an instance of Unrestricted Vertex Multicut we can obtain an instance of Vertex Multicut by adding, for each s_i, a new vertex s_i' adjacent only to s_i, and, for each t_i, a new vertex t_i' adjacent only to t_i. Each pair (s_i, t_i) is substituted by the new pair (s_i', t_i'). Solving Vertex Multicut in this instance is equivalent to solving Unrestricted Vertex Multicut in the original instance.

Both Vertex Multicut and Unrestricted Vertex Multicut might be of interest on their own. Garg, Vazirani and Yannakakis considered the weighted version of Vertex Multicut and proved that their algorithm in [10] achieves a performance ratio of $O(\log k)$ for the weighted version of Vertex Multicut in general graphs.

From now on, we refer to Multicut as EDGE MULTICUT, to avoid confusion. Let us mention some results we obtained for Vertex Multicut and Unrestricted Vertex Multicut. We have a proof that Vertex Multicut is NP-hard in bounded-degree trees. Unrestricted Vertex Multicut is easier: it is polynomially solvable in trees, but it becomes NP-hard in bounded-degree series-parallel graphs.

The tree-width notion (first introduced by Robertson and Seymour [19]) seems to often capture a property of the input graph which makes hard problems easy. Various NP-hard problems, like Clique or Coloring, have a polynomial-time algorithm (linear time in fact) if the input graph has bounded tree-width (see for example [2]). We will present the formal definition of tree-width in Section 2.

Bounded tree-width can also be used to obtain good approximation algorithms for those problems that remain NP-hard even if restricted to graphs of bounded tree-width. In our case, Unrestricted Vertex Multicut is NP-hard in graphs of tree-width at most two, since this class of graphs coincides with the series-parallel graphs (see for example [21]). We give a straightforward PTAS for Unrestricted Vertex Multicut in graphs of bounded tree-width.

We present an approximation-ratio preserving reduction from Edge Multicut to Unrestricted Vertex Multicut. If the Edge Multicut instance graph has bounded degree and bounded tree-width, the Unrestricted Vertex Multicut instance obtained by the reduction has bounded tree-width. Combining the reduction with the PTAS for Unrestricted Vertex Multicut in graphs of bounded tree-width, we obtain a PTAS for Unweighted Edge Multicut in graphs with bounded degree and bounded tree-width. This is the main result of the paper. Note that, according to [8, page 140, Theorem 6.8], a FPTAS cannot exist for this problem, unless P=NP.

We also have a linear-time implementation of our PTAS for Edge Multicut in bounded-degree trees. The running time of our implementation is $O((n + k)\lceil 1/\epsilon \rceil d^{d\lceil 1/\epsilon \rceil + 2})$, where n is the number of vertices of the tree, k is the number of (s_i, t_i) pairs, d is the maximum degree of the tree and $1 + \epsilon$ is the desired approximation ratio of the algorithm. The size of the input is $\Theta(n + k)$.

We show that Edge Multicut is still NP-hard for binary (degree bounded by three) trees. Thus, on the class of graphs of bounded degree and bounded tree-width, which contains binary trees, Edge Multicut is easier (there is a PTAS) than on general graphs, yet still NP-hard. Identifying this class is the main theoretical result of this paper.

Hardness results indicate why we cannot eliminate any of the three restrictions—unweighted, bounded degree and bounded tree-width—on the input graph and still obtain a PTAS. It is known [1] that for a Max SNP-hard problem, unless P=NP, no PTAS exists. We have already seen that Unweighted Edge Multicut is Max SNP-hard in stars [9], so letting the input graph have unbounded degree makes the problem harder. We show that Weighted Edge Multicut is Max

SNP-hard in binary trees, therefore letting the input graph be weighted makes the problem harder. Finally, we show that Unweighted Edge Multicut is Max SNP-hard if the input graphs are walls. Walls, to be formally defined in Section 4, have degree at most three and there are walls with tree-width as large as we wish. We conclude that letting the input graph have unbounded tree-width makes the problem significantly harder.

In Section 2 we present the polynomial-time algorithm for Unrestricted Vertex Multicut in trees and the polynomial-time approximation scheme for Unrestricted Vertex Multicut in bounded-tree-width graphs. In Section 3, we show the approximation-preserving reduction from Edge Multicut to Unrestricted Vertex Multicut. Finally, in Section 4 we present our hardness results.

2   Algorithms for Unrestricted Vertex Multicut

In this section we concentrate on Unrestricted Vertex Multicut. We present a polynomial-time algorithm for trees and a PTAS for graphs with bounded tree-width. Let us start defining tree-width.

Let G be a graph and Θ be a pair $(T, (X_w)_{w \in V(T)})$, which consists of a tree T and a multiset whose elements X_w, indexed by the vertices of T, are subsets of $V(G)$. For a vertex v of G, we denote by F_v the subgraph of T induced by those vertices w of T for which X_w contains v. Then Θ is called a *tree decomposition of G* if it satisfies the two conditions below:

(1) For every edge $e = xy$ of G, there is a vertex w of T such that $\{x, y\} \subseteq X_w$;

(2) For every vertex v of G, the subgraph F_v of T is a tree.

The *width of Θ* is the maximum, over all vertices w of T, of $|X_w| - 1$, and the *tree-width of G*, denoted by $tw(G)$, is the minimum of the widths of all tree decompositions of G.

Consider an instance of Unrestricted Vertex Multicut, that is, a graph $G = (V, E)$ and a set $\mathcal{C}$ of pairs (s_i, t_i) of vertices of G. We say a pair (s_i, t_i) in $\mathcal{C}$ is *disconnected by a set $S \subseteq V$* if s_i is disconnected from t_i in the subgraph of G induced by $V - S$. A set S is a *solution for G* if S disconnects all pairs (s_i, t_i) in $\mathcal{C}$. If S has minimum size (i.e., minimum number of vertices), then S is an *optimal solution for G*.

Now, let us describe the polynomial-time algorithm for trees. The input of the algorithm is a tree T and a set $\mathcal{C}$ of pairs (s_i, t_i) of vertices of T.

Consider the tree T rooted at an arbitrary vertex and consider also an arbitrary ordering of the children of each vertex (so that we can talk about postorder).

Algorithm:

> Input: a tree T.
> Start with $S = \emptyset$.
> Call a pair (s_i, t_i) in $\mathcal{C}$ *active* if it is not disconnected by S.
> Traverse the tree in postorder.

When visiting vertex v, if v is the least common ancestor of some active pair (s_i, t_i) in $\mathcal{C}$, then insert v into S and mark (s_i, t_i).
Output S.

Clearly the following invariant holds: all non-active pairs in $\mathcal{C}$ are disconnected by S. A pair in $\mathcal{C}$ that becomes non-active does not go back to active since we never remove vertices from S. At the end of the algorithm, no pair in $\mathcal{C}$ is active, meaning that S is a solution for the problem. For the minimality of S, note that the paths joining s_i to t_i in T for all marked pairs (s_i, t_i) form a pairwise disjoint collection of paths. Any solution should contain at least one vertex in each of these paths. But there are $|S|$ marked paths, meaning that any solution has at least $|S|$ vertices. This implies that $|S|$ is a minimum-size solution. Besides, it is not hard to see that the algorithm can be implemented in polynomial time.

2.1   Bounded-Tree-Width Graphs

Next we present a PTAS for Unrestricted Vertex Multicut in graphs with bounded tree-width. A PTAS consists of, for each $\epsilon > 0$, a polynomial-time algorithm for the problem with a performance ratio of at most $1 + \epsilon$. Let us describe such an algorithm.

The input of our algorithm is a graph $G = (V, E)$, a tree decomposition $\Theta = (T, (X_w)_{w \in V(T)})$ of G, and a set $\mathcal{C}$ of pairs of vertices of G.

Given a subgraph G' of G, denote by $\mathcal{C}(G')$ the set of pairs in $\mathcal{C}$ whose two vertices are in G', and by $G \setminus G'$ the subgraph of G induced by $V(G) \setminus V(G')$. For the description of the algorithm, all the instances we mention are on a subgraph G' of G and the set of pairs to be disconnected is $\mathcal{C}(G')$. So we will drop $\mathcal{C}(G')$ of the notation and refer to an instance only by the graph G'. Denote by $opt(G')$ the size (i.e., the number of vertices) of an optimal solution for G'.

Root the tree T (of the given tree decomposition) at a vertex r and consider an arbitrary ordering of the children of each vertex of T. For a vertex u of T, let $T(u)$ be the subtree of T rooted at u. Let $G(u)$ be the subgraph of G induced by the union of all X_w, $w \in V(T(u))$. Let $t = \lceil (tw(G) + 1)/\epsilon \rceil$.

Here is a general description of the algorithm: label the vertices of T in postorder. Find the lowest labeled vertex u such that an optimal solution for $G(u)$ has at least t vertices. If there is no such vertex, let u be the root. Find an approximate solution S_u for $G(u)$ such that $|S_u| \leq (1 + \epsilon)opt(G(u))$ and $X_u \subseteq S_u$. If u is the root of T, then output S_u. Otherwise, let $G' = G \setminus G(u)$ and let $\Theta' = (T', (X'_w)_{w \in V(T')})$ be the tree decomposition of G' where $T' = T \setminus T(u)$ and $X'_w = X_w \setminus V(G(u))$, for all $w \in V(T')$. Recursively get a solution S' for G'. Output $S = S' \cup S_u$.

Next we present a detailed description of the algorithm. It works in iterations. Iteration i starts with a subgraph G^{i-1} of G, a tree decomposition $\Theta^{i-1} = (T^{i-1}, (X_w^{i-1})_{w \in V(T^{i-1})})$ of G^{i-1} with T^{i-1} rooted at r, and a set S^{i-1} of vertices of G. Initially, $G^0 = G$, $\Theta^0 = \Theta$, $S^0 = \emptyset$ and $i = 1$. The algorithm halts when $G^{i-1} = \emptyset$. When G^{i-1} is nonempty, the algorithm starts calling a procedure

Get (u, A), which returns a vertex u of T^{i-1} and a solution A for $G^{i-1}(u)$ such that $|A| \leq (1 + \epsilon)opt(G^{i-1}(u))$ and $X_u^{i-1} \subseteq A$. Then the algorithm starts a new iteration with $G^i = G^{i-1} \setminus G^{i-1}(u)$, $\Theta^i = (T^i, (X_w^i)_{w \in V(T^i)})$, where $T^i = T^{i-1} \setminus T^{i-1}(u)$ and $X_w^i = X_w^{i-1} \setminus V(G^{i-1}(u))$, for all $w \in V(T^i)$, and $S^i = S^{i-1} \cup A$. The formal description of the algorithm appears in Figure 2.1.

Algorithm:

$G^0 \leftarrow G$;
$\Theta^0 \leftarrow \Theta$;
$S^0 \leftarrow \emptyset$;
$i \leftarrow 1$;
while $G^{i-1} \neq \emptyset$ do
    Get (u_i, A^i);   /* $|A^i| \leq (1 + \epsilon)opt(G^{i-1}(u_i))$ and $X_u^{i-1} \subseteq A^i$ */
    $G^i \leftarrow G^{i-1} \setminus G^{i-1}(u_i)$;
    $T^i \leftarrow T^{i-1} \setminus T^{i-1}(u_i)$;
    $X_w^i \leftarrow X_w^{i-1} \setminus V(G^{i-1}(u_i))$, for each $w \in V(T^i)$;
    $S^i \leftarrow S^{i-1} \cup A^i$;
    $i \leftarrow i + 1$;
endwhile;
$f \leftarrow i - 1$;
output S^f.

Fig. 1. The algorithm for Unrestricted Vertex Multicut in bounded-tree-width graphs.

We will postpone the description of *Get* (u, A) and, for now, assume that it works correctly and in polynomial time. The next lemma states a property of tree decompositions that we will use later.

Lemma 1. *Consider a graph G and a tree decomposition $\Theta = (T, (X_w)_{w \in V(T)})$ of G. Let u be a vertex of T, x be a vertex of $G(u)$ and y be a vertex of $G \setminus G(u)$. Then any path in G from x to y contains a vertex of X_u.*

Next we prove that the output of the algorithm is in fact a solution.

Lemma 2. *S^f is a solution for G.*

Proof. Let (s, t) be a pair in $\mathcal{C}$ and P be a path in G from s to t. We need to show that there is a vertex of P in S^f. Note that the vertex sets $V(G^{i-1}(u_i))$ define a partition of $V(G)$.

Let i be such that s is in $G^{i-1}(u_i)$. If all vertices of P lie in $G^{i-1}(u_i)$ then, in particular, both s and t are in $G^{i-1}(u_i)$, which means $(s, t) \in \mathcal{C}(G^{i-1}(u_i))$. Since S^f contains a solution for $G^{i-1}(u_i)$, S^f must contain a vertex of P.

If not all vertices of P lie in $G^{i-1}(u_i)$, let y be the first vertex of P that does not lie in $G^{i-1}(u_i)$. If y is in $G^{i-1} \setminus G^{i-1}(u_i)$ then, by Lemma 1, there is a vertex of $X_{u_i}^{i-1}$ in the segment of P from s to y. Since $X_{u_i}^{i-1} \subseteq S^f$, there is a vertex of P in S^f. If y is not in $G^{i-1} \setminus G^{i-1}(u_i)$, then y is not in G^{i-1}. This means y is in $G^{j-1}(u_j)$, for some $j < i$. Moreover, s is in $G^{j-1} \setminus G^{j-1}(u_j)$ (because this is a supergraph of G^{i-1}). Again by Lemma 1, there is a vertex of $X_{u_j}^{j-1} \subseteq S^f$ in P, concluding the proof of the lemma. ∎

The next lemma proves that the performance ratio of the algorithm is at most $1 + \epsilon$.

Lemma 3. $|S^f| \leq (1 + \epsilon)opt(G)$.

Proof. We have that

$$|S^f| = \sum_{i=1}^{f} |A^i| \leq \sum_{i=1}^{f} (1 + \epsilon)opt(G^{i-1}(u_i))$$

$$= (1 + \epsilon) \sum_{i=1}^{f} opt(G^{i-1}(u_i)) \leq (1 + \epsilon)opt(G),$$

because the subgraphs $G^{i-1}(u_i)$ are vertex disjoint. ∎

Now we proceed with the description of a straightforward polynomial-time implementation of $Get\,(u, A)$.

Search the vertices of the tree T^{i-1} in postorder. Stop if the vertex u being visited is either the root or is such that $opt(G^{i-1}(u)) \geq t$. Let us show how we check whether $opt(G^{i-1}(u)) \geq t$ in polynomial time.

If we are searching vertex u, it is because all children of u have been searched and have an optimal solution with less than t vertices. Compute an optimal solution for each child v of u. This can be done in $O(n^{t+1})$ time by brute force: check all subsets of $G(v)$ of size at most t. The time is polynomial, since $t = \lceil (tw(G) + 1)/\epsilon \rceil$ is fixed. Let s be the sum of the sizes of the solutions for the children of u.

Let us show that the optimum of $G^{i-1}(u)$ is at most $s + tw(G) + 1$. We do this by presenting a solution B for $G^{i-1}(u)$ of size at most $s + tw(G) + 1$. The set B is the union of X_u and an optimal solution for $G^{i-1}(v)$, for each child v of u. Thus $|B| \leq |X_u| + s \leq tw(G) + 1 + s$. Now we must prove that B is in fact a solution for $G^{i-1}(u)$. Let (s,t) be a pair in $\mathcal{C}(G^{i-1}(u))$ and P be a path in $G^{i-1}(u)$ from s to t. We need to show that there is a vertex of P in B. If there is a vertex of P in X_u, then clearly B contains a vertex of P. If, on the other hand, P contains no vertex of X_u, we must have all vertices of P in the same $G^{i-1}(v)$, for some child v of u, by Lemma 1. But B contains a solution for $G^{i-1}(v)$. Therefore, B contains a vertex of P. This completes the proof that B is a solution for $G^{i-1}(u)$, and so the optimum of $G^{i-1}(u)$ is at most $s + tw(G) + 1$.

Now, let us proceed with the description of $Get\,(u, A)$. If $s < t$, then $opt(G(u)) \leq s + tw(G) + 1 < t + tw(G) + 1$, and we can compute in polynomial time an optimal solution A_0 for $G(u)$ (by brute force in $O(n^{t+tw(G)+2})$ time, which is polynomial since $t = \lceil (tw(G) + 1)/\epsilon \rceil$). If $|A_0| < t$ then proceed to the next vertex in postorder. If $|A_0| \geq t$, then we output u and the

set $A = A_0 \cup X_u$. Note that in fact $opt(G^{i-1}(u)) = |A_0| \geq t$, $X_u \subseteq A$ and
$|A| \leq opt(G^{i-1}(u)) + (tw(G) + 1) \leq opt(G^{i-1}(u)) + t\epsilon \leq (1 + \epsilon)opt(G^{i-1}(u))$,
as desired. On the other hand, if $s \geq t$, then $t \leq s \leq opt(G^{i-1}(u)) \leq s +$
$tw(G) + 1 \leq s + t\epsilon \leq opt(G^{i-1}(u)) + opt(G^{i-1}(u))\epsilon = (1 + \epsilon)opt(G^{i-1}(u))$.
Thus B (from the previous paragraph) is a solution for $G^{i-1}(u)$ of size at most
$s + tw(G) + 1 \leq (1 + \epsilon)opt(G^{i-1}(u))$ that can be computed in polynomial time.
Moreover, $X_u \subseteq B$. So in this case we output u and $A = B$. This finishes the
description of $Get\ (u, A)$.

3   Edge Multicut

In this section we show that Edge Multicut can be reduced to Unrestricted
Vertex Multicut by a reduction that preserves approximability.

The reduction has the following property. If the instance of Edge Multicut is
a graph with bounded degree and bounded tree-width, then the corresponding
instance of Unrestricted Vertex Multicut has bounded tree-width.

Given a graph $G = (V, E)$, the *line graph of G* is the graph whose vertex set
is E and such that two of its vertices (edges of G) are adjacent if they share an
endpoint in G. In other words, the line graph of G is the graph (E, L), where
$L = \{ef : e, f \in E$ and e and f have a common endpoint$\}$.

Consider an instance of Edge Multicut, that is, a graph $G = (V, E)$ and a set $\mathcal{C}$
of pairs of distinct vertices of G. Let us describe the corresponding instance of
Unrestricted Vertex Multicut. The input graph for Unrestricted Vertex Multicut
is the line graph of G, denoted by G'. Now let us describe the set of pairs of
vertices of G'. For each pair (s, t) in $\mathcal{C}$, we have in $\mathcal{C}'$ all pairs (e, f) such that e
has s as endpoint and f has t as endpoint.

Clearly G' can be obtained from G in polynomial time. Note that $\mathcal{C}'$ has at
most $k\Delta^2$ pairs, where $k = |\mathcal{C}|$ and Δ is the maximum degree of G. Also $\mathcal{C}'$ can
be obtained from G and $\mathcal{C}$ in polynomial time.

The following theorem completes the reduction.

Theorem 4. *S is a solution for Edge Multicut in G if and only if S is a solution
for Unrestricted Vertex Multicut in G'.*

Proof. Consider a solution S of Edge Multicut in G, that is, a set S of edges of
G such that any pair in $\mathcal{C}$ is disconnected in $(V, E - S)$. Note that $S \subseteq E(G) =
V(G')$. Let us verify that the removal of S from G' disconnects all pairs in $\mathcal{C}'$.
For any pair (e, f) in $\mathcal{C}'$, there are s and t in $V(G)$ such that s is an endpoint
of e, t is an endpoint of f and the pair (s, t) is in $\mathcal{C}$. Moreover, a path P' in G'
from e to f corresponds to a path P in G from s to t whose edges are a subset of
the vertices in P' (which are edges of G). Since S is a solution of Edge Multicut
in G, there must be an edge of P in S, which means that there is a vertex of P'
in S. Hence S is a solution for Unrestricted Vertex Multicut in G'.

Conversely, let S be a solution for Unrestricted Vertex Multicut in G', that
is, S is a set of edges of G whose removal from G' disconnects all pairs of vertices
of G' in $\mathcal{C}'$. Let (s, t) be a pair in $\mathcal{C}$, and P a path in G from s to t. (Recall that,

by the description of Edge Multicut, $s \neq t$.) Let e be the first edge of P and f the last one (possibly e=f). Clearly s is incident to e, and t to f. Thus (e, f) is a pair in $\mathcal{C}'$. Corresponding to P, there is a path P' in G' from e to f containing as vertices all edges of P. Since S is a solution for Unrestricted Vertex Multicut in G' and (e, f) is in $\mathcal{C}'$, S must contain a vertex of P'. Therefore there is an edge of P in S, which implies that S is a solution of Edge Multicut in G. ∎

The next lemma shows the previously mentioned property of this reduction.

Lemma 5. *If G has bounded degree and bounded tree-width, then the line graph of G has bounded tree-width.*

Proof. Denote by G' the line graph of G. Let us present a tree decomposition of G' whose tree-width is at most $(tw(G)+1)\Delta$, where Δ is the maximum degree of G.

Let $\Theta = (T, (X_u)_{u \in V(T)})$ be a tree decomposition of G of width $tw(G)$. For each $u \in V(T)$, let Y_u be the set of edges of G incident to some vertex in X_u. First let us prove that $\Theta' = (T, (Y_u)_{u \in V(T)})$ is a tree decomposition of G'. For this, given an edge e of G, denote by T_e the subgraph of T induced by those vertices in T for which Y_u contains e. We shall prove that (1) any edge h of G' has both endpoints in Y_u, for some u in $V(T)$; and (2) that T_e is a tree for any edge e of G'.

The endpoints of an edge h of G' are two edges e and f of G with a common endpoint, say, v. But $v \in X_u$ for some $u \in V(T)$. This implies that both e and f belong to Y_u, proving (1). For (2), let e be a vertex of G', that is, an edge $e = xy$ of G. For any u such that $e \in Y_u$, we must have that either $x \in X_u$ or $y \in X_u$. Therefore $T_e = T_x \cup T_y$. We know that the subgraphs T_x and T_y of T are subtrees of T. Moreover, T_x and T_y have a vertex in common, because both x and y belong to the same X_u, for some $u \in V(T)$. Hence T_e is a subtree of T. This completes the proof that Θ' is a tree decomposition of G'.

To verify that the width of Θ' is at most $(tw(G) + 1)\Delta$, just note that $|Y_u| \leq |X_u|\Delta$, for all $u \in V(T)$. ∎

The next corollary is a consequence of the previous reduction and the PTAS given in Section 2.1.

Lemma 6. *There is a PTAS for Edge Multicut in bounded-degree graphs with bounded tree-width.*

In fact we know how to implement the PTAS given in Section 2.1, for Edge Multicut in bounded-degree trees, in time $O((n + k)\lceil 1/\epsilon \rceil d^{d\lceil 1/\epsilon \rceil + 2})$, where n is the number of vertices of the tree, k is the number of (s_i, t_i) pairs, d is the maximum degree of the tree and $1 + \epsilon$ is the desired approximation ratio of the algorithm. The size of the input is $\Theta(n + k)$. We omit the description of this linear-time implementation in this extended abstract.

4   Complexity Results

In this section, we examine the complexity of Edge, Vertex and Unrestricted Vertex Multicut. First we prove that Edge and Vertex Multicut are NP-hard in bounded-degree trees, while Unrestricted Vertex Multicut is NP-hard in series-parallel graphs of bounded degree. Second, we show that the Weighted Edge Multicut is Max SNP-hard in binary tree. Finally we prove that Edge, Vertex and Unrestricted Vertex Multicut are Max SNP-hard in walls (defined in Section 4).

Theorem 7. *Edge Multicut in binary trees is NP-hard.*

Proof.   The reduction is from 3-SAT, a well-known NP-complete problem [8].

Consider an instance Φ of 3-SAT, that is, a set of m clauses $C_1, C_2, \ldots, C_m$ on n variables $x_1, x_2, \ldots, x_n$, each clause with exactly three literals.

Let us construct an instance of Edge Multicut: a binary tree T and a set of pairs of distinct vertices of T. The tree T is built as follows. For each variable x_i, there is a gadget as depicted in Figure 2 (a). The gadget consists of a binary tree with three vertices: the root and two leaves, one labeled x_i and the other labeled $\overline{x}_i$. For each clause C_j, there is a gadget as depicted in Figure 2 (b). The gadget consists of a binary tree with five vertices: the root, one internal vertex and three leaves, each one labeled by one of the literals in C_j.

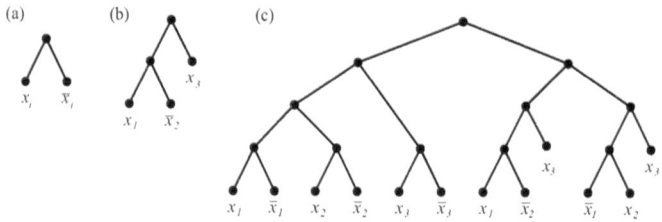

Fig. 2. (a) The gadget for variable x_i. (b) The gadget for clause $C_j = \{x_1, \overline{x}_2, x_3\}$. (c) Tree T built for the instance $\Phi = (x_1 \vee \overline{x}_2 \vee x_3) \wedge (\overline{x}_1 \vee x_2 \vee x_3)$, that is, $C_1 = \{x_1, \overline{x}_2, x_3\}$ and $C_2 = \{\overline{x}_1, x_2, x_3\}$.

The tree T is built from these $n + m$ gadgets by arbitrarily connecting them using new vertices to get a binary tree. See Figure 2 (c) for an example.

Next, we give the set of pairs of vertices of T in our instance. For each variable x_i, there is a pair with the vertices labeled x_i and $\overline{x}_i$ in its gadget. For each clause C_j, there are two pairs: one formed by the two leaves that are siblings and the other formed by the last leaf and the internal vertex. Finally, each vertex labeled $\tilde{x}_i$ in the gadget for a clause is paired to the vertex labeled $\tilde{x}_i$ in the gadget for the variable x_i, where $\tilde{x}_i \in \{x_i, \overline{x}_i\}$. This ends the construction of the instance for Edge Multicut. Note that all this can be done in polynomial time in the size of Φ.

The next lemma completes the proof of Theorem 7. ■

Lemma 8. *Φ is satisfiable if an only if there is a solution for T of size exactly n + 2m. Moreover, we can construct in polynomial time such a solution for T from a truth assignment for Φ and vice versa.*

Proof. Assume $Φ$ is satisfiable. Let us present a solution S for T of size exactly $n + 2m$. The edge set S consists of two types of edges:

1. For each variable x_i, S contains the edge in the gadget for x_i incident to the leaf labeled x_i if $x_i = TRUE$ or to the leaf labeled $\overline{x}_i$ if $x_i = FALSE$.
2. For each clause C_j, S contains two distinct edges in the gadget for C_j. These edges are such that (1) they disconnect the two pairs in the gadget, and (2) the only leaf that is still connected to the root of the gadget is a leaf with a label $\tilde{x}_i \in C_j$ such that $\tilde{x}_i = TRUE$. (The four possible choices for the two edges are shown in Figure 3.)

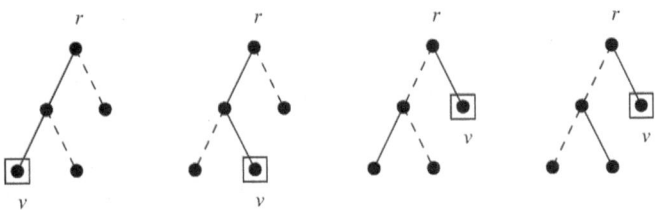

Fig. 3. Possible choices of two edges, the dashed edges, in the gadget for a clause that leave exactly one leaf (the marked leaf v) connected to the root r.

Clearly such set S has exactly $n + 2m$ edges and can be constructed in polynomial time from $Φ$. Let us prove that S is in fact a solution for T. It is easy to see that S disconnected the pairs for the variables, and the pairs for the clauses. The remaining pairs consist of two vertices labeled by a literal $\tilde{x}_i$, one in the variable gadget for x_i and the other in a clause gadget. If $\tilde{x}_i = TRUE$, then the edge in the variable gadget incident to the vertex labeled $\tilde{x}_i$ is in S, guaranteeing that the pair is disconnected. If $\tilde{x}_i = FALSE$, then the vertex labeled $\tilde{x}_i$ in the clause gadget is disconnected from the root of this gadget, and therefore, from the gadget for x_i. Thus S is a solution for T, and it has exactly $n + 2m$ edges.

Let us prove the inverse implication. Assume there is a solution S for T with exactly $n + 2m$ edges: one per variable and two per clause (one for each of the "disjoint" pairs). More specifically, S has exactly one edge in each variable gadget, and exactly two edges in each clause gadget in one of the configurations of Figure 3. Set $x_i = TRUE$ if the edge of S in the gadget for x_i is incident to the vertex labeled x_i; set $x_i = FALSE$ otherwise. Clearly, we can determine this truth assignment in polynomial time.

For each clause C_j, there is exactly one leaf v in the gadget for C_j that is connected to the root r of the gadget. Let $\tilde{x}_i \in \{x_i, \overline{x}_i\}$ be the label for this leaf. There is a pair formed by this leaf v and the leaf in the gadget for x_i whose label is $\tilde{x}_i$. In S, there must be an edge e in the path between these two leaves. Since leaf v is connected to the root r of the gadget for C_j and all edges in S are either in a variable gadget or in a clause gadget, this edge e has to be in the variable gadget. This means e is the edge incident to the leaf labeled $\tilde{x}_i$ in the gadget for x_i. Hence $\tilde{x}_i = TRUE$, and the clause is satisfied. Since this holds for all the clauses, the given assignment makes Φ $TRUE$, implying that Φ is satisfiable. ∎

Theorem 9. *Vertex Multicut in trees with maximum degree at most four is NP-hard.*

We omit the proof. The construction is similar to the one used in Theorem 7.

Theorem 10. *Unrestricted Vertex Multicut in series-parallel graphs with maximum degree at most three is NP-hard.*

We omit the proof. The construction is similar to the one used in Theorem 7.

Theorem 11. *Weighted Edge Multicut is Max SNP-hard in binary trees.*

Proof sketch. Let us reduce Edge Multicut in stars to Weighted Edge Multicut in binary trees. From an instance of the Unweighted Edge Multicut restricted to stars, we construct an instance of the Weighted Edge Multicut restricted to binary trees in the following way: for each leaf of the star S, there is a corresponding leaf in the binary tree T. The pairs are the same (we may assume there is no pair involving the root of the star). We connect the leaves of T arbitrarily into a binary tree. The edges in T incident to the leaves get weight one and all other edges of T get weight $2n + 1$, where n is the number of leaves in the star S (which is the same as the number of leaves in the tree T we construct). Any solution within twice the optimum for the Weighted Edge Multicut instance we constructed will contain only edges of T incident to the leaves, since any other edge is too heavy (removing all edges incident to the leaves, we get a solution of weight n). Then it is easy to see that any optimal solution for the Weighted Edge Multicut instance we constructed corresponds to an optimal solution for the original Unweighted Multicut star instance, and *vice versa*. Also approximability is preserved by this reduction. ∎

A *wall of height* h consists of $h + 1$ vertex disjoint paths $R_0, \ldots, R_h$, which we call *rows*, and $h + 1$ vertex disjoint paths $L_0, \ldots, L_h$, which we call *columns*. A wall of height six is depicted in Figure 4 (a). The reader should be able to complete the definition by considering Figure 4 (a). The formal definition is as follows. Each row is a path of $2h + 2$ vertices. Each column, a path with $2h + 2$ vertices. Column r contains the $(2r + 1)^{st}$ and the $(2r + 2)^{nd}$ vertices of all rows, as well as the edge between them. For $i < h$ and even, each L_r contains an edge between the $(2r + 2)^{nd}$ vertex of R_i and the $(2r + 2)^{nd}$ vertex of R_{i+1}. For $i < h$

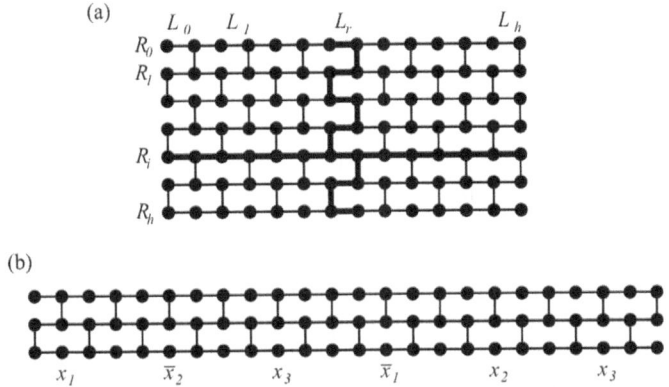

Fig. 4. (a) A wall of height six. The dark edges indicate row R_i and column L_r. (b) The three last rows of the wall built from $\Phi = (x_1 \vee \bar{x}_2 \vee x_3)(\bar{x}_1 \vee x_2 \vee x_3)$.

and odd, each L_r contains an edge between the $(2r+1)^{st}$ vertex of R_i and the $(2r+1)^{st}$ vertex of R_{i+1}. These are all the edges of the wall.

We prove that Edge, Vertex and Unrestricted Vertex Multicut are Max SNP-hard in walls. This means, by Arora et al. [1], that there is a constant $\epsilon > 0$ such that the existence of a polynomial-time approximation algorithm for any of the three versions of Multicut with performance ratio at most $1 + \epsilon$ implies that P=NP.

As in [18], we use the concept of *L-reduction*, which is a special kind of reduction that preserves approximability.

Let A and B be two optimization problems. We say A *L-reduces to* B if there are two polynomial-time algorithms f and g, and positive constants α and β, such that for each instance I of A,

1. Algorithm f produces an instance $I' = f(I)$ of B, such that the optima of I and I', of costs denoted $Opt_A(I)$ and $Opt_B(I')$ respectively, satisfy $Opt_B(I') \leq \alpha \cdot Opt_A(I)$, and
2. Given any feasible solution of I' with cost c', algorithm g produces a solution of I with cost c such that $|c - Opt_A(I)| \leq \beta \cdot |c' - Opt_B(I')|$.

Theorem 12. *Edge, Vertex and Unrestricted Vertex Multicut are Max SNP-hard in walls.*

Proof sketch. The reduction is from the well-known Max SNP-hard problem MAX 3-SAT [18]. We show the reduction for Unrestricted Vertex Multicut. The other two reductions are similar.

The first part of the *L*-reduction is the polynomial-time algorithm f and the constant α. Given any instance Φ of MAX 3-SAT, f produces an instance $W, \mathcal{C}$ of Unrestricted Vertex Multicut such that W is a wall. Also, the cost of the optimum of $W, \mathcal{C}$ in Unrestricted Vertex Multicut, denoted $Opt_{MC}(W, \mathcal{C})$, is at most α times the cost of the optimum of Φ in MAX 3-SAT, denoted by $Opt_{SAT}(\Phi)$, i.e., $Opt_{MC}(W, \mathcal{C}) \leq \alpha \cdot Opt_{SAT}(\Phi)$.

Consider an instance Φ of MAX 3-SAT, that is, a collection of m clauses on n variables $x_1, \ldots, x_n$, each consisting of exactly three literals. Let us describe the corresponding instance for Unrestricted Vertex Multicut. The wall W is a wall of height $6m$. To describe the collection $\mathcal{C}$ of pairs of vertices of W, consider the last row of W partitioned into m same length paths, each one associated to one of the clauses of Φ. Each path has length 12. Label the $2^{nd}, 6^{th}$ and 10th vertices in the jth path each with one of the literals in the jth clause. See Figure 4 (b) for an example. For each pair of vertices u, v in W, u labeled x_i and v labeled $\overline{x}_i$, include into $\mathcal{C}$ the pair u, v. For each clause, include three pairs. The three pairs formed by each two of the vertices labeled by its three literals. This ends the description of the instance of Unrestricted Vertex Multicut.

First note that W and $\mathcal{C}$ can be obtained in polynomial time in the size of Φ.

Lemma 13. $Opt_{MC}(W, \mathcal{C}) \leq 6 \cdot Opt_{SAT}(\Phi)$.

Proof sketch. $W, \mathcal{C}$ clearly has a solution of size $3m$. Also $Opt_{SAT}(\Phi) \geq m/2$. ∎

Lemma 14. *From a solution to Φ of size s, $0 \leq s \leq m$, we can obtain a solution to $W, \mathcal{C}$ of size $3m - s$ and vice versa.*

Proof sketch.   Given an assignment that satisfies s clauses of Φ, let S be the set of all labeled vertices of W except one labeled vertex per satisfied clause. Choose to not include in S a vertex labeled by a literal that is assigned TRUE. One can verify that this set S is a solution for $W, \mathcal{C}$ of size $3m - s$.

Now, consider a solution S for $W, \mathcal{C}$ of size $3m - s$. Since W has height $6m$, there is a row R_i of W which has no vertex of S. Set to TRUE any literal which appears as a label of a vertex of W that is connected to R_i after the removal of S. If some variable was not assigned a value by this rule, assign it an arbitrary value. Note that, since vertices labeled x_i are not connected to vertices labeled $\overline{x}_i$ after the removal of S, the assignment is well-defined. Consider the six columns of the wall corresponding to the jth clause of Φ. S should contain at least two vertices in these columns, otherwise there would be a path connecting at least two of the labeled vertices in these columns. This means that at least s clauses have only two vertices removed from their columns of W. Thus one of the labeled vertices is connected to row R_i, meaning that this clause is satisfied. ∎

The previous two lemmas can be used in an obvious way to show the reduction we presented is an L-reduction. ∎

Acknowledgments

The first two authors would like to thank Howard Karloff for suggesting the problem, and for some helpful discussions.

References

1. S. Arora, C. Lund, R. Motwani, M. Sudan, and M. Szegedy. Proof verification and hardness of approximation problems. *Proc. 33rd IEEE Symposium on Foundations of Computer Science*, pages 14–23, 1992.

2. S. Arnborg and J. Lagergren. Problems easy for tree-decomposable graphs. *Journal of Algorithms*, 12(2):308–340, 1991.

3. E. Dahlhaus, D. S. Johnson, C. H. Papadimitriou, P. D. Seymour, and M. Yannakakis. The complexity of multiterminal cuts. *SIAM Journal on Computing*, 23(4):864–894, 1994.

4. E. A. Dinic. Algorithm for solution of a problem of maximum flow in networks with power estimation. *Soviet Mathematics Doklady*, 11:1277–1280, 1970.

5. G. Even, J. S. Naor, B. Schieber, and L. Zosin. Approximating minimum subset feedback sets in undirected graphs with applications. *Proc. 4th Israel Symposium on Theory of Computing and Systems*, pages 78–88, 1996.

6. L. R. Ford and D. R. Fulkerson. Maximal flow through a network. *Canadian Journal of Mathematics*, 8:399–404, 1956.

7. L. R. Ford and D. R. Fulkerson. A suggested computation for maximal multicommodity network flows. *Management Science*, 5:97–101, 1958.

8. M. R. Garey and D. S. Johnson. *Computers and Intractability*. Freeman, 1979.

9. N. Garg, V. Vazirani, and M. Yannakakis. Approximate max-flow min-(multi)cut theorems and their applications. *SIAM Journal on Computing*, 25(2):235–251, 1996.

10. N. Garg, V. Vazirani, and M. Yannakakis. Primal-dual approximation algorithms for integral flow and multicut in trees. *Algorithmica*, 18(1)3–20, 1997.

11. T. C. Hu. Multicommodity network flows. *Operations Research*, 9:898–900, 1963.

12. A. Itai. Two-commodity flow. *Journal of ACM*, 25:596–611, 1978.

13. P. Klein, A. Agrawal, R. Ravi, and S. Rao. Approximation through multicommodity flow. *Proc. 31st IEEE Symposium on Foundations of Computer Science*, pages 726–737, 1990.

14. P. Klein, S. Plotkin, S. Rao, and E. Tardos. Approximation algorithms for Steiner and directed multicuts. *Journal of Algorithms*, 22(2):241–269, 1997.

15. F. T. Leighton and S. Rao. An approximate max-flow min-cut theorem for uniform multicommodity flow problems with application to approximation algorithms. *Proc. 29th IEEE Symposium on Foundations of Computer Science*, pages 422–431, 1988.

16. D. B. Shmoys. Computing near-optimal solutions to combinatorial optimization problems. In W. Cook and L. Lovász, editors, *Combinatorial Optimization, DIMACS Series in Discrete Mathematics and Theoretical Computer Science, Vol. 20*, pages 355–397. AMS Publications, 1995.

17. S. Plotkin and É. Tardos. Improved bounds on the max-flow min-cut ratio for multicommodityfFlows. *Proc. 25th Annual ACM Symp. on Theory of Computing*, pages 691–697, 1993.

18. C. H. Papadimitriou and M. Yannakakis. Optimization, approximation, and complexity classes. *Journal of Computer and System Sciences*, 43:425–440, 1991.

19. N. Robertson and P. Seymour. Graph minor II. Algorithmic aspects of tree-width. *Journal of Algorithms*, 7:309–322, 1986.

20. É. Tardos and V. V. Vazirani. Improved bounds for the max-flow min-multicut ratio for planar and $K_{r,r}$-free graphs. *Information Processing Letters*, 47(2):77-80, 1993.
21. J. van Leeuwen. Graph algorithms. *Handbook of Theoretical Computer Science, Vol. A*, chapter 10, pages 525–631. The MIT Press/Elsevier, 1990.

Approximating Disjoint-Path Problems Using Greedy Algorithms and Packing Integer Programs *

Stavros G. Kolliopoulos and Clifford Stein

Dartmouth College, Department of Computer Science
Hanover, NH 03755–3510, USA
{stavros, cliff}@@cs.dartmouth.edu

Abstract. The *edge* and *vertex-disjoint path* problems together with their *unsplittable flow* generalization are NP-hard problems with a multitude of applications in areas such as routing, scheduling and bin packing. Given the hardness of the problems, we study polynomial-time approximation algorithms with bounded performance guarantees. We introduce techniques which yield new algorithms for a wide range of disjoint-path problems. We use two basic techniques. First, we propose simple greedy algorithms for edge- and vertex-disjoint paths and second, we propose the use of a framework based on *packing integer programs* for more general problems such as unsplittable flow. As part of our tools we develop improved approximation algorithms for a class of packing integer programs, a result that we believe is of independent interest.

1 Introduction

This paper examines approximation algorithms for disjoint-path problems and their generalizations. In the *edge(vertex)-disjoint path* problem, we are given a graph $G = (V, E)$ and a set $\mathcal{T}$ of *connection requests*, also called *commodities*. Every connection request in $\mathcal{T}$ is a vertex pair (s_i, t_i), $1 \leq i \leq K$. The objective is to connect a maximum number of the pairs via edge(vertex)-disjoint paths. For the vertex-disjoint paths problem, the connection requests are assumed to be disjoint. We call the set of connected pairs *realizable*. A generalization of the edge-disjoint paths problem is *multiple-source unsplittable flow*. In this problem every commodity k in the set $\mathcal{T}$ has an associated demand ρ_k, and every edge e has a capacity u_e. The demand ρ_k must be routed on a single path from s_k to t_k. The objective is to maximize the sum of the demands that can be fully routed while respecting the capacity constraints. Wlog, we assume that $\max_k \rho_k = 1$, and following the standard definition of the problem in the literature, $u_e \geq 1$, $\forall e \in E$. When all demands and capacities are 1 in the multiple-source unsplittable

---

* Research partly supported by NSF Award CCR-9308701 and NSF Career Award CCR-9624828.

R. E. Bixby, E. A. Boyd, and R. Z. Ríos-Mercado (Eds.): IPCO VI
LNCS 1412, pp. 153–168, 1998. © Springer–Verlag Berlin Heidelberg 1998

flow problem we obtain the edge-disjoint path problem. (See [10,14] for further applications and motivation for unsplittable flow.) In all the above problems one can assign a weight $w_i \leq 1$ to each connection request and seek to find a realizable set of maximum total weight. In this paper we will state explicitly when we deal with the weighted version of a problem.

Both the edge- and vertex-disjoint path problems are fundamental, extensively studied (see e.g. [26,6,27,21,10,13,3]), NP-hard problems [9], with a multitude of applications in areas such as telecommunications, VLSI and scheduling. Despite the attention they have received, disjoint-path problems on general graphs remain notoriously hard in terms of approximation; even for edge-disjoint paths, no algorithm is known which can find even an $\omega(1/\sqrt{|E|})$ fraction of the realizable paths.

In approximating these problems, we use the traditional notion of a ρ-*approximation algorithm*, $\rho > 1$, which is one that outputs, in polynomial time, a realizable set of size at least $1/\rho$ times the optimum. We will also give and refer to algorithms which output a realizable set whose size is a non-linear function of the optimum OPT, such as $OPT^2/|E|$.

Overview of Previous Work. Two main approaches have been followed for approximation.

(i) The first approach, which we call the rounding approach, consists of solving a fractional relaxation and then use rounding techniques to obtain an integral solution. The fractional relaxation is typically multicommodity flow and the rounding techniques used to date involved sophisticated and non-standard use of randomized rounding [31]. The objective value of the resulting solution is compared to the *fractional optimum* y^*, which is an upper bound on the *integral optimum*, OPT. This approach has been the more successful one and recently yielded the first approximation algorithm for *uniform unsplittable flow* [31] which is the special case of unsplittable flow where all the capacities have the same value. Let d denote the *dilation* of the fractional solution, i.e. the maximum length of a flow path in the fractional relaxation. Bounds that rely on the dilation are particularly appealing for expander graphs where it is known that $d = O(polylog(n))$ [16,12]. The rounding approach yields, for unweighted uniform unsplittable flow (and thus for unweighted edge-disjoint paths as well) a realizable set of size $\Omega(\max\{(y^*)^2/|E|, y^*/\sqrt{|E|}, y^*/d\})$ and an $\Omega(\max\{(y^*)^2/|E|, y^*/d\})$ bound for the weighted version [31] . This approach is known to have limitations, e.g. it is known that a gap of $\Omega(\sqrt{|V|})$ exists between the fractional and integral optima for both the edge- and vertex-disjoint path problems on a graph with $|E| = \Theta(|V|)$ [7].

(ii) Under the second approach, which we call the routing approach, a commodity is never split, i.e. routed fractionally along more than one path during the course of the algorithm. In the analysis, the objective value of the solution is compared to an estimated upper bound on the OPT. This approach has found very limited applicability so far, one reason being the perceived hardness of deriving upper bounds on OPT without resorting to a fractional relaxation. The only example of this method we are aware of is the on-line Bounded Greedy Algorithm in

[10] whose approximation guarantee depends also on the diameter of the graph. The algorithm can be easily modified into an off-line procedure that outputs realizable sets of size $\Omega(OPT/\sqrt{|E|})$ $(\Omega(OPT/\sqrt{|V|}))$ for edge(vertex)-disjoint paths. The $\Omega(OPT/\sqrt{|V|})$ bound is the best known bound to date for vertex-disjoint paths.

Table 1. Known approximation bounds for edge-disjoint paths (EDP), uniform capacity unsplittable flow (UCUFP), and general unsplittable flow (UFP), Ω-notation omitted. E_o denotes the set of edges used by some path in an integral optimal solution and d_o the average length of the paths in the same solution. Results with no citation come from the present paper. Our $y^*/\sqrt{|E|}$ bound for the weighted EDP problem holds under the assumption that the number of connection requests $K = O(|E|)$.

	routing approach	rounding approach
unweighted EDP	$\frac{OPT}{\sqrt{|E|}}$ [10], $\frac{OPT}{\sqrt{|E_o|}}$, $\frac{OPT^2}{|E_o|}$, $\frac{OPT}{d_o}$	$\frac{y^*}{\sqrt{|E|}}$ [31], $\frac{(y^*)^2}{|E|}$ [31], $\frac{y^*}{d}$ [31]
weighted EDP	—	$\frac{y^*}{\sqrt{|E|}}$, $\frac{(y^*)^2}{|E|}$ [31], $\frac{y^*}{d}$ [31]
weighted UCUFP	—	$\frac{(y^*)^2}{|E|}$ [31], $\frac{y^*}{d}$ [31]
weighted UFP	—	$\frac{y^*}{\log|E|\sqrt{|E|}}$, $\frac{(y^*)^2}{|E|\log^3|E|}$, $\frac{y^*}{d}$

Our Contribution. In this paper we provide techniques for approximating disjoint-path problems that bear on both of the above approaches. Tables 1 and 2 summarize previous and new bounds for edge-, vertex-disjoint path and unsplittable flow problems.

Under the routing approach (approach (**ii**)) we give a simple deterministic greedy algorithm GREEDY_PATH for edge-disjoint paths that has performance guarantees comparable to those obtained by the multicommodity flow based algorithms. Greedy algorithms have been extensively studied in combinatorial optimization due to their elegance and simplicity. Our work provides another example of the usefulness of the greedy method. The underlying idea is that if one keeps routing commodities along sufficiently short paths the final number of commodities routed is lowerbounded with respect to the optimum.

GREEDY_PATH outputs a realizable set of size $\Omega(\max\{OPT^2/|E_o|,$ $OPT/\sqrt{|E_o|}\})$ for the edge-disjoint path problem. Here $E_o \subseteq E$ is the set of edges used by the paths in an optimal solution. Note that $OPT^2/|E_o|$ always dominates $OPT/\sqrt{|E_o|}$ in the unweighted case that we consider; we give both bounds to facilitate comparison with existing work and to conform to the traditional notion of a ρ-approximation algorithm. Our approximation existentially improves upon the multicommodity-flow based results when $|E_o| = o(|E|)$, i.e. when the optimal solution uses a small portion of the edges of the graph. Another bound can be obtained by noticing that $OPT^2/|E_o| = OPT/d_o$, where d_o denotes the average length of the paths in an optimal solution.

Essentially the same algorithm, GREEDY_VPATH, obtains for the vertex-disjoint path problem a realizable set of size $\Omega(\max\{OPT^2/|V_o|, OPT/\sqrt{|V_o|}\})$, where $V_o \subseteq V$ is the set of vertices used by the paths in an optimal solution. Recall that the best known bound to date is $t = \Omega(OPT/\sqrt{|V|})$. The realizable set output by our algorithm has size $\Omega(t^2)$ and potentially better than this when $|V_o| = o(|V|)$. This is a significant improvement when $OPT = \omega(\sqrt{|V|})$. For example, when $OPT = \Omega(|V|)$, we obtain a constant-factor approximation. Again an $\Omega(OPT/d_o)$ guarantee follows immediately.

Table 2. Known approximation bounds for vertex-disjoint paths, Ω-notation omitted. V_o denotes the set of vertices used by some path in an integral optimal solution and d_o the average length of the paths in the same solution. Results with no citation come from the present paper.

	routing approach				rounding approach		
unweighted	$\frac{OPT}{\sqrt{|V|}}$ [10],	$\frac{OPT}{\sqrt{|V_o|}}$,	$\frac{OPT^2}{|V_o|}$,	$\frac{OPT}{d_o}$	$\frac{y^*}{\sqrt{|V|}}$,	$\frac{(y^*)^2}{|V|}$,	$\frac{y^*}{d}$
weighted	—				$\frac{y^*}{\sqrt{|V|}}$,	$\frac{(y^*)^2}{|V|}$,	$\frac{y^*}{d}$

We turn to the rounding approach (approach **(i)**) to handle the weighted disjoint path and unsplittable flow problems. We propose the use of *packing integer programs* as a unifying framework that abstracts away the need for customized and complex randomized rounding schemes. A *packing integer program* is of the form maximize $c^T \cdot x$, subject to $Ax \leq b$, $A, b, c \geq 0$. We first develop, as part of our tools, an improved approximation algorithm for a class of packing integer programs, called *column restricted*, that are relevant to unsplittable flow problems. Armed with both this new algorithm and existing algorithms for general packing integer programs, we show how packing formulations both provide a unified and simplified derivation of many results from [31] and lead to new ones. In particular, we obtain the first approximation algorithm for weighted multiple-source unsplittable flow on networks with arbitrary demands and capacities and the first approximation algorithm for weighted vertex-disjoint paths. Further, we believe that our new algorithm for column-restricted packing integer programs is of independent interest. We now elaborate on our results under the rounding approach, providing further background as necessary.

1.1   Packing Integer Programs

Packing integer programs are a well-studied class of integer programs that can model several NP-complete problems, including independent set, hypergraph k-matching [19,1], job-shop scheduling [23,28,33,20] and many flow and path related problems. Many of these problems seem to be difficult to approximate, and not much is known about their worst-case approximation ratios. Following [30] a packing integer program (PIP) is defined as follows.

Definition 1. *Given $A \in [0,1]^{m \times n}$, $b \in [1, \infty)^m$ and $c \in [0,1]^n$ with $\max_j c_j = 1$, a PIP $\mathcal{P} = (A, b, c)$ seeks to maximize $c^T \cdot x$ subject to $x \in Z_+^n$ and $Ax \leq b$. Constraints of the form $0 \leq x_j \leq d_j$ are also allowed. If $A \in \{0,1\}^{m \times n}$, each entry of b is assumed integral. Let $B = \min_i b_i$, and α be the maximum number of non-zero entries in any column of A.*

The parameters B and α in the definition above appear in the approximation bounds. For convenience we call b_i the *capacity* of row i. The restrictions on the values of the entries of A, b, c are wlog; the values in an arbitrary packing program can be scaled to satisfy the above requirements [29]. We will state explicitly when some packing program in this paper deviates from these requirements. When $A \in \{0,1\}^{m \times n}$, we say that we have a *(0,1)-PIP*.

Previous Work on Packing Programs. The basic techniques for approximating packing integer programs have been the *randomized rounding* technique of Raghavan and Thompson [24,25] and the work of Plotkin, Shmoys and Tardos [23]. Let y^* denote the optimum value of the linear relaxation. Standard randomized rounding yields integral solutions of value $\Omega(y^*/m^{1/B})$ for general PIP's and $\Omega(y^*/m^{1/(B+1)})$ for (0,1)-PIP's [25] (see also [29].) Srinivasan [29,30] improved on the standard randomized rounding bounds and obtained bounds of $\Omega(y^*(y^*/m)^{1/(B-1)})$ and $\Omega(y^*/\alpha^{1/(B-1)})$ for general PIP's and $\Omega(y^*(y^*/m)^{1/B})$ and $\Omega(y^*/\alpha^{1/B})$ for (0,1)-PIP's.

New Results for Column-Restricted PIP's. The above results show that for various combinations of values for y^*, m and B, the bounds obtained for a (0,1)-PIP are significantly better than those for general PIP's. In fact they are always better when $y^* < m$. As another example, the approximation ratio $m^{1/(B+1)}$ obtained for a (0,1)-PIP is *polynomially better* than the approximation ratio of a PIP with the same parameters. Thus it is natural to ask whether we can bridge this gap. We make progress in this direction by defining a *column-restricted* PIP $\mathcal{P}_r$ as one where all non-zero entries of the j-th column of A have the same value $\rho_j \leq 1$. Column-restricted PIP's arise in applications such as unsplittable flow problems (see next section). We show how to obtain approximation guarantees for column-restricted PIP's that are similar to the ones obtained for (0,1)-PIP's. Let y_r^* denote the optimum of the linear relaxation of $\mathcal{P}_r$. We obtain an integral solution of value $\Omega(y_r^*/m^{1/(B+1)})$ and $\Omega(y_r^*/\alpha^{1/B})$. Letting $\sigma(y_r^*) = \Omega(y_r^*(y_r^*/m)^{1/B})$ we also obtain a bound that is at least as good as $\sigma(y_r^*)$ for $y_r^* < m \log n$ and in any case it is never worse by more than a $O(\log^{1/B} n)$ factor. Finally we show how to improve upon the stated approximations when $\max_j \rho_j$ is bounded away from 1. We develop the latter two results in a more complete version of this paper [15].

We now give an overview of our technique. First we find an optimum solution x^* to the linear relaxation of the column-restricted PIP $\mathcal{P}_r$. We partition the ρ_j's into a fixed number of intervals according to their values and generate a packing subproblem for each range. In a packing subproblem $\mathcal{P}^L$ corresponding to range L, we only include the columns of A with $\rho_j \in L$ and to each component of the b^L-vector we allocate only a fraction of the original b_i value, a fraction that is determined based on information from x^*. Next we find approximate solutions to each subproblem and combine them to obtain a solution to the original problem. Perhaps the key idea is in using the solution x^* to define the capacity allocation to the b^L-vector for subproblem $\mathcal{P}^L$. This generalizes previous work of the authors [14] on single-source unsplittable flow. The other key idea is that each subproblem can be approximated almost as well as a $(0, 1)$-PIP.

1.2   Applications of Packing to Approximation

We introduce a new framework for applying packing techniques to disjoint-path problems. First, we formulate an integer program (which is not necessarily a PIP) and solve a linear relaxation of this integer program to obtain a solution x. Typically this is a multicommodity flow problem. We then explicitly use the solution x to guide the formation of a column-restricted or $(0, 1)$ PIP. A related usage of a solution to the linear relaxation of integer programs in a different context can be found in [8,32]. An integral approximate solution to the created PIP will be an approximate solution to the original disjoint path problem (with possibly some small degradation in the approximation factor). This integral solution can be found using existing algorithms for approximating PIP's as a black box. Our algorithms apply to the case when there are weights on the commodities, and thus generalize those of Srinivasan for edge-disjoint paths. This approach yields four applications which we explain below.

Application 1: Weighted Unsplittable Flow. Let F_1, F_2, F_3 denote $(y^*)^2/|E|, y^*/d$ and $y^*/\sqrt{|E|}$ respectively. We obtain a realizable set of weight $\Omega(\max\{F_3/\log|E|, F_1/\log^3|E|, F_2\})$ for unsplittable flow with arbitrary demands and capacities. In fact we can give a better F_1-type bound for small enough y^*, whose analytical form is complicated. See [15] for further details. In the case where the number of commodities $K = O(|E|)$ we show how to obtain also an $\Omega(\max\{F_1/\log|E|, F_3\})$ bound. Notice that for the edge-disjoint path problem this is a natural assumption since at most $|E|$ connection requests can be feasibly routed. We also note that a ρ-approximation for y^* entails an $O(\rho\log|E|)$ approximation for the problem of *routing in rounds* [2,10]. We do not pursue any further the latter problem in this extended abstract.

Application 2: Weighted Vertex-Disjoint Paths. We give an algorithm that outputs a solution of value $\Omega(\max\{(y^*)^2/|V|, y^*/\sqrt{|V|}, y^*/d\})$. The algorithm relies on the observation that, after solving a fractional relaxation, the problem of rounding is essentially an instance of hypergraph matching; thus it can be formulated as a packing program with $|V|$ constraints. The algorithm is surprisingly simple but the performance guarantee matches the integrality gap known for the problem [7].

Application 3: Routing with Low Congestion. A problem that has received a lot of attention in the literature on routing problems (e.g. [25,17,23,22,10,14]) is that of minimizing *congestion,* i.e. the factor by which one is allowed to scale up capacities in order to achieve an optimal (or near-optimal) realizable set. In our usage of packing in the rounding algorithms we have assumed that the parameter B of the packing program is equal to 1. Allowing $B > 1$ is equivalent to allowing congestion B in the corresponding disjoint-path problem. Thus another advantage of the packing approach is that tradeoffs with the allowed congestion B can be obtained immediately by plugging in B in the packing algorithms that we use as a black box. For example the approximation for edge-disjoint paths becomes $\Omega(\max\{y^*(y^*/|E|\log|E|)^{1/B}, y^*/|E|^{1/(B+1)}, y^*/d^{1/B}\})$, when the number of connection requests is $O(|E|)$. Our congestion tradeoffs generalize previous work by Srinivasan [31] who showed the $\Omega(y^*/d^{1/B})$ tradeoff for uniform capacity unsplittable flow. We do not state the tradeoffs explicitly for the various problems since they can be obtained easily by simple modifications to the given algorithms.

Application 4: Independent Set in the Square of a Graph. Given a graph $G = (V, E)$ the k-th power $G^k = (V, E^k)$ of G is a graph where two vertices are adjacent if and only if they are at distance at most k in G. We further demonstrate the power of packing formulations by providing an $O(\sqrt{|V|})$ approximation algorithm for finding a maximum independent set in the square of a graph. We also give results that depend on the maximum vertex degree Δ in G. Our approximation ratio cannot be polynomially improved in the sense that no $(n/4)^{1/2-\varepsilon}$ approximation, for any fixed $\varepsilon > 0$, can be obtained in polynomial time unless $NP = ZPP$. Studying NP-hard problems in powers of graphs is a topic that has received some attention in the literature [5,34,18,4].

Independently of our work, Baveja and Srinivasan (personal communication) have obtained results similar to ours for approximating vertex-disjoint paths under the rounding approach, unsplittable flow and column-restricted packing integer programs. Their work builds on the methods in [31].

2   Approximating a Column-Restricted PIP

In this section we present the approximation algorithm for column-restricted PIP's. Let $\mathcal{P} = (A, b, c)$ be a column-restricted PIP. We call $\rho_j \leq 1$, the value of the non-zero entries of the j-th column, $1 \leq j \leq n$, the *value* of column j. Throughout this section we assume that there is a polynomial-time algorithm that given a $(0, 1)$-PIP with fractional optimum y^* outputs an integral solution of value at least $\sigma(m, B, \alpha, y^*)$ where m, B, α are the parameters of the packing program. For example a known σ is $\Omega(y^*/m^{1+B})$. We start by providing a subroutine for solving a column-restricted PIP when the column values are close in range.

Theorem 1. *Let $\mathcal{P} = (A, b, c)$ be a column-restricted PIP where all column values ρ_j are equal to ρ and where each $b_i = k_i \Gamma \rho$, $\Gamma \geq 1$, k_i positive integer, $1 \leq i \leq m$. Here $\min_i b_i$ is not necessarily greater than 1. Then we can find in polynomial time a solution of value at least $\sigma(m, k\Gamma, \alpha, y^*)$, where y^* denotes the optimum of the linear relaxation of $\mathcal{P}$ and $k = \min_i k_i$.*

Proof. Transform the given system $\mathcal{P}$ to a $(0,1)$-PIP $\mathcal{P}' = (A', b', c)$ where $b'_i = k_i \Gamma$, and $A'_{ij} = A_{ij}/\rho$. Every feasible solution (either fractional or integral) $\bar{x}$ to $\mathcal{P}'$ is a feasible solution to $\mathcal{P}$ and vice versa. Therefore the fractional optimum y^* is the same for both programs. Also the maximum number of non-zero entries on any column is the same for A and A'. Thus we can unambiguously use α for both. We have assumed that there is an approximation algorithm for $\mathcal{P}'$ returning a solution with objective value $\sigma(m, k\Gamma, \alpha, y^*)$. Invoking this algorithm completes the proof. □

The proof of the following lemma generalizes that of Lemma 4.1 in [11].

Lemma 1. *Let $\mathcal{P} = (A, b, c)$, be a column-restricted PIP with column values in the interval $(a_1, a_2]$, and $b_i \geq \Gamma a_2$, $\forall i$ and some number $\Gamma \geq 1$. Here $\min_i b_i$ is not necessarily greater than 1. There is an algorithm α_PACKING that finds a solution g to $\mathcal{P}$ of value at least $\sigma(m, \Gamma, \alpha, \frac{a_1}{2a_2} y^*)$, where y^* is the optimum of the fractional relaxation of $\mathcal{P}$. The algorithm runs in polynomial time.*

Proof sketch. We sketch the algorithm α_PACKING. Obtain a PIP $\mathcal{P}' = (A', b', c)$ from $\mathcal{P}$ as follows. Round down b_i to the nearest multiple of Γa_2 and then multiply it with a_1/a_2. Set b'_i equal to the resulting value. Every b'_i is now between $a_1/2a_2$ and a_1/a_2 times the corresponding b_i. Set A'_{ij} to a_1 if $A_{ij} \neq 0$ and to 0 otherwise. $\mathcal{P}'$ has thus a fractional solution of value at least $(a_1/2a_2)y^*$ that can be obtained by scaling down the optimal fractional solution of $\mathcal{P}$. Note that every b'_i is a multiple of Γa_1. Thus we can invoke Theorem 1 and find a solution g' to $\mathcal{P}'$ of value at least $\sigma(m, \Gamma, \alpha, (a_1/2a_2)y^*)$. Scaling up every component of g' by a factor of at most a_2/a_1 yields a vector g that is feasible for $\mathcal{P}$ and has value at least $\sigma(m, \Gamma, \alpha, \frac{a_1}{2a_2} y^*)$. □

Lemma 2. *Let $\mathcal{P} = (A, b, c)$, be a column-restricted PIP with column values in the interval $(a_1, a_2]$, and b_i a multiple of Γa_2, $\forall i$ and some number $\Gamma \geq 1$. Here $\min_i b_i$ is not necessarily greater than 1. There is an algorithm INTER-VAL_PACKING that finds a solution g to $\mathcal{P}$ of value at least $\sigma(m, \Gamma, \alpha, \frac{a_1}{a_2} y^*)$, where y^* is the optimum of the fractional relaxation of $\mathcal{P}$. The algorithm runs in polynomial time.*

Proof sketch. Similar to that of Lemma 1. Since all b_i's are already multiples of Γa_2, we don't pay the $1/2$ factor for the initial rounding. □

We now give the idea behind the full algorithm. The technique generalizes earlier work of the authors on single-source unsplittable flow [14]. Let x^* denote the optimum solution to the linear relaxation of $\mathcal{P}$. We are going to create packing subproblems $P^\lambda = (A^\lambda, b^\lambda, c^\lambda)$ where A^λ contains only the columns of A with values in some fixed range $(\alpha_{\lambda-1}, \alpha_\lambda]$. We will obtain our integral solution to $\mathcal{P}$ by combining approximate solutions to the subproblems. The crucial step is capacity allocation to subproblems. Consider a candidate for the b^λ-vector that we call the λ-th *fractional capacity vector*. In the fractional capacity vector the i-th entry is equal to $\sum_{j | \rho_j \in (\alpha_{\lambda-1}, \alpha_\lambda]} A_{ij} x_j^*$. In other words we allocate capacity to the λ-th subproblem by "pulling" out of b the amount of capacity used up in the solution x^* by the columns with values in $(\alpha_{\lambda-1}, \alpha_\lambda]$. The benefit of such a scheme would be that by using the relevant entries of x^* we could obtain a feasible solution to the linear relaxation of P^λ. However, to be able to benefit from Lemma 1 to find an approximate integral solution to each P^λ, we need all entries of b^λ to be larger than $B\alpha_\lambda$. This increases the required capacity for each subproblem potentially above the fractional capacity. Thus we resort to more sophisticated capacity allocation to ensure that (i) there is enough total capacity in the b-vector of $\mathcal{P}$ to share among the subproblems (ii) the subproblems can benefit from the subroutines in Lemmata 1, 2. In particular, we initially assign to each subproblem λ only $1/2$ of the fractional capacity vector; this has the asymptotically negligible effect of scaling down the fractional optimum for each subproblem by at most 2. We exploit the unused $(1/2)b$ vector of capacity to add an extra $B\alpha_\lambda$ units to the entries of b^λ, $\forall\lambda$.

Given α_i, α_j, let J^{α_i,α_j} be the set of column indices k for which $\alpha_i < \rho_k \le \alpha_j$. We then define, A^{α_i,α_j} to be the $m \times |J^{\alpha_i,\alpha_j}|$ submatrix of A consisting of the columns in J^{α_i,α_j}, and for any vector x, x^{α_i,α_j} to be the $|J^{\alpha_i,\alpha_j}|$-entry subvector x consisting of the entries whose indices are in J^{α_i,α_j}. We will also need to combine back together the various subvectors, and define $x^{\alpha_1,\alpha_2} \cup \cdots \cup x^{\alpha_{k-1},\alpha_k}$ to be the n-entry vector in which the various entries in the various subvectors are mapped back into their original positions. Any positions not in any of the corresponding index sets J^{α_i,α_j} are set to 0.

ALGORITHM COLUMN_PARTITION($\mathcal{P}$)

Step 1. Find the n-vector x_* that yields the optimal solution to the linear relaxation of $\mathcal{P}$.

Step 2a. Define a partition of the $(0,1]$ interval into $\xi = O(\log n)$ consecutive subintervals $(0, n^{-k}], \ldots, (4^{-\lambda}, 4^{-\lambda+1}], \ldots, (4^{-2}, 4^{-1}], (4^{-1}, 1]$ where k is a constant larger than 1. For $\lambda = 1 \ldots \xi - 1$ form subproblem $P^\lambda = (A^\lambda, b^\lambda, c^\lambda)$. A^λ and c^λ are the restrictions defined by $A^\lambda = A^{4^{-\lambda}, 4^{-\lambda+1}}$ and $c^\lambda = c^{4^{-\lambda}, 4^{-\lambda+1}}$. Define similarly $P^\xi = (A^\xi, b^\xi, c^\xi)$ such that $A^\xi = A^{0, n^{-k}}$, $c^\xi = c^{0, n^{-k}}$.

Step 2b. Let d_i^λ be the ith entry of $(A^\lambda \cdot x_*^\lambda)$. We define $b^\lambda = b$ when λ is 1 or ξ, and otherwise $b_i^\lambda = B(1/4)^{\lambda-1} + (1/2)d_i^\lambda$.

Step 3. Form solution $\dot{x}$ by setting x_j to 1 if $\rho_j \in (0, n^{-k}]$, and 0 otherwise.

Step 4. On each P^λ, $2 \le \lambda \le \xi - 1$, invoke α_PACKING to obtain a solution vector x^λ. Combine the solutions to subproblems 2 through $\xi - 1$ to form n-vector $\hat{x} = \cup_{2 \le \lambda \le \xi - 1} x^\lambda$.

Step 5. Invoke INTERVAL_PACKING on P^1 to obtain a solution vector x^1. Let $\bar{x} = \cup x^1$.

Step 6. Of the three vectors $\dot{x}$, $\hat{x}$ and $\bar{x}$, output the one, call it x, that maximizes $c^T \cdot x$.

Two tasks remain. First, we must show that the vector x output by the algorithm is a feasible solution to the original packing problem $\mathcal{P}$. Second, we must lower bound $c^T \cdot x$ in terms of the optimum $y_* = c^T \cdot x_*$ of the fractional relaxation of $\mathcal{P}$. Let $y_*^{r_1,r_2} = c^{r_1,r_2} x_*^{r_1,r_2}$, and if $r_1 = (1/4)^\lambda$ and $r_2 = (1/4)^{\lambda-1}$ then we abbreviate $y_*^{r_1,r_2}$ as y_*^λ. We examine first the vector $\hat{x}$.

Lemma 3. *Algorithm* COLUMN_PARTITION *runs in polynomial time and the n-vector $\hat{x}$ it outputs is a feasible solution to $\mathcal{P}$ of value at least $\sum_{\lambda=2}^{\lambda=\xi-1} \sigma(m, B, \alpha, (1/16)y_*^\lambda)$.*

Proof sketch. Let $\hat{y}^\lambda$ be the optimal solution to the linear relaxation of P^λ. By Lemma 1 the value of the solution x^λ, $2 \le \lambda \le \xi - 1$, found at Step 4 is at least $\sigma(m, B, \alpha, (1/2)(1/4)\hat{y}^\lambda)$. By the definition of b^λ in P_λ, $(1/2)x_*^\lambda$,i.e. the restriction of x_* scaled by $1/2$, is a feasible fractional solution for P_λ. Thus $\hat{y}^\lambda \geq (1/2)y_*^\lambda$. Hence the value of x^λ is at least $\sigma(m, B, \alpha, (1/16)y_*^\lambda)$. The claim on the value follows.

For the feasibility, we note that the aggregate capacity used by $\hat{x}$ on row i of A is the sum of the capacities used by x^λ, $2 \leq \lambda \leq \xi - 1$, on each subproblem. This sum is by Step 2b at most $(1/2)\sum_{\lambda=2}^{\lambda=\xi-1} d_i^\lambda + B\sum_{\lambda=2}^{\lambda=\xi-1}(1/4)^{\lambda-1}$. But $B\sum_{\lambda=2}^{\xi-1}(1/4)^{\lambda-1} < (1/2)B < (1/2)b_i$ and $\sum_{\lambda=2}^{\lambda=\xi-1} d_i^\lambda \leq b_i$. Thus the aggregate capacity used by $\hat{x}$ is at most $(1/2)b_i + (1/2)b_i = b_i$. □

It remains to account for $\bar{x}$ and $\dot{x}$. The following theorem is the main result of this section.

Theorem 2. *Let* $\mathcal{P} = (A, b, c)$ *be a column-restricted PIP and y^* be the optimum of the linear relaxation of $\mathcal{P}$. Algorithm* COLUMN_PARTITION *finds in polynomial time a solution g to $\mathcal{P}$ of value* $\Omega(\max\{y^*/m^{1/(B+1)}, y^*/\alpha^{1/B}, y^*(y^*/m\log n)^{1/B}\})$.

Proof. Each of the three vectors $\dot{x}, \hat{x}, \bar{x}$ solves a packing problem with column values lying in $(0, 1/n^k]$, $(1/n^k, 1/4]$ and $(1/4, 1]$ respectively. Let $\dot{\mathcal{P}}, \hat{\mathcal{P}}, \bar{\mathcal{P}}$ be the three induced packing problems. The optimal solution to the linear relaxation of at least one of them will have value at least $1/3$ of the optimal solution to the linear relaxation of $\mathcal{P}$. It remains to lower bound the approximation achieved by each of the three vectors on its corresponding domain of column values. Since m, B, and α are fixed for all subproblems, note that σ is a function of one variable, y^*. Vector $\dot{x}$ solves $\dot{\mathcal{P}}$ optimally. The solution is feasible since all column values are less than $1/n^k$ and thus the value of the left-hand side of any packing constraint cannot exceed 1. By Lemma 1, vector $\bar{x}$ outputs a solution to $\bar{\mathcal{P}}$ of value at least $\sigma(m, B, \alpha, (1/4)y_*^{1/4,1})$. For $\hat{\mathcal{P}}$, the value of the solution output is given by the sum $A = \sum_{\lambda=2}^{\xi-1}\sigma(m, B, \alpha, (1/16)y_*^\lambda)$ in Lemma 3. We distinguish two cases. If σ is a function linear in y_* then $A \geq \sigma(m, B, \alpha, (1/16)y_*^{n^{-k},4^{-1}})$. If σ is a function convex in y_* the sum A is minimized when all the terms y_*^λ are equal to $\Theta(y_*^{n^{-k},4^{-1}}/\log n)$. Instantiating $\sigma(y_*)$ with the function $\Omega(\max\{y_*/m^{1/(B+1)}, y_*/\alpha^{1/B}\})$ in the linear case and with $\Omega(y_*(y_*/m)^{1/B})$ in the convex case completes the proof. □

3   Applications of PIP's to Approximation

3.1   Weighted Multiple-Source Unsplittable Flow

Our approach for weighted unsplittable flow consists of finding first the optimum of the fractional relaxation, i.e. weighted multicommodity flow, which can be solved in polynomial time via linear programming. The relaxation consists of allowing commodity k to be shipped along more than one path. Call these paths the *fractional paths*. We round in two stages. In the first stage we select at most one of the fractional paths for each commodity, at the expense of congestion, i.e. some capacities may be violated. In addition, some commodities may not be routed at all. In the second stage, among the commodities routed during the first stage, we select those that will ultimately be routed while respecting the capacity constraints. It is in this last stage that a column-restricted PIP is used.

We introduce some terminology before giving the algorithm. A *routing* is a set of s_{k_i}-t_{k_i} paths P_{k_i}, used to route ρ_{k_i} amount of flow from s_{k_i} to t_{k_i} for each $(s_{k_i}, t_{k_i}) \in I \subseteq \mathcal{T}$. Given a routing g, the flow g_e through edge e is equal to $\sum_{P_i \in g, P_i \ni e} \rho_i$. A routing g for which $g_e \leq u_e$ for every edge e is an *unsplittable flow*. A *fractional routing* is one where for commodity k (i) the flow is split along potentially many paths (ii) demand $f_k \leq \rho_k$ is routed. A fractional routing corresponds thus to standard multicommodity flow. A *fractional single-path routing* is one where the flow for a commodity is shipped on one path if at all, but only a fraction $f_k \leq \rho_k$ of the demand is shipped for commodity k. The *value* of a routing g is the weighted sum of the demands routed in g.

ALGORITHM MAX_ROUTING($G = (V, E, u), \mathcal{T}$)

Step 1. Find an optimum fractional routing f by invoking a weighted multicommodity flow algorithm. Denote by $\alpha^f(\mathcal{T})$ the value of f.

Step 2. Scale up all capacities by a factor of $\Theta(\log|E|)$ to obtain network G' with capacity function u'. Invoke Raghavan's algorithm [24] on G' to round f to a routing g'.

Step 3. Scale down the flow on every path of g' by a factor of at most $\Theta(\log|E|)$ to obtain a fractional single-path routing g that is feasible for G.

Step 4. Construct a column-restricted PIP $\mathcal{P} = (A, b, c)$ as follows. Let $k_1, k_2, \ldots, k_\lambda$ be the set of commodities shipped in g, $\lambda \leq K$. A has λ columns, one for each commodity in g, and $|E|$ rows, one for each edge of G. $A_{ij} = \rho_{k_j}$ if the path P_{k_j} in g for commodity k_j goes through edge e_i and 0 otherwise. The cost vector c has entry c_j set to $w_{k_j}\rho_{k_j}$ for each commodity k_j shipped in g. Finally, $b_i = u_{e_i}$, $1 \leq i \leq |E|$.

Step 5. Invoke algorithm COLUMN_PARTITION to find an integral solution $\hat{g}$ to $\mathcal{P}$. Construct an unsplittable flow g'' by routing commodity k_j on path P_{k_j} if and only if $\hat{g}_j = 1$.

Theorem 3. *Given a weighted multiple-source unsplittable flow problem $(G = (V, E), \mathcal{T})$, algorithm MAX_ROUTING finds in polynomial time an unsplittable flow of value at least $\Omega(\max\{\alpha^f(\mathcal{T})/(\log|E|\sqrt{|E|}), (\alpha^f(\mathcal{T}))^2/(|E|\log^3|E|)\})$, where $\alpha^f(\mathcal{T})$ is the value of an optimum fractional routing f.*

Proof sketch. First note that since routing g' is feasible for G', after scaling down the flow at Step 3, routing g is feasible for G. The rounded solution to $\mathcal{P}$ maintains feasibility. For the value, we can easily extend the analysis of Raghavan's algorithm to show that even with weights, it routes in G' a constant fraction of $\alpha^f(\mathcal{T})$ [24]. Set z_{k_j} to be equal to the amount of flow that is routed in g for commodity k_j divided by ρ_{k_j}. The λ-vector z is a feasible fractional solution to $\mathcal{P}$ of value $\Omega(\alpha^f(\mathcal{T})/\log|E|)$. The theorem follows by using Theorem 2 to lower bound the value of the integral solution $\hat{g}$ to $\mathcal{P}$. □

We now give a bound that depends on the dilation d of the fractional routing.

Theorem 4. *Given a weighted multiple-source unsplittable flow problem* $(G = (V, E), \mathcal{T})$, *there is a polynomial-time algorithm that finds an unsplittable flow of value at least* $\Omega(\alpha^f(\mathcal{T})/d)$, *where* $\alpha^f(\mathcal{T})$ *is the value and* d *the dilation of an optimum fractional routing* f.

The construction in the proof of Theorem 4 can also be used to give an $\Omega(\max\{\alpha^f(\mathcal{T})/(\sqrt{|E|}), (\alpha^f(\mathcal{T}))^2/(|E|\log|E|)\})$, bound in the case where the number of commodities $|\mathcal{T}| = O(|E|)$. We omit the details in this version.

3.2 Weighted Vertex-Disjoint Paths

A relaxation of the problem is integral multicommodity flow where every commodity has an associated demand of 1. We can express vertex-disjoint paths as an integer program $\mathcal{I}$ that has the same constraints as multicommodity flow together with additional "bandwidth" constraints on the vertices such that the total flow through a vertex is at most 1. Let $\mathcal{LP}$ be the linear relaxation of $\mathcal{I}$. The optimal solution f to $\mathcal{LP}$ consists of a set of *fractional flow paths*. Our algorithm relies on the observation that f gives rise (and at the same time is a fractional solution) to a PIP. The particular PIP models a 1-matching problem on a hypergraph H with vertex set V and a hyperedge (subset of V) for every path in the fractional solution. In other words, the paths in f may be viewed as sets of vertices *without any regard for the flow through the edges of* G. We proceed to give the full algorithm.

ALGORITHM PATH_PACKING$(G = (V, E), \mathcal{T})$
Step 1. Formulate the linear relaxation $\mathcal{LP}$ and find an optimal solution f to it. Using flow decomposition express f as a set of paths $P_1, P_2, \ldots, P_\lambda$ each connecting a pair of terminals and carrying $z_i \leq 1$, $1 \leq i \leq \lambda$, units of flow.
Step 2. Construct a $(0, 1)$-PIP $\mathcal{P} = (A, b, c)$ as follows. A is a $|V| \times \lambda$ matrix; A_{ij} is 1 if path P_j includes vertex i and 0 otherwise. b is a vector of ones. c_j is equal to w_k such that path P_j connects terminals s_k and t_k.
Step 3. Find an integral solution g to $\mathcal{P}$ and output the corresponding set of paths $P(g)$.

Theorem 5. *Given a weighted vertex-disjoint paths problem* $(G = (V, E), \mathcal{T})$, *algorithm* PATH_PACKING *finds in polynomial time a solution of value* $\Omega(\max\{y^*/\sqrt{|V|}, (y^*)^2/|V|, y^*/d\})$, *where* d *is the dilation and* y^* *is the value of an optimum solution to the linear relaxation* $\mathcal{LP}$.

Proof sketch. We show first that $P(g)$ is a feasible solution to the problem. Clearly the constraints of the packing program $\mathcal{P}$, constructed at Step 2 ensure that the paths in $P(g)$ are vertex disjoint. This excludes the possibility that more than one (s_k, t_k)-path is present in $P(g)$ for some k. The optimal value of the linear relaxation of $\mathcal{P}$ is at least y^* since setting x_j equal to z_j, $1 \leq j \leq \lambda$, yields a feasible fractional solution to $\mathcal{P}$. By applying either standard randomized rounding [25] or Srinivasan's algorithms [29,30] at Step 3, we obtain the claimed bounds on the objective value $c^T \cdot g$.   □

3.3   An Application to Independent Set

In this section we show how a packing formulation leads to an $O(\sqrt{|V|})$ approximation for the following problem: find a maximum weight independent set in the square of the graph $G = (V, E)$.

Theorem 6. *Given a graph $G = (V, E)$ and $c \in [0,1]^{|V|}$ a weight vector on the vertices, there exists a polynomial-time algorithm that outputs an independent set in the square $G^2 = (V, E^2)$ of G of weight $\Omega(\max\{y^*/\sqrt{|V|}, (y^*)^2/|V|, y^*/\Delta\})$. Here y^* denotes the optimum of a fractional relaxation and Δ is the maximum vertex degree in G.*

A hardness of approximation result for the problem of finding a maximum independent set in the k-th power of a graph follows.

Theorem 7. *For the problem of finding a maximum independent set in the k-th power $G^k = (V, E^k)$ of a graph $G = (V, E)$ for any fixed integer $k > 0$, there is no ρ-approximation with $\rho = (\frac{|V|}{k+2})^{1/2-\varepsilon}$, for any fixed $\varepsilon > 0$, unless $NP = ZPP$.*

4   Greedy Algorithms for Disjoint Paths

In this section we turn to the routing approach for the unweighted edge- and vertex-disjoint path problems.

ALGORITHM GREEDY_PATH(G, T)
Step 1. Set $\mathcal{A}$ to $\emptyset$.
Step 2. Let (s_*, t_*) be the commodity in T such that the shortest path P_* in G from s_* to t_* has minimum length. If no such path exists halt and output $\mathcal{A}$.
Step 3. Add P_* to $\mathcal{A}$ and remove the edges of P_* from G. Remove (s_*, t_*) from T. Goto Step 2.

We begin by lowerbounding the approximation ratio of the algorithm.
Theorem 8. *Algorithm GREEDY_PATH runs in polynomial time and outputs a solution to an edge-disjoint paths problem $(G = (V, E), T)$ of size at least $1/(\sqrt{|E_o|} + 1)$ times the optimum, where $E_o \subseteq E$ is the set of edges used by the paths in an optimal solution.*

We now give improved bounds on the size of the output realizable set.

Theorem 9. *Algorithm GREEDY_PATH outputs a solution to an edge-disjoint path problem $(G = (V, E), T)$ of size $\Omega(OPT^2/|E_o|)$, where $E_o \subseteq E$ is the set of edges used by the paths in an optimal solution.*

Proof. Let t be the total number of iterations of GREEDY_PATH and $\mathcal{A}_i$ be the set $\mathcal{A}$ at the end of the i-th iteration. Let $\mathcal{O}$ be an optimal set of paths. We say that a path P_x *hits* a path P_y, if P_x and P_y share an edge. We define the set $\mathcal{O} \ominus \mathcal{A}$ as the paths in $\mathcal{O}$ that correspond to commodities not routed in $\mathcal{A}$. Let P_i be the path added to $\mathcal{A}$ at the i-th iteration of the algorithm. If P_i hits k_i paths in $\mathcal{O} \ominus \mathcal{A}_i$ that are not hit by a path in $\mathcal{A}_{i-1}$, then P_i must have length at least k_i. In turn each of the paths hit has length at least k_i otherwise it would have been selected by the algorithm instead of P_i. Furthermore all paths in $\mathcal{O}$ are edge-disjoint with total number of edges $|E_o|$. Therefore $\sum_{i=1}^t k_i^2 \leq |E_o|$. Applying the Cauchy-Schwarz inequality on the left-hand side we obtain that $(\sum_{i=1}^t k_i)^2/t \leq |E_o|$. But $\sum_{i=1}^t k_i = |\mathcal{O} \ominus \mathcal{A}_t|$ since upon termination of the algorithm all paths in $\mathcal{O} \ominus \mathcal{A}_t$ must have been hit by some path in $\mathcal{A}_t$. We obtain $\frac{|\mathcal{O} \ominus \mathcal{A}_t|^2}{t} \leq |E_o|$. Wlog we can assume that $|\mathcal{A}_t| = o(|\mathcal{O}|)$, since otherwise GREEDY_PATH obtains a constant-factor approximation. It follows that $t = \Omega(|\mathcal{O}|^2)/|E_o|) = \Omega(OPT^2/|E_o|)$.   □

Corollary 1. *Algorithm* GREEDY_PATH *outputs a solution to an edge-disjoint path problem* $(G = (V, E), \mathcal{T})$ *of size* $\Omega(OPT/d_o)$, *where* d_o *is the average length of the paths in an optimal solution.*

Algorithm GREEDY_PATH gives a solution to vertex-disjoint paths with the following modification at Step 3: remove the vertices of P_* from G. Call the resulting algorithm GREEDY_VPATH. The analogues of the results above can be seen to hold for GREEDY_VPATH as well.

Theorem 10. *Algorithm* GREEDY_VPATH *outputs a solution to a vertex-disjoint path problem* $(G = (V, E), \mathcal{T})$ *of size* $\Omega(\max\{OPT/|V_o|, OPT^2/|V_o|, OPT/d_o\})$, *where* $V_o \subseteq V$ *is the set of vertices used by the paths in an optimal solution and* d_o *is the average length of the paths in an optimal solution.*

Acknowledgments. We wish to thank Jon Kleinberg and Aravind Srinivasan for valuable discussions. We also thank Javed Aslam for helpful discussions.

References

1. R. Aharoni, P. Erdős, and N. Linial. Optima of dual integer linear programs. *Comb.*, 8:13–20, 1988.
2. Y. Aumann and Y. Rabani. Improved bounds for all-optical routing. In *Proc. 6th ACM-SIAM Symp. on Discrete Algorithms*, pages 567–576, 1995.
3. A. Z. Broder, A. M. Frieze, and E. Upfal. Static and dynamic path selection on expander graphs: a random walk approach. In *Proc. 29th Ann. ACM Symp. on Theory of Computing*, 531–539, 1997.
4. C. Cooper. The thresold of hamilton cycles in the square of a random graph. *Random Structures and Algorithms*, 5:25–31, 1994.
5. H. Fleischner. The square of every two-connected graph is hamiltonian. *J. of Combinatorial Theory B*, 16:29–34, 1974.
6. A. Frank. Packing paths, cuts and circuits – A survey. In B. Korte, L. Lovász, H. J. Prömel, and A. Schrijver, editors, *Paths, Flows and VLSI-Layout*, pages 49–100. Springer-Verlag, Berlin, 1990.
7. N. Garg, V. Vazirani, and M. Yannakakis. Primal-dual approximation algorithms for integral flow and multicut in trees. *Algorithmica*, 18:3–20, 1997.
8. R. M. Karp, F. T. Leighton, R. L. Rivest, C. D. Thompson, U. V. Vazirani, and V. V. Vazirani. Global wire routing in two-dimensional arrays. *Algorithmica*, 2:113–129, 1987.

9. R. M. Karp. On the computational complexity of combinatorial problems. *Networks*, 5:45–68, 1975.

10. J. M. Kleinberg. *Approximation algorithms for disjoint paths problems*. PhD thesis, MIT, Cambridge, MA, May 1996.

11. J. M. Kleinberg. Single-source unsplittable flow. In *Proc. 37th Ann. Symp. on Foundations of Computer Science*, pages 68–77, October 1996.

12. J. M. Kleinberg and R. Rubinfeld. Short paths in expander graphs. In *Proc. 37th Ann. Symp. on Foundations of Computer Science*, pages 86–95, 1996.

13. J. M. Kleinberg and É. Tardos. Disjoint paths in densely-embedded graphs. In *Proc. 36th Ann. Symp. on Foundations of Computer Science*, pages 52–61, 1995.

14. S. G. Kolliopoulos and C. Stein. Improved approximation algorithms for unsplittable flow problems. In *Proc. 38th Ann. Symp. on Foundations of Computer Science*, pages 426–435, 1997.

15. S. G. Kolliopoulos and C. Stein. Approximating disjoint-path problems using greedy algorithms and packing integer programs. Technical Report PCS TR97–325, Department of Computer Science, Dartmouth College, 1997.

16. F. T. Leighton and S. B. Rao. Circuit switching: A multi-commodity flow approach. In *Workshop on Randomized Parallel Computing*, 1996.

17. T. Leighton and S. Rao. An approximate max-flow min-cut theorem for uniform multicommodity flow problems with applications to approximation algorithms. In *Proc. 29th Ann. Symp. on Foundations of Computer Science*, pages 422–431, 1988.

18. Y.-L. Lin and S. E. Skiena. Algorithms for square roots of graphs. *SIAM J. on Discrete Mathematics*, 8(1):99–118, 1995.

19. L. Lovász. On the ratio of optimal and fractional covers. *Discrete Mathematics*, 13:383–390, 1975.

20. P. Martin and D. B. Shmoys. A new approach to computing optimal schedules for the job-shop scheduling problem. In *Proc. 5th Conference on Integer Programming and Combinatorial Optimization*, pages 389–403, 1996.

21. D. Peleg and E. Upfal. Disjoint paths on expander graphs. *Comb.*, 9:289–313, 1989.

22. S. Plotkin. Competitive routing of virtual circuits in ATM networks. *IEEE J. Selected Areas in Comm.*, 1128–1136, 1995.

23. S. Plotkin, D. B. Shmoys, and E. Tardos. Fast approximation algorithms for fractional packing and covering problems. *Math. of Oper. Res.*, 20:257–301, 1995.

24. P. Raghavan. Probabilistic construction of deterministic algorithms: approximating packing integer programs. *J. of Computer and System Sciences*, 37:130–143, 1988.

25. P. Raghavan and C. D. Thompson. Randomized rounding: a technique for provably good algorithms and algorithmic proofs. *Comb.*, 7:365–374, 1987.

26. N. Robertson and P. D. Seymour. Outline of a disjoint paths algorithm. In B. Korte, L. Lovász, H. J. Prömel, and A. Schrijver, editors, *Paths, Flows and VLSI-Layout*. Springer-Verlag, Berlin, 1990.

27. A. Schrijver. Homotopic routing methods. In B. Korte, L. Lovász, H. J. Prömel, and A. Schrijver, editors, *Paths, Flows and VLSI-Layout*. Springer, Berlin, 1990.

28. D. B. Shmoys, C. Stein, and J. Wein. Improved approximation algorithms for shop scheduling problems. *SIAM J. on Computing*, 23(3):617–632, 1994.

29. A. Srinivasan. Improved approximations of packing and covering problems. In *Proc. 27th Ann. ACM Symp. on Theory of Computing*, pages 268–276, 1995.

30. A. Srinivasan. An extension of the Lovász Local Lemma and its applications to integer programming. In *Proc. 7th ACM-SIAM Symp. on Discrete Algorithms*, pages 6–15, 1996.

31. A. Srinivasan. Improved approximations for edge-disjoint paths, unsplittable flow and related routing problems. In *Proc. 38th Ann. Symp. on Foundations of Computer Science*, pages 416–425, 1997.
32. A. Srinivasan and C.-P. Teo. A constant-factor approximation algorithm for packet routing and balancing local vs. global criteria. In *Proc. 29th Ann. ACM Symp. on Theory of Computing*, pages 636–643, 1997.
33. C. Stein. *Approximation Algorithms for Multicommodity Flow and Shop Scheduling Problems*. PhD thesis, MIT, Cambridge, MA, August 1992.
34. P. Underground. On graphs with hamiltonian squares. *Disc. Math.*, 21:323, 1978.

Approximation Algorithms for the Mixed Postman Problem

Balaji Raghavachari and Jeyakesavan Veerasamy

Computer Science Program
The University of Texas at Dallas
Richardson, TX 75083, USA
{rbk, veerasam}@@utdallas.edu

Abstract. The mixed postman problem, a generalization of the Chinese postman problem, is that of finding a shortest tour that traverses each edge of a given mixed graph (a graph containing both undirected and directed edges) at least once. The problem is solvable in polynomial time either if the graph is undirected or if the graph is directed, but NP-hard in mixed graphs. An approximation algorithm with a performance ratio of 3/2 for the postman problem on mixed graphs is presented.

1   Introduction

Problems of finding paths and tours on graphs are of fundamental importance and find many practical applications. The Traveling salesman problem (TSP) is a well-known and widely studied problem. The objective is to find a shortest tour that visits all vertices of a given graph exactly once. The problem is known to be NP-hard. Postman problems are similar to TSP at first glance, but are quite different in terms of the complexity of the problems. Given a graph G, the Chinese postman problem (CPP) is to find a minimum cost tour covering all edges of G at least once [9]. It is the optimization version of the Euler tour problem, which asks if there is a tour that traverses every edges of a graph exactly once. Edmonds and Johnson [3] showed that the problem is solvable in polynomial time. They also showed that the problem is solvable in polynomial time if G is a directed graph.

The **Mixed Postman Problem** (MPP) is a generalization of the Chinese Postman Problem and is listed as Problem ND25 by Garey and Johnson [8]. In the mixed postman problem, the input graph may contain both undirected edges and arcs (directed edges). The objective is to find a tour that traverses every edge at least once, and which traverses directed edges only in the direction of the arc. Even though both undirected and directed versions of the Chinese postman problem are polynomially solvable, Papadimitriou [13] showed that MPP is NP-hard. There are other related problems, such as the Rural postman problem and the Windy postman problem, which are also NP-hard [4,5]. Many practical

R. E. Bixby, E. A. Boyd, and R. Z. Ríos-Mercado (Eds.): IPCO VI
LNCS 1412, pp. 169–179, 1998. © Springer–Verlag Berlin Heidelberg 1998

applications like mail delivery, snow removal, and trash pick-up can be modeled as instances of MPP, and hence it is important to design good approximation algorithms for this problem.

Previous Work: Numerous articles have appeared in the literature over the past three decades about the mixed postman problem. Edmonds and Johnson [3], and Christofides [2] presented the first approximation algorithms. Frederickson [7] showed that the algorithm of [3] finds a tour whose length is at most 2 times the length of an optimal tour (i.e., approximation ratio of 2). He also presented a mixed strategy algorithm, which used the solutions output by two different heuristics, and then selected the shorter of the two tours. He proved that the approximation ratio of the mixed strategy algorithm is $\frac{5}{3}$. Comprehensive surveys are available on postman problems [1,4]. Integer and linear programming formulations of postman problems have generated a lot of interest in recent years [10,12,16]. Ralphs [16] showed that a linear relaxation of MPP has optimal solutions that are half-integral. One could use this to derive a 2-approximation algorithm for the problem. Improvements in implementation are discussed in [12,15]. It is interesting to note that Nobert and Picard [12] state that their implementation has been used for scheduling snow removal in Montreal. Several other articles have appeared on generalized postman problems, such as k-CPP [14] and the Windy Postman problem [4].

Our Results: Even though numerous articles have appeared in the literature on MPP after Frederickson's paper in 1979, his result has been the best approximation algorithm for MPP in terms of proven worst-case ratio until now. In this paper, we present an improved approximation algorithm for MPP with an approximation ratio of $\frac{3}{2}$. We study the properties of feasible solutions to MPP, and derive a new lower bound on the cost of an optimal solution. Our algorithm uses a subtle modification of Frederickson's algorithm, and the improved performance ratio is derived from the new lower bound. We present examples showing that our analysis is tight.

2   Preliminaries

Problem Statement: The input graph $G = (V, E, A)$ consists of a set of vertices V, a multi-set of edges E, and a multi-set of arcs (directed edges) A. A non-negative cost function C is defined on edges and arcs. We extend the definition of C to graphs (multisets of edges and arcs), by taking the sum of the costs of its edges and arcs. We assume that the graph is strongly connected, i.e., there exists a path from any vertex u to any other vertex v, since the problem is clearly infeasible for graphs that are not strongly connected. The output is a tour which may travel each edge or arc several times. Therefore, except for traversing each edge/arc once, the traversal could always use a shortest path between any two nodes. Hence we assume that the weights of the edges satisfy the triangle inequality.

Definitions: A *cut* is a partition of the vertex set into S and $V - S$. It is called *nontrivial* if neither side is empty. An edge *crosses* the cut if it connects a vertex in S to a vertex in $V - S$. Let $outdegree(v)$ be the number of outgoing arcs from v. Similarly, $indegree(v)$ is the number of incoming arcs into v. Let $degree(v)$ be the total number of edges and arcs incident to v. We say v has even degree if $degree(v)$ is even. A graph has even degree if all its vertices have even degree. Let $surplus(v) = outdegree(v) - indegree(v)$. If $surplus(v)$ is negative, we may call it as a *deficit*. The definition of surplus can be extended to sets of vertices S, by finding the difference between outgoing and incoming edges that cross the cut $(S, V - S)$.

Properties of Eulerian Graphs: A graph is called *Eulerian* if there is a tour that traverses each edge of the graph exactly once. It is known that an undirected, connected graph is Eulerian if and only if the degree of each vertex is even. For a directed graph to be Eulerian, the underlying graph must be connected and for each vertex v, $outdegree(v) = indegree(v)$. In other words, for each vertex v, $surplus(v) = 0$. A mixed graph $G = (V, E, A)$ is Eulerian if the graph is strongly connected and satisfies the following properties [6,11]:

- **Even Degree Condition:** Every vertex v is incident to an even number of edges and arcs, i.e., $degree(v)$ is even. This condition implies that the number of edges and arcs crossing any cut $(S, V - S)$ is even.
- **Balanced Set Condition:** For every nontrivial cut $S \subset V$, the absolute surplus of S must be less than or equal to the number of undirected edges crossing the cut $(S, V - S)$.

The above conditions can be checked in polynomial time using algorithms for the maximum flow problem [6]. In other words, we can decide in polynomial time whether a given mixed graph is Eulerian. The problem we are interested in is to find set of additional edges and arcs of total minimum cost that can be added to G to make it Eulerian, and this problem is NP-hard. In the process of solving mixed postman problem, arcs and edges may be duplicated. For convenience, we may also orient some undirected edges by giving them a direction. The output of our algorithm is an Eulerian graph H that contains the input graph G as a subgraph. So each edge of H can be classified either as an original edge or as a duplicated edge. Also, each arc of H is either an original arc, or a duplicated arc, or an oriented edge, or a duplicated & oriented edge.

3   Frederickson's MIXED Algorithm

Frederickson defined the following algorithms as part of his solution to MPP:

- EVENDEGREE: augment a mixed graph G by duplicating edges and arcs such that the resulting graph has even degree. A minimum-cost solution is obtained by disregarding the directions of the arcs (i.e., by taking the underlying undirected graph) and solving CPP by adding a minimum-weight matching of odd-degree nodes.
- INOUTDEGREE: augment a mixed graph G by duplicating edges and arcs, and orienting edges such that in the resulting graph, for each vertex v,

$surplus(v) = 0$. We will refer to this as the INOUT problem. A minimum-cost solution G_{IO} is obtained by formulating a flow problem and solving it optimally. The *augmentation cost*, $C_{IO}(G)$ is defined as the cost of additional arcs and edges that were added to G to get G_{IO} by INOUTDEGREE.

 - EVENPARITY: applied to the output of INOUTDEGREE on an even-degree graph; restores even degree to all nodes without increasing the cost, while retaining the property that *indegree = outdegree* at all nodes. Edmonds and Johnson [3] indicated that INOUTDEGREE can be applied to an even-degree graph in such a way that the resulting graph has even degree and hence Eulerian. Frederickson [7] showed a simple linear-time algorithm to perform the task. The basis of this algorithm is that a suitably defined subgraph of undirected edges and duplicated edges/arcs forms a collection of Eulerian graphs.
 - LARGECYCLES: similar to EVENDEGREE except that only edges are allowed to be duplicated, and arcs are not considered.

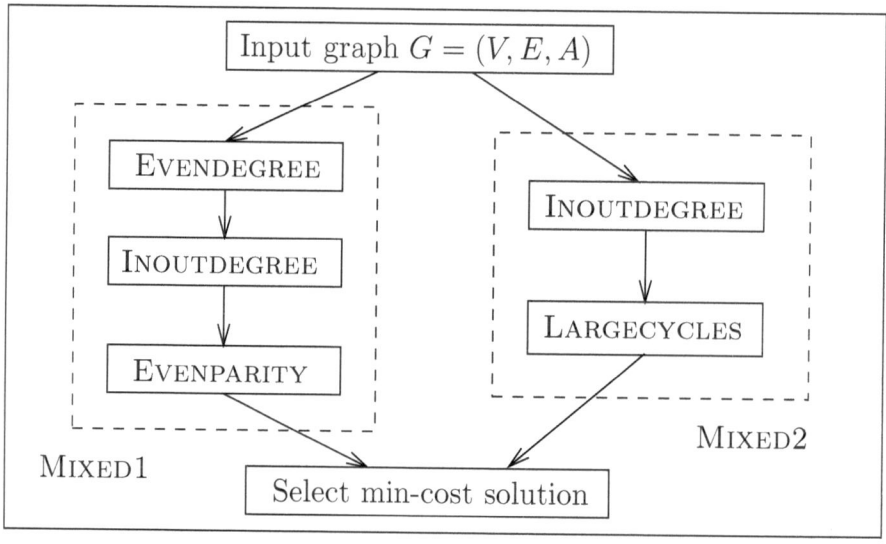

Fig. 1. MIXED algorithm

Figure 1 shows an approximation algorithm for MPP presented by Frederickson [7], called the MIXED algorithm. It comprises of two heuristics called MIXED1 and MIXED2. Heuristic MIXED1 first runs EVENDEGREE to make the degree of all nodes even. Then it runs INOUTDEGREE to make *indegree = outdegree* for all vertices. Finally, EVENPARITY restores even degree to all the nodes without increasing the cost, and the graph becomes Eulerian. Heuristic MIXED2 first calls INOUTDEGREE, and then makes the graph Eulerian by calling LARGECY-

CLES. Since LARGECYCLES disregards the arcs of the graph, no further steps are needed. The MIXED algorithm outputs the best solution of the two heuristics, and Frederickson showed its performance ratio is at most $\frac{5}{3}$.

4   Structure of INOUT Problem

In this section, we will identify a few critical properties of the INOUT problem, which we use improve the MIXED algorithm. Let G be a mixed graph, and let G_b and G_r (blue graph and red graph respectively) be two augmentations of G for the INOUT problem, i.e., G_b and G_r are two INOUT solutions.

Let $G_{br} = G_b \oplus G_r$ denote the *symmetric difference* of G_b and G_r, the multigraph containing edges and arcs in one graph, but not the other. Since we are dealing with multigraphs, an edge/arc (u, v), occurs in G_{br} as many times as the difference between the number of copies of (u, v) in G_b and G_r. A set of arcs C in G_{br} is an *alternating cycle* if the underlying edges of C form a simple cycle, and in any walk around the cycle, the blue arcs of C are directed one way, and the red arcs of C are directed in the opposite direction (see Figure 2).

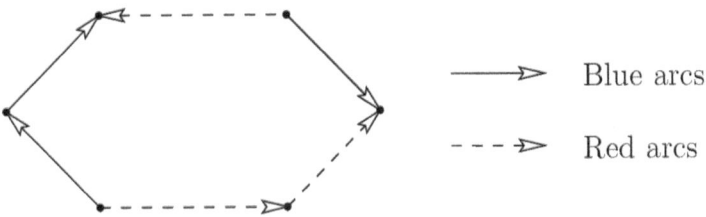

Fig. 2. An alternating cycle

The following lemmas show how alternating cycles of G_{br} can be used to switch between different INOUT solutions.

Lemma 1. *The arcs of G_{br} can be decomposed into alternating cycles.*

Proof. This proof is based on the properties of Eulerian directed graphs. Since they are INOUT solutions, each vertex of G_b and G_r satisfies the condition, *in_degree = out_degree*. However, this rule does not apply to G_{br}, since all common arcs and edges have been removed. Since only common arcs are removed, it can be verified that for each vertex in G_{br}, $surplus_b = surplus_r$. In other words, at each vertex of G_{br}, the net surplus or deficit created by blue arcs is equal to the net surplus or deficit created by red arcs. So at each vertex, an incoming blue arc can be associated with either an outgoing blue arc, or an incoming red arc. Similar statement is valid for red arcs also.

Consider a directed walk starting at an arbitrary vertex in G_{br}, in which blue edges are traversed in the forward direction and red edges are traversed in the reverse direction. Whenever any vertex is revisited during the walk, an alternating cycle is removed and output. When the start vertex is re-visited and there are no more arcs incident to it, restart the walk from another vertex. Continue the process until all arcs of G_{br} have been decomposed into alternating cycles. An alternative proof can be given by showing that reversing all red arcs makes G_{br} satisfy $indegree = outdegree$ at each node and hence the arcs can be decomposed into cycles.

Each alternating cycle AC in G_{br} is composed of one or more blue paths and red paths. In each cycle, all blue paths go in one direction and all red paths go in the opposite direction. Let us call blue arcs of AC as B and red arcs as R.

Define the operation of removing a set of arcs B from an INOUT solution G_{IO}, denoted by $G_{IO} - B$, as follows. If an arc of B is an additional arc or edge added during the augmentation, then it can be removed. On the other hand, if the arc was an undirected edge in G that was oriented during the augmentation, then we remove the orientation but leave the edge in the graph. The cost of such an operation is the total cost of the arcs/edges that are removed. Similarly, we can define the addition operation $G_{IO} + B$, where a set of arcs B are added to G_{IO}.

Lemma 2. *Consider two INOUT solutions G_b and G_r, and their symmetric difference G_{br}. Let AC be an alternating cycle in G_{br}, with blue component B and red component R. Then, $G_b - B + R$ and $G_r - R + B$ are also INOUT solutions.*

Proof. Clearly, adding or deleting edges of AC does not affect the deficit/surplus of the nodes that are not in AC. For all nodes in AC, the net deficit/surplus created by blue paths of AC is same as the net deficit/surplus created by red paths of AC. So, the net deficit/surplus created by removal of B from G_b is compensated by the addition of R. Therefore, $G_b - B + R$ is an INOUT solution, as each node of $G_b - B + R$ has net deficit/surplus of zero. By symmetry, $G_r - R + B$ is also an INOUT solution.

Lemma 3. *Let $d^-(R)$ be the cost of removing R from G_r and let $d^+(R)$ be the cost of adding R to G_b. Similarly, let $d^-(B)$ be the cost of removing B from G_b and let $d^+(B)$ be the cost of adding B to G_r. Then $d^+(R) \leq d^-(R)$ and $d^+(B) \leq d^-(B)$.*

Proof. We prove that the cost of adding R to G_b does not exceed the cost of removing R from G_r. Arcs of R in G_r can be either additional arc, oriented edge, or additional oriented edge. Additional arcs and additional oriented edges contribute at most the same cost to $d^+(R)$. If an oriented edge is in R, then either it is not currently oriented in G_b or it is oriented in the opposite direction in G_b. To add this oriented edge to G_b, we can either orient the undirected edge, or remove its opposite orientation; either way there is no additional cost. Therefore R can be added to G_b without incurring additional cost. Hence, $d^+(R) \leq d^-(R)$. By symmetry, $d^+(B) \leq d^-(B)$.

5   Improved Lower Bound for Mixed Postman Problem

Consider a mixed graph $G = (V, E, A)$, with edges E and arcs A. Let C^* be the weight of an optimal postman tour of G. Suppose, given G as input, IN-OUTDEGREE outputs $G_{IO} = (V, U, M)$. $U \subseteq E$ are the edges of G that were not oriented by INOUTDEGREE. $M \supseteq A$ are arcs that satisfy $indegree = outdegree$ at each vertex. Let C_M and C_U be the total weight of M and U respectively. Note that $C_M + C_U = C_{IO}(G) + C(G)$. M may contain several disjoint directed components. Suppose we shrink each directed component of G_{IO} into a single vertex and form an undirected graph UG. If we run EVENDEGREE on UG, it adds a minimum-weight matching (i.e., a subset of U) to make the degree of each vertex in UG even. Let the weight of this matching be C_X. Frederickson [7] used a lower bound of $C_M + C_U$ on C^*. We show the following improved lower bound on C^*, which allows us to improve the approximation ratio.

Lemma 4. $C_M + C_U + C_X \le C^*$.

In order to prove the above lemma, we first show that when we find an optimal INOUT solution, if some undirected edges are not oriented (edges in the set U) by INOUTDEGREE, then adding additional copies of edges in U (when one runs EVENDEGREE in UG) to G does not decrease $C_{IO}(G)$, the augmentation cost of INOUTDEGREE.

Lemma 5. *Let $G = (V, E, A)$ be a mixed graph. Let $G_{IO} = (V, U, M)$ be an optimal INOUT solution computed by INOUTDEGREE algorithm. Let $U = \{u_1, ..., u_k\}$ be the undirected edges of G that were not oriented by INOUTDEGREE, even though the algorithm could orient these edges without incurring additional cost. Adding additional copies of edges in U to G does not decrease the augmentation cost of an optimal INOUT solution.*

Proof. We give a proof by contradiction. Let H be the graph G to which n_i copies of u_i $(i = 1, \ldots, k)$ are added such that the augmentation cost of INOUTDEGREE is less for H than G, i.e., $C_{IO}(H) < C_{IO}(G)$. In addition, let H be a minimal supergraph of G with this property, and if we add any fewer copies of the edges of U, then $C_{IO}(H) = C_{IO}(G)$. Clearly, some edges of U must be oriented in H_{IO}, otherwise an INOUT solution of H would also be an INOUT solution of G. Assigning $G_b = G_{IO}$ and $G_r = H_{IO}$, we will prove that $d^-(R) < d^-(B)$ for every cycle AC that uses at least one edge of U. Recall that, $d^-(R)$ is the cost of removing R from the red INOUT solution H_{IO}, and $d^-(B)$ is the cost of removing B from the blue INOUT solution G_{IO}.

Consider any alternating cycle AC that contains an edge of U. Let B and R be its blue and red arcs respectively. By Lemma 3, $d^+(B) \le d^-(B)$. If $d^-(R) \ge d^-(B)$ then,

$$C(H_{IO} - R + B) = C(H_{IO}) - d^-(R) + d^+(B) \le C(H_{IO}).$$

Therefore, $H_{IO} - R + B$ is an optimal INOUT solution of H, and it uses fewer copies of edges of U than H. This contradicts the minimality assumption of

H. Therefore $d^-(R) < d^-(B)$ for all cycles AC that involve at least one edge from U. Choose any such cycle AC. We already know that $d^+(R) \le d^-(R)$ and $d^-(R) < d^-(B)$. Combining these, we get,

$$C(G_{IO} - B + R) = C(G_{IO}) - d^-(B) + d^+(R) < C(G_{IO}).$$

Note that $G_{IO} - B + R$ uses at most one copy of each edge from U and its cost is less than $C(G_{IO})$. In other words, $G_{IO} - B + R$ is an INOUT solution of G, and $C(G_{IO} - B + R) < C(G_{IO})$. This contradicts the assumption that G_{IO} is an optimal INOUT solution of G. Therefore, adding additional copies of edges from U does not decrease the augmentation cost of an INOUT solution.

Proof of Lemma 4. Let G^* be an optimal solution, whose cost is C^*. Consider the nodes of UG, the nodes corresponding to the directed components of M. G^* needs to have additional matching edges (subset of U) between these components to satisfy the even-degree condition for each component. This matching costs at least C_X. By Lemma 5, we know that additional edges of U do not decrease the cost of optimal INOUT solution. Therefore the augmentation cost of the INOUT problem is still $C_{IO}(G)$. Hence the total cost of G^* is at least $C_X + C_{IO}(G) + C(G)$. Substituting $C_M + C_U$ for $C_{IO}(G) + C(G)$, we get, $C_M + C_U + C_X \le C^*$.

6   Modified MIXED Algorithm

Figure 3 describes the modified MIXED algorithm. First, we run algorithm IN-OUTDEGREE on input graph G and obtain $G_{IO} = (V, U, M)$. Before running EVENDEGREE of MIXED1 algorithm, reset the weights of all arcs and edges used by M to 0, forcing EVENDEGREE to duplicate edges/arcs of M whenever possible, as opposed to duplicating edges of U. Use the actual weights for the rest of the MIXED1 algorithm. There are no changes made in MIXED2 algorithm.
Remark: The cost of arcs and edges of M need not be set to zero. In practice, the cost of each edge in U could be scaled up by a big constant. This ensures that the total cost of edges from U in the the minimum-cost matching is minimized.

6.1   Analysis of Modified MIXED Algorithm

Lemma 6. *Let C_M be the cost of arcs in M and let C^* be the cost of an optimal postman tour of G. The cost of the tour generated by Modified MIXED1 algorithm is at most $C^* + C_M$.*

Proof. Consider the components induced by the arcs of M. In original graph G, the arcs of M correspond to arcs and oriented edges, possibly duplicated. Since the original edges and arcs of G corresponding to M are made to have zero-cost when EVENDEGREE is run, any component that has an even number of odd-degree vertices can be matched using the arcs of M at zero cost. Therefore, the algorithm adds a minimum-cost matching of cost C_X. Let H be the graph

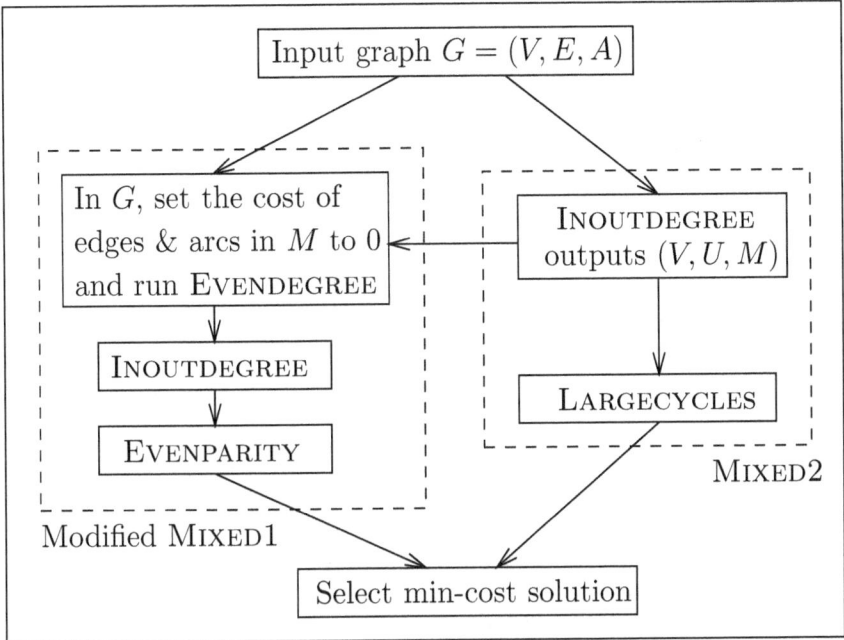

Fig. 3. Modified MIXED algorithm

at this stage. Note that EVENDEGREE duplicates each arc of M at most once to form H. We follow Frederickson's analysis [7] for the rest of the proof: Let M_1 be the multiset of arcs such that there are two arcs in M_1 for each arc in M. Clearly, M and M_1 both satisfy INOUT property. Hence, the union of U, X, and M_1 forms a INOUT solution containing H, whose cost is $C_U + C_X + 2 * C_M$. Since INOUTDEGREE is an optimal algorithm for INOUT problem, it is guaranteed to find an INOUT solution of cost at most $C_U + C_X + 2 * C_M$. This is at most $C^* + C_M$ by Lemma 4. Finally, EVENPARITY does not change the cost of the solution.

Lemma 7 (Frederickson [7]). *Algorithm* MIXED2 *finds a tour whose cost is at most* $2C^* - C_M$.

Theorem 1. *Algorithm Modified* MIXED *produces a tour whose cost is at most* $\frac{3}{2} C^*$.

Proof. By Lemma 6, Modified MIXED1 outputs a solution whose cost is at most $C_M + C^*$, which is at most $\frac{3}{2} C^*$, if $C_M \leq C^*/2$. On the other hand, if $C_M > C^*/2$, then by Lemma 7, MIXED2 outputs a solution whose cost is at most $2C^* - C_M$, which is at most $\frac{3}{2} C^*$.

7 Conclusion

We have presented an algorithm and analysis to achieve approximation ratio of 3/2 for the mixed postman problem. Improvement in the performance ratio is achieved by proving an improved lower bound on the cost of an optimal postman tour. The performance ratio is tight as shown by Frederickson's examples.

Acknowledgments

The research of the first author was supported in part by a grant from the National Science Foundation under Research Initiation Award CCR-9409625. The second author gratefully acknowledges the support of his employer, Samsung Telecommunications America Inc.

References

1. P. Brucker. The Chinese postman problem for mixed networks. In *Proceedings of the International Workshop on Graph Theoretic Concepts in Computer Science, LNCS, Vol. 100*, pages 354–366. Springer, 1980.
2. N. Christofides, E. Benavent, V. Campos, A. Corberan, and E. Mota. An optimal method for the mixed postman problem. In *System Modelling and Optimization, Notes in Control and Information Sciences, Vol. 59*. Springer, 1984.
3. J. Edmonds and E. L. Johnson. Matching, Euler tours and the Chinese postman. *Math. Programming*, 5:88–124, 1973.
4. H. A. Eiselt. Arc routing problems, Part I: The Chinese postman problem. *Operations Research*, 43:231–242, 1995.
5. H. A. Eiselt. Arc routing problems, Part II: The rural postman problem. *Operations Research*, 43:399–414, 1995.
6. L. R. Ford and D. R. Fulkerson. *Flows in Networks*. Princeton U. Press, Princeton, NJ, 1962.
7. G. N. Frederickson. Approximation algorithms for some postman problems. *J. Assoc. Comput. Mach.*, 26:538–554, 1979.
8. M. R. Garey and D. S. Johnson. *Computers and Intractability: A Guide to the Theory of NP-Completeness*. Freeman, New York, 1979.
9. M. Guan. Graphic programming using odd and even points. *Chinese Math.*, 1:273–277, 1962.
10. C. H. Kappauf and G. J. Koehler. The mixed postman problem. *Discret. Appl. Math.*, 1:89–103, 1979.
11. Y. Nobert and J. C. Picard. An optimal algorithm for the mixed Chinese postman problem. Publication # 799. Centre de recherche sur les transports, Montreal, Canada, 1991.
12. Y. Nobert and J. C. Picard. An optimal algorithm for the mixed Chinese postman problem. *Networks*, 27:95–108, 1996.
13. C. H. Papadimitriou. On the complexity of edge traversing. *J. Assoc. Comput. Mach.*, 23:544–554, 1976.
14. W. L. Pearn. Solvable cases of the k-person Chinese postman problem. *Operations Research Letters*, 16:241–244, 1994.

15. W. L. Pearn and C. M. Liu. Algorithms for the Chinese postman problem on mixed networks. *Computers & Operations Research*, 22:479–489, 1995.
16. T. K. Ralphs. On the mixed Chinese postman problem. *Operations Research Letters*, 14:123–127, 1993.

Improved Approximation Algorithms for Uncapacitated Facility Location

Fabián A. Chudak *

School of Operations Research & Industrial Engineering
Cornell University
Ithaca, NY 14853, USA
chudak@@cs.cornell.edu

Abstract. We consider the uncapacitated facility location problem. In this problem, there is a set of locations at which facilities can be built; a fixed cost f_i is incurred if a facility is opened at location i. Furthermore, there is a set of demand locations to be serviced by the opened facilities; if the demand location j is assigned to a facility at location i, then there is an associated service cost of c_{ij}. The objective is to determine which facilities to open and an assignment of demand points to the opened facilities, so as to minimize the total cost. We assume that the service costs c_{ij} are symmetric and satisfy the triangle inequality. For this problem we obtain a $(1 + 2/e)$-approximation algorithm, where $1 + 2/e \approx 1.736$, which is a significant improvement on the previously known approximation guarantees.

The algorithm works by rounding an optimal fractional solution to a linear programming relaxation. Our techniques use properties of optimal solutions to the linear program, randomized rounding, as well as a generalization of the decomposition techniques of Shmoys, Tardos, and Aardal.

1 Introduction

The study of the location of facilities to serve clients at minimum cost has been one of the most studied themes in the field of Operations Research (see, e.g., the textbook edited by Mirchandani and Francis [9]). In this paper, we focus on one of its simplest variants, *the uncapacitated facility location problem*, also known as *the simple plant location problem*, which has been extensively treated in the literature (see the chapter by Cornuéjols, Nemhauser, and Wolsey in [9]). This problem can be described as follows. There is a set of potential facility locations $\mathcal{F}$; building a facility at location $i \in \mathcal{F}$ has an associated nonnegative fixed cost f_i, and any open facility can provide an unlimited amount of certain commodity. There also is a set of clients or demand points $\mathcal{D}$ that require service; client $j \in \mathcal{D}$

* Research partially supported by NSF grants DMS-9505155 and CCR-9700029 and by ONR grant N00014-96-1-00500.

has a positive demand of commodity d_j that must be shipped from one of the open facilities. If a facility at location $i \in \mathcal{F}$ is used to satisfy the demand of client $j \in \mathcal{D}$, the service or transportation cost incurred is proportional to the distance from i to j, c_{ij} . The goal is to determine a subset of the set of potential facility locations at which to open facilities and an assignment of clients to these facilities so as to minimize the overall total cost, that is, the fixed costs of opening the facilities plus the total service cost. We will only consider the *metric* variant of the problem in which the distance function c is nonnegative, symmetric and satisfies the triangle inequality. Throughout the paper, a *ρ-approximation algorithm* is a polynomial-time algorithm that is guaranteed to deliver a feasible solution of objective function value within a factor of ρ of optimum. The main result of this paper is a $(1 + 2/e)$-approximation algorithm for the metric uncapacitated facility location problem, where $1 + 2/e \approx 1.736$.

Notice that our result is based on *worst case* analysis, that is, our solutions will be within a factor of $(1 + 2/e)$ of optimum for *any* instance of the problem. Such a strong assurance can have salient practical implications: these algorithms often outperform algorithms whose design was not grounded by the mathematical understanding required for proving performance guarantees. We have corroborated this assertion for our algorithm through a few computational experiments, which will be reported in a follow-up paper.

In contrast to the uncapacitated facility location problem, Cornuéjols, Fisher, and Nemhauser [3] studied the problem in which the objective is to maximize the difference between assignment and facility costs. They showed that with this objective, the problem can be thought of as a bank account location problem. Notice that even though these two problems are equivalent from the point of view of optimization, they are not from the point of view of approximation. Interestingly, Cornuéjols, Fisher, and Nemhauser showed that for the maximization problem, a greedy procedure that iteratively tries to open the facility that most improves the objective function yields a solution of value within a constant factor of optimum.

The metric uncapacitated facility location problem is known to be NP-hard (see [4]). Very recently, Guha and Khuller [7] have shown that it is Max SNP-hard. In fact, they have also shown that the existence of a ρ-approximation algorithm for $\rho < 1.463$ implies that NP $\subseteq$ TIME($n^{O(\log \log n)}$) (see also Feige [5]).

We briefly review previous work on approximation algorithms for the metric uncapacitated facility location problem. The first constant factor approximation algorithm was given by Shmoys, Tardos, and Aardal [11], who presented a 3.16-approximation algorithm, based on rounding an optimal solution of a classical linear programming relaxation for the problem. This bound was subsequently improved by Guha and Khuller [7], who provided a 2.408-approximation algorithm. Guha and Khuller's algorithm requires a stronger linear programming relaxation. They add to the relaxation of [11] a facility budget constraint that bounds the total fractional facility cost by the optimal facility cost. After running the algorithms of [11], they use a greedy procedure (as in [3]) to improve the quality of the solution: iteratively, open one facility at a time if it improves the cost of the solution. Since they can only guess the optimal facility cost to within a factor of $(1 + \epsilon)$, they are in fact solving a weakly polynomial number of linear programs. In contrast, the 1.736-approximation algorithm presented in this paper requires the solution of just one linear program, providing as a by-product further evidence of the strength of this linear programming relaxation.

Without loss of generality we shall assume that the sets $\mathcal{F}$ and $\mathcal{D}$ are disjoint; let $\mathcal{N} = \mathcal{F} \cup \mathcal{D}$, $n = |\mathcal{N}|$. Even though all our results hold for the case of arbitrary demands, for sake of simplicity of the exposition, we will assume that each demand d_j is 1 ($j \in \mathcal{D}$); thus, the cost of assigning a client j to an open facility at location i is c_{ij}. The distance between any two points $k, \ell \in \mathcal{N}$ is $c_{k\ell}$. We assume that the $n \times n$ distance matrix $(c_{k\ell})$ is nonnegative, symmetric ($c_{k\ell} = c_{\ell k}$, for all $k, \ell \in \mathcal{N}$) and satisfies the triangle inequality, that is, $c_{ij} \leq c_{ik} + c_{kj}$, for all $i, j, k \in \mathcal{N}$. The simplest linear programming relaxation (due to Balinski, 1965 [2]), which we will refer to as P, is as follows:

$$
\text{Min} \qquad \sum_{j \in \mathcal{D}} \sum_{i \in \mathcal{F}} c_{ij} x_{ij} + \sum_{i \in \mathcal{F}} f_i y_i
$$

(P) subject to
$$
\sum_{i \in \mathcal{F}} x_{ij} = 1 \quad \text{for each } j \in \mathcal{D} \qquad (1)
$$
$$
x_{ij} \leq y_i \text{ for each } i \in \mathcal{F}, \, j \in \mathcal{D} \qquad (2)
$$
$$
x_{ij} \geq 0 \text{ for each } i \in \mathcal{F}, \, j \in \mathcal{D} \; . \qquad (3)
$$

Any 0-1 feasible solution corresponds to a feasible solution to the uncapacitated facility location problem: $y_i = 1$ indicates that a facility at location $i \in \mathcal{F}$ is open, while $x_{ij} = 1$ means that client $j \in \mathcal{D}$ is serviced by the facility built at location $i \in \mathcal{F}$. Inequalities (1) state that each demand point $j \in \mathcal{D}$ must be assigned to some facility, while inequalities (2) say that clients can only be assigned to open facilities. Thus the linear program P is indeed a relaxation of the problem. Given a feasible fractional solution $(\hat{x}, \hat{y})$, we will say that $\sum_{i \in \mathcal{F}} f_i \hat{y}_i$ and $\sum_{j \in \mathcal{D}} \sum_{i \in \mathcal{F}} c_{ij} \hat{x}_{ij}$ are, respectively, its fractional facility and service cost.

Given a feasible solution to the linear programming relaxation P, the algorithm of Shmoys, Tardos, and Aardal first partitions the demand points into clusters and then for each cluster opens exactly one facility, which services all the points in it. In their analysis, they show that the resulting solution has the property that the total facility cost is within a constant factor of the fractional facility cost and the total service cost is within a constant factor of the fractional service cost. The main drawback of this approach is that *most of the time* the solution is *unbalanced*, in the sense that the first constant is approximately three times smaller than the second.

One of the simplest ways to round an optimal solution (x^*, y^*) to the linear program P is to use the randomized rounding technique of Raghavan and Thompson [10] as proposed by Sviridenko [12] for the special case in which all the distances are 1 or 2. Essentially, the 1.2785-approximation algorithm of [12] opens a facility at location $i \in \mathcal{F}$ with probability y_i^*; and then assigns each demand point to its nearest facility. This matches the lower bound of Guha and

Khuller for this special case. Independently of our work, Ageev and Sviridenko [1] have recently shown that the randomized rounding analysis for the maximum satisfiability problem of Goemans and Williamson [6] can be adapted to obtain improved bounds for the maximization version of the problem.

The following simple ideas enable us to develop a rounding procedure for the linear programming relaxation P with an improved performance guarantee. We explicitly exploit optimality conditions of the linear program, and in particular, we use properties of the optimal dual solution and complementary slackness. A key element to our improvement is the use of randomized rounding in conjunction with the approach of Shmoys, Tardos, and Aardal. To understand our approach, suppose that for each location $i \in \mathcal{F}$, independently, we open a facility at i with probability y_i^*. The difficulty arises when attempting to estimate the expected service cost: the distance from a given demand point to the closest open facility might be too large. However, we could always use the routing of the algorithm of Shmoys, Tardos, and Aardal if we knew that each cluster has a facility open. In essence, rather than opening each facility independently with probability y_i^*, we instead open *one* facility in each cluster with probability y_i^*. The algorithm is not much more complicated, but the most refined analysis of it is not quite so simple. Our algorithms are randomized, and can be easily derandomized using the method of conditional expectations. The main result of this paper is the following.

Theorem 1. *There is a polynomial-time algorithm that rounds an optimal solution to the linear programming relaxation* P *to a feasible integer solution whose value is within* $(1 + 2/e) \approx 1.736$ *of the optimal value of the linear programming relaxation* P.

Since the optimal LP value is a lower bound on the integer optimal value, the theorem yields a 1.736-approximation algorithm, whose running time is dominated by the time required to solve the linear programming relaxation P. As a consequence of the theorem, we obtain the following corollary on the quality of the value of the linear programming relaxation.

Corollary 1. *The optimal value of the linear programming relaxation* P *is within a factor of 1.736 of the optimal cost.*

This improves on the previously best known factor of 3.16 presented in [11].

2   A Simple 4-Approximation Algorithm

In this section we present a new simple 4-approximation algorithm. Even though the guarantees we will prove in the next section are substantially better, we will use most of the ideas presented here. Next we define the *neighborhood* of a demand point $k \in \mathcal{D}$.

Definition 1. *If* $(\overline{x}, \overline{y})$ *is a feasible solution to the linear programming relaxation* P, *and* $j \in \mathcal{D}$ *is any demand point, the* neighborhood of j, $\mathsf{N}(j)$, *is the set of facilities that fractionally service* j, *that is,* $\mathsf{N}(j) = \{i \in \mathcal{F} : \overline{x}_{ij} > 0\}$.

The following definition was crucial for the algorithm of Shmoys, Tardos, and Aardal [11].

Definition 2. *Suppose that $(\overline{x}, \overline{y})$ is a feasible solution to the linear programming relaxation* P *and let $g_j \geq 0$, for each $j \in \mathcal{D}$. Then $(\overline{x}, \overline{y})$ is g-close if $\overline{x}_{ij} > 0$ implies that $c_{ij} \leq g_j$ $(j \in \mathcal{D}, i \in \mathcal{F})$.*

Notice that if $(\overline{x}, \overline{y})$ is g-close and $j \in \mathcal{D}$ is any demand point, j is fractionally serviced by facilities inside the ball of radius g_j centered at j. The following lemma is from [11].

Lemma 1. *Given a feasible g-close solution $(\overline{x}, \overline{y})$, we can find, in polynomial time, a feasible integer $3g$-close solution $(\hat{x}, \hat{y})$ such that*

$$\sum_{i \in \mathcal{F}} f_i \hat{y}_i \leq \sum_{i \in \mathcal{F}} f_i \overline{y}_i .$$

We briefly sketch the proof below. The algorithm can be divided into two steps: a clustering step and a facility opening step. The clustering step works as follows (see Table 1). Let $\mathcal{S}$ be the set of demand points that have not yet been assigned to any cluster; initially, $\mathcal{S} = \mathcal{D}$. Find the unassigned demand point j_0 with smallest g_j-value and create a new cluster *centered* at j_0. Then all of the unassigned demand points that are fractionally serviced by facilities in the neighborhood of j_0 (that is, all the demand points $k \in \mathcal{S}$ with $N(k) \cap N(j_0) \neq \emptyset$) are assigned to the cluster centered at j_0; the set $\mathcal{S}$ is updated accordingly. Repeat the procedure until all the demand points are assigned to some cluster (i.e., $\mathcal{S} = \emptyset$). We will use $\mathcal{C}$ to denote the set of centers of the clusters.

Table 1. The clustering construction of Shmoys, Tardos, and Aardal.

```
1. S ← D,   C ← ∅
2. while S ≠ ∅
3.    choose j₀ ∈ S with smallest gⱼ value (j ∈ S)
4.    create a new cluster Q centered at j₀, C ← C ∪ {j₀}
5.    Q ← {k ∈ S : N(k) ∩ N(j₀) ≠ ∅}
6.    S ← S − Q
```

The following fact follows easily from the clustering construction and the definition of neighborhood, and is essential for the success of the algorithm.

Fact 1. *Suppose that we run the clustering algorithm of Table 1, using any g-close solution $(\overline{x}, \overline{y})$, then:*
(a) neighborhoods of distinct centers are disjoint (i.e., if j and k are centers, $j \neq k \in \mathcal{C}$, then $N(j) \cap N(k) = \emptyset$),
(b) for every demand point $k \in \mathcal{D}$, $\sum_{i \in N(k)} \overline{x}_{ik} = 1$.

After the clustering step, the algorithm of [11] opens exactly one facility per cluster. For each center $j \in C$ we open the facility i_o in the neighborhood of j, $N(j)$, with smallest fixed cost f_i and assign all the demand points in the cluster of j to facility i_o. Observe that by inequalities (2), $\sum_{i \in N(j)} \overline{y}_i \geq 1$, thus $f_{i_o} \leq \sum_{i \in N(j)} f_i \overline{y}_i$. Using Fact 1(a), the total facility cost incurred by the algorithm is never more than the total fractional facility cost $\sum_{i \in \mathcal{F}} f_i \overline{y}_i$. Next consider any demand point $k \in \mathcal{D}$ and suppose it belongs to the cluster centered at j_o; let $\ell \in N(k) \cap N(j_o)$ be a common neighbor and let i be the open facility in the neighborhood of j_o (see Figure 1). Then, the distance from k to i can be

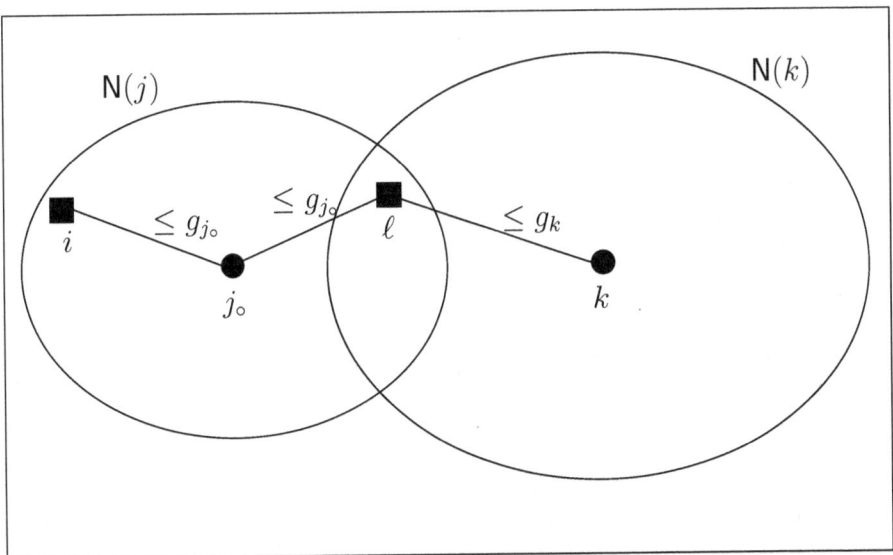

Fig. 1. Bounding the service cost of k (4-approximation algorithm). The circles (●) are demand points, whereas the squares (■) are facility locations.

bounded by the distance from k to ℓ (which is at most g_k) plus the distance from ℓ to j_o (which is at most g_{j_o}) plus the distance from j_o to i (which is at most g_{j_o}). Thus, the distance from k to an opened facility is at most $2g_{j_o} + g_k$, which is at most $3g_k$, since j_o was the remaining demand point with minimum g-value. Hence the total service cost can be bounded by $3\sum_{k \in \mathcal{D}} g_k$.

Shmoys, Tardos, and Aardal used the filtering technique of Lin & Vitter [8] to obtain g-close solutions and then applied Lemma 1 to obtain the first constant factor approximation algorithms for the problem. However, a simpler g-close solution is directly obtained by using the optimal solution to the dual linear program of P. More precisely, the dual of the linear program P is given by

$$\text{Max} \qquad \sum_{j \in \mathcal{D}} v_j \qquad\qquad\qquad (4)$$

$$\text{(D) subject to} \quad \sum_{j \in \mathcal{D}} w_{ij} \le f_i \quad \text{for each } i \in \mathcal{F} \qquad\qquad (5)$$

$$v_j - w_{ij} \le c_{ij} \quad \text{for each } i \in \mathcal{F}, j \in \mathcal{D} \qquad (6)$$

$$w_{ij} \ge 0 \quad \text{for each } i \in \mathcal{F}, j \in \mathcal{D} \ . \qquad (7)$$

Fix an optimal primal solution (x^*, y^*) and an optimal dual solution (v^*, w^*), and let LP^* be the optimal LP value. Complementary slackness gives that $x_{ij}^* > 0$ implies that $v_j^* - w_{ij}^* = c_{ij}$; since $w_{ij}^* \ge 0$, we get the following lemma.

Lemma 2. *If (x^*, y^*) is an optimal solution to the primal linear program* P *and (v^*, w^*) is an optimal solution to the dual linear program* D*, then (x^*, y^*) is v^*-close.*

By applying Lemma 1 to the optimal v^*-close solution (x^*, y^*), we obtain a feasible solution for the problem with total facility cost at most $\sum_{i \in \mathcal{F}} f_i y_i^*$ and with total service cost bounded by $3 \sum_{j \in \mathcal{D}} v_j^* = 3\,\mathsf{LP}^*$. We can bound the sum of these by $4\,\mathsf{LP}^*$; thus we have a 4-approximation algorithm. Note the inbalance in bounding facility and service costs.

3   A Randomized Algorithm

After solving the linear program P, a very simple randomized algorithm is the following: open a facility at location $i \in \mathcal{F}$ with probability y_i^* *independently* for every $i \in \mathcal{F}$, and then assign each demand point to its closest open facility. Notice that the expected facility cost is just $\sum_{i \in \mathcal{F}} f_i y_i^*$, the same bound as in the algorithm of Section 2. Focus on a demand point $k \in \mathcal{D}$. If it happens that one of its neighbors has been opened, then the service cost of k would be bounded by the optimal dual variable v_k^*. However, if we are unlucky and this is not the case (an event that can easily be shown to occur with probability at most $1/e \approx 0.368$, where the bound is tight), the service cost of k could be very large. On the other hand, suppose that we knew, for instance, that for the clustering computed in Section 2, k belongs to a cluster centered at j, where one of the facilities in $\mathsf{N}(j)$ has been opened. Then in this unlucky case we could bound the service cost of k using the routing cost of the 4-approximation algorithm.

Our algorithm is also based on randomized rounding and the expected facility cost is $\sum_{i \in \mathcal{F}} f_i y^*$. However, we weaken the randomized rounding step and do *not* open facilities *independently* with probability y_i^*, but rather in a dependent way to ensure that each cluster center has *one* of its neighboring facilities opened.

Even though the algorithms presented in this section work for any g-close feasible solution, for sake of simplicity we will assume as in the end of Section 2, that we have a fixed optimal primal solution (x^*, y^*) and a fixed optimal dual solution (v^*, w^*), so that (x^*, y^*) is v^*-close. It is easy to see that we can assume that $y_i^* \le 1$ for each potential facility location $i \in \mathcal{F}$. To motivate the following definition, fix a demand location $j \in \mathcal{D}$, and suppose without loss of generality that the neighbors of j (that is, the facilities i for which $x_{ij}^* > 0$) are $\{1, \ldots, d\}$ with $c_{1j} \le c_{2j} \le \ldots \le c_{dj}$. Then it is clear that we can assume without loss of generality that j is assigned "as much as possible" to facility 1, then to facility 2 and so on; that is, $x_{1j}^* = y_1^*$, $x_{2j}^* = y_2^*$, $\ldots$, $x_{d-1,j}^* = y_{d-1}^*$ (but maybe $x_{dj}^* < y_d^*$).

Definition 3. *A feasible solution $(\overline{x}, \overline{y})$ to the linear programming relaxation* P *is complete if $\overline{x}_{ij} > 0$ implies that $\overline{x}_{ij} = \overline{y}_i$, for every $i \in \mathcal{F}$, $j \in \mathcal{D}$.*

Thus the optimal solution (x^*, y^*) is "almost" complete, in the sense that for every $j \in \mathcal{D}$ there is at most one $i \in \mathcal{F}$ with $0 < x_{ij}^* < y_i^*$.

Lemma 3. *Suppose that $(\overline{x}, \overline{y})$ is a feasible solution to the linear program* P *for a given instance of the uncapacitated facility location problem $\mathcal{I}$. Then we can find in polynomial time an equivalent instance $\tilde{\mathcal{I}}$ and a complete feasible solution $(\tilde{x}, \tilde{y})$ to its linear programming relaxation with the same fractional facility and service costs as $(\overline{x}, \overline{y})$, that is, $\sum_{i \in \mathcal{F}} f_i \tilde{y}_i = \sum_{i \in \mathcal{F}} f_i \tilde{y}_i$ and $\sum_{j \in \mathcal{D}} \sum_{i \in \mathcal{F}} c_{ij} \overline{x}_{ij} = \sum_{j \in \mathcal{D}} \sum_{i \in \mathcal{F}} c_{ij} \tilde{x}_{ij}$. Moreover, if $(\overline{x}, \overline{y})$ is g-close, so is $(\tilde{x}, \tilde{y})$.* □*

Proof. Pick any facility $i \in \mathcal{F}$ for which there is a demand point $j \in \mathcal{D}$ with $0 < \overline{x}_{ij} < \overline{y}_i$ (if there is no such a facility, the original solution $(\overline{x}, \overline{y})$ is complete and we are done). Among the demand points $j \in \mathcal{D}$ for which $\overline{x}_{ij} > 0$, let $j_\circ$ be the one with smallest $\overline{x}_{ij}$ value. Next create a new facility location i' which is an exact copy of i (i.e. same fixed cost and in the same location), and set $\tilde{y}_{i'} = \overline{y}_i - \overline{x}_{ij_\circ}$ and reset $\tilde{y}_i$ equal to $\overline{x}_{ij_\circ}$. Next for every $j \in \mathcal{D}$ with $\overline{x}_{ij} > 0$, reset $\tilde{x}_{ij} = \overline{x}_{ij_\circ} = \tilde{y}_i$, and set $\tilde{x}_{i'j} = \overline{x}_{ij} - \overline{x}_{ij_\circ}$ (which is nonnegative by the choice of $j_\circ$). All the other components of $\overline{x}$ and $\overline{y}$ remain unchanged. Clearly $(\tilde{x}, \tilde{y})$ is a feasible solution to the linear programming relaxation of the new instance; if $(\overline{x}, \overline{y})$ is g-close so is $(\tilde{x}, \tilde{y})$. It is straightforward to verify that the new instance is equivalent to the old one (because there are no capacity restrictions) and that the fractional facility and service costs of the solutions $(\overline{x}, \overline{y})$ and $(\tilde{x}, \tilde{y})$ are the same. Since the number of pairs (k, j) for which $0 < \overline{x}_{kj} < \overline{y}_k$ has decreased at least by one, and initially there can be at most $|\mathcal{D}||\mathcal{F}| \le n^2$ such pairs, n^2 iterations suffice to construct a new instance with a complete solution. □

By Lemma 3, we can assume that (x^*, y^*) is complete. To understand some of the crucial points of our improved algorithm we will first consider the following RANDOMIZED ROUNDING WITH CLUSTERING. Suppose that we run the clustering procedure exactly as in Table 1, and let $\mathcal{C}$ be the set of cluster centers. We partition the facility locations into two classes, according to whether they are in the neighborhood of a cluster center or not.

Definition 4. *The set of central facility locations $\mathcal{L}$ is the set of facility locations that are in the neighborhood of some cluster center, that is, $\mathcal{L} = \cup_{j \in \mathcal{C}} \mathsf{N}(j)$; the remaining set of facility locations $\mathcal{R} = \mathcal{F} - \mathcal{L}$ are noncentral facility locations.*

The algorithm opens facilities in a slightly more complicated way that the simplest randomized rounding algorithm described in the beginning of the section. First we open exactly one central facility per cluster as follows: independently for each center $j \in \mathcal{C}$, open neighboring facility $i \in \mathsf{N}(j)$ at random with probability x_{ij}^* (recall Fact 1(b)). Next independently we open each noncentral facility $i \in \mathcal{R}$ with probability y_i^*. The algorithm then simply assigns each demand point to its closest open facility.

Lemma 4. *For each facility location* $i \in \mathcal{F}$, *the probability that a facility at location* i *is open is* y_i^*.

Proof. If i is a noncentral facility ($i \in \mathcal{R}$), we open a facility at i with probability y_i^*. Suppose next that i is a central facility ($i \in \mathcal{L}$), and assume that $i \in \mathsf{N}(j)$, for a center $j \in \mathcal{C}$. A facility will be opened at location i only if the center j chooses it with probability x_{ij}^*; but $x_{ij}^* = y_i^*$ since (x^*, y^*) is complete.   □

Corollary 2. *The expected total facility cost is* $\sum_{i \in \mathcal{F}} f_i y_i^*$.

For each demand point $k \in \mathcal{D}$, let $\overline{C}_k$ denote the fractional service cost of k, that is, $\overline{C}_k = \sum_{i \in \mathcal{F}} c_{ik} x_{ik}^*$. The expected service cost of $k \in \mathcal{D}$ is bounded in the following lemma whose proof is presented below.

Lemma 5. *For each demand point* $k \in \mathcal{D}$, *the expected service cost of* k *is at most* $\overline{C}_k + (3/e) v_k^*$.

Overall, since $\sum_{k \in \mathcal{D}} v_k^* = \mathsf{LP}^*$, the expected total service cost can be bounded as follows.

Corollary 3. *The expected total service cost is at most* $\sum_{k \in \mathcal{D}} \overline{C}_k + (3/e) \, \mathsf{LP}^*$.

By combining Corollaries 2 and 3, and noting that $\sum_{k \in \mathcal{D}} \overline{C}_k + \sum_{i \in \mathcal{F}} f_i y_i^* = \mathsf{LP}^*$, we obtain the following.

Theorem 2. *The expected total cost incurred by* RANDOMIZED ROUNDING WITH CLUSTERING *is at most* $(1 + 3/e)\mathsf{LP}^*$.

Proof of Lemma 5. Fix a demand point $k \in \mathcal{D}$. For future reference, let j_o be the center of the cluster to which k belongs; notice that j_o always has a neighboring facility i_o ($i_o \in \mathsf{N}(j_o)$) opened, hence its service cost is never greater than $v_{j_o}^*$. To gain some intuition behind the analysis, suppose first that each center in $\mathcal{C}$ shares at most one neighbor with k; that is, $|\mathsf{N}(j) \cap \mathsf{N}(k)| \leq 1$, for each center $j \in \mathcal{C}$. Each neighbor $i \in \mathsf{N}(k)$ is opened with probability $y_i^* = x_{ik}^*$ *independently* in this special case. For notational simplicity suppose that $\mathsf{N}(k) = \{1, \dots, d\}$, with $c_{1k} \leq \dots \leq c_{dk}$. Let q be the probability that none of the facilities in $\mathsf{N}(k)$ is open. Note that $q = \prod_{i=1}^d (1 - y_i^*) = \prod_{i=1}^d (1 - x_{ik}^*)$. One key observation is that q is "not too big": since $1 - x \leq e^{-x}$ ($x > 0$), using Fact 1(b),

$$q = \prod_{i=1}^d (1 - x_{ik}^*) \leq \prod_{i=1}^d e^{-x_{ik}^*} = \exp\left(-\sum_{i=1}^d x_{ik}^*\right) = \frac{1}{e},$$

where $\exp(x) = e^x$. We will bound the expected service cost of k by considering a clearly worse algorithm: assign k to its closest open neighbor; if none of the neighbors of k is open, assign k to the open facility $i_o \in N(j_o)$ (exactly as in Section 2). If facility 1 is open, an event which occurs with probability y_1^*, the service cost of k is c_{1k}. If, on the other hand, facility 1 is closed, but facility 2 is open, an event which occurs with probability $(1 - y_1^*)y_2^*$, the service cost of k is c_{2k}, and so on. If all the facilities in the neighborhood of k are closed, which occurs with probability q, then k is assigned to the open facility $i_o \in N(j_o)$. But in this case, k is serviced by i_o, so the service cost of k is at most $2v_{j_o}^* + v_k^* \le 3v_k^*$ exactly as in Figure 1 (Section 2); in fact, this *backup routing* gives a deterministic bound: the service cost of k is *always* no more than $3v_k^*$. Thus the expected service cost of k is at most

$$c_{1k}\, y_1^* + c_{2k}\, y_2^*(1 - y_1^*) + \cdots + c_{dk}\, y_d^*(1 - y_1^*) \ldots (1 - y_{d-1}^*) + 3v_k^*\, q$$

$$\le \sum_{i=1}^{d} c_{ik}x_{ik}^* + \frac{1}{e}\, 3v_k^* = \overline{C}_k + \frac{3}{e}\, v_k^* \; ,$$

which concludes the proof of the lemma in this special case.

Now we return to the more general case in which there are centers in $\mathcal{C}$ that can share more than one neighbor with k. We assumed that this was not the case in order to ensure that the events of opening facilities in $N(k)$ were independent, but now this is no longer true for facilities $i, i' \in N(k)$ that are neighbors of the same center. However, if one of i or i' is closed, the probability that the other is open increases; thus the dependencies are favorable for the analysis. The key idea of the proof is to group together those facilities that are neighbors of the same cluster center, so that the independence is retained and the proof of the special case above still works. A more rigorous analysis follows.

Let $\hat{\mathcal{C}}$ be the subset of centers that share neighbors with k. For each center $j \in \mathcal{C}$, let $S_j = N(j) \cap N(k)$, and so $\hat{\mathcal{C}} = \{j \in \mathcal{C} : S_j \ne \emptyset\}$. We have already proved the lemma when $|S_j| \le 1$, for each center $j \in \mathcal{C}$. For each center $j \in \hat{\mathcal{C}}$, let E_j be the event that at least one common neighbor of j and k is open (see Figure 2). To follow the proof, for each $j \in \hat{\mathcal{C}}$, it will be convenient to think of the event of choosing facility i in S_j as a sequence of two events: first j chooses to "open" S_j with probability $p_j = \sum_{i \in S_j} x_{ik}^*$ (i.e., event E_j occurs); and then if S_j is open, j chooses facility $i \in S_j$ with probability x_{ij}^*/p_j (which is the conditional probability of opening i given event E_j). Now let $\overline{C}_j = \sum_{i \in S_j} c_{ik}x_{ik}^*/p_j$; that is, $\overline{C}_j$ is the conditional expected distance from k to S_j, given the event E_j. For example, if $S_j = \{r, s, t\}$ are the common neighbors of j and k, the event E_j occurs when one of r, s or t is open, $p_j = x_{rk}^* + x_{sk}^* + x_{tk}^*$ and $\overline{C}_j = c_{rk}x_{rk}^*/p_j + c_{sk}x_{sk}^*/p_j + c_{tk}x_{tk}^*/p_j$. Notice that by Fact 1(a), the events E_j ($j \in \hat{\mathcal{C}}$) are independent. This completes the facility grouping. Consider the neighbors of k that are noncentral facility locations; that is, locations $i \in N(k) \cap \mathcal{R}$. For each each noncentral neighbor $i \in N(k) \cap \mathcal{R}$, let E_i be the event in which facility i is open, $\overline{C}_i$ be the distance c_{ik}, and $p_i = x_{ik}^*$. Next notice that *all* of the events E_ℓ are independent. It follows easily from the definitions that $\sum_\ell p_\ell = \sum_{i \in \mathcal{F}} \overline{x}_{ik} = 1$ and $\sum_\ell \overline{C}_\ell p_\ell = \overline{C}_k$.

Now we can argue essentially as in the simple case when $|S_j| \leq 1$ for each center $j \in \mathcal{C}$. Assume that there are d events E_ℓ, and for notational simplicity, they are indexed by $\ell \in \{1, \ldots, d\}$, with $\overline{C}_1 \leq \ldots \leq \overline{C}_d$. Let D be the event that none of $E_1, \ldots, E_d$ occurs; that is, D is precisely the event in which all the facilities in the neighborhood of k, $\mathsf{N}(k)$, are closed; let q be the probability of event D. Note that, as in the simple case, the service cost of k is never greater that its backup routing cost $3v_k^*$, in particular, this bound holds even conditioned on D. As before, we will analyze the expected service cost of a worse algorithm: k is assigned to the open neighboring facility with smallest $\overline{C}_\ell$; and if all the neighbors are closed, k is assigned through its backup routing to the open facility $i_\circ \in \mathsf{N}(j_\circ)$. If the event E_1 occurs (with probability p_1), the expected

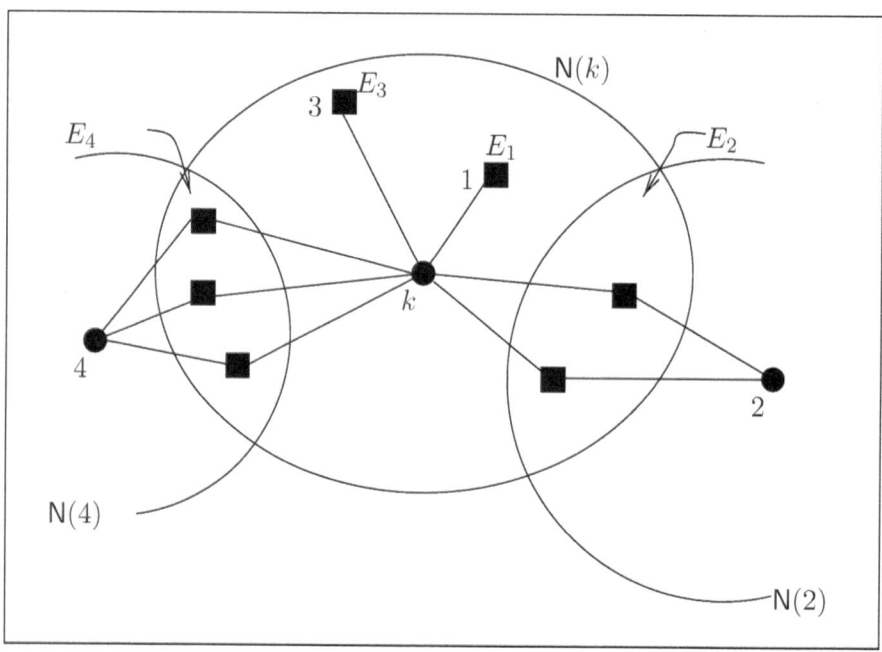

Fig. 2. Estimating the expected service cost of k. Here the centers that share a neighbor with k are demand locations 2 and 4 ($\mathcal{C}_k = \{2, 4\}$). The neighbors of k that are noncentral locations are 1 and 3. Event E_2 (respectively E_4) occurs when a facility in $\mathsf{N}(k) \cap \mathsf{N}(2)$ (respectively $\mathsf{N}(k) \cap \mathsf{N}(4)$) is open, while event E_1 (respectively E_3) occurs when facility 1 (respectively 3) is open. Though there are dependencies among the neighbors of a fixed center, the events E_1, E_2, E_3 and E_4 are independent.

service cost of k is $\overline{C}_1$. If event E_1 does not occur, but event E_2 occurs (which happens with probability $(1 - p_1)p_2$), the expected service cost of k is $\overline{C}_2$, and so on. If we are in the complementary space D, which occurs with probability $q = \prod_{\ell=1}^{d}(1 - p_\ell)$, the service cost of k is never greater than its backup service cost $3v_k^*$. Thus the expected service cost of k can be bounded by

$$\overline{C}_1\, p_1 + \overline{C}_2\, (1 - p_1)p_2 + \cdots + \overline{C}_d\, (1 - p_1)\ldots(1 - p_{d-1})p_d + 3v_k^*\, q \;. \qquad (8)$$

To prove the lemma we bound the first d terms of (8) by $\overline{C}_k$, and q by $1/e$. $\quad\square$

Notice than even though the clustering construction is deterministic, the backup service cost of k (that is, the distance between k and the facility open in $N(j_\circ)$) is a random variable B. In the proof above, we used the upper bound $B \leq 3v_k^*$. In fact, the proof of Lemma 5 shows that the expected service cost of k is no more than $\overline{C}_k + q\,\mathsf{E}[B|D]$. As it can be easily seen, the upper bound used for equation (8) is not tight. In fact, we can get an upper bound of $(1 - q)\overline{C}_k + q\,\mathsf{E}[B|D]$ as follows. First note the following simple probabilistic interpretation of the first d terms of (8). Let Z_ℓ $(\ell = 1,\ldots,d)$ be independent 0-1 random variables with $\mathsf{Prob}\{Z_\ell = 1\} = p_\ell$. Consider the set of indices for which Z_ℓ is 1, and let Z be the minimum $\overline{C}_\ell$ value in this set of indices; if all of the Z_ℓ are 0, Z is defined to be 0. Then the expected value of Z is exactly equal to the first d terms of (8). Given a set of numbers S, we will use $\min_\circ (S)$ to denote the smallest element of S if S is nonempty, and 0 if S is empty, so that $Z = \min_{\circ\{Z_\ell=1\}} \overline{C}_\ell Z_\ell$. The following intuitive probability lemma provides a bound on the first d terms of (8).

Lemma 6. *Suppose that* $0 \leq \overline{C}_1 \leq \ldots \leq \overline{C}_d$, $p_1,\ldots,p_d > 0$, *with* $\sum_{\ell=1}^{d} p_\ell = 1$. *Let* $Z_1,\ldots,Z_d$ *be 0-1 independent random variables, with* $\mathsf{Prob}\{Z_\ell = 1\} = p_\ell$. *Let* $\overline{C} = \sum \overline{C}_\ell p_\ell$. *Then*

$$\mathsf{E}\left[\min_{Z_\ell=1}\!{}_\circ\, \overline{C}_\ell Z_\ell + \prod_{\ell=1}^{d}(1 - Z_\ell)\overline{C}\right] \leq \overline{C} \;. \qquad\qquad \square$$

Applying the lemma to the first d terms of equation (8), since $\mathsf{E}[\prod_{\ell=1}^{d}(1 - Z_\ell)] = \prod_{\ell=1}^{d}(1 - p_\ell) = q$, we have that

$$\overline{C}_1 p_1 + \overline{C}_2(1 - p_1)p_2 + \cdots + \overline{C}_d(1 - p_1)\ldots(1 - p_{d-1})p_d \leq \overline{C}_k(1 - q) \;. \quad (9)$$

Thus we have proved the following.

Lemma 7. *For each demand point* $k \in D$, *the expected service cost of* k *is at most* $(1 - q)\overline{C}_k + q\,\mathsf{E}[B|D]$.

Finally we introduce the last idea that leads to the $(1+2/e)$-approximation algorithm. In Figure 1, we have bounded the distance from center $j_\circ$ to the open facility i, $c_{ij_\circ}$, by $v_{j_\circ}^*$. However, i is selected (by the center $j_\circ$) with probability $x_{ij_\circ}^*$ and, thus, the expected length of this leg of the routing is $\sum_{i\in\mathcal{F}} c_{ij_\circ} x_{ij_\circ}^* = \overline{C}_{j_\circ}$, which in general is smaller than the estimate $v_{j_\circ}^*$ used in the proof of Lemma 5. Thus, to improve our bounds, we slightly modify the clustering procedure by changing line 3 of the procedure of Table 1 to

> 3'.   choose $j_o \in S$ with smallest $v_j^* + \overline{C}_j$ value $(j \in S)$

We will call the modified algorithm RANDOMIZED ROUNDING WITH IMPROVED CLUSTERING. Notice that Lemmas 4 and 7 are unaffected by this change. We will show that the modified rule 3' leads to the bound $\mathsf{E}[B|D] \leq 2v_k^* + \overline{C}_k$, improving on the bound of $3v_k^*$ we used in the proof of Lemma 5.

Lemma 8. *If we run* RANDOMIZED ROUNDING WITH IMPROVED CLUSTERING, *the conditional expected backup service cost of* k, $\mathsf{E}[B|D]$, *is at most* $2v_k^* + \overline{C}_k$.

Proof. Suppose that the clustering partition assigned k to the cluster with center j_o. Deterministically, we divide the proof into two cases.

Case 1. Suppose that there is a facility $\ell \in \mathsf{N}(k) \cap \mathsf{N}(j_o)$, such that $c_{\ell j_o} \leq \overline{C}_{j_o}$ (see Figure 3(a)). Let i be the facility in $\mathsf{N}(j_o)$ that was opened by j_o; notice that $c_{ij_o} \leq v_{j_o}^*$ (because $(\overline{x}, \overline{y})$ is v^*-close). Then the service cost of k is at most $c_{ik} \leq c_{k\ell} + c_{\ell j_o} + c_{j_o i}$, which using again that (x^*, y^*) is v^*-close, is at most $v_k^* + c_{\ell j_o} + v_{j_o}^* \leq v_k^* + \overline{C}_{j_o} + v_{j_o}^* \leq \overline{C}_k + 2v_k^*$, where the last inequality follows from the fact that the center has the minimum $(\overline{C}_j + v_j^*)$ value. In this case, we have a (deterministic) bound, $B \leq \overline{C}_k + 2v_k^*$.

Case 2. Assume that $c_{\ell j_o} > \overline{C}_{j_o}$ for every $\ell \in \mathsf{N}(k) \cap \mathsf{N}(j_o)$ (see Figure 3(b)). First note that when we do not condition on D (i.e., that no facility in $\mathsf{N}(k)$ is open), then the expected length of the edge from j_o to the facility that j_o has selected is $\overline{C}_{j_o}$. However, we are given that all the facilities in the neighborhood of k are closed, but in this case, all of these facilities that contribute to the expected service cost of j_o (the facilities in $\mathsf{N}(k) \cap \mathsf{N}(j_o)$) are at distance greater than the average $\overline{C}_{j_o}$. Thus the conditional expected service cost of j_o is at most the unconditional expected service cost of j_o, $\overline{C}_{j_o}$. It follows then that if $\ell \in \mathsf{N}(k) \cap \mathsf{N}(j_o)$, the conditional expected service cost of k is at most $\overline{C}_{j_o} + c_{j_o \ell} + c_{\ell k} \leq \overline{C}_{j_o} + v_{j_o}^* + v_k^* \leq \overline{C}_k + 2v_k^*$. Hence, $\mathsf{E}[B|D] \leq \overline{C}_k + 2v_k^*$ in this case too.   □

Thus, using Lemmas 7 and 8, the expected service cost of k can be bounded by

$$\overline{C}_k(1 - q) + q\,(2v_k^* + \overline{C}_k) = \overline{C}_k + 2q\,v_k^* \leq \overline{C}_k + \frac{2}{e}v_k^* ,$$

where once again we bound q by $1/e$.

Corollary 4. *The expected total service cost of* RANDOMIZED ROUNDING WITH IMPROVED CLUSTERING *is at most* $\sum_{k \in D} \overline{C}_k + (2/e) \sum_{k \in D} v_k^*$.

Combining Corollaries 2 and 4, RANDOMIZED ROUNDING WITH IMPROVED CLUSTERING produces a feasible solution with expected cost no greater than

$$\sum_{i \in \mathcal{F}} f_i y_i^* + \sum_{k \in D} \overline{C}_k + \frac{2}{e} \sum_{k \in D} v_k^* = \left(1 + \frac{2}{e}\right) \mathsf{LP}^* \approx 1.736\,\mathsf{LP}^*.$$

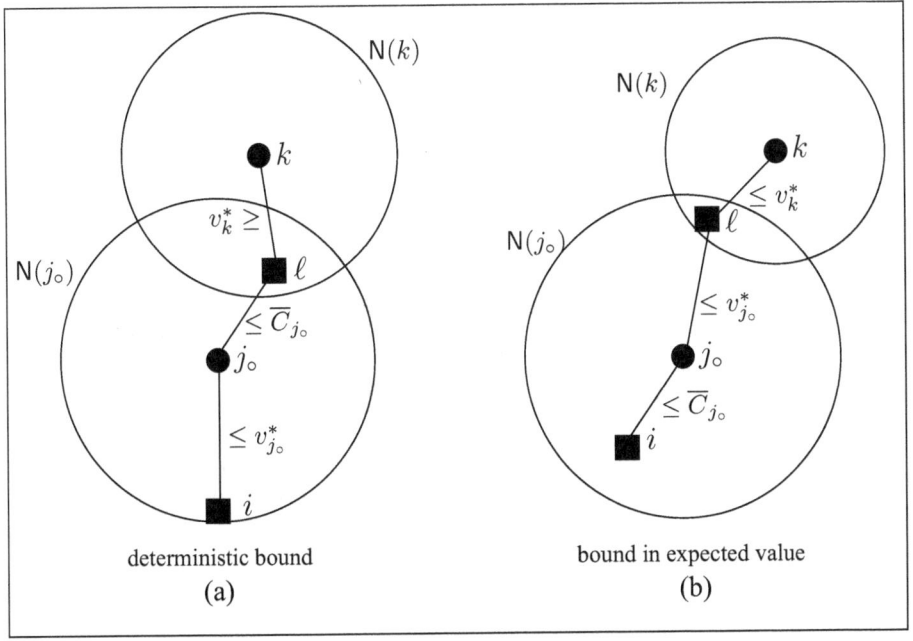

deterministic bound

(a)

bound in expected value

(b)

Fig. 3. Bounding the backup service cost of k.

Thus we have proved the following theorem.

Theorem 3. *There is a polynomial-time randomized algorithm that finds a feasible solution to the uncapacitated facility location problem with expected cost at most $(1 + 2/e)\mathsf{LP}^*$.*

To finish the proof of Theorem 1 we can show that the algorithm of Theorem 3 can be derandomized using the standard method of conditional expectations.

4   Discussion

We conclude with a few remarks concerning the algorithm of Section 3. A standard technique to improve the guarantees of randomized rounding is to use a fixed parameter, say $\gamma \geq 0$, to boost the probabilities. For instance, the simplest randomized rounding algorithm would open facility $i \in \mathcal{F}$ with probability $\min\{\gamma y_i^*, 1\}$. This technique can be also applied to our randomized algorithm in a very simple fashion. The bounds of Theorem 3 are thus parameterized for each γ. Even though this does not lead to an overall improvement in the performance guarantee, it allows us to improve the performance guarantee for some values of ρ, where $\rho \in [0, 1]$ is defined by $\rho \mathsf{LP}^* = \sum_{i \in \mathcal{F}} f_i y_i^*$. In particular, we can show that if $\rho \leq 2/e \approx 0.736$, there is a variant of our algorithm with performance guarantee $\rho \ln(2/\rho) + 1$. This bound gets better as ρ approaches 0. Another algorithm is obtained by applying Theorem 3 to the filtering g-close solution proposed by Shmoys, Tardos, and Aardal. This new algorithm performs better when ρ approaches 1. Thus for "most" values of ρ, we can provide a better guarantee than the one of Theorem 1. For example, the worst possible guarantee 1.736 is achieved only when ρ is "close" to 0.70.

Another interesting issue concerns the proof of Theorem 3. In particular, for the bound $q \leq 1/e$ to be tight, it requires having many terms (d has to be relatively large) and all the x values have to be the same. This bad instance seems very unlikely to occur. Therefore for a particular solution, a more careful bookkeeping of the constants can provide a substantially improved guarantee.

Finally, our work also indicates that it might be worthwhile to investigate the of randomization for more complex location problems.

Acknowledgments. The author is grateful to David Shmoys for providing great assistance and numerous ideas, that led to some crucial points of the results in this paper. We also wish to thank Gena Samorodnitsky and Mike Todd for helpful comments.

References

1. A. A. Ageev and M. I. Sviridenko. An approximation algorithm for the uncapacitated facility location problem. Manuscript, 1997.
2. M. L. Balinski. Integer programming: Methods, uses, computation. *Management Science*, 12(3):253–313, 1965.
3. G. Cornuéjols, M. L. Fisher, and G. L. Nemhauser. Location of bank accounts to optimize float: An analytic study of exact and approximate algorithms. *Management Science*, 23(8):789–810, 1977.
4. G. Cornuéjols, G. L. Nemhauser, and L. A. Wolsey. The uncapacitated facility location problem. In P. Mirchandani and R. Francis, editors, *Discrete Location Theory*, pages 119–171. John Wiley and Sons, Inc., New York, 1997.
5. U. Feige. A threshold of ln n for approxiamting set-cover. In *28th ACM Symposium on Theory of Computing*, pages 314–318, 1996.
6. M. X. Goemans and D. P. Williamson. New 3/4-approximation algorithms for max-sat. *SIAM Journal on Discrete Mathematics*, 7:656–666, 1994.
7. S. Guha and S. Khuller. Greedy strikes back: improved facility location algorithms. In *Ninth Annual ACM-SIAM Symposium on Discrete Algorithms (SODA)*, 1998. To appear.
8. J. H. Lin and J. S. Vitter. ϵ-Approximation with minimum packing constraint violation. In *Proceedings of the 24th Annual ACM Symposium on Theory of Computing*, pages 771–782, 1992.
9. P. Mirchandani and R. Francis, editors. *Discrete Location Theory*. John Wiley and Sons, Inc., New York, 1990.
10. P. Raghavan and C. D. Thompson. Randomized rounding. *Combinatorica*, 7:365–374, 1987.
11. D. B. Shmoys, É. Tardos, and K. Aardal. Approximation algorithms for facility location problems. In *29th ACM Symposium on Theory of Computing*, pages 265–274, 1997.
12. M. I. Sviridenko. Personal communication, July 1997.

The Maximum Traveling Salesman Problem Under Polyhedral Norms

Alexander Barvinok[1*], David S. Johnson[2], Gerhard J. Woeginger[3**], and
Russell Woodroofe[1***]

[1] University of Michigan, Dept. of Mathematics
Ann Arbor, MI 48109-1009, USA
barvinok@@math.lsa.umich.edu
[2] AT&T Labs – Research, Room C239
Florham Park, NJ 07932-0971, USA
dsj@@research.att.com
[3] Institut für Mathematik, TU Graz
Steyrergasse 30, A-8010 Graz, Austria
gwoegi@@opt.math.tu-graz.ac.at

Abstract. We consider the traveling salesman problem when the cities
are points in R^d for some fixed d and distances are computed according
to a polyhedral norm. We show that for any such norm, the problem of
finding a tour of *maximum* length can be solved in polynomial time. If
arithmetic operations are assumed to take unit time, our algorithms run
in time $O(n^{f-2} \log n)$, where f is the number of facets of the polyhedron
determining the polyhedral norm. Thus for example we have $O(n^2 \log n)$
algorithms for the cases of points in the plane under the Rectilinear and
Sup norms. This is in contrast to the fact that finding a *minimum* length
tour in each case is NP-hard.

1 Introduction

In the *Traveling Salesman Problem* (TSP), the input consists of a set C of *cities*
together with the distances $d(c, c')$ between every pair of distinct cities $c, c' \in C$.
The goal is to find an ordering or *tour* of the cities that minimizes (Minimum
TSP) or maximizes (Maximum TSP) the total tour length. Here the length of a
tour $c_{\pi(1)}, c_{\pi(2)}, \ldots, c_{\pi(n)}$ is

$$\sum_{i=1}^{n-1} d(c_{\pi(i)}, c_{\pi(i+1)}) + d(c_{\pi(n)}, c_{\pi(1)}).$$

---

[*] Supported by an Alfred P. Sloan Research Fellowship and NSF grant DMS 9501129.
[**] Supported by the START program Y43-MAT of the Austrian Ministry of Science.
[***] Supported by the NSF through the REU Program.

R. E. Bixby, E. A. Boyd, and R. Z. Ríos-Mercado (Eds.): IPCO VI
LNCS 1412, pp. 195–201, 1998. © Springer–Verlag Berlin Heidelberg 1998

Of particular interest are *geometric* instances of the TSP, in which cities correspond to points in R^d for some $d \geq 1$, and distances are computed according to some geometric norm. Perhaps the most popular norms are the Rectilinear, Euclidean, and Sup norms. These are examples of what is known as an "L^p norm" for $p = 1$, 2, and ∞. In general, the distance between two points $x = (x_1, x_2, \ldots, x_d)$ and $y = (y_1, y_2, \ldots, y_d)$ under the L^p norm, $p \geq 1$, is

$$d(x, y) = \left(\sum_{i=1}^{d} |x_i - y_i|^p \right)^{1/p}$$

with the natural asymptotic interpretation that distance under the L^∞ norm is

$$d(x, y) = \max_{i=1}^{d} |x_i - y_i|.$$

This paper concentrates on a second class of norms which also includes the Rectilinear and Sup norms, but can only approximate the Euclidean and other L^p norms. This is the class of *polyhedral norms*. Each polyhedral norm is determined by a *unit ball* which is a centrally-symmetric polyhedron P with the origin at its center. To determine $d(x, y)$ under such a norm, first translate the space so that one of the points, say x, is at the origin. Then determine the unique factor α by which one must rescale P (expanding if $\alpha > 1$, shrinking if $\alpha < 1$) so that the other point (y) is on the boundary of the polyhedron. We then have $d(x, y) = \alpha$.

Alternatively, and more usefully for our purposes, we can view a polyhedral norm as follows. If P is a polyhedron as described above and has F facets, then F is divisible by 2 and there is a set $H_P = \{h_1, \ldots, h_{F/2}\}$ of points in R^d such that P is the intersection of a collection of half-spaces determined by H_P:

$$P = \left(\bigcap_{i=1}^{F/2} \{x : \ x \cdot h_i \leq 1\} \right) \cap \left(\bigcap_{i=1}^{F/2} \{x : \ x \cdot h_i \geq -1\} \right)$$

Then we have

$$d(x, y) = \max \left\{ \left| (x - y) \cdot h_i \right| : \ 1 \leq i \leq F/2 \right\}$$

Note that for the Rectilinear norm in the plane we can take $H_P = \{(1, 1), (-1, 1)\}$ and for the Sup norm in the plane we can take $H_P = \{(1, 0), (0, 1)\}$.

For the Minimum TSP on geometric instances, all of the key complexity questions have been answered. As follows from results of Itai, Papadimitriou, and Swarcfiter [5], the Minimum TSP is NP-hard for any fixed dimension d and any L^p or polyhedral norm. On the other hand, recent results of Arora [1,2] and Mitchell [8] imply that in all these cases a polynomial-time approximation scheme (PTAS) exists, i.e., a sequence of polynomial-time algorithms A_k, $1 \leq k < \infty$, where A_k is guaranteed to find a tour whose length is within a ratio of $1 + (1/k)$ of optimal.

The situation for geometric versions of the Maximum TSP is less completely resolved. Barvinok [3] has shown that once again polynomial-time approximation schemes exist for all fixed dimensions d and all L^p or polyhedral norms (and in a sense for *any* fixed norm; see [3]). Until now, however, the complexity of the optimization problems themselves when d is fixed has remained open: For no fixed dimension d and L^p or polyhedral norm was the problem of determining the maximum tour length known either to be NP-hard or to be polynomial-time solvable. In this paper, we resolve the question for polyhedral norms, showing that, in contrast to the case for the Minimum TSP, the Maximum TSP is solvable in polynomial time for any fixed dimension d and any polyhedral norm:

Theorem 1. *Let dimension d be fixed, and let $\| \cdot \|$ be a fixed polyhedral norm in R^d whose unit ball is a polyhedron P determined by a set of f facets. Then for any set of n points in R^d, one can construct a traveling salesman tour of maximum length with respect to $\| \cdot \|$ in time $O(n^{f-2} \log n)$, assuming arithmetic operations take unit time.*

As an immediate consequence of Theorem 1, we get relatively efficient algorithms for the Maximum TSP in the plane under Rectilinear and Sup norms:

Corollary 2. *The Maximum TSP for points in R^2 under the L^1 and L^∞ norms can be solved in $O(n^2 \log n)$ time, assuming arithmetic operations take unit time.*

The restriction to unit cost arithmetic operations in Theorem 1 and Corollary 2 is made primarily to simplify the statements of the conclusions, although it does reflect that fact that our results hold for the *real number RAM* computational model. Suppose on the other hand that one assumes, as one typically must for complexity theory results, that the components of the vectors in H_P and the coordinates of the cities are all rationals. Let U denote the maximum absolute value of any of the corresponding numerators and denominators. Then the conclusions of the Theorem and Corollary hold with running times multiplied by $n \log(U)$. If the components/coordinates are all integers with maximum absolute value U, the running times need only be multiplied by $\log(nU)$. For simplicity in the remainder of this paper, we shall stick to the model in which numbers can be arbitrary reals and arithmetic operations take unit time. The reader should have no trouble deriving the above variants.

The paper is organized as follows. Section 2 introduces a new special case of the TSP, the *Tunneling TSP*, and shows how the Maximum TSP under a polyhedral norm can be reduced to the Tunneling TSP with the same number of cities and $f/2$ tunnels. Section 3 sketches how the latter problem can be solved in $O(n^{f+1})$ time, a slightly weaker result than that claimed in Theorem 1. The details of how to improve this running time to $O(n^{f-2} \log n)$ will be presented in the full version of this paper, a draft of which is available from the authors. Section 4 concludes by describing some related results and open problems.

2   The Tunneling TSP

The *Tunneling TSP* is a special case of the Maximum TSP in which distances are determined by what we shall call a *tunnel system* distance function. In such a distance function we are given a set $T = \{t_1, t_2, \ldots, t_k\}$ of auxiliary objects that we shall call *tunnels*. Each tunnel is viewed as a bidirectional passage having a front and a back end. For each pair c, t of a city and a tunnel we are given real-valued *access distances* $F(c, t)$ and $B(c, t)$ from the city to the front and back ends of the tunnel respectively. Each potential tour edge $\{c, c'\}$ must pass through some tunnel t, either by entering the front end and leaving the back (for a distance of $F(c, t) + B(c', t)$), or by entering the back end and leaving the front (for a distance of $B(c, t) + F(c', t)$). Since we are looking for a tour of maximum length, we can thus define the distance between cities c and c' to be

$$d(c, c') = \max \left\{ F(c, t_i) + B(c', t_i), \ B(c, t_i) + F(c', t_i) : 1 \le i \le k \right\}$$

Note that this distance function, like our geometric norms, is symmetric.

It is easy to see that Maximum TSP remains NP-hard when distances are determined by arbitrary tunnel system distance functions. However, for the case where $k = |T|$ is fixed and not part of the input, we will show in the next section that Maximum TSP can be solved in $O(n^{2k-1})$ time. We are interested in this special case because of the following lemma.

Lemma 3. *If $\| \cdot \|$ is a polyhedral norm determined by a set H_P of k vectors in R^d, then for any set C of points in R^d one can in time $O(dk|C|)$ construct a tunnel system distance function with k tunnels that yields $d(c, c') = \| c - c' \|$ for all $c, c' \in C$.*

Proof. The polyhedral distance between two cities $c, c' \in R^d$ is

$$\| c - c' \| = \max \left\{ \left| (c - c') \cdot h_i \right| : 1 \le i \le k \right\}$$

$$= \max \left\{ (c - c') \cdot h_i, \ (c' - c) \cdot h_i : 1 \le i \le k \right\}$$

Thus we can view the distance function determined by $\| \cdot \|$ as a tunnel system distance function with set of tunnels $T = H_P$ and $F(c, h) = c \cdot h$, $B(c, h) = -c \cdot h$ for all cities c and tunnels h. $\qquad\square$

3   An Algorithm for Bounded Tunnel Systems

This section is devoted to the proof of the following lemma, which together with Lemma 3 implies that the Maximum TSP problem for a fixed polyhedral norm with f facets can be solved in $O(n^{f+1})$ time.

Lemma 4. *If the number of tunnels is fixed at k, the Tunneling TSP can be solved in time $O(n^{2k+1})$, assuming arithmetic operations take unit time.*

Proof. Suppose we are given an instance of the Tunneling TSP with sets $C = \{c_1, \ldots, c_n\}$ and $T = \{t_1, \ldots, t_k\}$ of cities and tunnels, and access distances $F(c, t)$, $B(c, t)$ for all $c \in C$ and $t \in T$. We begin by transforming the problem to one about subset construction.

Let $G = (C \cup T, E)$ be an edge-weighted, bipartite multigraph with four edges between each city c and tunnel t, denoted by $e_i[c, t, X]$, $i \in \{1, 2\}$ and $X \in \{B, F\}$. The weights of these edges are $w(e_i[c, t, F]) = F(c, t)$ and $w(e_i[c, t, B]) = B(c, t)$, $i \in \{1, 2\}$. For notational convenience, let us partition the edges in E into sets $E[t, F] = \{e_i[c, t, F] : c \in C, i \in \{1, 2\}\}$ and $E[t, B] = \{e_i[c, t, B] : c \in C, i \in \{1, 2\}\}$, $t \in T$. Each tour for the TSP instance then corresponds to a subset E' of E that has $\sum_{e \in E'} w(e)$ equal to the tour length and satisfies

(T1) Every city is incident to exactly two edges in E'.

(T2) For each tunnel $t \in T$, $|E' \cap E[t, F]| = |E' \cap E[t, B]|$.

(T3) The set E' is connected.

To construct the multiset E', we simply represent each tour edge $\{c, c'\}$ by a pair of edges from E that connect in the appropriate way to the tunnel that determines $d(c, c')$. For example, if $d(c, c') = F(c, t) + B(c', t)$, and c appears immediately before c' when the tour is traversed starting from $c_{\pi(1)}$, then the edge (c, c') can be represented by the two edges $e_2[c, t, F]$ and $e_1[c', t, B]$. Note that there are enough (city,tunnel) edges of each type so that all tour edges can be represented, even if a given city uses the same tunnel endpoint for both its tour edges. Also note that if $d(c, c')$ can be realized in more than one way, the multiset E' will not be unique. However, any multiset E' constructed in this fashion will still have $\sum_{e \in E'} w(e)$ equal to the tour length.

On the other hand, any set E' satisfying (T1) – (T3) corresponds to one (or more) tours having length at least $\sum_{e \in E'} w(e)$: Let $T' \in T$ be the set of tunnels t with $|E' \cap E[t, F]| > 0$. Then $G' = (C \cup T', E')$ is a connected graph all of whose vertex degrees are even by (T1) – (T3). By an easy result from graph theory, this means that G' contains an Euler tour that by (T1) passes through each city exactly once, thus inducing a TSP tour for C. Moreover, by (T2) one can construct such an Euler tour with the additional property that if $e_i[c, t, x]$ and $e_j[c', t, y]$ are consecutive edges in this tour, then $x \neq y$, i.e, either $x = F, y = B$ or $x = B, y = F$. Thus we will have $w(e_i[c, t, x]) + w(e_j[c', t, y]) \leq d(c, c')$, and hence the length of the TSP tour will be at least $\sum_{e \in E'} w(e)$, as claimed.

Thus our problem is reduced to finding a maximum weight set of edges $E' \subseteq E$ satisfying (T1) – (T3). We will now sketch our approach to solving this latter problem; full details are available from the authors and will appear in the journal version of this paper. The basic idea is to divide into $O(n^{2k-2})$ subproblems, each of which can be solved in linear time. Each subcase corresponds to a choice of a degree sequence d for the tunnels, for which there are $O(n^{k-1})$ possibilities, and a choice, for that degree sequence, of a canonical-form sequence s of edges that connects together those tunnels that have positive degree, for which there are again $O(n^{k-1})$ possibilities.

Having chosen d and s, what we are left with is a maximum weight bipartite b-matching problem: starting with a set consisting of the edges specified by s, each tunnel end must have its degree augmented up to that specified by d and each city must have its degree augmented up to 2. This b-matching problem can be solved in $O(n^3)$ time by the standard technique that converts it to an assignment problem on an expanded graph. The overall running time for the algorithm is thus $O(n^{k-1}n^{k-1}n^3) = O(n^{2k+1})$, as claimed.   □

In the full paper we show how two additional ideas enable us to reduce our running times to $O(n^{2k-2}\log n)$, as needed for the proof of Theorem 1. The first idea is to view each b-matching problem as a transportation problem with a bounded number of customer locations. This latter problem can be solved in linear time by combining ideas from [7,4,11]. The second idea is to exploit the similarities between the transportation instances we need to solve. Here a standard concavity result implies that one dimension of our search over degree sequences can be handled by a binary search. In the full paper we also discuss how the constants involved in our algorithms grow with k.

4   Conclusion

We have derived a polynomial time algorithm for the Maximum TSP when the cities are points in R^d for some fixed d and when the distances are measured according to some polyhedral norm. The complexity of the Maximum TSP with *Euclidean distances* and fixed d remains unsettled, however, even for $d = 2$. Although the Euclidean norm can be approximated arbitrarily closely by polyhedral norms (and hence Barvinok's result [3] that Maximum TSP has a PTAS), it is not itself a polyhedral norm.

A further difficulty with the Euclidean norm (one shared by both the Minimum and Maximum TSP) is that we still do not know whether the TSP is in NP under this norm. Even if all city coordinates are rationals, we do not know how to compare a tour length to a given rational target in less than exponential time. Such a comparison would appear to require us to evaluate a sum of n square roots to some precision, and currently the best upper bound known on the number of bits of precision needed to insure a correct answer remains exponential in n. Thus even if we were to produce an algorithm for the Euclidean Maximum TSP that ran in polynomial time when arithmetic operations (and comparisons) take unit time, it might not run in polynomial time on a standard Turing machine.

Another set of questions concerns the complexity of the Maximum TSP when d is *not* fixed. It is relatively easy to show that the problem is NP-hard for all L^p norms (the most natural norms that are defined for all $d > 0$). For the case of L^∞ one can use a transformation from Hamiltonian Circuit in which each edge is represented by a separate dimension. For the L^p norms, $1 \le p < \infty$, one can use a transformation from the Hamiltonian Circuit problem for cubic graphs, with a dimension for each *non*-edge. However, this still leaves open the question of whether there might exist a PTAS for any such norm when d is not fixed. Trevisan [10] has shown that the Minimum TSP is Max SNP-hard for any such norm, and so cannot have such PTAS's unless P = NP. We can obtain a similar result for the Maximum TSP under L^∞ by modifying our NP-hardness transformation so that the source problem is the Minimum TSP with all edge lengths in $\{1, 2\}$, a special case that was proved Max SNP-hard by Papadimitriou and Yannakakis [9]. The question remains open for L^p, $1 \leq p < \infty$, although we conjecture that these cases are Max SNP-hard as well.

Finally, we note that our results can be extended in several ways. For instance, one can get polynomial-time algorithms for asymmetric versions of the Maximum TSP in which distances are computed based on non-symmetric unit balls. Also, algorithmic approaches analogous to ours can be applied to geometric versions of other NP-hard maximization problems: For example, consider the *Weighted 3-Dimensional Matching Problem* that consists in partitioning a set of $3n$ elements into n triples of maximum total weight. The special case where the elements are points in $\mathbf{R}^d$ and where the weight of a triple equals the perimeter of the corresponding triangle measured according to some fixed polyhedral norm can be solved in polynomial time.

Acknowledgment. We thank Arie Tamir for helpful comments on a preliminary version of this paper, and in particular for pointing out that a speedup of $O(n^2)$ could be obtained by using the transportation problem results of [7] and [11]. Thanks also to Mauricio Resende and Peter Shor for helpful discussions.

References

1. S. Arora. Polynomial-time approximation schemes for Euclidean TSP and other geometric problems. *Proc. 37th IEEE Symp. on Foundations of Computer Science*, pages 2–12. IEEE Computer Society, Los Alamitos, CA, 1996.
2. S. Arora. Nearly linear time approximation schemes for Euclidean TSP and other geometric problems. *Proc. 38th IEEE Symp. on Foundations of Computer Science*, pages 554–563. IEEE Computer Society, Los Alamitos, CA, 1997.
3. A. I. Barvinok. Two algorithmic results for the traveling salesman problem. *Math. of Oper. Res.*, 21:65–84, 1996.
4. D. Gusfield, C. Martel, and D. Fernandez-Baca. Fast algorithms for bipartite network flow. *SIAM J. Comput.*, 16:237–251, 1987.
5. A. Itai, C. Papadimitriou, and J. L. Swarcfiter. Hamilton paths in grid graphs. *SIAM J. Comput.*, 11:676–686, 1982.
6. E. L. Lawler, J. K. Lenstra, A. H. G. Rinnooy Kan, and D. B. Shmoys. *The Traveling Salesman Problem*, Wiley, Chichester, 1985.
7. N. Megiddo and A. Tamir. Linear time algorithms for some separable quadratic programming problems. *Oper. Res. Lett.*, 13:203–211, 1993.
8. J. Mitchell. Guillotine subdivisions approximate polygonal subdivisions: Part II – A simple PTAS for geometric k-MST, TSP, and related problems. Preliminary manuscript, April 1996.
9. C. H. Papadimitriou and M. Yannakakis. The traveling salesman problem with distances one and two. *Math. of Oper. Res.*, 18:1–11, 1993.
10. L. Trevisan. When Hamming meets Euclid: The approximability of geometric TSP and MST. *Proc. 29th Ann. ACM Symp. on Theory of Computing*, pages 21–29. ACM, New York, 1997.
11. E. Zemel. An $O(n)$ algorithm for the linear multiple choice knapsack problem and related problems. *Inf. Proc. Lett.*, 18:123–128, 1984.

Polyhedral Combinatorics of Benzenoid Problems

Hernán Abeledo[1] and Gary Atkinson[2]

[1] Department of Operations Research
The George Washington University
Washington, DC 20052, USA
abeledo@@seas.gwu.edu
[2] Bell Laboratories, Lucent Technologies
Holmdel, NJ 07733, USA
atkinson@@lucent.com

Abstract. Many chemical properties of benzenoid hydrocarbons can be understood in terms of the maximum number of mutually resonant hexagons, or Clar number, of the molecules. Hansen and Zheng (1994) formulated this problem as an integer program and conjectured, based on computational evidence, that solving the linear programming relaxation always yields integral solutions. We establish their conjecture by showing that the constraint matrices of these problems are unimodular.

Previously, Hansen and Zheng (1992) showed that a certain minimum weight cut cover problem defined for benzenoids yields an upper bound for the Clar number and conjectured that equality always holds. We prove that strong duality holds by formulating a network flow problem that can be used to compute the Clar number. We show that our results extend to generalizations of the Clar number and cut cover problems defined for plane graphs that are bipartite and 2-connected.

1 Introduction

In this paper we study optimization problems defined for maps of graphs that are bipartite and 2-connected. These problems are generalizations of ones that arise in chemical graph theory, specifically those encountered in the analysis of benzenoid hydrocarbons.

Benzenoid hydrocarbons are organic molecules composed of carbon and hydrogen atoms organized into connected (hexagonal) benzene rings. The structure of such a molecule is usually represented by a *benzenoid system*, a bipartite and two-connected plane graph whose interior faces are all regular hexagons. Each node of a benzenoid system represents a carbon atom, and each edge corresponds to a single or double bond between a pair of carbon atoms. The hydrogen atoms are not explicitly represented since their location is immediately deduced from the configuration of the carbon atoms.

R. E. Bixby, E. A. Boyd, and R. Z. Ríos-Mercado (Eds.): IPCO VI
LNCS 1412, pp. 202–212, 1998. © Springer–Verlag Berlin Heidelberg 1998

Generally, there is more than one way to arrange n carbon atoms into h hexagonal rings where each arrangement has $n - 2(h - 1)$ hydrogen atoms. The chemical formula $C_nH_{n-2(h-1)}$ represents a series of benzenoid *isomers*, structurally distinct molecules having the same atomic makeup. However, not all possible arrangements correspond to actual molecules. For benzenoid molecules, empirical evidence [4,5] shows that their graphs always have a perfect matching.

In chemical graph theory, perfect matchings are called *Kekulé structures*. A benzenoid system will usually have many Kekulé structures; analyzing them is of interest to chemists and gives rise to interesting combinatorial optimization problems [8]. For example, the Clar number of a benzenoid system is the optimal value of a particular maximization problem over its set of Kekulé structures and is a key concept in the aromatic sextet theory developed by E. Clar [3] to explain benzenoid phenomenology. It has been observed that as the Clar number increases within an isomeric series of benzenoid hydrocarbons, (1) isomeric stability increases, so chemical reactivity decreases, and (2) the absorption bands shift towards shorter wavelengths, so the isomer colors change from dark blue-green to red, yellow or white [5]. It has also been demonstrated that the Clar number provides a rough estimate of the Dewar-type resonance energy [1].

Hansen and Zheng [7] formulated the computation of the Clar number as an integer (linear) program and conjectured, based on empirical results, that solving its linear programming relaxation yields integer solutions for all benzenoids. We prove here that the constraint matrix of this integer program is always unimodular. In particular, this establishes that the relaxation polytope is integral since the linear program is in standard form.

Interestingly, these unimodular constraint matrices coming from an applied problem constitute an unusual case: they are not, in general, totally unimodular as often occurs with optimization problems on graphs that give rise to integral polyhedra. However, for the Clar problem, we remark that a subset of them are (also) totally unimodular, namely those constraint matrices corresponding to a natural subset of benzenoids called catacondensed benzenoid hydrocarbons.

In a previous paper, Hansen and Zheng [6] considered a minimization problem for benzenoid systems that we call here the minimum weight cut cover problem. They showed that the optimal value of the cover problem is an upper bound for the Clar number and conjectured that equality always holds. We prove this is the case by formulating the minimum weight cut cover problem as a network flow problem and using its dual to construct a solution for the Clar number problem. As a consequence, the Clar number and the minimum weight cut cover problems can be solved in strongly polynomial time using a minimum cost network flow algorithm. Our results are established for the generalized versions of the Clar number and cut cover problems that are defined here.

2   Preliminaries and Background

In this paper we consider combinatorial optimization problems defined for plane graphs that are bipartite and 2-connected. The undefined graph theory terminology we use is standard and may be found in [2] or [9]. Here (V, E, F) will always denote the plane map with set of finite faces F that results from embedding a bipartite and 2-connected planar graph $G = (V, E)$. We will further assume that the nodes of V are colored black or white so that all edges connect nodes of different color.

Since G is bipartite and 2-connected, the boundary of each face $f \in F$ is an even cycle which can be perfectly matched in two different ways. Each of these two possible matchings is called a *frame* of the face. A frame of a face f is *clockwise oriented* if each frame edge is drawn from a black node to white node in a clockwise direction along the boundary of f. Otherwise, the frame is *counterclockwise oriented*. A *framework* for (V, E, F) is a mapping $\varphi : F \mapsto E$ such that $\varphi(f)$ is a frame for each $f \in F$. A framework φ is *simple* if $\varphi(f) \cap \varphi(f') = \emptyset$ whenever $f \neq f'$. An *oriented* framework has all frames with the same orientation. It follows that oriented frameworks are simple.

A set of node-disjoint faces $F' \subseteq F$ is called a *resonant set* for (V, E, F) if there exists a perfect matching for G that simultaneously contains a frame for each face in F'. We denote by $\mathrm{res}(V, E, F)$ the maximum cardinality of a resonant set for (V, E, F). The *maximum resonant set problem* seeks to determine $\mathrm{res}(V, E, F)$. It can be viewed as a set partitioning problem where nodes in a plane map must be covered by faces or edges, and the objective is to maximize the number of faces in the partition. Of course, this problem is infeasible if G does not have a perfect matching.

When (V, E, F) is a benzenoid system, $\mathrm{res}(V, E, F)$ is called its *Clar number*. A *Clar structure* is a representation of an optimal solution to this problem where the (hexagonal) faces and edges in the partition have inscribed circles and double lines, respectively (see Fig. 1).

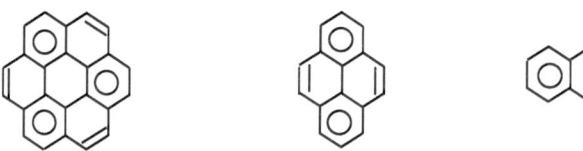

Fig. 1. Clar structures of different benzenoids

We recall that a *basis* of a matrix is a maximal linearly independent column submatrix. An integer matrix is called *unimodular* if, for each basis, the determinants of the largest submatrices are relatively prime. A matrix is *totally unimodular* if the determinant of every square submatrix is 0, 1, or -1. Totally unimodular matrices are unimodular. The relationship between unimodular matrices and integral polyhedra is stated in a theorem established by Veinott and Dantzig [13] and extended by Truemper [12] to matrices that do not have full row rank. To simplify the presentation of results, we consider an empty polyhedron to be integral.

Theorem 1 ([13], [12]). *An integer matrix A is unimodular if and only if the polyhedron $\{x : Ax = b, \; x \geq 0\}$ is integral for any integer vector b.*

An integer matrix M is *column eulerian* if $M1 = 0 \pmod 2$, i.e., if the sum of its columns is a vector with all entries even. Truemper gave the following sufficient condition for a matrix to be unimodular.

Theorem 2 ([12]). *An integer matrix A is unimodular if:*

1. *$A = BC$, with B a unimodular basis of A and each entry of C is 0, 1 or -1.*
2. *No column eulerian column submatrix of A is linearly independent.*

Throughout this paper we assume familiarity with basic properties of totally unimodular matrices as can be found in [10] or [11].

3   Two Useful Propositions

We say two subgraphs of a plane graph are *intrafacial* if they are edge-disjoint and each one is contained in the closure of a face of the other. Let G_1 and G_2 be two intrafacial subgraphs. We say G_1 is *interior* to G_2 and G_2 is *exterior* to G_1 if G_1 is contained in the closure of an interior face of G_2. We say G_1 and G_2 are *mutually exterior* if each subgraph is contained in the exterior face of the other one.

Proposition 3. *A plane graph with all even degree nodes can be decomposed into pairwise intrafacial cycles.*

Proof. We outline here an iterative procedure that yields the desired decomposition. Note that the blocks of a plane graph with all even degree nodes are eulerian, 2-connected and pairwise intrafacial. Since a block is 2-connected, the boundary of each of its faces is a cycle. Removing such a cycle from the original graph results in a new graph whose blocks continue to be eulerian, 2-connected and pairwise intrafacial. Furthermore, the removed cycle is intrafacial with respect to the new graph.   □

An (even) cycle is *signed* if each edge is alternatively assigned a value of 1 or -1. The signing of a plane cycle is *clockwise* if edges drawn from a black to a white node in a clockwise direction along the cycle are given a value of 1; otherwise the signing is *counterclockwise*. For brevity in the following proof, we say a face and a cycle are *frame adjacent* if the face's frame and the cycle have common edges.

Proposition 4. *Let $G' = (V', E')$ be a plane subgraph of (V, E, F) that is decomposed into intrafacial cycles. Then the intrafacial cycles of G' can be signed such that, for any framework φ for (V, E, F) and for every $f \in F$, all edges in $\varphi(f) \cap E'$ have the same sign.*

Proof. All cycles referred to in this proof are those of the G' decomposition. Since G is bipartite, all cycles are signable. We will show that the following rule yields the desired signing: if a cycle lies in the interior of an even number of cycles (not counting itself), sign it clockwise; otherwise sign it counterclockwise.

Consider any face $f \in F$. Then f is interior or exterior to any cycle. Let n denote the total number of cycles that contain f and assume, without loss of generality, that $\varphi(f)$ is clockwise oriented. If f is not frame adjacent to any cycle, the proposition holds vacuously for f. Next, we consider the cases when f is interior or exterior to a frame adjacent cycle.

Let c be a frame adjacent cycle that contains f in its interior. Then c must be the only such cycle and must also be the innermost cycle that contains f in its interior. Therefore, c must be interior to $n - 1$ cycles. For n odd (even), c is signed clockwise (counterclockwise). Since f is interior to c and $\varphi(f)$ is clockwise oriented, its frame edges in c are drawn from a black to a white node in the clockwise direction along c. Hence, the frame edges in c are all assigned a $+1$ if n is odd and a -1 if n is even.

Next, let $\mathcal{C}$ denote the set cycles exterior and frame adjacent to f. All cycles in $\mathcal{C}$ are signed with the same orientation since each must be interior to the same number of cycles as f. Since $\varphi(f)$ is clockwise oriented, the frame edges that f shares with a cycle in $\mathcal{C}$ are drawn from a black node to white node in the counterclockwise direction along the cycle. Thus, the frame edges of f in the exterior cycles are assigned a $+1$ if n is odd and a -1 if n is even.

In both cases, edges in $\varphi(f) \cap E'$ are assigned a $+1$ if n is odd and a -1 if n is even. □

4   Formulations of the Maximum Resonant Set Problem

The following integer program (IP1) is a formulation of the maximum resonant set problem that was proposed by Hansen and Zheng [7] to compute the Clar number of a benzenoid system.

$$\max \{1^T y : Kx + Ry = 1, \ x \geq 0, \ y \geq 0, \ x \in \mathbb{Z}^E, \ y \in \mathbb{Z}^F\}$$

where

K is the $V \times E$ node-edge incidence matrix of $G = (V, E)$, and
R is the $V \times F$ node-face incidence matrix of (V, E, F).

To show that the linear programming relaxation of IP1 yields an integral polytope, it is convenient to consider a reformulation of this integer program. Given a framework φ for (V, E, F), we define its edge-frame incidence matrix U to be an $E \times F$ matrix such that, for each face $f \in F$, its corresponding column of U is the edge incidence vector of the frame $\varphi(f)$. The incidence matrix of a framework yields the following useful factorization of the matrix R defined above.

Lemma 5. *Let U be the incidence matrix of a framework. Then, $R = KU$.*

Introducing the vector of variables $z = x + Uy$ yields an alternative formulation of the maximum resonant set problem (IP2).

$$\max \ \{1^T y : Kz = 1, \ x + Uy - z = 0, \ x \geq 0, y \geq 0, z \geq 0, \ y \in \mathbb{Z}^F, \ x, z \in \mathbb{Z}^E\}.$$

All feasible solutions to the above integer program are necessarily binary vectors. Hence, constraints $Kz = 1$ express that z is the incidence vector of a perfect matching for G. Constraints $x + Uy - z = 0$ partition the edges in the perfect matching represented by z between those in faces considered resonant (represented by Uy) and those identified by x.

Note that this alternative formulation of the maximum resonant set problem is valid for the incidence matrix U of any framework of (V, E, F). To facilitate the subsequent analysis of this integer program, it is advantageous to assume that U is the incidence matrix of a simple framework. Recall this occurs whenever the framework is oriented.

5 Unimodularity of the Constraint Matrices

In this section we prove that the constraint matrices of both formulations of the maximum resonant set problem are unimodular. Therefore, their linear relaxation polytopes are integral and the optimization problems can be solved directly as linear programs.

First, we prove that the constraint matrix of the integer program IP2 is unimodular when U is the incidence matrix of a simple framework. The general form of this matrix is

$$\begin{bmatrix} 0 & 0 & K \\ I & U & -I \end{bmatrix}$$

The incidence matrix of a simple framework has the following useful property.

Lemma 6. *Let U be the incidence matrix of a simple framework. Then, each row of U has at most one entry of 1.*

We are now ready for the main result of this section.

Theorem 7. *Let U be the incidence matrix of a simple framework φ. Then, the constraint matrix of the integer program IP2 is unimodular.*

Proof. Since unimodularity is preserved when rows and columns are multiplied by -1, it is equivalent to show that the following matrix is unimodular.

$$A = \begin{bmatrix} 0 & 0 & K \\ I & U & I \end{bmatrix}$$

To show that matrix A satisfies the first condition of Theorem 2, it suffices to choose as basis B a column submatrix

$$B = \begin{bmatrix} 0 & K_T \\ I & I_T \end{bmatrix}$$

where K_T and I_T are the column submatrices of K and I, respectively, corresponding to the edges of a spanning tree of G. Since G is bipartite, K is totally unimodular. This implies that B is totally unimodular and, in particular, B is unimodular. It can easily be shown that if A is factored as $A = BC$, then all entries of C are 0, 1 or -1.

Next, we show that the second condition of Theorem 2 is satisfied. Let A' be a column eulerian column submatrix of A. Then

$$A' = \begin{bmatrix} 0 & 0 & K_{E'} \\ I_{D'} & U_{F'} & I_{E'} \end{bmatrix}$$

where $D' \subseteq E$, $E' \subseteq E$ and $F' \subseteq F$ are edge and face subsets indexing the columns of A'. Since A' is column eulerian and each row of U has at most one entry of 1, it follows that the nonzero rows of $[I_{D'} \ U_{F'} \ I_{E'}]$ have exactly two entries of 1. To prove that A' is linearly dependent, we must show there are vectors $u \in \mathbb{R}^{D'}$, $v \in \mathbb{R}^{F'}$, and $w \in \mathbb{R}^{E'}$, not all zero, such that

$$\begin{bmatrix} 0 & 0 & K_{E'} \\ I_{D'} & U_{F'} & I_{E'} \end{bmatrix} \begin{bmatrix} u \\ v \\ w \end{bmatrix} = 0 \tag{1}$$

If E' is empty, we can set all entries of u equal to 1 and all entries of v equal to -1. Otherwise, let G' denote the plane subgraph of $G = (V, E)$ induced by E'. Since $K_{E'}$ is column eulerian, G' is a plane graph with all even degree nodes. By Proposition 3, we can assume that G' is decomposed into intrafacial cycles. Since these cycles are even and edge-disjoint, each cycle can be signed independently of the others. Note that in any signing, if we set w_e equal to the value given to edge e in the cycle signing, for each $e \in E'$, then $K_{E'}w = 0$. In particular, we choose to sign the cycles so that, for each face $f \in F$, $w_e = w_{e'}$ for all pairs of edges e and e' in $\varphi(f) \cap E'$. We showed in Proposition 4 that such a signing exists.

To assign values to the entries of u and v we proceed as follows. For each $f \in F'$, if $\varphi(f) \cap E' \neq \emptyset$, we let $v_f = -w_e$ for $e \in \varphi(f) \cap E'$. Otherwise, if $\varphi(f) \cap E' = \emptyset$, v_f can be chosen arbitrarily, say $v_f = 1$. Finally for each $e \in D'$, either $e \in E'$ or $e \in \varphi(f)$ for some $f \in F'$. In the first case, we assign $u_e = -w_e$ and, in the second case, we let $u_e = -v_f$. It can now be easily seen that the u, v, w vectors so defined are a solution to the homogeneous system (1), establishing that the columns of A' are linearly dependent. $\qquad\square$

For any given framework incidence matrix U, let $P_2(b')$ denote the polyhedron defined by the system of equations with constraint matrix of problem IP2 and integer right hand side vector b', together with nonnegativity constraints on all variables. Similarly, let $P_1(b)$ denote the polyhedron defined by the system of equations with constraint matrix of problem IP1 with integer right hand side vector b, and nonnegativity constraints on all variables. Proving the following lemma is straightforward.

Lemma 8. *Let b be an integer vector such that $P_1(b)$ is nonempty. Then, there exists an integer vector b' such that (x, y) is an extreme point of $P_1(b)$ if and only if $(x, y, x + Uy)$ is an extreme point of $P_2(b')$.*

Note Lemma 8 holds for the incidence matrix U of any framework. In particular, if we choose the framework to be simple, then combining Theorem 7, Lemma 8, and Theorem 1 establishes the unimodularity of the constraint matrix of problem IP1.

Corollary 9. *The constraint matrix of problem IP1 is unimodular.*

In conclusion, we have established that the constraint matrices of both formulations of the maximum resonant set problem are unimodular. In particular, for IP2 we showed here that the result holds when the framework is simple. Although it is not necessary for our development, it can be shown with additional effort that Theorem 7 is also valid for an arbitrary framework.

6 Minimum Weight Cut Cover Problem

Hansen and Zheng [6] proposed a class of cuts for benzenoids systems and defined an optimal covering problem based on these cuts. In this section we apply their definitions to a larger class of plane graphs. The proofs of results established by Hansen and Zheng [6] extend directly to the more general version considered here, so we will omit them. As before, (V, E, F) continues to denote the plane map of a bipartite and 2-connected planar graph $G = (V, E)$, where the node set is bipartitioned as $V = V_1 \cup V_2$. We also assume in this section that G has a perfect matching.

Definition 10. *A cut of (V, E, F) is a simple (possibly closed) plane curve c such that*

(1) c intersects G only at some edge subset E_c of E, (i.e., c does not go through any nodes);
(2) the subgraph $G_c = (V, E \setminus E_c)$ is disconnected such that the nodes of each edge in E_c belong to different components of G_c; and
(3) all nodes of edges in E_c in any component of G_c belong to either V_1 or V_2.

Definition 11. *A cut cover of (V, E, F) is a set of cuts C such that each face in F is intersected by at least one cut in C.*

Definition 12. *Let $M \subseteq E$ be a perfect matching for (V, E, F). The weight m_c, with respect to M, of a cut c is $m_c = |E_c \cap M|$ and the weight $m(C)$ of a cut cover C is $m(C) = \sum_{c \in C} m_c$.*

Hansen and Zheng [6] proved the following two theorems.

Theorem 13 ([6]). *The weight of a cut is the same for all perfect matchings.*

Thus, for any cut cover C of (V, E, F), its weight $m(C)$ is also independent of the perfect matching. Let $cov(V, E, F)$ denote the value of a minimum weight cut cover for (V, E, F). Then

Theorem 14 ([6]). *For all (V, E, F), $res(V, E, F) \leq cov(V, E, F)$.*

Hansen and Zheng [6] conjectured that the above inequality is satisfied as equation by all benzenoid systems. In what follows, we sketch our proof that this conjecture holds true for the class of plane maps we consider in this paper. To accomplish this, we will define a minimum cost network flow problem associated with the minimum weight cut cover problem. The directed graph of this flow problem is based on the geometric dual graph of the plane map (V, E, F).

Let $G^* = (F \cup \{t\}, E^*)$ denote the dual graph of (V, E, F), where t denotes the external face of G. We now transform G^* into a directed graph. Let φ denote the clockwise oriented framework of (V, E, F) and let $\{f, i\}$ be an edge in E^*. Without loss of generality we can assume $f \in F$. If the edge that separates faces f and i in (V, E, F) belongs to $\varphi(f)$, then $\{f, i\}$ becomes arc (f, i). Otherwise, $\{f, i\}$ becomes the arc (i, f). Next, for each node $f \in F$, we perform a node splitting transformation to obtain nodes $f_1 \in F_1$ and $f_2 \in F_2$ connected by the arc $(f_1, f_2) \in A_F$, where F_1 and F_2 are two copies of F and A_F is the set of resulting arcs connecting nodes in these two sets. Outgoing (incoming) arcs of node f become outgoing (incoming) arcs of node f_2 (f_1). Let $D = (F_1 \cup F_2 \cup \{t\}, A_E \cup A_F)$ denote the resulting directed graph, where A_E is the set of arcs connecting nodes in different faces. Observe that each directed cycle in $D = (F_1 \cup F_2 \cup \{t\}, A_E \cup A_F)$ corresponds to a cut of (V, E, F).

We now define a minimum cost circulation problem on the directed graph D. A lower bound of 1 is imposed on the flow value of each arc in A_F (this enforces that all faces in F are covered by cuts). Flow on all remaining arcs is required to be nonnegative. The cost assigned to each arc in A_F is zero. Finally, let $p \in \{0, 1\}^E$ be the incidence vector of a perfect matching for $G = (V, E)$. The cost of arcs in A_E is given by the corresponding entry of the vector p. Let γ denote the optimal value of this network flow problem. Since the right hand side data for this network flow problem is integer, there exists an optimal solution to the problem that is integer. As integer solutions can be interpreted as cut covers, it follows that $cov(V, E, F) \leq \gamma$. We next prove that $\gamma \leq res(V, E, F)$.

Let $\overline{U}$ be the incidence matrix of the counterclockwise oriented framework of (V, E, F) and let $a \in \{-1, 0, 1\}^E$ be defined as $a = (U - \overline{U})1$. We note that the nonzero entries of a correspond to the edges in the perimeter of (V, E, F) and they alternate in sign along the perimeter.

The dual of the network flow problem can be written as follows:

$$\text{minimize} \quad 1^T y$$
$$\text{subject to} \quad Uu - \overline{U}v - aw + x = p \tag{2}$$
$$y - u + v = 0 \tag{3}$$
$$y, u, v, w, x \geq 0 \tag{4}$$

where $y, u, v \in \mathbb{R}^F$, $w \in \mathbb{R}$, and $x \in \mathbb{R}^E$.

The proof is completed by showing that the x, y components of any feasible solution to the above dual problem are also feasible for the linear relaxation of problem IP1. The key observation here is that premultiplying constraints (2) by K (the node-edge incidence matrix of G) and rearranging terms gives

$$Kx + R(u - v) = Kp = 1$$

Using equations (3), we substitute y for $(u - v)$ and obtain $Kx + Ry = 1$, the equation constraints of problem IP1. Thus, the strong duality result follows.

Theorem 15. *For all (V, E, F), $res(V, E, F) = cov(V, E, F)$.*

The arguments in the proof of Theorem 15 also show that $res(V, E, F)$ and $cov(V, E, F)$ can be computed in strongly polynomial time using a network flow algorithm.

References

1. J. Aihara. On the number of aromatic sextets in a benzenoid hydrocarbon. *Bulletin of the Chemical Society of Japan*, 49:1429–1430, 1976.
2. B. Bollobás. *Graph Theory: An Introductory Course*. Springer-Verlag, New York, 1979.
3. E. Clar. *The Aromatic Sextet*. John Wiley & Sons, London, 1972.
4. S. J. Cyvin and I. Gutman. *Kekulé Structures in Benzenoid Hydrocarbons*. Springer-Verlag, Berlin, 1988.
5. I. Gutman and S. J. Cyvin. *Introduction to the Theory of Benzenoid Hydrocarbons*, Springer-Verlag, Berlin, 1989.
6. P. Hansen and M. Zheng. Upper bounds for the Clar number of benzenoid hydrocarbons. *Journal of the Chemical Society, Faraday Transactions*, 88:1621–1625, 1992.
7. P. Hansen and M. Zheng. The Clar number of a benzenoid hydrocarbon and linear programming. *Journal of Mathematical Chemistry*, 15:93–107, 1994.
8. P. Hansen and M. Zheng. Numerical bounds for the perfect matching vectors of a polyhex. *Journal of Chemical Information and Computer Sciences*, 34:305–308, 1994.
9. F. Harary. *Graph Theory*. Addison-Wesley, Boston, 1969.
10. G. L. Nemhauser and L.A. Wolsey. *Integer and Combinatorial Optimization*. John Wiley & Sons, New York, 1988.
11. A. Schrijver. *Theory of Integer and Linear Programming*. John Wiley & Sons, New York, 1986.

12. K. Truemper. Algebraic characterizations of unimodular matrices. *SIAM Journal of Applied Mathematics*, 35(2):328–332, 1978.
13. A. F. Veinott and G. B. Dantzig. Integral extreme points. *SIAM Review*, 10(3):371–372, 1968.

Consecutive Ones and a Betweenness Problem in Computational Biology

Thomas Christof, Marcus Oswald, and Gerhard Reinelt

Institut für Angewandte Mathematik
Universität Heidelberg
Im Neuenheimer Feld 293/294
D-69120 Heidelberg, Germany
{Thomas.Christof, Marcus.Oswald, Gerhard.Reinelt}@@IWR.Uni-Heidelberg.De

Abstract. In this paper we consider a variant of the betweenness problem occurring in computational biology. We present a new polyhedral approach which incorporates the solution of consecutive ones problems and show that it supersedes an earlier one. A particular feature of this new branch-and-cut algorithm is that it is not based on an explicit integer programming formulation of the problem and makes use of automatically generated facet-defining inequalities.

1 Introduction

The general *Betweenness Problem* is the following combinatorial optimization problem. We are given a set of n objects $1, 2, \ldots, n$, a set $\mathcal{B}$ of betweenness conditions, and a set $\overline{\mathcal{B}}$ of non-betweenness conditions. Every element of $\mathcal{B}$ (of $\overline{\mathcal{B}}$) is a triple (i, j, k) (a triple $\overline{(i, j, k)}$) requesting that object j should be placed (should not be placed) between objects i and k. The task is to find a linear order of all objects such that as few betweenness and non-betweenness conditions as possible are violated, resp. to characterize all orders that achieve this minimum. If violations are penalized by weights, we call the problem of finding a linear order minimizing the sum of weights of violations *Weighted Betweenness Problem*. This problem is $\mathcal{NP}$-hard in general.

In this paper we consider a special variant of this problem occurring in computational molecular biology, namely in the physical mapping problem with end probes. For the purpose of this paper we do not elaborate on the biological background, but refer to [4]. We define the problem simply as follows. We are given a set of m so-called *clones* $i \in \{1, 2, \ldots, m\}$ to each of which two so-called *end probes* i^t and i^h are associated. These $n = 2m$ probes are numbered such that $i^t = 2i - 1$ and $i^h = 2i$. Depending on the data, we have for every pair of a clone i and a probe $j \in \{1, 2, \ldots, n\} \setminus \{i^t, i^h\}$ either a betweenness condition (i^t, j, i^h) or a non-betweenness condition $\overline{(i^t, j, i^h)}$. Violation of a betweenness condition is penalized with cost c_ρ, and violation of a non-betweenness constraint receives a

R. E. Bixby, E. A. Boyd, and R. Z. Ríos-Mercado (Eds.): IPCO VI
LNCS 1412, pp. 213–228, 1998. © Springer–Verlag Berlin Heidelberg 1998

penalty of c_μ. The problem is then to find a linear order of the probes minimizing the sum of penalties for violated constraints.

This Weighted Betweenness Problem can also be stated in a different version: A 0/1 matrix $A \in \{0,1\}^{m \times n}$ has the *consecutive ones property* (for rows) if the columns of A can be permuted so that the 1's in each row appear consecutively. For a 0/1 matrix $B \in \{0,1\}^{m \times n}$ having the consecutive ones property let n_ρ^B (n_μ^B) denote the number of 1's (of 0's) that have to be switched to transform A into B. For given nonnegative numbers c_ρ and c_μ, we define the *Weighted Consecutive Ones Problem* as the task to find a matrix B with the consecutive ones property minimizing $c_\rho n_\rho^B + c_\mu n_\mu^B$. This problem is known to be $\mathcal{NP}$-hard [2]. All column permutations π of a feasible matrix B so that the 1's in each row of B^π appear consecutively can be found in time linear in the number of 1's in B by a so called *PQ-tree algorithm* [1].

For our discussion, we assume that the data is given as clone × probe 0/1-matrix A where a_{i,i^t} and a_{i,i^h} are fixed to 1. The other entries are obtained from some experiment where an entry $a_{ij} = 1$ gives rise to a betweenness constraint (i^t, j, i^h) and an entry $a_{ij} = 0$ corresponds to a non-betweenness constraint $\overline{(i^t, j, i^h)}$. A solution of the Weighted Betweenness Problem corresponds then to a solution of the Weighted Consecutive Ones Problem with the additional constraint that in some column permuted matrix B^π (in which the 1's in each row appear consecutively) the first and the last 1 in each row i correspond to the end probes of clone i.

The Weighted Consecutive Ones Problem models the biological situation when the information that the probes are the ends of clones is missing [8]. Note that by introducing artificial variables the Weighted Consecutive Ones Problem can easily be transformed to a Weighted Betweenness Problem.

The paper is organized as follows. Section 2 discusses our previous approach which could already be applied successfully. An improved approach is presented in section 3 leading to the definition of the betweenness polytope. This polytope is then studied in the following section. Separation in the context of a branch-and-cut algorithm to solve the betweenness problem to optimality is the topic of section 5. Computational results conclude this paper.

2   A First IP Model

Our previous computational approach to the Weighted Betweenness Problem was based on a different model. We state it here mainly for a comparison and for introducing some notations. In the following we will introduce 0/1 variables indicating whether a betweenness or non-betweenness constraint is violated. Since a betweenness condition (i, j, k) is violated if and only if the non-betweenness condition $\overline{(i, j, k)}$ is satisfied, we can just complement variables and will speak only about betweenness conditions in the following. The objective function coefficients of the variables will then give the preference if the condition should be satisfied or violated.

Let m be the number of clones and n be the number of probes. In our special setup we have $n = 2m$ and we have $n - 2$ betweenness constraints (i^t, j, i^h) for every clone i. We write, for short, (i, j) for the betweenness constraint (i^t, j, i^h), and call (i, j) a *clone-probe pair*. The set $\mathcal{B}_m$ of all possible clone-probe pairs is

$$\mathcal{B}_m := \{(i, j) \mid 1 \le i \le m, 1 \le j \le n, j \ne i^t, j \ne i^h\}.$$

Obviously, $|\mathcal{B}_m| = 2m(m - 1)$.

We will develop an integer programming formulation with two types of variables. For every ordered pair (i, j) of probes, we introduce a 0/1 variable y_{ij} which has value 1 if and only if i precedes j in the order π. With every clone-probe pair $(i, j) \in \mathcal{B}_m$, we associate the 0/1 variable x_{ij} which has value 1 if and only if constraint (i^t, j, i^h) is met in the order π. Equivalently, $x_{ij} = 0$ if and only if the non-betweenness constraint $\overline{(i^t, j, i^h)}$ is met.

To ensure that the variables y_{ij} encode a linear order π of the probes, the constraints of the IP formulation of the linear ordering problem have to be met, i.e., they have to satisfy

$$y_{ij} + y_{ji} = 1, \quad \text{for all } 1 \le i < j \le n,$$
$$y_{ij} + y_{jk} + y_{ki} \le 2, \quad \text{for all triples } 1 \le i, j, k \le n.$$

To ensure that the x_{ij} count violations of the betweenness and nonbetweenness constraints, we have to add further inequalities. To force a $x_{ij}, (i, j) \in \mathcal{B}_m$ to be 0 if and only if (i^t, j, i^h) is violated (or $\overline{(i^t, j, i^h)}$ is satisfied), we add

$$x_{ij} \le y_{ji^h} + y_{ji^t},$$
$$x_{ij} \le y_{i^h j} + y_{i^t j},$$
$$x_{ij} \ge -y_{i^t j} - y_{ji^h} + 1,$$
$$x_{ij} \ge y_{i^t j} + y_{ji^h} - 1.$$

Thus, x_{ij} is 1 if and only if $y_{i^t j} = y_{ji^h}$.

We do not discuss the objective function here. Due to the positive parameters c_ρ and c_μ we can omit some of these inequalities depending on whether x_{ij} corresponds to an original betweenness or non-betweenness condition, and moreover, we only have to require

$$0 \le x_{ij} \le 1.$$

The objective function will force these variables to have integer values if the linear ordering variables are integer. Note that the objective function is zero on the linear ordering variables.

With every feasible solution of the problem, corresponding to a permutation π of the probes, we associate 0/1-vectors $\psi^\pi \in \{0, 1\}^{n(n-1)}$ and $\chi^\pi \in \{0, 1\}^{|\mathcal{B}_m|}$ with

$$\psi^\pi_{ij} = \begin{cases} 1 & \text{if } i \text{ precedes } j \text{ in the order } \pi, \\ 0 & \text{otherwise.} \end{cases}$$

and

$$\chi_{ij}^{\pi} = \begin{cases} 1 & \text{if constraint } (i^t, j, i^h) \text{ is met in the order } \pi, \\ 0 & \text{otherwise.} \end{cases}$$

The polytope P_{LOBW}^m associated with the instance of the Weighted Betweenness Problem is the convex hull of all possible vectors $\binom{\psi^\pi}{\chi^\pi}$,

$$P_{LOBW}^m = \text{conv}\left(\left\{ \binom{\psi^\pi}{\chi^\pi} \mid \pi \text{ is a permutation of the probes}\right\}\right).$$

Previous computations were based on partial descriptions of this polytope (see [4]).

3   Modelling Without Linear Ordering Variables

The formulation of the preceding section is somewhat redundant. Namely, let $\binom{\psi}{\chi} \in P_{LOBW}^m \cap \{0,1\}^{6m^2-4m}$ be an incidence vector of the Weighted Betweenness Problem with $n = 2m$ probes and the set $\mathcal{B}_m$ of betweenness constraints. Obviously χ can be retrieved from ψ, because

$$\chi_{ij} = 1 \Leftrightarrow \psi_{i^t j} = \psi_{j i^h}.$$

Conversely, for a given χ there exist one or more feasible settings of ψ. These settings cannot be obtained directly but in linear time by application of the PQ-tree algorithm [1]. For a given χ we define for every clone i three sets in the following way.

$$S_{i_1} = \{j \mid \chi_{ij} = 1\} \cup \{i^t\},$$
$$S_{i_2} = \{j \mid \chi_{ij} = 1\} \cup \{i^h\},$$
$$S_{i_3} = \{j \mid \chi_{ij} = 1\} \cup \{i^t\} \cup \{i^h\}.$$

The feasible settings of the linear ordering variables ψ correspond to all permutations of $\{1, \dots, n\}$ where the elements of every set introduced above occur in consecutive order.

We now define the projection of P_{LOBW}^m onto the χ variables

$$P_{BW}^m = \text{conv}\left(\left\{ \chi \mid \text{ there exists } \psi \text{ such that } \binom{\psi}{\chi} \in P_{LOBW}^m \right\}\right).$$

It can easily be tested if a 0/1-vector χ is contained in P_{BW}^m. Namely if not, i.e. if $\chi \in \{0,1\}^{2m(m-1)}, \chi \notin P_{BW}^m$, then with the sets defined above for the clones, the application of the PQ-tree algorithm would yield the result that no permutation exists in which the elements of the sets appear as consecutive subsequences. Otherwise, if $\chi \in P_{BW}^m$, then the PQ-tree provides all consistent permutations.

Because a feasible linear ordering can be derived from a $\chi \in P_{BW}^m$ and the objective function is zero for the linear ordering variables, they can be omitted. A solution of the Weighted Betweenness Problem for physical mapping is obtained by solving

$$\max c^T x$$
$$x \in P_{BW}^m$$
$$x \in \{0,1\}^{2m(m-1)}$$

Now we want to derive an integer programming formulation of this problem from an integer programming formulation of the Weighted Consecutive Ones Problem.

For each χ the sets S_{i_k} can be written as rows of a matrix $M^\chi \in \{0,1\}^{3m \times 2m}$. For $1 \le i \le m$ and $1 \le j \le n$ and $1 \le k \le 3$ an entry $m^\chi_{3(i-1)+k,j}$ of M^χ has the value 1 if and only if $j \in S_{i_k}$.

Proposition 1. $\chi \in P_{BW}^m$ if and only if M^χ has the consecutive ones property for rows.

Proof. Clear because of the construction of the sets S_{i_k}.

Corollary 2. *Each integer programming formulation of the Weighted Consecutive Ones Problem leads to an integer programming formulation of the Weighted Betweenness Problem for physical mapping.*

Proof. Observe that the entries of M^χ are constant or equal to some corresponding entries of χ. Hence, it exists a (simple) linear transformation from χ to M^χ which can be used to substitute M^χ in an IP formulation of the Weighted Consecutive Ones Problem, yielding an IP formulation of the Weighted Betweenness Problem.

An IP formulation of the Weighted Consecutive Ones Problem can be derived from a theorem of Tucker [15]. In this paper it is shown that the 0/1 matrix M has the consecutive ones property for rows if and only if no submatrix of M is a member of $\{M_i\}$, where $\{M_i\}$ is a given set of 0/1 matrices. Now for each submatrix of M and each matrix M_i with the same size it is easy to derive an inequality, which is violated if and only if the submatrix is equal to M_i. The set of all these inequalities gives an IP formulation of the Weighted Consecutive Ones Problem.

Using the linear transformation from χ to M^χ we obtain an IP formulation of the Weighted Betweenness Problem for physical mapping. For $m = 2$ the nontrivial inequalities of the formulation are

$$2x_{12^t} + x_{21^t} + x_{21^h} \le 3,$$
$$2x_{12^t} - 2x_{12^h} - x_{21^t} - x_{21^h} \le 1,$$

and further inequalities which can be obtained by the symmetry operations described in section 4.2. Note that these two classes of inequalities do not define facets of P_{BW}^2.

4   The Polytope P_{BW}^m

In this section we will investigate some properties of the polytope P_{BW}^m (see also [13] for a more detailed discussion). In particular we are interested in exhibiting classes of facet-defining inequalities.

4.1   Dimension and Lifting

We first determine the dimension of P_{BW}^m and address the question of trivial (node) lifting of facets.

Proposition 3. *Let $m > \overline{m} \geq 2$. An inequality $g^T x \leq g_0$ for P_{BW}^m, obtained by trivial lifting from an inequality $f^T x \leq f_0$ which is valid for $P_{BW}^{\overline{m}}$ is valid for P_{BW}^m.*

Proof. Let a binary vector $\chi \in P_{BW}^m$ be given. Since by removing the components χ_{ij} with $(i,j) \in \mathcal{B}_m \setminus \mathcal{B}_{\overline{m}}$ from χ one obtains a vector $\overline{\chi}$ which is a feasible incidence vector of $P_{BW}^{\overline{m}}$, the proposition follows.

We can also compute the dimension of a face of P_{BW}^m which is induced by an inequality resulting from trivial lifting of an inequality of $P_{BW}^{\overline{m}}$.

Theorem 4. *Let $f^T x \leq f_0$ be valid for $P_{BW}^{\overline{m}}$, and let $F = \{x \in P_{BW}^{\overline{m}} \mid f^T x = f_0\}$ be the induced face with $\dim F \geq 0$. Let $m > \overline{m}$ and let $g^T x \leq g_0$ be obtained from $f^T x \leq f_0$ by trivial lifting. Let $G = \{x \in P_{BW}^m \mid g^T x = g_0\}$. Then*

$$\dim G - \dim F = |\mathcal{B}_m| - |\mathcal{B}_{\overline{m}}| = 2m(m-1) - 2\overline{m}(\overline{m}-1).$$

Proof. Obviously, it is sufficient to show the theorem for $m = \overline{m} + 1$. Let $m = \overline{m} + 1$ and $k := \dim F \geq 0$. Then there exist $k+1$ permutations $\pi_0, \ldots, \pi_k$ and corresponding incidence vectors χ^{π_i} with $\chi^{\pi_i} \in F$ for $i = 0, \ldots, k$ and $\mathrm{arank}\{\chi^{\pi_0}, \ldots, \chi^{\pi_k}\} = k+1$. W.l.o.g. by symmetry considerations we may assume that

$$\pi_0^{-1}(i^t) < \pi_0^{-1}(i^h) \text{ for } 1 \leq i \leq \overline{m},$$
$$\pi_0^{-1}(i^t) < \pi_0^{-1}(j^t) \text{ for } 1 \leq i < j \leq \overline{m}.$$

We now construct $k+1+|\mathcal{B}_m|-|\mathcal{B}_{\overline{m}}| = k+1+4\overline{m}$ permutations $\rho_0, \ldots, \rho_{k+4\overline{m}}$ with $\chi^{\rho_i} \in G$ for $i = 0, \ldots, k+4\overline{m}$ and $\mathrm{arank}\{\chi^{\rho_0}, \ldots, \chi^{\rho_{k+4\overline{m}}}\} = k+4\overline{m}+1$.
For $0 \leq l \leq k$ set

$$\rho_l = (\pi_l(1), \ldots, \pi_l(2\overline{m}), m^t, m^h)$$

The remaining permutations $\rho_{k+1}, \ldots, \rho_{k+4\overline{m}}$ are obtained from π_0 by inserting m^t and m^h at different positions. For $1 \leq l \leq \overline{m}$ we define the permutations as

$$\rho_{k-2+3l} = (\pi_0(1), \ldots \pi_0(\pi_0^{-1}(l^t)-1), m^t, l^t, m^h, \pi_0(\pi_0^{-1}(l^t)+1), \ldots \pi_0(2\overline{m})),$$
$$\rho_{k-1+3l} = (\pi_0(1), \ldots \pi_0(\pi_0^{-1}(l^t)-1), m^h, l^t, m^t, \pi_0(\pi_0^{-1}(l^t)+1), \ldots \pi_0(2\overline{m})),$$
$$\rho_{k+3l} = (\pi_0(1), \ldots \pi_0(\pi_0^{-1}(l^t)-1), l^t, m^t, m^h, \pi_0(\pi_0^{-1}(l^t)+1), \ldots \pi_0(2\overline{m})),$$
$$\rho_{k+3\overline{m}+l} = (\pi_0(1), \ldots \pi_0(\pi_0^{-1}(l^h)-1), m^t, l^h, m^h, \pi_0(\pi_0^{-1}(l^h)+1), \ldots \pi_0(2\overline{m})).$$

The following properties of $\chi_{ij}^{\rho_l}, k+1 \leq l \leq k+4\overline{m}, (i,j) \in \mathcal{B}_m \setminus \mathcal{B}_{\overline{m}}$ are needed to show that the vectors χ^{ρ_l} are affinely independent. They result in a straightforward way from the definition of $\rho_l, k+1 \leq l \leq k+4\overline{m}$.

For $l < r \leq \overline{m}$ it holds that $\chi_{rm^t}^{\rho_k-2+3l} = 0$, since $\pi_0^{-1}(m^t) < \pi_0^{-1}(l^t) < \pi_0^{-1}(r^t) < \pi_0^{-1}(r^h)$.

By similar arguments, it holds for $l < r \leq \overline{m}$ that

$$\chi_{ij}^{\rho_k-2+3l} = \chi_{ij}^{\rho_k-1+3l} = \chi_{ij}^{\rho_k+3l} = 0,$$

for

$$(i,j) \in \{(r,m^t), (r,m^h), (m,r^t)\}.$$

Moreover, for $1 \leq l \leq \overline{m}$, we have

$$\chi_{lm^t}^{\rho_k-2+3l} = \chi_{lm^h}^{\rho_k-1+3l} = \chi_{ml^t}^{\rho_k+3l} = 0,$$

and

$$\chi_{lm^h}^{\rho_k-2+3l} = \chi_{ml^t}^{\rho_k-2+3l} = 1,$$
$$\chi_{lm^t}^{\rho_k-1+3l} = \chi_{ml^t}^{\rho_k-1+3l} = 1,$$
$$\chi_{lm^t}^{\rho_k+3l} = \chi_{lm^h}^{\rho_k+3l} = 1.$$

In addition, for $1 \leq r, l \leq \overline{m}$

$$\chi_{mr^h}^{\rho_k+3\overline{m}+l} = \begin{cases} 1 & \text{for } l = r, \\ 0 & \text{otherwise,} \end{cases}$$

and, for $1 \leq r, l \leq \overline{m}$,

$$\chi_{mr^h}^{\rho_k-2+3l} = \chi_{mr^h}^{\rho_k-1+3l} = \chi_{mr^h}^{\rho_k+3l} = 0.$$

In the following table, the rows are the incidence vectors $(\chi^{\rho_{k+i}})^T, 1 \leq i \leq 4\overline{m}$, restricted to the variables in $\mathcal{B}_m \setminus \mathcal{B}_{\overline{m}}$. The vectors are partitioned into $\overline{m}$ blocks of coefficients $\chi_{lm^t}^{\rho_{k+i}}, \chi_{lm^h}^{\rho_{k+i}}, \chi_{ml^t}^{\rho_{k+i}}$, and the remaining coefficients $\chi_{m1^h}^{\rho_{k+i}}, \ldots, \chi_{mm^h}^{\rho_{k+i}}$.

	$\chi_{1m^t}^{\rho_{k+i}}$ $\chi_{1m^h}^{\rho_{k+i}}$ $\chi_{m1^t}^{\rho_{k+i}}$	$\cdots$	$\chi_{\overline{m}m^t}^{\rho_{k+i}}$ $\chi_{\overline{m}m^h}^{\rho_{k+i}}$ $\chi_{m\overline{m}^t}^{\rho_{k+i}}$	$\chi_{m1^h}^{\rho_{k+i}}$ $\cdots$ $\chi_{m\overline{m}^h}^{\rho_{k+i}}$
$i=1$				
$i=2$	$\mathbf{1}-I$	$\mathbf{0}$	$\mathbf{0}$	$\mathbf{0}$
$i=3$				
$\vdots$	$*$	$\ddots$ $\mathbf{0}$	$\mathbf{0}$	$\mathbf{0}$
$i=3\overline{m}-2$				
$i=3\overline{m}-1$	$*$	$*$	$\mathbf{1}-I$	$\mathbf{0}$
$i=3\overline{m}$				
$i=3\overline{m}+1$				
$\vdots$	$*$	$*$	$*$	I
$i=4\overline{m}$				

The system has full rank and since it holds that, for $0 \leq l \leq k$,

$$\chi_{ij}^{\rho_l} = \begin{cases} \chi_{ij}^{\pi_l} & \text{for } (i,j) \in \mathcal{B}_{\overline{m}}, \\ 0 & \text{for } (i,j) \in \mathcal{B}_m \setminus \mathcal{B}_{\overline{m}}, \end{cases}$$

the vectors $\chi^{\rho_0}, \ldots, \chi^{\rho_{k+4\overline{m}}}$ are affinely independent.

It is easy to see that $\chi^{\rho_i} \in G$ for $i = 0, \ldots, k + 4\overline{m}$. Hence, $\dim G \geq k + 4\overline{m}$. On the other hand, $\dim G \leq k + 4\overline{m}$, because by the lifting operation only $4\overline{m}$ variables are added (for $m = \overline{m} + 1$). Therefore, $\dim G = \dim F + 4\overline{m}$.

From these results we obtain that trivial lifting preserves the facet-defining property and that P_{BW}^m is full-dimensional.

Corollary 5. *Let $m > \overline{m} \geq 2$. An inequality $g^T x \leq g_0$ for P_{BW}^m, obtained by trivial lifting from a facet-defining inequality $f^T x \leq f_0$ of $P_{BW}^{\overline{m}}$ is facet-defining.*

Corollary 6. *For $m \geq 2$*

$$\dim P_{BW}^m = 2m(m-1),$$

i.e., P_{BW}^m is full-dimensional.

Proof. P_{BW}^2 is full-dimensional. By application of Theorem 4 with $F = P_{BW}^2$ and $G = P_{BW}^m$ the corollary follows.

4.2   Small Instance Relaxations

Because facet-defining inequalities of P_{BW}^m are trivially liftable, we can obtain from linear descriptions of polytopes associated with small instances of the problem, say for $2 \leq m \leq 4$, relaxations of P_{BW}^m for any m. We call such a relaxation *small instance relaxation* [3,6].

In order to characterize the symmetry properties of P_{BW}^m we use the following notation. Let $v \in \mathbb{R}^{|\mathcal{B}_m|}$. The entries of v can be placed in an m by m *clone-clone matrix* $\tilde{v}$ of 2-dimensional vectors. The rows and columns of the matrix correspond to the clones, and an entry $\tilde{v}_{ij} \in \mathbb{R}^2$ of $\tilde{v}$ is defined for $i \neq j$ as $\tilde{v}_{ij} = \binom{v_{ij^t}}{v_{ij^h}}$, i.e., an entry ij refers to the relations of the end probes obtained from clone j with clone i.

Now, the set of incidence vectors in P_{BW}^m can be partitioned into equivalence classes with respect to the following two operations. First, for $\chi \in P_{BW}^m$ an arbitrary permutation of the clones is allowed. This corresponds to a simultaneous permutation of the rows and columns of the associated clone-clone matrix $\tilde{\chi}$. Second, for an arbitrary clone j its end probes can be reversed. This corresponds to reversing in column j of the clone-clone matrix the entries of $\binom{\chi_{ij^t}}{\chi_{ij^h}}$ to $\binom{\chi_{ij^h}}{\chi_{ij^t}}$ for all $1 \leq i \leq m, i \neq j$.

We used the algorithms for facet enumeration discussed in [5] to get more insight into the facet structure of P_{BW}^m for $m \leq 4$. It is clear that also the facet-defining inequalities of P_{BW}^m can be assigned to equivalence classes.

P_{BW}^2 has 7 vertices and is completely described by 3 equivalence classes of facets. P_{BW}^3 has 172 vertices and 241 facets in total which can be partitioned into 16 equivalence classes of facets. They are displayed in Table 1 ($\tilde{f}^i$ denotes the clone-clone matrix representation of f^i). Observe that the (lifted) facets of P_{BW}^2 are among the facets of P_{BW}^3 ($f^{1T}x \leq f_0^1$ to $f^{3T}x \leq f_0^3$).

The computation of the complete linear description of P_{BW}^4 which has 9,197 vertices was not possible in reasonable time. However, by an algorithm for parallel facet enumeration (see [5]) we found more than $1,16 \cdot 10^7$ (!) different equivalence classes of facet-defining inequalities, yielding a lower bound of $4.4 \cdot 10^9$ for the number of facets of P_{BW}^4.

4.3   Cycle Inequalities

We use the following notation. For $j \in \{1, \ldots, n\}$, $c(j)$ denotes the clone from which the probe j is extracted, i.e., $j = c(j)^t$ or $j = c(j)^h$, and by $q(j)$ we denote the second probe $q(j)$ which the clone $c(j)$ defines. Of course, we have $q(q(j)) = j$.

A class of valid inequalities can be motivated by the following observation. Suppose we have $n = 2m$ betweenness constraints $(i_1{}^t, 1, i_1{}^h), \ldots, (i_n{}^t, n, i_n{}^h)$, such that every probe j, $j \in \{1, \ldots, n\}$, is exactly once the probe between the ends of a clone $i_j \neq c(j)$. Then for any permutation π of the probes at least two betweenness constraints are violated, because the first probe $\pi(1)$ and the last probe $\pi(2m)$ are not between other probes. Hence, the inequality

$$\sum_{j=1}^{n} x_{i_j j} \leq n - 2, \quad i_j \neq c(j) \tag{1}$$

is valid for P_{BW}^m. The following cycle inequalities are a special case of these inequalities. In particular, they are facet-defining and they generalize the facet $f^{9T}x \leq f_0^9$ of Table 1.

Theorem 7. *Let D_n be the complete directed graph on n nodes corresponding to all probes, and let R be a cycle in D_n with the property that if k in $V(R)$, then $q(k) \notin V(R)$. Then*

$$\sum_{(k,l)\in R} x_{c(k)l} + x_{c(l)q(k)} \leq 2|R| - 2 \tag{2}$$

defines a facet of P_{BW}^m.

Proof. Let us assume that

$$a^T x = \sum_{(i,j)\in \mathcal{B}_m} a_{ij} x_{ij} \leq b \tag{3}$$

Table 1. Facet-defining inequalities $f^{i^T} x \leq f_0^i$ of P_{BW}^3.

$$\tilde{f}^1 = \begin{pmatrix} * & \binom{1}{1} & \binom{0}{0} \\ \binom{1}{1} & * & \binom{0}{0} \\ \binom{0}{0} & \binom{0}{0} & * \end{pmatrix}, \; f_0^1 = 2 \qquad \tilde{f}^2 = \begin{pmatrix} * & \binom{0}{0} & \binom{0}{0} \\ \binom{0}{-1} & * & \binom{0}{0} \\ \binom{0}{0} & \binom{0}{0} & * \end{pmatrix}, \; f_0^2 = 0$$

$$\tilde{f}^3 = \begin{pmatrix} * & \binom{1}{-1} & \binom{0}{0} \\ \binom{-1}{-1} & * & \binom{0}{0} \\ \binom{0}{0} & \binom{0}{0} & * \end{pmatrix}, \; f_0^3 = 0 \qquad \tilde{f}^4 = \begin{pmatrix} * & \binom{2}{0} & \binom{-1}{-1} \\ \binom{0}{0} & * & \binom{0}{0} \\ \binom{-1}{-1} & \binom{2}{0} & * \end{pmatrix}, \; f_0^4 = 2$$

$$\tilde{f}^5 = \begin{pmatrix} * & \binom{2}{0} & \binom{-1}{-1} \\ \binom{0}{0} & * & \binom{0}{0} \\ \binom{1}{1} & \binom{-2}{0} & * \end{pmatrix}, \; f_0^5 = 2 \qquad \tilde{f}^6 = \begin{pmatrix} * & \binom{1}{1} & \binom{-1}{-1} \\ \binom{-1}{-1} & * & \binom{1}{1} \\ \binom{1}{1} & \binom{-1}{-1} & * \end{pmatrix}, \; f_0^6 = 2$$

$$\tilde{f}^7 = \begin{pmatrix} * & \binom{1}{1} & \binom{-1}{-1} \\ \binom{1}{1} & * & \binom{-1}{-1} \\ \binom{1}{1} & \binom{-1}{-1} & * \end{pmatrix}, \; f_0^7 = 2 \qquad \tilde{f}^8 = \begin{pmatrix} * & \binom{2}{0} & \binom{1}{-1} \\ \binom{0}{0} & * & \binom{-1}{1} \\ \binom{1}{1} & \binom{-1}{1} & * \end{pmatrix}, \; f_0^8 = 4$$

$$\tilde{f}^9 = \begin{pmatrix} * & \binom{1}{0} & \binom{1}{0} \\ \binom{1}{0} & * & \binom{0}{1} \\ \binom{0}{1} & \binom{0}{1} & * \end{pmatrix}, \; f_0^9 = 4 \qquad \tilde{f}^{10} = \begin{pmatrix} * & \binom{1}{1} & \binom{-1}{-1} \\ \binom{1}{-1} & * & \binom{0}{0} \\ \binom{1}{1} & \binom{-2}{-2} & * \end{pmatrix}, \; f_0^{10} = 2$$

$$\tilde{f}^{11} = \begin{pmatrix} * & \binom{2}{0} & \binom{0}{-2} \\ \binom{0}{-2} & * & \binom{-1}{1} \\ \binom{-2}{0} & \binom{1}{-1} & * \end{pmatrix}, \; f_0^{11} = 2 \qquad \tilde{f}^{12} = \begin{pmatrix} * & \binom{1}{-1} & \binom{1}{-1} \\ \binom{1}{-1} & * & \binom{0}{-2} \\ \binom{1}{-1} & \binom{0}{-2} & * \end{pmatrix}, \; f_0^{12} = 2$$

$$\tilde{f}^{13} = \begin{pmatrix} * & \binom{1}{-3} & \binom{1}{-3} \\ \binom{1}{-3} & * & \binom{1}{-3} \\ \binom{1}{-3} & \binom{1}{-3} & * \end{pmatrix}, \; f_0^{13} = 2 \qquad \tilde{f}^{14} = \begin{pmatrix} * & \binom{1}{-1} & \binom{1}{-1} \\ \binom{1}{-1} & * & \binom{-2}{0} \\ \binom{1}{-1} & \binom{-2}{0} & * \end{pmatrix}, \; f_0^{14} = 2$$

$$\tilde{f}^{15} = \begin{pmatrix} * & \binom{3}{-1} & \binom{3}{-1} \\ \binom{1}{-1} & * & \binom{-3}{1} \\ \binom{1}{-1} & \binom{-3}{1} & * \end{pmatrix}, \; f_0^{15} = 6 \qquad \tilde{f}^{16} = \begin{pmatrix} * & \binom{1}{0} & \binom{0}{-1} \\ \binom{0}{-1} & * & \binom{0}{1} \\ \binom{0}{1} & \binom{-1}{0} & * \end{pmatrix}, \; f_0^{16} = 2$$

is facet-defining for P_{BW}^m, such that the face induced by (2) is contained in the facet induced by (3). We show that (3) is a positive multiple of (2).

W.l.o.g. we set $R = \{(1^t, 2^t), (2^t, 3^t), \ldots, (m^t, 1^t)\}$. Then (2) becomes

$$\sum_{i=1}^{m-1} (x_{i(i+1)^t} + x_{(i+1)i^h}) + x_{m1^t} + x_{1m^h} \leq 2m - 2 \tag{4}$$

Figure 1 displays the graph belonging to (4) for $m = 4$.

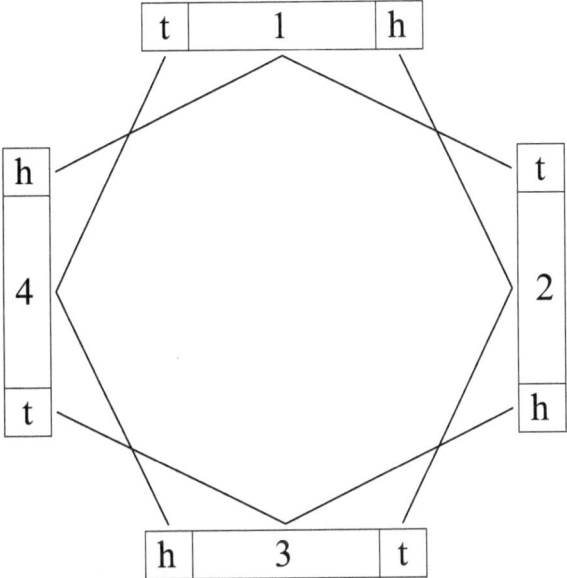

Fig. 1. Cycle inequality with 4 clones

First we show that all coefficients a_{ij} are constant for the clone-probe pairs (i, j) corresponding to the nonzero coefficients in (4).

Let

$$\pi_1 = (1^h, 2^t, 3^t, 2^h, \ldots, i^t, (i-1)^h, \ldots, m^t, (m-1)^h, m^h, 1^t),$$
$$\pi_2 = (2^t, 1^h, 3^t, 2^h, \ldots, i^t, (i-1)^h, \ldots, m^t, (m-1)^h, m^h, 1^t).$$

It is easy to see that χ^{π_1} and χ^{π_2} satisfy (4) as equation. Thus $a^T \chi^{\pi_1} = a^T \chi^{\pi_2}$. This yields $a_{12^t} = a_{21^h}$.

Because R is invariant under the cyclic permutation $1 \to 2 \to \cdots m \to 2 \to 1$, the same operations applied to π_1 and π_2 imply that

$$a_{23^t} = a_{32^h}, \ldots, a_{m1^t} = a_{1m^h}. \tag{5}$$

Analogously, using the permutations

$$\pi_3 = (1^t, 2^t, 1^h, 3^t, 2^h, \ldots, i^t, (i-1)^h, \ldots, m^t, (m-1)^h, m^h),$$
$$\pi_4 = (2^t, 3^t, 2^h, \ldots, i^t, (i-1)^h, \ldots, m^t, (m-1)^h, 1^t, m^h, 1^h)$$

we obtain

$$a_{12^t} + a_{21^h} = a_{23^t} + a_{32^h}, \ldots, a_{m-1m^t} + a_{m(m-1)^h} = a_{m1^t} + a_{1m^h}. \qquad (6)$$

With (5) and (6) we have shown the first claim.

Now we show that $a_{ij} = 0$, for all remaining coefficients. For this purpose we define the following $m-1$ permutations, which only differ in the position of 1^h.

$$\rho_1 = (1^t, 2^t, 3^t, \ldots, (m-1)^t, m^t, 1^h, 2^h, \ldots (m-1)^h, m^h),$$
$$\rho_2 = (1^t, 2^t, 3^t, \ldots, (m-1)^t, 1^h, m^t, 2^h, \ldots (m-1)^h, m^h),$$

$$\vdots$$

$$\rho_{m-1} = (1^t, 2^t, 1^h, 3^t, \ldots, (m-1)^t, m^t, 2^h, \ldots (m-1)^h, m^h).$$

Again, the associated incidence vectors satisfy (4) as equation. The equations $a^T \chi^{\rho_i} = a^T \chi^{\rho_{i+1}}$, for $1 \le i \le m-2$, give

$$a_{1(m+1-i)^t} = -a_{(m+1-i)1^h}, \quad \text{for } 1 \le i \le m-2. \qquad (7)$$

With the permutations (which only differ in the position of 1^h)

$$\tau_1 = (2^t, 3^t, \ldots, (m-1)^t, m^t, 1^h, 1^t, 2^h, \ldots (m-1)^h, m^h),$$
$$\tau_2 = (2^t, 3^t, \ldots, (m-1)^t, 1^h, m^t, 1^t, 2^h, \ldots (m-1)^h, m^h),$$

$$\vdots$$

$$\tau_{m-1} = (2^t, 1^h, 3^t, \ldots, (m-1)^t, m^t, 1^t, 2^h, \ldots (m-1)^h, m^h),$$

we obtain analogously that

$$a_{1(m+1-i)^t} = a_{(m+1-i)1^h}, \quad \text{for } 1 \le i \le m-2. \qquad (8)$$

With (7) and (8) we have shown that $a_{13^t} = a_{31^h} = a_{14^t} = a_{41^h} = \ldots = a_{1m^t} = a_{m1^h} = 0$.

Applying cyclic permutations to ρ_i and τ_i, for $1 \le i \le m-1$, we finally obtain that all coefficients a_{ij} are zero for all zero-coefficients in (4).

Hence, the left hand side of (3) is a multiple of the left hand side of (4). By the properties of (3) and (4), (4) induces that same facet as (3).

5   Separation Procedures

In our computations, we do not use the IP formulation presented above. because on the one hand it is fairly complicated and on the other hand the inequalities do not define facets in general. Rather, we proceed as follows. As discussed before, we can check feasibility of an integer vector x^* by making use of the PQ-tree algorithm. If x^* is feasible, then the PQ-tree algorithm also generates all optimal solutions π^*. Otherwise, if x^* is not feasible (but binary), then one can construct a cutting plane, which is satisfied by all 0/1-vectors different from x^*. Let $P = \{i \mid x_i^* = 1\}$ and $Z = \{i \mid x_i^* = 0\}$, then

$$\sum_{i \in P} x_i - \sum_{i \in Z} x_i \leq |P| - 1$$

is a cutting plane with the desired properties.

5.1   Separation of Cycle Inequalities

The separation of the cycle inequalities can be done by a shortest path algorithm. Let x^* be an LP solution and $D_n = (V_n, A_n)$ be the complete directed graph on n nodes with edge weights

$$w_{ij} = 2 - x^*_{c(i)j} - x^*_{c(j)q(i)}.$$

Then it is easy to see by the following transformation that a cycle with weight less than 2 corresponds to a violated inequality. Let $y^* = 1 - x^*$. Then x^* violates a cycle inequality (4) if and only if y^* violates an inequality

$$\sum_{(i,j) \in R} x_{c(i)j} + x_{c(j)q(i)} \geq 2$$

which is true if and only $w(R) < 2$ for the cycle R in $D_n = (V_n, A_n)$.

5.2   Separation of Small Instance Relaxations

We also implemented separation heuristics for inequalities of small instance relaxations.

Let an LP solution $x^* \in \mathbb{R}^{|\mathcal{B}_m|}$ and a facet-defining inequality $g^T x \leq g_0$ be given. For a permutation σ of $\{1, \ldots, m\}$ we define

$$C(\sigma) = \sum_{i=1}^{m} \sum_{j=1, j \neq i}^{m} \begin{pmatrix} x^*_{ijt} \\ x^*_{ijh} \end{pmatrix}^T \begin{pmatrix} g_{\sigma(i)\sigma(j)t} \\ g_{\sigma(i)\sigma(j)h} \end{pmatrix}.$$

It is clear that the set $\{\sigma \mid C(\sigma) > g_0\}$ gives all violated inequalities among all inequalities which are (by means of relabeling of the clones) equivalent to $g^T x \leq g_0$. Thus the separation problem for equivalent inequalities (relative to relabeling of the clones) is a quadratic assignment problem in which all entries are 2-dimensional vectors.

Because an inequality $g^T x \leq g_0$ of a small instance relaxation results from trivial node lifting of an inequality $f^T x \leq f_0$ which is facet-defining for $P_{BW}^{\overline{m}}$, $\overline{m} < m$ we have

$$
C(\sigma) = \sum_{i=1}^{\overline{m}} \sum_{j=1, j \neq i}^{\overline{m}} \begin{pmatrix} x^*_{\sigma^{-1}(i)\sigma^{-1}(j)^t} \\ x^*_{\sigma^{-1}(i)\sigma^{-1}(j)^h} \end{pmatrix}^T \begin{pmatrix} f_{ij^t} \\ f_{ij^h} \end{pmatrix}.
$$

Since $\overline{m}$ is rather small ($\overline{m} \leq 4$) the quadratic assignment problem is computationally tractable and can effectively be solved by heuristics; in our implementation we use a Grasp procedure similar to [12].

In order to separate all inequalities of one equivalence class this separation procedure has to be executed for at most $2^{\overline{m}}$ different inequalities $f_i^T x \leq f_0^i$ which are equivalent relative to the end probe reversing symmetry.

6   Computational Results

In our branch-and-cut computations we used ABACUS [14,9,10] with LP solver CPLEX [7] and an implementation of the PQ-tree algorithm [11].

In [4] we used a set of 150 randomly generated problem instances with 42 to 70 probes which resemble instances occurring in practical applications. To generate the data the clones were randomly distributed across the chromosome. We used a coverage (which gives the average number of how often a single point along the chromosome is covered by a clone) varying from 3 to 5, a false negative rate of 10%, and a false positive rate varying from 0% to 5%. To create a false positive or false negative, a coin was flipped at each entry with the given probability. Across experiments with varying coverage, the clone length was held constant.

Our computational experiments show that the new approach clearly supersedes our previous approach. Table 2 compares our previous approach (using linear ordering variables) with the new model described here. The table displays the average CPU-times t_{tot} (in hh:mm:ss, on a Sun Sparc IPX), the average number of nodes of the branch-and-cut tree n_{sub}, the average number of cutting planes which are added to the linear program n_{cut} and the average number of LP reoptimizations n_{lp} to solve the problem instances.

Table 2. Average values for 150 problems.

Model	t_{tot}	n_{sub}	n_{cut}	n_{lp}
Linear ordering based	0:44:32	3.2	22197.8	201.5
PQ-tree based	0:01:28	1.1	1022.0	6.3

In our new approach the following strategy for separating a fractional LP solution turned out to be favorable. If by exact enumeration of the nontrivial facet classes of P^2_{BW} not enough cutting planes are found, we execute the heuristics of section 5.2 for a small subset (less than 15) of all equivalence classes of facet-defining inequalities of P^4_{BW}. If this separation fails, then we separate the cycle inequalities.

To improve the performance of our branch-and-cut algorithm we developed algorithms for executing the separation procedures for different classes of facets in parallel. In addition, we studied by computational experiments two important questions in detail:

- Which and how many classes of facets should be used for separation?
- Given a large set of cutting planes generated by the separation heuristics, which ones should be selected and should be added to the LP?

Concerning the first problem, we observed that facet classes with a large number of roots (feasible incidence vectors contained in the facet) are the most important ones. The best solution to the second problem is to use cutting planes whose normal vectors have a small angle with the objective function gradient. Detailed computational experiments using small instance relaxations in parallel branch-and-cut algorithms are reported in [3,6].

For the Weighted Betweenness Problem we tested 25 different strategies on 8 randomly generated instances with the following characteristics: 125 clones (=250 probes), a coverage of 4, a false positive rate of 5% and a false negative rate of 10%. With our best strategy we could solve the 8 instances on a parallel machine with 8 Motorola Power PC processors (7 of them reserved for executing the separation heuristics) in 2:01:34h total time.

References

1. K. Booth and G. Lueker. Testing for the consecutive ones property, interval graphs, and graph planarity using PQ-tree algorithms. *Journal of Computer and System Sciences*, 13:335–379, 1976.
2. K. S. Booth. *PQ-Tree Algorithms*. PhD thesis, University of California, Berkeley, 1975.
3. T. Christof. *Low-Dimensional 0/1-Polytopes and Branch-and-Cut in Combinatorial Optimization*. Aachen: Shaker, 1997.
4. T. Christof, M. Jünger, J. Kececioglu, P. Mutzel, and G. Reinelt. A branch-and-cut approach to physical mapping of chromosomes by unique end-probes. *Journal of Computational Biology*, 4(4):433–447, 1997.
5. T. Christof and G. Reinelt. Efficient parallel facet enumeration for 0/1 polytopes. Technical report, University of Heidelberg, Germany, 1997.
6. T. Christof and G. Reinelt. Algorithmic aspects of using small instance relaxations in parallel branch-and-cut. Technical report, University of Heidelberg, Germany, 1998.
7. CPLEX. *Using the CPLEX Callable Library*. CPLEX Optimization, Inc, 1997.
8. M. Jain and E. W. Myers. Algorithms for computing and integrating physical maps using unique probes. In *First Annual International Conference on Computational Molecular Biology*, pages 84–92. ACM, 1997.

9. M. Jünger and S. Thienel. The design of the branch-and-cut system ABACUS. Technical Report 95.260, Universität zu Köln, Germany, 1997.
10. M. Jünger and S. Thienel. Introduction to ABACUS – A Branch-And-CUt system. Technical Report 95.263, Universität zu Köln, Germany, 1997.
11. S. Leipert. PQ-trees, an implementation as template class in C++. Technical Report 97.259, Universität zu Köln, Germany, 1997.
12. Y. Li, P. M. Pardalos, and M. G. C. Resende. A greedy randomized adaptive search procedure for the quadratic assignment problem. In P. M. Pardalos and H. Wolkowicz, editors, *DIMACS Series in Discrete Mathematics and Theoretical Computer Science, Vol. 16*, pages 237–261. American Mathematical Society, 1994.
13. M. Oswald. PQ-Bäume im Branch & Cut-Ansatz für das Physical-Mapping-Problem mit Endprobes. Master's thesis, Universität Heidelberg, Germany, 1997.
14. S. Thienel. *ABACUS A Branch-And-CUt System*. PhD thesis, Universität zu Köln, 1995.
15. A. Tucker. A structure theorem for the consecutive 1's property. *Journal of Combinatorial Theory*, 12:153–162, 1972.

Solving a Linear Diophantine Equation with Lower and Upper Bounds on the Variables

Karen Aardal[1*], Cor Hurkens[2], and Arjen K. Lenstra[3**]

[1] Department of Computer Science, Utrecht University
aardal@@cs.ruu.nl
[2] Department of Mathematics and Computing Science
Eindhoven University of Technology
wscor@@win.tue.nl
[3] Emerging Technology, Citibank N.A.
arjen.lenstra@@citicorp.com

Abstract. We develop an algorithm for solving a linear diophantine equation with lower and upper bounds on the variables. The algorithm is based on lattice basis reduction, and first finds short vectors satisfying the diophantine equation. The next step is to branch on linear combinations of these vectors, which either yields a vector that satisfies the bound constraints or provides a proof that no such vector exists. The research was motivated by the need for solving constrained linear diophantine equations as subproblems when designing integrated circuits for video signal processing. Our algorithm is tested with good result on real-life data.

Subject classification: Primary: 90C10. Secondary: 45F05, 11Y50.

1 Introduction and Problem Description

We develop an algorithm for solving the following integer feasibility problem:

$$\text{does there exist a vector } \mathbf{x} \in \mathbb{Z}^n \text{ such that } \mathbf{ax} = a_0, \; \mathbf{0} \le \mathbf{x} \le \mathbf{u}? \quad (1)$$

We assume that $\mathbf{a}$ is an n-dimensional row vector, $\mathbf{u}$ is an n-dimensional column vector, and that a_0 is an integer scalar. This is an NP-complete problem; in the absence of bound constraints, it can be solved in polynomial time. The research was motivated by a need for solving such problems when designing integrated circuits (ICs) for video signal processing, but several other problems

* Research partially supported by ESPRIT Long Term Research Project No. 20244 (Project ALCOM-IT: *Algorithms and Complexity in Information Technology*), and by NSF grant CCR-9307391 through David B. Shmoys, Cornell University.
** Research partially supported by ESPRIT Long Term Research Project No. 20244 (Project ALCOM-IT: *Algorithms and Complexity in Information Technology*).

R. E. Bixby, E. A. Boyd, and R. Z. Ríos-Mercado (Eds.): IPCO VI
LNCS 1412, pp. 229–242, 1998. © Springer–Verlag Berlin Heidelberg 1998

can be viewed as problem (1), or generalizations of (1). One such example is the *Frobenius* problem that was recently considered by Cornuéjols, Urbaniak, Weismantel and Wolsey [4]. The instances related to video signal processing were difficult to tackle by linear programming (LP) based branch-and-bound due to the characteristics of the input. In order to explain the structure of these instances we briefly explain the origin of the problem below. We also tested our algorithm with good results on the Frobenius instances of Cornuéjols et al.

In one of the steps of the design of ICs for video signal processing one needs to assign so-called *data streams* to processors. A data stream is a repetitive set of arithmetic operations. The attributes of a data stream are the starting time of the first execution, the number of repetitions, and the period of repetition. One can view a data stream as a set of nested loops. The outer loop has an iterator $i_0 : 0 \le i_0 \le I_0$. The following loop has iterator $i_1 : 0 \le i_1 \le I_1$, and so forth. The periodicity corresponding to a loop is the time interval between two consecutive iterations. When constructing an assignment of the streams to the processors the following *conflict detection problem* occurs: check whether there is any point in time at which operations of two different streams are carried out. If such a point in time exists, then the streams should not be assigned to the same processor. Consider an arbitrary data stream f. Let $\mathbf{i}_f = (i_{f0}, i_{f1}, \ldots, i_{fm})^T$ be the *iterator vector* of the stream. The iterator vector satisfies upper and lower bounds, $\mathbf{0} \le \mathbf{i}_f \le \mathbf{I}_f$. Let $\mathbf{p}_f$ denote the *period vector* and s_f the *starting time* of the stream. The point in time at which execution $\mathbf{i}_f$ of data stream f takes place is expressed as $t(\mathbf{i}_f) = s_f + \mathbf{p}_f^T \mathbf{i}_f$. The conflict detection problem can be formulated mathematically as the following integer feasibility problem: given data streams f and g, do there exist iterator vectors $\mathbf{i}_f$ and $\mathbf{i}_g$ such that

$$s_f + \mathbf{p}_f^T \mathbf{i}_f = s_g + \mathbf{p}_g^T \mathbf{i}_g, \text{ and such that } \mathbf{0} \le \mathbf{i}_f \le \mathbf{I}_f, \quad \mathbf{0} \le \mathbf{i}_g \le \mathbf{I}_g?$$

The Frobenius problem is defined as follows: given nonnegative integers $(a_1, \ldots, a_n)$ with $\gcd(a_1, \ldots, a_n) = 1$, find the largest integer a_0 that cannot be expressed as a nonnegative integer combination of $a_1, \ldots, a_n$. The number a_0 is called the Frobenius number. The instances considered by Cornuéjols et al. [4] were also hard to solve using LP-based branch-and-bound. They developed a test set approach that was successful on their instances.

When solving a feasibility problem such as (1) by LP-based branch-and-bound, two difficulties may arise. First, the search tree may become large due to the magnitude of the upper bounds on the variables, and second, round-off errors may occur. The size of the branch-and-bound tree may also be sensitive to the objective function that is used. For our problem (1) an objective function does not have any meaning since it is a feasibility problem; as long as we either find a feasible vector, or are able to verify that no feasible vector exists, the objective function as such does not matter. The problem is that one objective function may give an answer faster than another, but which one is best is hard to predict. An objective function also introduces an aspect to the problem that is not natural. Round-off errors occur quite frequently for the instances related to the conflict detection problem, since the coefficients of some of the variables are very large ($\approx 10^7$). The special characteristics of these instances – some very large and some relatively small coefficients and a very large right-hand side value a_0 – are due to the difference in periodicity of the nested loops. This difference is explained by the composition of a television screen image. Such an image consist of 625 lines, and each line is composed of 720 pixels. Every second 25 pictures are shown on the screen, so the time between two pictures is 40 ms. The time between two lines and between two pixels are 64 μs and 74 ns respectively. Since the output rate of the signals has to be equal to the input rate, we get large differences in periodicity when the data stream corresponds to operations that have to be repeated for all screens, lines and pixels. Due to the large difference in the magnitude of the coefficients we often observe that the LP-based branch-and-bound algorithm terminates with a solution in which for instance variable x_j takes value 4.999999, simply because the hardware does not allow for greater precision. If one would round x_j to $x_j = 5.0$, then one would obtain a vector $\mathbf{x}$ such that $\mathbf{a}\mathbf{x} \neq a_0$. It is obviously a serious drawback that the algorithm terminates with an infeasible solution.

To overcome the mentioned deficiencies we have developed an algorithm based on the L^3 basis reduction algorithm as developed by Lenstra, Lenstra and Lovász [10]. The motivation behind choosing basis reduction as a core of our algorithm is twofold. First, basis reduction allows us to work directly with integers, which avoids the round-off problems. Second, basis reduction finds short, nearly orthogonal vectors belonging to the lattice described by the basis. Given the lower and upper bounds on the variables, we can interpret problem (1) as checking whether there exists a *short vector* satisfying a given linear diophantine equation. It is easy to find an initial basis that describes the lattice containing all vectors of interest to our problem. This initial basis is not "good" in the sense that it contains very long vectors, but it is useful as we can prove structural properties of the reduced basis obtained by applying the L^3 algorithm to it. It is important to note that basis reduction does not change the lattice, it only derives an alternative way of spanning it. Furthermore, our algorithm is designed for feasibility problems. Once we have obtained the vectors given by the reduced basis, we use them as input to a heuristic that tries to find a feasible vector fast or, in case the heuristic fails, we call an algorithm that branches on linear combinations of vectors and yields either a vector satisfying the bound constraints, or a proof that no such vector exists.

In Sect. 2 we give a short description of the L^3 basis reduction algorithm and a brief review of the use of basis reduction in integer programming. In Sect. 3 we introduce a lattice that contains all interesting vectors for our problem (1), and provide an initial basis spanning that lattice. We also derive structural properties of the reduced basis. Our algorithm is outlined in Sect. 4, and we report on our computational experience in Sect. 5.

We are indebted to Laurence Wolsey for stimulating discussions and for numerous suggestions on how to improve the exposition.

2   Basis Reduction and Its Use in Integer Programming

We begin by giving the definition of a lattice and a reduced basis.

Definition 1. *A subset $L \subset \mathbb{R}^n$ is called a* lattice *if there exists linearly independent vectors $\mathbf{b}_1, \mathbf{b}_2, \ldots, \mathbf{b}_l$ in $\mathbb{R}^n$ such that*

$$L = \{\textstyle\sum_{j=1}^{l} \alpha_j \mathbf{b}_j : \ \alpha_j \in \mathbb{Z}, \ 1 \le j \le l\}. \tag{2}$$

The set of vectors $\mathbf{b}_1, \mathbf{b}_2, \ldots, \mathbf{b}_l$ is called a lattice basis.

Gram-Schmidt orthogonalization is an algorithm for deriving orthogonal vectors $\mathbf{b}_j^*$, $1 \le j \le l$ from independent vectors $\mathbf{b}_j$, $1 \le j \le l$. The vectors $\mathbf{b}_j^*$, $1 \le j \le l$ and the real numbers μ_{jk}, $1 \le k < j \le l$ are defined inductively by:

$$\mathbf{b}_j^* = \mathbf{b}_j - \textstyle\sum_{k=1}^{j-1} \mu_{jk} \mathbf{b}_k^* \tag{3}$$

$$\mu_{jk} = (\mathbf{b}_j)^T \mathbf{b}_k^* / (\mathbf{b}_k^*)^T \mathbf{b}_k^* \tag{4}$$

Let $\| \ \|$ denote the Euclidean length in $\mathbb{R}^n$. Lenstra, Lenstra and Lovász [10] used the following definition of a reduced basis:

Definition 2. *A basis $\mathbf{b}_1, \mathbf{b}_2, \ldots, \mathbf{b}_l$ is* reduced *if*

$$|\mu_{jk}| \le \tfrac{1}{2} \ for \ 1 \le k < j \le l \tag{5}$$

and

$$\|\mathbf{b}_j^* + \mu_{j,j-1} \mathbf{b}_{j-1}^*\|^2 \ge \tfrac{3}{4} \|\mathbf{b}_{j-1}^*\|^2 \ for \ 1 < j \le l. \tag{6}$$

The vector $\mathbf{b}_j^*$ is the projection of $\mathbf{b}_j$ on the orthogonal complement of $\sum_{k=1}^{j-1} \mathbb{R}\mathbf{b}_k$, and the vectors $\mathbf{b}_j^* + \mu_{j,j-1} \mathbf{b}_{j-1}^*$ and $\mathbf{b}_{j-1}^*$ are the projections of $\mathbf{b}_j$ and $\mathbf{b}_{j-1}$ on the orthogonal complement of $\sum_{k=1}^{j-2} \mathbb{R}\mathbf{b}_k$. Inequality (5) states that the vectors $1 \le j \le l$ are "reasonably" orthogonal. Inequality (6) can be interpreted as follows. Suppose that vectors $\mathbf{b}_{j-1}$ and $\mathbf{b}_j$ are interchanged. Then the vectors $\mathbf{b}_{j-1}^*$ and $\mathbf{b}_j^*$ will change as well. More precisely, the new vector $\mathbf{b}_{j-1}^*$ is the vector $\mathbf{b}_j^* + \mu_{j,j-1} \mathbf{b}_{j-1}^*$. Inequality (6) therefore says that the length of vector $\mathbf{b}_{j-1}^*$ does not decrease too much if vectors vectors $\mathbf{b}_{j-1}$ and $\mathbf{b}_j$ are interchanged. The constant $\tfrac{3}{4}$ in inequality (6) is arbitrarily chosen and can be replaced by any fixed real number $\tfrac{1}{4} < y < 1$. Lenstra et al. [10] developed a polynomial time algorithm for obtaining a reduced basis for a lattice given an initial basis. The algorithm consists of a sequence of size reductions and interchanges as described below. For the precise algorithm we refer to [10].

Size reduction: If for any pair of indices $j, k : \ 1 \le k < j \le l$ condition (5) is violated, then replace $\mathbf{b}_j$ by $\mathbf{b}_j - \lceil \mu_{jk} \rfloor \mathbf{b}_k$, where $\lceil \mu_{jk} \rfloor = \lceil \mu_{jk} - \tfrac{1}{2} \rceil$.

Interchange: If condition (6) is violated for an index j, $1 < j \le l$, then interchange vectors $\mathbf{b}_{j-1}$ and $\mathbf{b}_j$.

Basis reduction was introduced in integer programming by H.W. Lenstra, Jr. [11], who showed that the problem of determining if there exists a vector $\mathbf{x} \in \mathbb{Z}^n$ such that $\mathbf{A}\mathbf{x} \leq \mathbf{d}$ can be solved in polynomial time when n is fixed. Before this result was published, only the cases $n = 1, 2$ were known to be polynomially solvable. The idea behind Lenstra's algorithm can be explained considering a two-dimensional convex body. Suppose that this body is "thin" as illustrated in Fig. 1. If it extends arbitrarily far in both directions, as indicated in the figure, then an LP-based branch-and-bound tree will become arbitrarily deep before concluding that no feasible solution exists. It is easy to construct a similar example in which a feasible vector does exist. So, even if $n = 2$, an LP-based branch-and-bound algorithm may require exponentially many iterations in terms of the dimension. What Lenstra observed was the following. Assume

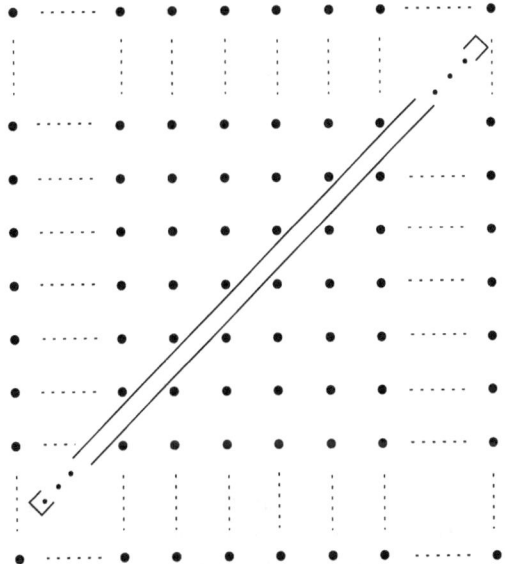

Fig. 1. A thin convex body in $\mathbb{Z}^2$.

that we start with the full-dimensional bounded convex body $X \in \mathbb{R}^n$ and that we consider the lattice $\mathbb{Z}^n$. The problem is to determine whether there exists a vector $\mathbf{x} \in (X \cap \mathbb{Z}^n)$. We refer to this problem as problem P. We use $\mathbf{b}_j = \mathbf{e}_j$, $1 \leq j \leq n$ as a basis for the lattice $\mathbb{Z}^n$, where $\mathbf{e}_j$ is the vector where all elements of the vector are equal to zero, except element j that is equal to one. To avoid having a convex body that is thin we apply a linear transformation τ to X to make it appear "regular". Problem P is equivalent to the problem of determining whether there exists a vector $\mathbf{x} \in (\tau X \cap \tau \mathbb{Z}^n)$. The new convex body τX has a regular shape but the basis vectors $\tau \mathbf{e}_j$ are not necessarily orthogonal any longer, so from the point of view of branching the difficulty is still present.

We can view this as having shifted the problem we had from the convex body to the lattice. This is where basis reduction proves useful. By applying the L^3 algorithm to the basis vectors τe_j, we obtain a new basis $\hat{b}_1, \ldots, \hat{b}_n$ spanning the same lattice, $\tau \mathbb{Z}^n$, but having short, nearly-orthogonal vectors. In particular it is possible to show that the distance d between any two consecutive hyperplanes $H + k\hat{b}_n$, $H + (k+1)\hat{b}_n$, where $H = \sum_{j=1}^{n-1} \mathbb{R}b_j$ and $k \in \mathbb{Z}$, is not too short, which means that if we branch on these hyperplanes, then there cannot be too many of them. Each branch at a certain level of the search tree corresponds to a subproblem with dimension one less than the dimension of its predecessor. In Fig. 2 we show how the distance between hyperplanes $H + k\hat{b}_n$ increases if we use a basis with orthogonal vectors instead of a basis with long non-orthogonal ones.

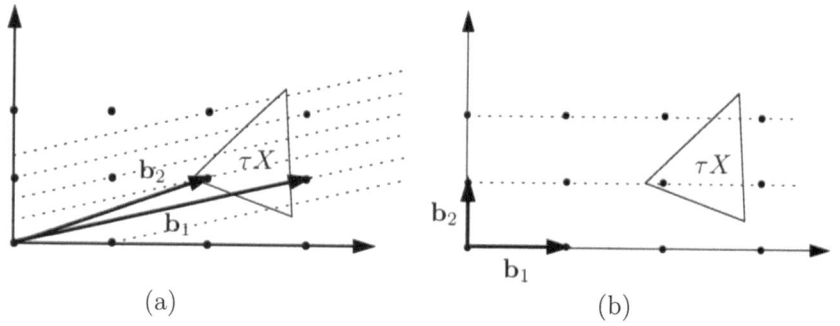

(a)   (b)

Fig. 2. (a) Non-orthogonal basis. (b) Orthogonal basis.

Due to the special structure of our instances we do not use a transformation τ in our algorithm. We can simply write down an initial basis for a lattice that contains all vectors of interest for our problem, and then apply the L^3 algorithm directly to this basis. Our initial basis resembles the bases used by several authors for finding feasible solutions to subset sum problems that arise in certain cryptographic systems, see further Sect. 3.

For the integer programming problem P, Lovász and Scarf [13] developed an algorithm that, as Lenstra's algorithm, uses branching on hyperplanes. Instead of using a transformation τ to transform the convex body and the initial basis vectors, and then applying a basis reduction algorithm, their algorithm produces a "Lovász-Scarf-reduced" basis by measuring the width of the considered convex body in different independent directions. Lovász and Scarf's definition of a reduced basis is a generalization of the definition given by Lenstra et al. [10]. Cook, Rutherford, Scarf and Shallcross [2] report on a successful implementation of the Lovász-Scarf algorithm. Cook et al. were able to solve some integer programming problems arising in network design that could not be solved by traditional LP-based branch-and-bound.

Integer programming is not the only application of lattice basis reduction. A prominent application is factoring polynomials with rational coefficients. Lenstra et al. [10] developed a polynomial-time algorithm based on basis reduction for finding a decomposition into irreducible factors of a non-zero polynomial in one variable with rational coefficients. In cryptography, basis reduction has been used to solve subset sum problems arising in connection with certain cryptosystems, see for instance [5], [9], [15], [16]. A recent application in cryptography is due to Coppersmith [3] who uses basis reduction to find small integer solutions to a polynomial in a single variable modulo N, and to a polynomial in two variables over the integers. This has applications to some RSA-based cryptographic schemes. In extended g.c.d.-computations, basis reduction is used by for instance Havas, Majewski and Matthews [8]. Here, the aim is to find a short multiplier vector $\mathbf{x}$ such that $\mathbf{a}\mathbf{x} = a_0$, where $a_0 = \gcd(a_1, a_2, \ldots, a_n)$.

3   Structure of Initial and Reduced Basis

Here we consider a lattice that contains all vectors of interest to our problem (1):

does there exist a vector $\mathbf{x} \in \mathbb{Z}^n$ such that $\mathbf{a}\mathbf{x} = a_0, \quad \mathbf{0} \le \mathbf{x} \le \mathbf{u}$?

Without loss of generality we assume that $\gcd(a_1, a_2, ..., a_n) = 1$. We formulate an initial basis $\mathbf{B}$ that generates this lattice and derive structural properties of the reduced basis $\mathbf{RB}$ obtained after applying the L^3 algorithm to $\mathbf{B}$.

Let $\mathbf{x}_0$ denote the vector $(x_1, \ldots, x_n, x_0)^T$. We refer to $\mathbf{x}_0$ as the *extended* $\mathbf{x}$-vector. Here, x_0 is a variable that we associate with the right-hand side coefficient a_0. Given the bounds on the variables, we tackle problem (1) by trying to find a short vector $\mathbf{x}$ satisfying $\mathbf{a}\mathbf{x} = a_0$. This can be done by finding short vectors in the lattice L_s containing the vectors

$$(x_1, \ldots, x_n, x_0, (a_1 x_1 + \cdots + a_n x_n - a_0 x_0))^T, \qquad (7)$$

where $x_1, \ldots, x_n, x_0 \in \mathbb{Z}$. In particular, we want to find integral extended $\mathbf{x}$-vectors lying in the null-space of $\mathbf{a}_0 = (a_1, \ldots, a_n, a_0)$, denoted by $N(\mathbf{a}_0)$, i.e., vectors $\mathbf{x}_0 \in \mathbb{Z}^{n+1}$ that satisfy $\mathbf{a}\mathbf{x} - a_0 x_0 = 0$. Moreover, if possible, we want $\mathbf{x}$ to satisfy the upper and lower bound constraints, and x_0 to be equal to one. Below we will show how we can use basis reduction to find such a vector if one exists.

The lattice L spanned by the basis $\mathbf{B}$ given by

$$\mathbf{B} = \begin{pmatrix} 1 & 0 & 0 & \cdots & & 0 \\ 0 & 1 & 0 & \cdots & & 0 \\ \vdots & & \ddots & & & \vdots \\ \vdots & & & 1 & & 0 \\ 0 & 0 & \cdots & 0 & & N_1 \\ N_2 a_1 & \cdots & & \cdots & N_2 a_n & -N_2 a_0 \end{pmatrix}$$

contains the vectors

$$(\mathbf{x}, N_1 x_0, h)^T = (x_1, \ldots, x_n, N_1 x_0, N_2(-a_0 x_0 + a_1 x_1 + \cdots + a_n x_n))^T, \quad (8)$$

where N_1 and N_2 are large integral numbers. Note that the basis vectors are given columnwise, and that the basis consists of $n+1$ vectors $\mathbf{b}_j = (b_{1j}, \ldots, b_{n+2,j})^T$, $1 \leq j \leq n+1$. A vector $(\mathbf{x}_0, 0)^T$ in the lattice L_s that belongs to $N(\mathbf{a}_0, 0)$ corresponds to a vector $(\mathbf{x}, N_1 x_0, 0)^T$ in the lattice L. We will show below that by choosing the multipliers N_1 and N_2 large enough the reduced basis will contain a vector $(\mathbf{x}_d, N_1, 0)^T$, and vectors $(\mathbf{x}_j, 0, 0,)^T$, $1 \leq j \leq n-1$. The vector $\mathbf{x}_d$ is a solution to the equation $\mathbf{a}\mathbf{x} = a_0$, and the vectors $\mathbf{x}_j$ are solutions to $\mathbf{a}\mathbf{x} = 0$, i.e., they belong to the null-space $N(\mathbf{a})$. Since the vectors $\mathbf{x}_d$ and $\mathbf{x}_j$, $1 \leq j \leq n-1$ belong to the reduced basis we can expect them to be relatively short.

Lemma 1 (Lenstra, Lenstra, Lovász [10]). *Let $\Lambda \subset \mathbb{R}^n$ be a lattice with reduced basis $\mathbf{b}_1, \mathbf{b}_2, \ldots, \mathbf{b}_l \in \mathbb{R}^n$. Let $\mathbf{y}_1, \mathbf{y}_2, \ldots, \mathbf{y}_t \in \Lambda$, $t \leq l$ be linearly independent. Then,*

$$||\mathbf{b}_j||^2 \leq 2^{n-1} \max\{||\mathbf{y}_1||^2, ||\mathbf{y}_2||^2, ..., ||\mathbf{y}_t||^2\} \text{ for } 1 \leq j \leq t. \quad (9)$$

Let $\hat{\mathbf{b}}_j = (\hat{b}_{1j} \ldots, \hat{b}_{n+2,j})^T$, $1 \leq j \leq n+1$, denote the vectors of the reduced basis, **RB**, obtained by applying L^3 to **B**.

Theorem 1. *There exists numbers N_{01} and N_{02} such that if $N_1 > N_{01}$, and if $N_2 > N_1 N_{02}$, then the vectors $\hat{\mathbf{b}}_j \in \mathbb{Z}^{n+2}$, of the reduced basis **RB** have the following properties:*

1. $\hat{b}_{n+1,j} = 0$ for $1 \leq j \leq n-1$,
2. $\hat{b}_{n+2,j} = 0$ for $1 \leq j \leq n$,
3. $\hat{b}_{n+1,n} = N_1$.

Proof. Without loss of generality we assume that $a_n \neq 0$. Consider the following n vectors in $\mathbb{Z}^{n+1}$: $\mathbf{v}_j = (a_n \mathbf{e}_j - a_j \mathbf{e}_n, 0)^T$, $1 \leq j \leq n-1$, $\mathbf{v}_n = (a_0 \mathbf{e}_n, a_n)^T$. Next, define vectors $\mathbf{z}_1, \ldots, \mathbf{z}_n$ as follows:

$$\mathbf{z}_j = \mathbf{B}\mathbf{v}_j = (\mathbf{v}_j, 0)^T, \quad 1 \leq j \leq n-1, \qquad (10)$$
$$\mathbf{z}_n = \mathbf{B}\mathbf{v}_n = (a_0 \mathbf{e}_n, N_1 a_n, 0)^T. \qquad (11)$$

The vectors $\mathbf{z}_j$, $1 \leq j \leq n$ belong to the lattice L.
Choose N_{01} such that

$$N_{01}^2 > 2^{n+1} \max\{||\mathbf{z}_1||^2, \ldots, ||\mathbf{z}_{n-1}||^2\} = 2^{n+1} \max\{||\mathbf{v}_1||^2, \ldots, ||\mathbf{v}_{n-1}||^2\}.$$

From Lemma 1 we have that

$$||\hat{\mathbf{b}}_j||^2 \leq 2^{n+1} \max\{||\mathbf{z}_1||^2, \ldots, ||\mathbf{z}_{n-1}||^2\} \quad \text{for } 1 \leq j \leq n-1.$$

Suppose that $\hat{b}_{n+1,j} \neq 0$ for some $j: 1 \leq j \leq n-1$. Then $||\hat{\mathbf{b}}_j||^2 \geq \hat{b}_{n+1,j}^2 \geq N_1^2$ as N_1 divides $\hat{b}_{n+1,j}$. As a consequence, $||\hat{\mathbf{b}}_j||^2 > N_{01}^2$, which contradicts the outcome of Lemma 1. We therefore have that $\hat{b}_{n+1,j} = 0$ for $1 \leq j \leq n-1$.

Next, select N_{02} such that

$$N_{02}^2 > 2^{n+1} \max\{||\mathbf{v}_1||^2, \ldots, ||\mathbf{v}_n||^2\}.$$

Due to Lemma 1, the following holds for the reduced basis vectors $\hat{\mathbf{b}}_j, 1 \le j \le n$:

$$
\begin{aligned}
||\hat{\mathbf{b}}_j||^2 &\le 2^{n+1} \max\{||\mathbf{z}_1||^2, \ldots, ||\mathbf{z}_n||^2\} = \\
&\quad 2^{n+1} \max\{||\mathbf{z}_1||^2, \ldots, ||\mathbf{z}_{n-1}||^2, a_0^2 + N_1^2 a_n^2\} = \\
&\quad 2^{n+1} \max\{||\mathbf{v}_1||^2, \ldots, ||\mathbf{v}_{n-1}||^2, a_0^2 + N_1^2 a_n^2\} \le \\
&\quad 2^{n+1} \max\{||\mathbf{v}_1||^2, \ldots, ||\mathbf{v}_{n-1}||^2, N_1^2 ||\mathbf{v}_n||^2\} \le \\
&\quad N_1^2 \, 2^{n+1} \max\{||\mathbf{v}_1||^2, \ldots, ||\mathbf{v}_{n-1}||^2, ||\mathbf{v}_n||^2\} < N_1^2 N_{02}^2.
\end{aligned}
$$

If $\hat{b}_{n+2,j} \ne 0$ for some $j : 1 \le j \le n$, then $||\hat{\mathbf{b}}_j||^2 \ge \hat{b}_{n+2,j}^2 \ge N_2^2$ since N_2 divides $\hat{b}_{n+2,j}$. This implies that $||\hat{\mathbf{b}}_j||^2 > N_1^2 N_{02}^2$, which yields a contradiction. We can therefore conclude that $\hat{b}_{n+2,j} = 0$ for $1 \le j \le n$.

Finally, we prove Property 3. The equation $\mathbf{a}x = a_0$ has a feasible solution since we assume that $\gcd(a_1, \ldots, a_n) = 1$ and since a_0 is integral. Let $\mathbf{x}_d$ be a solution to $\mathbf{a}x = a_0$. This implies that the vector $(\mathbf{x}_d, N_1, 0)^T = \mathbf{B}(\mathbf{x}_d, 1)^T$ belongs to L spanned by $\mathbf{B}$. The lattice L is also spanned by the reduced basis $\mathbf{RB}$, and hence the vector $(\mathbf{x}_d, N_1, 0)^T$ can be obtained as $(\mathbf{x}_d, N_1, 0)^T = \mathbf{RB}(\lambda_1, \ldots, \lambda_{n+1})^T$. Properties 1 and 2 imply the following:

$$0\lambda_1 + \cdots + 0\lambda_{n-1} + \hat{b}_{n+1,n}\lambda_n + \hat{b}_{n+1,n+1}\lambda_{n+1} = N_1 \tag{12}$$

$$0\lambda_1 + \cdots + 0\lambda_n + \hat{b}_{n+2,n+1}\lambda_{n+1} = 0 \tag{13}$$

From equation (13) we obtain $\lambda_{n+1} = 0$ as $\hat{b}_{n+2,n+1}$ is equal to an integer multiple of N_2. If we use $\lambda_{n+1} = 0$ in equation (12) we obtain $\hat{b}_{n+1,n}\lambda_n = N_1$, which implies that $\hat{b}_{n+1,n}$ divides N_1. Moreover, N_1 divides $\hat{b}_{n,n+1}$ (cf. the proof of Property 1). We can therefore conclude that $\hat{b}_{n+1,n} = N_1$. □

Example 1. Consider the following instance of problem (1): does there exist a vector $\mathbf{x} \in \mathbb{Z}^5$ such that:

$$8,400,000x_1 + 4,000,000x_2 + 15,688x_3 + 6,720x_4 + 15x_5 = 371,065,262;$$
$$0 \le x_1 \le 45; \ 0 \le x_2 \le 39; \ 0 \le x_3 \le 349; \ 0 \le x_4 \le 199; \ 0 \le x_5 \le 170?$$

Let $N_1 = 1,000$ and $N_2 = 10,000$. The initial basis $\mathbf{B}$ looks as follows:

$$
\mathbf{B} = \begin{pmatrix}
1 & 0 & 0 & 0 & 0 & 0 \\
0 & 1 & 0 & 0 & 0 & 0 \\
0 & 0 & 1 & 0 & 0 & 0 \\
0 & 0 & 0 & 1 & 0 & 0 \\
0 & 0 & 0 & 0 & 1 & 0 \\
0 & 0 & 0 & 0 & 0 & N_1 \\
8,400,000N_2 & 4,000,000N_2 & 15,688N_2 & 6,720N_2 & 15N_2 & -371,065,262N_2
\end{pmatrix}
$$

After applying L^3 to **B** we obtain:

$$\mathbf{RB} = \begin{pmatrix} -10 & 0 & 1 & 5 & 36 & 2 \\ 21 & 0 & -2 & -10 & 17 & -4 \\ 0 & 15 & -25 & -95 & 39 & -42 \\ 0 & -35 & -1 & -76 & 8 & -21 \\ 0 & -8 & -72 & 72 & -22 & 1 \\ 0 & 0 & 0 & 0 & 1,000 & 0 \\ 0 & 0 & 0 & 0 & 0 & 10,000 \end{pmatrix}$$

Notice that the sixth $((n+1)$-st$)$ element of the first four $(n-1)$ basis vectors of **RB** are equal to zero, and that the last element $((n+2)$-nd$)$ of the first five (n) basis vectors are equal to zero. We also note that $\hat{b}_{n+1,n} = \hat{b}_{6,5} = N_1$ and that the elements of the first $n = 5$ rows of **RB** are quite small. $\qquad\square$

We conclude the section with a brief comparison between our initial basis **B** and various bases used when trying to find solutions to subset sum problems that arise in cryptography. In the cryptography application the message that the "sender" wants to transmit to the "receiver" is represented by a sequence $\mathbf{x} \in \{0,1\}^n$ of "bits". The receiver knows a sequence of numbers $a_1, \ldots, a_n$, and instead of sending the actual message $\mathbf{x}$, the sender sends a number $a_0 = \mathbf{ax}$. Once the receiver knows a_0 he can recover the message $\mathbf{x}$ by solving the subset sum problem $\mathbf{ax} = a_0$. Here, the equation $\mathbf{ax} = a_0$ is *known* to have a solution. Lagarias and Odlyzko [9] considered the following basis to generate the lattice containing vectors $(\mathbf{x}, (-\mathbf{ax} + a_0 x_0))^T$:

$$\mathbf{B} = \begin{pmatrix} \mathbf{I}^{(n)} & \mathbf{0}^{(n \times 1)} \\ -\mathbf{a} & a_0 \end{pmatrix}$$

Here, $\mathbf{I}^{(n)}$ denotes the n-dimensional identity matrix, and $\mathbf{0}^{(n \times 1)}$ denotes the $n \times 1$-matrix consisting of zeros only. Lagarias and Odlyzko proposed a polynomial time algorithm based on basis reduction. There is no guarantee that the algorithm finds a feasible vector $\mathbf{x}$, but the authors show that for "almost all" instances of *density* $d = n/\log_2(max_j \, a_j) < 0.645$, a feasible vector is found. Several similar bases have been considered later as input to algorithms for trying to find solutions to subset sum problems. Schnorr and Euchner [15] for instance used the basis

$$\mathbf{B} = \begin{pmatrix} \text{diag}(2)^{(n \times n)} & \mathbf{1}^{(n \times 1)} \\ n\mathbf{a} & na_0 \\ \mathbf{0}^{(1 \times n)} & 1 \end{pmatrix}$$

where $\text{diag}(2)^{(n \times n)}$ is the $n \times n$-matrix with twos along the main diagonal and zeros otherwise. Here, a lattice vector $\mathbf{v} \in \mathbb{Z}^{n+2}$ that satisfies $|v_{n+2}| = 1$, $v_{n+1} = 0$ and $v_j \in \{\pm 1\}$ for $0 \le j \le n$, corresponds to a feasible vector $x_j = \frac{1}{2}|v_j - v_{n+2}|$, $0 \le j \le n$. Schnorr and Euchner [15] proposed an algorithm that for "almost all" subset sum problem instances with $d < 0.9408$ finds a feasible vector. The algorithm uses the above basis as input. For further details on finding feasible solutions to subset sum problems arising in cryptography we refer to the above references and to the references given in the introduction.

The important differences between the approach described above and our approach are the following. First the question that is posed is different. We do not know a priori whether our instances have a feasible solution or not, and we want to *solve* the feasibility problem, i.e., if a feasible solution exists we want to find one, and if no feasible solution exists this should be verified. Hence, we propose a branching algorithm as described in the following section. We also use two large constants N_1 and N_2 to "force" the L^3 algorithm to find interesting vectors. Moreover, we have shown, see [1], that our algorithm can be generalized to handle systems of linear diophantine equations.

4   The Algorithm

Here we discuss how we can use the properties stated in Theorem 1 to design an algorithm for solving the feasibility problem (1). Since we are only interested in vectors belonging to the null-space $N(\mathbf{a}_0)$ we consider only the first $n+1$ elements of the first n vectors of $\mathbf{RB}$. We now have obtained a basis for $\mathbb{Z}^{n+1} \cap N(\mathbf{a}, \frac{a_0}{N_1})$ with the following structure:

$$\mathbf{RB}' = \begin{pmatrix} \mathbf{X} & \mathbf{x}_d \\ \mathbf{0}^{(1 \times (n-1))} & N_1 \end{pmatrix}$$

In our example above the basis $\mathbf{RB}'$ corresponds to

$$\begin{pmatrix} -10 & 0 & 1 & 5 & 36 \\ 21 & 0 & -2 & -10 & 17 \\ 0 & 15 & -25 & -95 & 39 \\ 0 & -35 & -1 & -76 & 8 \\ 0 & -8 & -72 & 72 & -22 \\ 0 & 0 & 0 & 0 & 1{,}000 \end{pmatrix}$$

The last column of the matrix $\mathbf{RB}'$, $(\mathbf{x}_d, N_1)^T$ is a solution to the equation $\mathbf{ax} - \frac{a_0}{N_1}x_0 = 0$, which implies that the vector $\mathbf{x}_d$ is a solution to the equation $\mathbf{ax} = a_0$. All other columns of $\mathbf{RB}'$ are solutions to the equation $\mathbf{ax} - 0 = 0$, i.e., the columns of the submatrix $\mathbf{X}$ all lie in the null-space $N(\mathbf{a})$.

In our algorithm we first check whether the vector $\mathbf{x}_d$ satisfies the lower and upper bounds. If yes, we are done. If any of the bounds is violated we search for a vector that is feasible, or for a proof that no feasible vector exists, by branching on linear integer multiples of the vectors in $N(\mathbf{a})$. Note that by adding any linear integer combination of vectors in $N(\mathbf{a})$, $\mathbf{x}_\lambda$, to $\mathbf{x}_d$ we obtain a vector $\mathbf{x}_d + \mathbf{x}_\lambda$ that satisfies $\mathbf{ax}_d + \mathbf{x}_\lambda = a_0$. For the feasible instances the search for a feasible vector turned out to be easy. To speed up the algorithm for most of these instances we developed a heuristic as follows. Let $\mathbf{X}_j$ be the j-th column of the submatrix $\mathbf{X}$ of $\mathbf{RB}'$. Suppose that we are at iteration t of the heuristic and that an integer linear combination of $t_0 < t$ vectors of $N(\mathbf{a})$ has been added to vector $\mathbf{x}_d$. The vector obtained in this way is called the "current vector". For simplicity we assume that only variable x_k of the current vector violates one of its bound constraints. At iteration t we add or subtract an integer multiple λ_t of the column vector $\mathbf{X}_t$ if the violation of variable x_k's bound constraint is reduced and if no other bound constraints becomes violated. As soon as the value of x_k satisfies its bounds, we do not consider any larger values of λ_t. If the heuristic does not find any feasible solution, we call an exact branching algorithm that branches on linear combinations of vectors in $N(\mathbf{a})$. A summary of the complete algorithm is given in Fig. 3.

procedure main$(\mathbf{a}, a_0, \mathbf{u})$
begin
    store initial basis $\mathbf{B}$;
    $\mathbf{RB} = L^3(\mathbf{B})$;
    extract $\mathbf{RB'}$ from $\mathbf{RB}$;
    if $0 \leq \mathbf{x}_d \leq \mathbf{u}$ **then** return $\mathbf{x}_d$;
    heuristic$(\mathbf{RB})$;
    if heuristic fails **then**
        branch on linear combinations of columns $j = 1, \ldots, n-1$ of the submatrix $\mathbf{X}$;
    return feasible vector $\mathbf{x}$, or a proof that no such vector exists;
end

Fig. 3. Algorithm 1.

In our example we can see that by subtracting the third column of $\mathbf{X}$, $\mathbf{X}_3 = (1, -2, -25, -1, -72)^T$, from $\mathbf{x}_d = (36, 17, 39, 8, -22)^T$, we obtain a vector $\mathbf{x} = (35, 19, 64, 9, 50)^T$ that satisfies the equation as well as all the bounds.

5   Computational Experience

We solved thirteen instances of problem (1). Eight of the instances were feasible and five infeasible. The instances starting with "P" in Table 1 were obtained from Philips Research Labs. The instances starting with "F" are the Frobenius instances of Cornuéjols et al. [4]. Here we used the Frobenius number as right-hand side a_0. The two other instances, starting with "E", were derived from F3 and F4. The information in Table 1 is interpreted as follows. In the first two columns, "Instance" and "n", the instance names and the dimension of the instances are given. An "F" in column "Type" means that the instance is feasible, and an "N" that it is not feasible. In the two columns of LP-based branch-and-bound, "LP B&B", the number of nodes and the computing time are given. In the "# Nodes" column, $500,000^*$ means that we terminated the search after 500,000 nodes without reaching a result. Two asterisks after the number of nodes indicate that a rounding error occurred, i.e., that the rounded solution given by the algorithm did not satisfy the diophantine equation. In both

Table 1. Results of the computational experiments.

Instance	n	Type	LP B&B		Algorithm		
			# Nodes	Time (s)	Heur.	# Nodes	Time (s)
P1	5	F	420**	–	1		$< 10^{-5}$
P2	5	F	327	0.09	1		$< 10^{-5}$
P3	4	F	75	0.05	0		$< 10^{-5}$
P4	5	F	313**	–	1		$< 10^{-5}$
P5	5	F	231	0.11	2		$< 10^{-5}$
P6	5	F	313**	–	1		$< 10^{-5}$
E1	6	F	3,271	0.97	0		$< 10^{-5}$
E2	7	F	500,000*	–	0		$< 10^{-5}$
F1	5	N	500,000*	–	–	1	$< 10^{-3}$
F2	6	N	500,000*	–	–	5	0.01
F3	6	N	500,000*	–	–	1	$< 10^{-3}$
F4	7	N	500,000*	–	–	1	0.01
F5	8	N	500,000*	–	–	5	0.01

cases we do not report on the computing times since no result was obtained. In the three columns corresponding to our algorithm, "Algorithm", the column "Heur." gives the number of vectors belonging to N(**a**) that was used in the integer linear combination of vectors added to the vector $\mathbf{x}_d$ by the heuristic in order to obtain a feasible solution. Notice that for every feasible instance, the heuristic found a feasible solution. A zero in column "Heur." therefore means that the vector $\mathbf{x}_d$ was feasible. For the infeasible instances the heuristic obviously failed, and therefore the sign "–" is given in the column. In that case we turn to the branching phase. Here, a one in the column "# Nodes" means that we solved the problem in the root node by using logical implications. The computing times are given in seconds on a 144MHz Sun Ultra-1. For the LP-based branch-and-bound we used CPLEX version 4.0.9 [6], and in our algorithm we used LiDIA, a library for computational number theory [12], for computing the reduced basis.

Our results indicate that the instances are rather trivial once they are represented in a good way. Using the basis $\mathbf{e}_j$ and branching on variables as in LP-based branch-and-bound is clearly not a good approach here, but it is the standard way of tackling integer programs. Using basis reduction seems to give a more natural representation of the problem. For our instances the computing times were very short, and, contrary to LP-based branch-and-bound, we avoid round-off errors. It is also worth noticing that the infeasibility of instances F1–F5 was particularly quickly verified using our algorithm.

References

1. K. Aardal, A. K. Lenstra, and C. A. J. Hurkens. An algorithm for solving a diophantine equation with upper and lower bounds on the variables. Report UU-CS-97-40, Department of Computer Science, Utrecht University, 1997. ftp://ftp.cs.ruu.nl/pub/RUU/CS/techreps/CS-1997/

2. W. Cook, T. Rutherford, H. E. Scarf, and D. Shallcross. An implementation of the generalized basis reduction algorithm for integer programming. *ORSA Journal on Computing*, 5:206–212, 1993.

3. D. Coppersmith. Small solutions to polynomial equations, and low exponent RSA vulnerability. *Journal of Cryptology*, 10:233–260, 1997.

4. G. Cornuéjols, R. Urbaniak, R. Weismantel, and L. Wolsey. Decomposition of integer programs and of generating sets. In R. Burkard and G. Woeginger, editors, *Algorithms – ESA '97, LNCS, Vol. 1284*, pages 92–103. Springer-Verlag, 1997.

5. M. J. Coster, A. Joux, B. A. LaMacchia, A. M. Odlyzko, and C. P. Schnorr. Improved low-density subset sum algorithms. *Computational Complexity*, 2:111–128, 1992.

6. CPLEX Optimization Inc. *Using the CPLEX Callable Library*, 1989.

7. B. de Fluiter. *A Complexity Catalogue of High-Level Synthesis Problems*. Master's thesis, Department of Mathematics and Computing Science, Eindhoven University of Technology, 1993.

8. G. Havas, B. S. Majewski, and K. R. Matthews. Extended gcd and Hermite normal form algorithms via lattice basis reduction. Working paper, Department of Mathematics, The University of Queensland, Australia, 1996.

9. J. C. Lagarias and A. M. Odlyzko. Solving low-density subset sum problems. *Journal of the Association for Computing Machinery*, 32:229–246, 1985.

10. A. K. Lenstra, H. W. Lenstra, Jr., and L. Lovász. Factoring polynomials with rational coefficients. *Mathematische Annalen*, 261:515–534, 1982.

11. H. W. Lenstra, Jr. Integer programming with a fixed number of variables. *Mathematics of Operations Research*, 8:538–548, 1983.

12. LiDIA – A library for computational number theory. TH Darmstadt / Universität des Saarlandes, Fachbereich Informatik, Institut für Theoretische Informatik. http://www.informatik.th-darmstadt.de/pub/TI/LiDIA

13. L. Lovász and H. E. Scarf. The generalized basis reduction algorithm. *Mathematics of Operations Research*, 17:751–764, 1992.

14. M. C. McFarland, A. C. Parker, and R. Camposano. The high-level synthesis of digital systems. *Proceedings of the IEEE, Vol. 78*, pages 301–318, 1990.

15. C. P. Schnorr and M. Euchner. Lattice basis reduction: Improved practical algorithms and solving subset sum problems. *Mathematical Programming*, 66:181–199, 1994.

16. C. P. Schnorr and H. H. Hörner. Attacking the Chor-Rivest Cryptosystem by improved lattice reduction. In L. C. Guillou and J.-J. Quisquater, editors, *Advances in Cryptology – EUROCRYPT '95, LNCS, Vol. 921*, pages 1–12. Springer Verlag, 1995.

17. W. F. J. Verhaegh, P. E. R. Lippens, E. H. L. Aarts, J. H. M. Korst, J. L. van Meerbergen, and A. van der Werf. Modeling periodicity by PHIDEO steams. *Proceedings of the Sixth International Workshop on High-Level Synthesis*, pages 256–266. ACM/SIGDA, IEEE/DATC, 1992.

The Intersection of Knapsack Polyhedra and Extensions

Alexander Martin and Robert Weismantel *

Konrad-Zuse-Zentrum Berlin
Takustr. 7
D-14195 Berlin, Germany
{martin, weismantel}@@zib.de

Abstract. This paper introduces a scheme of deriving strong cutting planes for a general integer programming problem. The scheme is related to Chvátal-Gomory cutting planes and important special cases such as odd hole and clique inequalities for the stable set polyhedron or families of inequalities for the knapsack polyhedron. We analyze how relations between covering and incomparability numbers associated with the matrix can be used to bound coefficients in these inequalities. For the intersection of several knapsack polyhedra, incomparabilities between the column vectors of the associated matrix will be shown to transfer into inequalities of the associated polyhedron. Our scheme has been incorporated into the mixed integer programming code SIP. About experimental results will be reported.

1 Introduction

LP-based branch-and-bound algorithms are currently the most important tool to deal with real-world general mixed integer programming problems computationally. Usually, the LP relaxations that occur during the execution of such methods are strengthened by cutting planes. Cutting planes for integer programs may be classified with regard to the question whether their derivation requires knowledge about the structure of the underlying constraint matrix. Examples of families of cutting planes that do not exploit the structure of the constraint matrix are Chvátal-Gomory cuts [6], [4], [12] or lift-and-project cuts [1]. An alternative approach to obtain cutting planes for an integer program follows essentially the scheme to derive relaxations associated with certain substructures of the underlying constraint matrix, and tries to find valid inequalities for these relaxations. Crowder, Johnson and Padberg [5] applied this methodology by interpreting each single row of the constraint matrix as a knapsack relaxation and strengthened the integer program by adding violated knapsack inequalities. An

---

* Supported by a "Gerhard Hess-Forschungsförderpreis" of the German Science Foundation (DFG).

R. E. Bixby, E. A. Boyd, and R. Z. Ríos-Mercado (Eds.): IPCO VI
LNCS 1412, pp. 243–256, 1998. © Springer–Verlag Berlin Heidelberg 1998

analysis of other important relaxations of an integer program allows to incorporate odd hole and clique inequalities for the stable set polyhedron [8] or flow cover inequalities for certain mixed integer models [11]. Further recent examples of this second approach are given in [3], [7].

Our paper follows the methodology to obtain cutting planes for an integer program by investigating relaxations of it. We try to go one step further and investigate the intersection of two or more knapsack polyhedra. We describe a general family of valid inequalities for an integer program that are associated with its feasible solutions. Usually such inequalities must be lifted in order to induce high dimensional faces. We derive lower and upper bounds on the exact lifting coefficients of such an inequality and discuss special cases when these bounds can be computed in polynomial time. We also relate our family of inequalities to Chvátal-Gomory cuts and discuss in detail the special case where only two knapsacks are involved. The use of feasible set inequalities within an implementation for the solution of general mixed integer programming problems is investigated.

Consider some matrix $A \in \mathbb{R}^{m \times n}$, vectors $b \in \mathbb{R}^m$, $u \in \mathbb{R}^n$, and the polytope

$$P := \text{conv}\{x \in \mathbb{Z}^n : Ax \leq b, 0 \leq x \leq u\},$$

that is the convex hull of all integral vectors x satisfying $Ax \leq b$ and $0 \leq x \leq u$. Set $N := \{1, \ldots, n\}$. For $S \subseteq N$, define $P_S := P \cap \{x \in \mathbb{R}^n : x_i = 0 \text{ for } i \in N \setminus S\}$, where $P_N = P$, and, for $x \in \mathbb{R}^n$, denote by $x|_S := (x_i)_{i \in S}$ the vector restricted to the components in S.

Definition 1. *Let $T \subseteq N$ such that $\sum_{i \in T} A_{.i} v_i \leq b$ for all $v \in \mathbb{R}^T$ with $v \leq u|_T$. T is called a* feasible set. *Let $w : T \mapsto \mathbb{Z}$ be some* weighting *of the elements of T. For $j \in N \setminus T$ with $\sum_{i \in T} A_{.i} u_i + A_{.j} u_j \not\leq b$, the inequality*

$$\sum_{i \in T} w_i x_i + w_j x_j \leq \sum_{i \in T} w_i u_i \qquad (1)$$

is called a feasible set inequality *associated with T (and $\{j\}$) if*

$$w_j \leq \min_{l=1,\ldots,u_j} \frac{1}{l} \min_x \sum_{i \in T} w_i x_i$$

$$\sum_{i \in T} A_{.i} x_i \geq A_{.j} - r(T) \qquad (2)$$

$$0 \leq x_i \leq u_i, x_i \in \mathbb{Z}, i \in T,$$

where $r(T) := b - \sum_{i \in T} A_{.i} u_i$.

Theorem 2. *Feasible set inequalities are valid for $P_{T \cup \{j\}}$.*

Proof. Let $\gamma = \sum_{i \in T} w_i u_i$ and $\gamma_l := \max\{\sum_{i \in T} w_i x_i : \sum_{i \in T} A_{.i} x_i + A_{.j} l \leq b, 0 \leq x_i \leq u_i, x_i \in \mathbb{Z}, i \in T\}$. After complementing variables x_i to $u_i - x_i$ for $i \in T$ we obtain that the right hand side of (2) is $\min_{l=1,\ldots,u_j} \frac{1}{l}(\gamma - \gamma_l)$. For some feasible solution $\bar{x} \in P$ we have $\sum_{i \in T} w_i \bar{x}_i + w_j \bar{x}_j \leq \gamma_{\bar{x}_j} + \bar{x}_j \min_{l=1,\ldots,u_j} \frac{1}{l}(\gamma - \gamma_l) \leq \gamma_{\bar{x}_j} + (\gamma - \gamma_{\bar{x}_j}) = \gamma$. $\square$

Examples of feasible set inequalities include $(1, k)$-configuration and minimal-cover inequalities that are known for the knapsack polytope $K := \text{conv}\{x \in \{0,1\}^n : a^T x \leq \alpha\}$ with $a \in \mathbb{R}_+^n, \alpha > 0$. Let $S \subseteq N$ be a minimal cover, i.e., $a(S) > \alpha$ and $a(S \setminus \{i\}) \leq \alpha$ for all $i \in S$, and partition S into T and $\{j\}$ for some $j \in S$. Set $w_i := 1$ for all $i \in T$. The feasible set inequality reads $\sum_{i \in T} x_i + w_j x_j \leq |T| = |S| - 1$ with $w_j \leq \min\{|V| : V \subseteq T, \sum_{i \in V} a_i \geq a_j - r(T)\}$. Since $\sum_{i \in T} a_i + r(T) = \alpha$ and $\sum_{i \in S} a_i > \alpha$, this minimum is greater than or equal to one. Therefore, under the regularity condition imposed in [10], the feasible set inequality is always a $(1, k)$-configuration inequality. In case the coefficient happens to be one we get a minimal cover inequality, see, for instance, [13].

Theorem 2 states the validity of the feasible set inequality for $P_{T \cup \{j\}}$. To obtain a (strong) valid inequality for P we resort to *lifting*, see [9]. Consider some permutation

$\pi_1, \ldots, \pi_{n-|T|-1}$ of the set $N \setminus (T \cup \{j\})$. For $k = 1, \ldots, n - |T| - 1$ and $l = 1, \ldots, u_{\pi_k}$ let

$$
\gamma(k, l) = \max \sum_{i \in T \cup \{j\}} w_i x_i + \sum_{i \in \{\pi_1, \ldots, \pi_{k-1}\}} w_i x_i
$$
$$
\sum_{i \in T \cup \{j\}} A_{\cdot i} x_i + \sum_{i \in \{\pi_1, \ldots, \pi_{k-1}\}} A_{\cdot i} x_i + A_{\cdot \pi_k} l \leq b \qquad (3)
$$
$$
0 \leq x_i \leq u_i, x_i \in \mathbb{Z} \text{ for } i \in T \cup \{\pi_1, \ldots, \pi_{k-1}\}.
$$

Let $\gamma = \sum_{i \in T} w_i u_i$, the lifting coefficients are

$$
w_{\pi_k} := \min_{l=1,\ldots,u_{\pi_k}} \frac{\gamma - \gamma(k, l)}{l}. \qquad (4)
$$

The following statement is immediate.

Theorem 3. *The (lifted) feasible set inequality $w^T x \leq \sum_{i \in T} w_i u_i$ is valid for P.*

Note that the right hand side of (2) coincides with (4) applied to variable j if we substitute in (3) the set $T \cup \{j\}$ by T. In other words, a lifted feasible set inequality associated with T and $\{j\}$, where the variables in $N \setminus (T \cup \{j\})$ are lifted according to the sequence $\pi_1, \ldots, \pi_{n-|T|-1}$, coincides with the inequality associated with T, where j is lifted first, and the remaining variables $N \setminus (T \cup \{j\})$ are lifted in the same order $\pi_1, \ldots, \pi_{n-|T|-1}$. Thus, instead of speaking of a feasible set inequality associated with T and $\{j\}$, we speak in the sequel of a feasible set inequality associated with T and view j as the variable that is lifted first.

Odd hole- and clique inequalities for the set packing polytope are examples of lifted feasible set inequalities. Consider the set packing polytope $P = \text{conv}\{x \in \{0,1\}^n : Ax \leq \mathbb{1}\}$ for some 0/1 matrix $A \in \{0,1\}^{m \times n}$. Let $G_A = (V, E)$ denote the associated column intersection graph whose nodes correspond to the columns of A and nodes i and j are adjacent if and only if the columns associated with i and j intersect in some row. Let $Q \subseteq V$ be a clique in G_A, then the clique inequality $\sum_{i \in Q} x_i \leq 1$ is valid for P. To see that this inequality is a lifted feasible set inequality, let $T = \{i\}$ for some $i \in Q$. The feasible set inequality $x_i \leq 1$ is valid for $P_{\{i\}}$. Lifting the remaining variables $k \in Q \setminus \{i\}$ by applying formula (4) yields $w_k = 1$, and the clique inequality follows.

2 Bounds on the Lifting Coefficients

For a feasible set inequality associated with T, the calculation of the lifting coefficients for the variables in $N \setminus T$ requires the solution of an integer program. In this section we study lower and upper bounds for these coefficients. It will turn out that these bounds are sometimes easier to compute. We assume throughout the section that $A \geq 0$ and $w_i \geq 0$ for $i \in T$.

Definition 4. *Let $T \subseteq N$ be a feasible set and $w : T \mapsto \mathbb{R}_+^T$ a weighting of T. For $v \in \mathbb{R}^m$ we define the*

Covering Number

$$\phi^{\geq}(v) := \min \{\sum_{i \in T} w_i x_i : \sum_{i \in T} A_{\cdot i} x_i \geq v, \, 0 \leq x_i \leq u_i, x_i \in \mathbb{Z}, i \in T\},$$

$\geq$-Incomparability Number

$$\phi^{\not\geq}(v) := \min \{\sum_{i \in T} w_i x_i : \sum_{i \in T} A_{\cdot i} x_i \not\geq v, \, 0 \leq x_i \leq u_i, x_i \in \mathbb{Z}, i \in T,$$
$$\wedge \exists j \in T, x_j < u_j : \sum_{i \in T} A_{\cdot i} x_i + A_{\cdot j} \geq v\},$$

$\leq$-Incomparability Number

$$\phi^{\not\leq}(v) := \min \{\sum_{i \in T} w_i x_i : \sum_{i \in T} A_{\cdot i} x_i \not\leq v, \, 0 \leq x_i \leq u_i, x_i \in \mathbb{Z}, i \in T\},$$

where we set $\phi^{\not\geq}(v) := 0$ and $\phi^{\not\leq}(v) := 0$ for $v \leq 0$.

Consider a (lifted) feasible set inequality $w^T x \leq \sum_{i \in T} w_i u_i$ associated with T, where the variables in $N \setminus T$ are lifted in the sequence $\pi_1, \ldots, \pi_{n-|T|}$. The following proposition gives upper bounds for the lifting coefficients derived from the covering number.

Proposition 5.

(a) $w_{\pi_1} = \min_{l=1,\ldots,u_{\pi_1}} \frac{1}{l} \phi^{\geq}(A_{\cdot \pi_1} l - r(T))$.
(b) $w_{\pi_k} \leq \min_{l=1,\ldots,u_{\pi_k}} \frac{1}{l} \phi^{\geq}(A_{\cdot \pi_k} l - r(T))$, for $k = 2, \ldots, n - |T|$.

Proof. (a) directly follows from Theorem 2. To see (b), it suffices to show that $\gamma - \gamma(k, l) \leq \phi^{\geq}(A_{.\pi_k} l - r(T))$ for $l = 1, \ldots, u_{\pi_k}$, see (4). This relation is obtained by

$$
\begin{aligned}
\gamma - \gamma(k, l) &= \gamma - \max \; \begin{array}{l} \sum_{i \in T} w_i x_i + \sum_{i \in \{\pi_1, \ldots, \pi_{k-1}\}} w_i x_i \\ \sum_{i \in T} A_{.i} x_i + \sum_{i \in \{\pi_1, \ldots, \pi_{k-1}\}} A_{.i} x_i + A_{.\pi_k} l \leq \gamma, \\ 0 \leq x \leq u|_{T \cup \{\pi_1, \ldots, \pi_{k-1}\}}, x \in \mathbb{Z}^{T \cup \{\pi_1, \ldots, \pi_{k-1}\}} \end{array} \\[2mm]
&= \min \; \begin{array}{l} \sum_{i \in T} w_i x_i - \sum_{i \in \{\pi_1, \ldots, \pi_{k-1}\}} w_i x_i \\ \sum_{i \in T} A_{.i} x_i - \sum_{i \in \{\pi_1, \ldots, \pi_{k-1}\}} A_{.i} x_i \geq A_{.\pi_k} l - r(T), \\ 0 \leq x \leq u|_{T \cup \{\pi_1, \ldots, \pi_{k-1}\}}, x \in \mathbb{Z}^{T \cup \{\pi_1, \ldots, \pi_{k-1}\}} \end{array} \\[2mm]
&\leq \min \; \begin{array}{l} \sum_{i \in T} w_i x_i \\ \sum_{i \in T} A_{.i} x_i \geq A_{.\pi_k} l - r(T), \\ 0 \leq x \leq u|_T, x \in \mathbb{Z}^T \end{array} \\[2mm]
&= \phi^{\geq}(A_{.\pi_k} l - r(T)),
\end{aligned}
$$

where the second equation follows by complementing variables $x_i, i \in T$. □

To derive lower bounds on the lifting coefficients we need the following relations.

Lemma 6. *For $v_1, v_2 \in \mathbb{R}^m$ with $v_1, v_2 \geq 0$ holds:*

(a) $\phi^{\geq}(v_1) \geq \phi^{\geqq}(v_1)$ and $\phi^{\geq}(v_1) \geq \phi^{\lessgtr}(v_1)$.
(b) $\phi^{\geq}, \phi^{\geqq}$, and $\phi^{\lessgtr}$ are monotonically increasing, that is for $v_1 \geq v_2$, $\phi^{\geq}(v_1) \geq \phi^{\geq}(v_2)$, $\phi^{\geqq}(v_1) \geq \phi^{\geqq}(v_2)$, and $\phi^{\lessgtr}(v_1) \geq \phi^{\lessgtr}(v_2)$.
(c) $\phi^{\geq}(v_1 + v_2) \geq \phi^{\geqq}(v_1) + \phi^{\lessgtr}(v_2)$.
(d) $\phi^{\geq}(v_1 + v_2) + \max\{w_i : i \in T\} \geq \phi^{\lessgtr}(v_1) + \phi^{\geq}(v_2)$.
(e) $\phi^{\lessgtr}(v_1 + v_2) + \max\{w_i : i \in T\} \geq \phi^{\lessgtr}(v_1) + \phi^{\lessgtr}(v_2)$.

Proof. (a) and (b) are obvious, the proofs of (c) and (e) follow the same line. We show exemplarily (c). Let $\bar{x} \in \mathbb{R}^T$ with $0 \leq \bar{x} \leq u|_T$ and $\sum_{i \in T} A_{.i} \bar{x}_i \geq v_1 + v_2$ such that $\sum_{i \in T} w_i \bar{x}_i = \phi^{\geq}(v_1 + v_2)$. If $v_1 = v_2 = 0$ the statement is trivial. Otherwise, suppose w.l.o.g. $v_1 > 0$. Let $z \in \mathbb{R}^T$ with $z \leq \bar{x}$ such that $\sum_{i \in T} A_{.i} z_i \geq v_1$ and $\sum_{i \in T} w_i z_i$ is minimal. Since $v_1 > 0$ there exists some $i_0 \in T$ with $z_{i_0} > 0$, and since z was chosen to be minimal, we have that $\sum_{i \in T} A_{.i} z_i - A_{.i_0} \not\geq v_1$. This implies that $\sum_{i \in T} A_{.i} (\bar{x}_i - z_i) + A_{.i_0} \not\geq v_2$. Summing up, we get $\phi^{\geq}(v_1 + v_2) = \sum_{i \in T} w_i \bar{x}_i = (\sum_{i \in T} A_{.i} z_i - A_{.i_0}) + (\sum_{i \in T} A_{.i} (\bar{x}_i - z_i) + A_{.i_0}) \geq \phi^{\geqq}(v_1) + \phi^{\lessgtr}(v_2)$. Finally, (d) directly follows from (c) and the fact that $\phi^{\geqq}(v_1) + \max\{w_i : i \in T\} \geq \phi^{\geq}(v_1)$. □

With the help of Lemma 6 we are able to bound the lifting coefficients from below.

Theorem 7. *For $k = 1, \ldots, n - |T|$ we have*

$$
w_{\pi_k} \geq \min_{l=1, \ldots, u_{\pi_k}} \frac{\phi^{\lessgtr}(A_{.\pi_k} l - r(T))}{l} - \max\{w_i : i \in T\}. \tag{5}
$$

Proof. Let $c_{\pi_k} := \min_{l=1,\ldots,u_{\pi_k}} \frac{\phi^{\leq}(A._{\pi_k}l - r(T))}{l} - \max\{w_i : i \in T\}$ denote the right hand side of (5), for $k = 1, \ldots, n - |T|$. We show by induction on k that the inequality $\sum_{i \in T} w_i x_i + \sum_{i=1}^{k} c_{\pi_i} x_i \leq \sum_{i \in T} w_i u_i$ is valid. For $k = 1$, the statement follows from Proposition 5 (a) and Lemma 6 (a). Now let $k \geq 2$ and suppose the statement is true for all $l < k$. Let $\bar{x}$ be an optimal solution of

$$\max \sum_{i \in T} w_i x_i + \sum_{i=1}^{k} c_{\pi_i} x_{\pi_i}$$

$$\sum_{i \in T} A._i x_i + \sum_{i=1}^{k} A._{\pi_i} x_{\pi_i} \leq b$$

$$0 \leq x_i \leq u_i, x_i \in \mathbb{Z} \text{ for } i \in T \cup \{\pi_1, \ldots, \pi_k\}.$$

We must show that $\sum_{i \in T} w_i \bar{x}_i + \sum_{i=1}^{k} c_{\pi_i} \bar{x}_{\pi_i} \leq \sum_{i \in T} w_i u_i$. First note that the inequality is valid if $\bar{x}_{\pi_k} = 0$. This is equivalent to saying

$$\phi^{\geq}\left(\sum_{i=1}^{k-1} A._{\pi_i} x_{\pi_i} - r(T)\right) \geq \sum_{i=1}^{k-1} c_{\pi_i} x_{\pi_i},$$

for all $x \in \mathbb{Z}^{\{\pi_1,\ldots,\pi_{k-1}\}}$ with $\sum_{i=1}^{k-1} A._{\pi_i} x_{\pi_i} \leq b, 0 \leq x_{\pi_i} \leq u_{\pi_i}, i = 1, \ldots, k-1$. Applying Lemma 6 (d) and (b) we obtain with $w_{max} := \max\{w_i : i \in T\}$

$$\phi^{\geq}\left(\sum_{i=1}^{k} A._{\pi_i} \bar{x}_{\pi_i} - r(T)\right) \geq \phi^{\geq}\left(\sum_{i=1}^{k-1}\left(A._{\pi_i} \bar{x}_{\pi_i} - r(T)\right)\right) + \phi^{\leq}(A._{\pi_k} \bar{x}_{\pi_k}) - w_{max}$$

$$\geq \sum_{i=1}^{k-1} c_{\pi_i} \bar{x}_{\pi_i} + \phi^{\leq}(A._{\pi_k} \bar{x}_{\pi_k} - r(T)) - w_{max}$$

$$\geq \sum_{i=1}^{k-1} c_{\pi_i} \bar{x}_{\pi_i}$$

$$+ \bar{x}_{\pi_k} \min_{l=1,\ldots,u_k} \frac{\phi^{\leq}(A._{\pi_k}l - r(T))}{l} - w_{max}$$

$$= \sum_{i=1}^{k} c_{\pi_i} \bar{x}_{\pi_i}.$$

On account of $\sum_{i \in T} w_i(u_i - \bar{x}_i) \geq \phi^{\geq}(\sum_{i=1}^{k} A._{\pi_i} \bar{x}_{\pi_i} - r(T))$ the statement follows. □

Theorem 7 applies, in particular, if we set the coefficient of the first lifted variable w_{π_1} to the upper bound of Proposition 5 (a). The subsequent example shows that in this case the lower bounds given in Theorem 7 may be tight.

Example 8. Let $A = \begin{bmatrix} 1 & 1 & 6 & 6 & 8 & 6 & 1 & 3 \\ 4 & 6 & 1 & 1 & 3 & 9 & 2 & 1 \end{bmatrix}$ and $b = \begin{bmatrix} 14 \\ 12 \end{bmatrix}$. The set $T = \{1, 2, 3, 4\}$ is feasible for the 0/1 program $\max\{c^T x : x \in P\}$ with $P := \text{conv}\{x \in \{0,1\}^8 :$

$Ax \leq b\}$. For $w_i = 1$, $i \in T$, we obtain $\phi^{\geq}\binom{8}{3} = 3$. Moreover, $\phi^{\lessgtr}\binom{6}{9} = 2$, because $\binom{6}{9} \geq A_{\cdot i}$ for $i \in T$. Accordingly we get $\phi^{\lessgtr}\binom{1}{2} = 1 = \phi^{\lessgtr}\binom{3}{1}$. The inequality $x_1 + x_2 + x_3 + x_4 + \phi^{\geq}\binom{8}{3}x_5 + \sum_{i=6}^{8}(\phi^{\lessgtr}(A_{\cdot i}) - 1)x_i \leq 4$ reads $x_1 + x_2 + x_3 + x_4 + 3x_5 + x_6 \leq 4$. It defines a facet of P.

The question remains, whether the values $\phi^{\geq}, \phi^{\lessgtr}$ and $\phi^{\lessgtr}$ are easier to compute than the exact lifting coefficient. Indeed, they sometimes are. Suppose $w_i = 1$ for all $i \in T$ and consider the *comparability digraph* $G = (V, E)$ that is obtained by introducing a node for each column and arcs (i, j) if $A_{\cdot i} \geq A_{\cdot j}$ and $A_{\cdot i} \neq A_{\cdot j}$ or if $A_{\cdot i} = A_{\cdot j}$ and $i > j$, for $i, j \in \{1, \ldots, n\}$ (where transitive arcs may be deleted). Let r denote the number of nodes with indegree zero, i.e., $\delta^{-}(i) = 0$. Then, $\phi^{\geq}, \phi^{\lessgtr}$ and $\phi^{\lessgtr}$ can be computed in time $O(n^r + \alpha)$, where α is the time to construct the comparability digraph. For example, in case of one knapsack inequality the comparability digraph turns out to be a path, and thus $\phi^{\geq}, \phi^{\lessgtr}$ and $\phi^{\lessgtr}$ can be computed in time $O(n + n \log n) = O(n \log n)$.

3   Connection to Chvátal-Gomory Cuts

So far we have been discussing feasible set inequalities for general integer programs. With the exception that with $u|_T$ also every vector $v \leq u|_T$ is valid for P we have not subsumed any assumptions on the matrix A. Thus a comparison to Chvátal-Gomory cutting planes that do not rely on any particular structure of A is natural. Recall that Chvátal-Gomory inequalities for the system $Ax \leq b, 0 \leq x \leq u, x \in \mathbb{Z}^n$ are cutting planes $d^T x \leq \delta$ such that $d_i = \lfloor \lambda^T \hat{A}_{\cdot i} \rfloor$, $i = 1, \ldots, n$, and

$$\delta = \lfloor \lambda^T \hat{b} \rfloor \text{ for some } \lambda \in \mathbb{R}_+^{m+n}, \text{ where } \hat{A} = \begin{bmatrix} A \\ I \end{bmatrix} \text{ and } \hat{b} = \begin{bmatrix} b \\ u \end{bmatrix}.$$

Consider a (lifted) feasible set inequality $w^T x \leq \sum_{i \in T} w_i u_i$ associated with T, whose remaining variables $N \setminus T$ are lifted in the sequence $\pi_1, \ldots, \pi_{n-|T|}$. This lifted feasible set inequality is compared to Chvátal-Gomory inequalities resulting from multipliers $\lambda \in \mathbb{R}_+^{m+n}$ that satisfy $\lfloor \lambda^T \hat{A}_{\cdot i} \rfloor = w_i$ for $i \in T$.

Proposition 9.

(a) $\lfloor \lambda^T \hat{b} \rfloor \geq \sum_{i \in T} u_i w_i$.
(b) If $\lfloor \lambda^T \hat{b} \rfloor = \sum_{i \in T} u_i w_i$, *let j be the smallest index with* $\lfloor \lambda^T \hat{A}_{\cdot \pi_j} \rfloor \neq w_{\pi_j}$. *Then,* $\lfloor \lambda^T \hat{A}_{\cdot \pi_j} \rfloor < w_{\pi_j}$.

Proof. Since T is a feasible set, (a) is obviously true. To see (b) suppose the contrary and let j be the first index with $\lfloor \lambda^T \hat{A}_{\cdot \pi_j} \rfloor \neq w_{\pi_j}$ and $\lfloor \lambda^T \hat{A}_{\cdot \pi_j} \rfloor > w_{\pi_j}$. Set $\gamma = \sum_{i \in T} u_i w_i$ and consider an optimal solution $\bar{x} \in \mathbb{Z}^{T \cup \{\pi_1, \ldots, \pi_j\}}$ of (3) such that $w_{\pi_j} = \frac{\gamma - \gamma(j, \bar{x}_j)}{\bar{x}_j}$. Obviously, $\bar{x}$ can be extended to a feasible solution $\tilde{x}$ of P by setting $\tilde{x}_i = \bar{x}_i$, if $i \in T \cup \{\pi_1, \ldots, \pi_j\}$, $\tilde{x}_i = 0$, otherwise.

This solution satisfies the feasible set inequality with equality, since $w^T \tilde{x} = \sum_{i \in T \cup \{\pi_1, \ldots, \pi_{j-1}\}} w_i \tilde{x}_i + w_{\pi_j} \tilde{x}_j = \gamma(j, \tilde{x}_j) + \frac{\gamma - \gamma(j, \tilde{x}_j)}{\tilde{x}_j} \tilde{x}_j = \gamma$. On the other hand, j is the first index where the Chvátal-Gomory and the feasible set coefficient differ and we conclude $\sum_{i \in N} \lfloor \lambda^T \hat{A}_{\cdot i} \rfloor \tilde{x}_i = \sum_{i \in T \cup \{\pi_1, \ldots, \pi_{j-1}\}} \lfloor \lambda^T \hat{A}_{\cdot i} \rfloor \tilde{x}_i + \lfloor \lambda^T \hat{A}_{\cdot \pi_j} \rfloor \tilde{x}_j > \sum_{i \in T \cup \{\pi_1, \ldots, \pi_{j-1}\}} w_i \tilde{x}_i + w_{\pi_j} \tilde{x}_j = \gamma = \lfloor \lambda^T \hat{b} \rfloor$, contradicting the validity of the Chvátal-Gomory inequality.   □

As soon as the first two coefficients differ, for $k \in \{\pi_{j+1}, \ldots, \pi_{n-|T|}\}$, no further statements on the relations of the coefficients are possible, in general.

Example 10. For $b \in \mathbb{N}$, let $P(b)$ be the convex hull of all 0/1 solutions that satisfy the knapsack inequality

$$2x_1 + 6x_2 + 8x_3 + 9x_4 + 9x_5 + 21x_6 + 4x_7 \le b.$$

One Chvátal-Gomory cutting plane for $P(b)$ reads

$$x_1 + 3x_2 + 4x_3 + 4x_4 + 4x_5 + 10x_6 + 2x_7 \le \lfloor b/2 \rfloor.$$

Let $b = 25$. The set $T := \{1, 2, 3, 4\}$ is a feasible set. Choosing coefficients $1, 3, 4, 4$ for the items in T and lifting the items $5, 6, 7$ in this order we obtain the lifted feasible set inequality that is valid for $P(25)$:

$$x_1 + 3x_2 + 4x_3 + 4x_4 + 4x_5 + 11x_6 + x_7 \le 12.$$

The right hand side of the Chvátal-Gomory cutting plane and the lifted feasible set inequality coincide. With respect to the lifting order $5, 6, 7$ the coefficient of item 6 is the first one in which the two inequalities differ. This coefficient is 11 in the feasible set inequality and 10 in the Chvátal-Gomory cutting plane. For item 7 the coefficient in the feasible set inequality is then smaller than the corresponding one in the Chvátal-Gomory cutting plane. For $b = 27$ we obtain the lifted feasible set inequality that is valid for $P(27)$:

$$x_1 + 3x_2 + 4x_3 + 4x_4 + 4x_5 + 9x_6 + x_7 \le 12.$$

The right hand side of this inequality is by one smaller than the right hand side of the corresponding Chvátal-Gomory cutting plane for $\lambda = \frac{1}{2}$. However, the coefficients of the items 6 and 7 are smaller than the corresponding coefficients of the Chvátal-Gomory cutting plane.

Under certain conditions a feasible set- and a Chvátal-Gomory cutting plane coincide.

Theorem 11. *Let $A \in \mathbb{N}^{m \times n}$ and T be a feasible set of the integer program* $\max \{c^T x : Ax \le b, \ 0 \le x \le \mathbb{1}, \ x \in \mathbb{Z}^n\}$. *Let $\lambda \in \mathbb{R}_+^m$ such that $\lfloor \lambda^T b \rfloor = \sum_{i \in T} \lfloor \lambda^T A_{\cdot i} \rfloor$. If, for all $j \in N \setminus T$, column vector $A_{\cdot j}$ is a 0/1-combination of elements from $\{A_{\cdot i} : i \in T\}$, then*

$$\lfloor \lambda^T A_{\cdot j} \rfloor = \phi^{\ge}(A_{\cdot j} - r(T)).$$

Proof. We first show that $\lfloor \lambda^T A_{\cdot j} \rfloor \geq \phi^\geq(A_{\cdot j} - r(T))$. Since $A_{\cdot j}$ is a 0/1-combination of elements from $\{A_{\cdot i} : i \in T\}$, there exist $\sigma \in \{0,1\}^T$ such that $\sum_{i \in T} \sigma_i A_{\cdot i} = A_{\cdot j}$. Thus, $\lfloor \lambda^T A_{\cdot j} \rfloor = \lfloor \lambda^T (\sum_{i \in T} \sigma_i A_{\cdot i}) \rfloor \geq \sum_{i \in T} \sigma_i \lfloor \lambda^T A_{\cdot i} \rfloor = \sum_{i \in T} \sigma_i w_i \geq \phi^\geq(A_{\cdot j} - r(T))$.

To show the opposite relation, we know by Proposition 5 that the coefficient of the feasible set inequality w_j satisfies $w_j \leq \phi^\geq(A_{\cdot j} - r(T))$. By Proposition 9 (b), however, we know that, if w_j does not coincide with $\lfloor \lambda^T A_{\cdot j} \rfloor$ for all $j \in N \setminus T$, there is at least one index j with $w_j > \lfloor \lambda^T A_{\cdot j} \rfloor$. This together with the first part of the proof implies $\phi^\geq(A_{\cdot j} - r(T)) \leq \lfloor \lambda^T A_{\cdot j} \rfloor < w_j \leq \phi^\geq(A_{\cdot j} - r(T))$, a contradiction. Thus, $w_j = \lfloor \lambda^T A_{\cdot j} \rfloor$ for all j, and the claim follows by Proposition 5. $\square$

By Proposition 5 the expression $\phi^\geq(A_{\cdot j} - r(T))$ is an upper bound on the exact coefficient of an item $j \in N \setminus T$ in any lifted feasible set inequality associated with the feasible set T and the weighting $\lfloor \lambda^T A_{\cdot i} \rfloor$, $i \in T$. On the other hand, the Chvátal-Gomory cutting plane $\sum_{j \in N} \lfloor \lambda^T A_{\cdot j} \rfloor \leq \lfloor \lambda^T b \rfloor$ is valid for P. Therefore, this Chvátal-Gomory cutting plane must coincide with any lifted feasible set inequality $\sum_{i \in T} \lfloor \lambda^T A_{\cdot i} \rfloor x_i + \sum_{j \in N \setminus T} w_j x_j \leq \lfloor \lambda^T b \rfloor$ independent on the sequence in which the lifting coefficients w_j for the items in $N \setminus T$ are computed.

4   Consecutively Intersecting Knapsacks

So far we have been discussing a framework with which one can define and explain families of cutting planes for a general integer program. On the other hand, from a practical point of view the cutting plane phase of usual codes for integer programming relies in particular on valid inequalities for knapsack polyhedra. From both a theoretical and a practical point of view it would be desirable to understand under what conditions facets of single knapsack polyhedra define or do not define strong cutting planes of an integer program when several knapsack constraints intersect. This question is addressed in this section. In fact, here we study a special family of 0/1 programs that

arises when $A \in \mathbb{N}^{m \times n}$ and $Ax \leq b$ defines a system of *consecutively intersecting knapsack constraints*. Throughout this section we assume $u = \mathbb{1}$, $A \geq 0$ and integral. For $i = 1, \ldots, m$, let $N_i := \mathrm{supp}(A_{i\cdot})$ and $P^i := \mathrm{conv}\{x \in \{0,1\}^{N_i} : \sum_{j \in N_i} A_{ij} x_j \leq b_i\}$.

Definition 12. *A system of linear inequalities $Ax \leq b$ is called a system of consecutively intersecting knapsack constraints if $A \in \mathbb{N}^{m \times n}$ and $N_i \cap N_l = \emptyset$ for all $i, l \in \{1, \ldots, m\}$, $|i - l| \geq 2$.*

A natural question when one starts investigating the intersection of several knapsack polyhedra is when this polyhedron inherits all the facets of the single knapsack polyhedra.

Proposition 13. *Let $A \in \mathbb{N}^{m \times n}$ and $Ax \leq b$ be a system of consecutively intersecting knapsack constraints. Let $i \in \{1, \ldots, m\}$. If $N_i \cap N_l \cup \{k\}$ is not a* *cover for all $l \in \{i-1, i+1\}$ and $k \notin N_i$, then every facet-defining inequality of the knapsack polyhedron P^i defines a facet of P.*

Proof. Suppose that $\sum_{j \in N_i} c_j x_j \leq \gamma$ defines a facet of P^i. Let $\sum_{j \in N} d_j x_j \leq \gamma$ define a facet F of P such that

$$F_c := \{x \in P : \sum_{j \in N_i} c_j x_j = \gamma\} \subseteq F.$$

Let $x^\circ \in \mathbb{Z}^n \cap F_c$ be a vector such that $x_j^\circ = 0$ for all $j \notin N_i$.

Consider some $j \notin N_i$, and let $j \in N_l$ for some $l \neq i$. If $|l - i| \geq 2$, obviously $x^\circ + e_j \in F_c$. In the other case, we know $N_i \cap N_l \cup \{j\}$ is not a cover (on account of the conditions in the proposition), and thus $x^\circ + e_j \in F_c$ as well. This implies that $d_j = 0$. Because $\sum_{j \in N_i} c_j x_j \leq \gamma$ defines a facet of P^i we obtain

$$\dim(F_c) \geq \dim(P^i) - 1 + |N| - |N_i| = |N| - 1 \geq \dim(P) - 1.$$

Therefore, F_c defines a facet of P that coincides with F.   □

The condition that, for every $k \notin N_i$, $l = i - 1, i + 1$, the set $(N_i \cap N_l) \cup \{k\}$ is not a cover, is essential for the correctness of the proposition as the following example shows.

Example 14. For $b \in \mathbb{N} \setminus \{0\}$ let A be the matrix $\begin{bmatrix} \frac{b}{3}+1 & \frac{b}{3}+1 & \frac{b}{3}+1 & 0 \\ 0 & \frac{b}{3}+1 & \frac{b}{3}+1 & b \end{bmatrix}$
and consider $P = \mathrm{conv}\,\{x \in \{0,1\}^4 : Ax \leq b\}$. Then $N_1 \cap N_2 = \{2, 3\}$, and the set $\{2, 3, 4\}$ defines a cover.

The inequality $x_1 + x_2 + x_3 \leq 2$ defines a facet of the knapsack polyhedron

$$\mathrm{conv}\,\{x \in \{0,1\}^3 : (\frac{b}{3}+1)x_1 + (\frac{b}{3}+1)x_2 + (\frac{b}{3}+1)x_3 \leq b\}.$$

On the other hand, the inequality $x_1 + x_2 + x_3 \leq 2$ is not facet-defining for P, since the face $F = \{x \in P : x_1 + x_2 + x_3 \leq 2\}$ is strictly contained in the face induced by the inequality $x_1 + x_2 + x_3 + x_4 \leq 2$.

In certain cases, a complete description of P may even be derived from the description of the single knapsack polyhedra.

Theorem 15. *Let $m = 2$ and $A \in \mathbb{N}^{2 \times n}$. Let $Ax \leq b$ be a system of consecutively intersecting knapsack constraints such that every pair of items from $N_1 \cap N_2$ is a cover. For $i = 1, 2$ let $C^i x \leq \gamma^i$ be a system of inequalities that describes the single knapsack polyhedron P^i.*

Then, P is described by the system of inequalities

$$\sum_{j \in N_1 \cap N_2} x_j \leq 1$$
$$C^i x \qquad \leq \gamma^i \text{ for } i = 1, 2.$$

Proof. Note that the system of linear inequalities given in the theorem is valid for P. To see that it suffices to describe P, let $c^T x \leq \gamma$ be a non-trivial facet-defining inequality of P that is not a positive multiple of the inequality $\sum_{j \in N_1 \cap N_2} x_j \leq 1$. W.l.o.g. we assume that $\text{supp}(c) \cap N_1 \neq \emptyset$. Let $Z = \{z_1, \ldots, z_t\} = N_1 \cap N_2$. We claim that $(\text{supp}(c) \setminus Z) \cap N_2 = \emptyset$. Suppose the contrary. We define natural numbers $\gamma^0, \gamma^1, \ldots, \gamma^t$ by

$$\gamma^0 := \max\{ \sum_{j \in N_1 \setminus Z} c_j x_j : \sum_{j \in N_1 \setminus Z} A_{1j} x_j \leq b_1, \; x \in \{0,1\}^{N_1 \setminus Z}\},$$

$$\gamma^i := \max\{ \sum_{j \in N_1 \setminus Z} c_j x_j : \sum_{j \in N_1 \setminus Z} A_{1j} x_j \leq b_1 - A_{1z_i}, \; x \in \{0,1\}^{N_1 \setminus Z}\},$$

and claim that the face induced by the inequality $c^T x \leq \gamma$ is contained in the face induced by the inequality

$$\sum_{j \in N_1 \setminus Z} c_j x_j + \sum_{i=1}^{t} (\gamma^0 - \gamma^i) x_{z_i} \leq \gamma^0. \tag{6}$$

This inequality is valid for P by definition of the values γ^i, $i = 0, \ldots, t$ and because $\sum_{i \in Z} x_i \leq 1$.

Let $x \in P \cap \mathbb{Z}^n$ such that $c^T x = \gamma$. If $\sum_{i \in Z} x_i = 0$, then $c^T x = \gamma^0$, since otherwise we obtain a contradiction to the validity of $c^T x \leq \gamma$. If $\sum_{i \in Z} x_i > 0$, then $\sum_{i \in Z} x_i = 1$. Let $z_i \in Z$ with $x_{z_i} = 1$. Then $c^T x = \gamma$ implies that $x - e^{z_i}$ is an optimal solution of the program

$$\max\{ \sum_{j \in N_1 \setminus Z} c_j w_j : w \in \{0,1\}^{N_1 \setminus Z} : A_1 . w \leq b_1 - A_{1z_i}\}.$$

This shows that in this case x satisfies inequality (6) as an equation, too. We obtain that $c^T x \leq \gamma$ must be a facet of the knapsack polyhedron P^1. This completes the proof. □

The correctness of the theorem strongly relies on the fact that every pair of items from the intersection $N_1 \cap N_2$ is a cover. If this condition is not satisfied, then Example 14 demonstrates that the facet-defining inequalities of the two single knapsack polyhedra do not suffice in general to describe the polyhedron associated with the intersection of the two knapsack constraints. Geometrically, the fact that we intersect two knapsack constraints generates incomparabilities between the column vectors of the associated matrix. These incomparabilities give rise to cutting planes that do not define facets of one of the two single knapsack polyhedra that we intersect. In fact, incomparabilities between column vectors in a matrix make it possible to "melt" inequalities from different knapsack polyhedra. A basic situation to which the operation of melting applies is

Proposition 16. *Let* $m = 2$, $A \in \mathbb{N}^{2 \times n}$ *and* $Ax \leq b$ *be a system of two consecutively intersecting knapsack constraints. Let* $\sum_{i \in N_1 \setminus N_2} c_i x_i + \sum_{i \in N_1 \cap N_2} c_i x_i \leq \gamma$

be a valid inequality for P^1, and let $\sum_{i\in N_2\setminus N_1} c_i x_i + \sum_{i\in N_1\cap N_2} c_i x_i \leq \gamma$ be a valid inequality for P^2. Setting $\Theta := \sum_{i\in N_1\setminus N_2} c_i$, the melted inequality

$$\sum_{i\in N_1\setminus N_2} c_i x_i + \sum_{i\in N_1\cap N_2} c_i x_i + \sum_{i\in N_2\setminus N_1} (c_i - \Theta)^+ x_i \leq \gamma$$

is valid for P.

Proof. Let $x \in P \cap \mathbb{Z}^n$. If $x_i = 0$ for all $i \in N_2 \setminus N_1$ with $c_i - \Theta > 0$, the inequality is satisfied because $\sum_{i\in N_1\setminus N_2} c_i x_i + \sum_{i\in N_1\cap N_2} c_i x_i \leq \gamma$ is valid for P^1. Otherwise, $\sum_{i\in N_2\setminus N_1} (c_i - \Theta)^+ x_i \leq \sum_{i\in N_2\setminus N_1} c_i x_i - \Theta$, and we obtain

$$\sum_{i\in N_1\setminus N_2} c_i x_i \quad + \sum_{i\in N_1\cap N_2} c_i x_i \quad + \sum_{i\in N_2\setminus N_1} (c_i - \Theta)^+ x_i$$

$$\leq \sum_{i\in N_1\setminus N_2} c_i x_i - \Theta \quad + \sum_{i\in N_1\cap N_2} c_i x_i + \sum_{i\in N_2\setminus N_1} c_i x_i$$

$$\leq \sum_{i\in N_1\cap N_2} c_i x_i + \sum_{i\in N_2\setminus N_1} c_i x_i$$

$$\leq \gamma.$$

$\square$

Proposition 16 can be extended to general upper bounds $u \in \mathbb{N}^n$. Often the inequality that results from melting valid inequalities from knapsack polyhedra as described in Proposition 16 does not define a facet of P. In such situations there is a good chance of strengthening the melted inequality by determining lifting coefficients for the items in $(N_1 \setminus N_2) \cup (N_2 \setminus N_1)$ with respect to a given order.

Example 17. Let A be the matrix $\begin{bmatrix} 1\,4\,5\,6\,0\,0 \\ 0\,0\,5\,6\,1\,4 \end{bmatrix}$ and $b = \begin{bmatrix} 15 \\ 15 \end{bmatrix}$. The inequality $x_1 + 4x_2 + 5x_3 + 6x_4 \leq 15$ defines a facet of $P^1 := \mathrm{conv}\,\{x \in \mathbb{Z}^4 : x_1 + 4x_2 + 5x_3 + 6x_4 \leq 15, 0 \leq x_i \leq 3, i \in \{1,2,3,4\}\}$. For $\alpha \in \{0,1,2,3\}$ the inequality $5x_3 + 6x_4 + \alpha x_6 \leq 15$ is valid for $P^2 := \mathrm{conv}\,\{x \in \mathbb{Z}^{\{3,4,5,6\}} : 5x_3 + 6x_4 + x_5 + 4x_6 \leq 15, 0 \leq x_i \leq 3, i \in \{3,4,5\}, 0 \leq x_6 \leq 1\}$. Setting $\alpha = 1$ and applying Proposition 16 and the succeeding remark we obtain that $0x_1 + 3x_2 + 5x_3 + 6x_4 + x_6 \leq 15$ is valid for $P := \mathrm{conv}\,\{x \in \mathbb{Z}^6 : Ax \leq b, 0 \leq x_i \leq 3, i \in \{1,2,3,4,5\}, 0 \leq x_6 \leq 1\}$. This inequality can be strengthened by lifting to yield the facet-defining inequality $x_1 + 3x_2 + 5x_3 + 6x_4 + x_6 \leq 15$. For $\alpha = 2$, we end up with the melted inequality $0x_1 + 2x_2 + 5x_3 + 6x_4 + 2x_6 \leq 15$. For $\alpha = 3$ we obtain the inequality $0x_1 + x_2 + 5x_3 + 6x_4 + 3x_6 \leq 15$ that can again be strengthened to yield a facet-defining inequality of P, $x_2 + 5x_3 + 6x_4 + x_5 + 3x_6 \leq 15$.

It turns out that sometimes the feasible set inequalities for two consecutively intersecting knapsacks can be interpreted in terms of melting feasible set inequalities for the associated single knapsack polytopes.

Example 18. Consider $A = \begin{bmatrix} 15 & 3 & 5 & 13 & 0 & 0 \\ 0 & 17 & 18 & 17 & 19 & 20 \end{bmatrix}$ and $b = \begin{bmatrix} 20 \\ 35 \end{bmatrix}$. The set $T = \{2,3\}$ is feasible, and the inequality $x_2 + x_3 + x_4 + 2x_5 + 2x_6 \leq 2$ is a feasible set inequality for $P^2 := \text{conv}\left\{x \in \{0,1\}^{\{2,3,4,5,6\}} : 17x_2 + 18x_3 + 17x_4 + 19x_5 + 20x_6 \leq 35\right\}$, where the variables not in T are lifted in the sequence 4, 5, 6. In addition, $x_1 + x_2 + x_3 + x_4 \leq 2$ is a feasible set inequality for $P^1 := \text{conv}\left\{x \in \{0,1\}^{\{1,2,3,4\}} : 15x_1 + 3x_2 + 5x_3 + 13x_4 \leq 20\right\}$, with respect to the same feasible set T and the lifting sequence 4, 1. Now $\Theta = \sum_{i \in N_1 \setminus N_2} c_i = 1$ and the melted inequality reads $x_1 + x_2 + x_3 + x_4 + x_5 + x_6 \leq 2$, which is facet-defining for $P := \{x \in \{0,1\}^6 : Ax \leq b\}$. Note that the melted inequality is also a feasible set inequality with respect to T and the lifting sequence 4, 1, 5, 6.

Note that Example 18 also shows that under certain conditions the operation of melting feasible set inequalities produces a facet-defining inequality for P.

5   Computational Experience

In this section we briefly report on our computational experience with a separation algorithm for the feasible set inequalities. We have incorporated this algorithm in a general mixed integer programming solver, called SIP, and tested it on instances from the MIPLIB 3.0 ([2]). Details of our separation algorithm such as how to determine a feasible set T, how to weight the variables in T, how to perform the lifting, and for which substructures of the underlying constraint matrix the separation algorithm should be called, will be given in a forthcoming paper. We compared SIP with and without using feasible set inequalities. The time limit for our runs was 3600 CPU seconds on a Sun Enterprise 3000. It turns out that for 14 out of 59 problems (*air05, fiber, gesa2, gesa2_o, gesa3_o, misc03, misc07, p0033, p0201, p2756, qnet1, seymour, stein27, stein45*) we find feasible set inequalities or our separation routines uses more than 1% of the total running time.

Table 1. Comparison of SIP with and without feasible set (FS) inequalities.

Example	B&B Nodes	Cuts Others	FS	Time FS	Total	Gap %
SIP without FS	733843	1241	0	0.0	18413.6	17.38
SIP with FS	690812	1450	1712	561.8	18766.2	9.92

Table 1 summarizes our results over these 14 problem instances. The first column gives the name of the problem, Column 2 the number of branch-and-bound nodes. The following two columns headed *Cuts* give the number of cutting planes found, *Others* are those that are found by the default separation routines in SIP, and *FS* shows the number of feasible set inequalities added. Columns 5 and 6 present the time spent for separating feasible set inequalities and the total time. The last column gives the sum of the gaps ($\frac{\text{upper bound - lower bound}}{\text{lower bound}}$) in percentage between the lower and upper bounds. Table 1 shows that the time slightly increases (by 2%), but the quality of the solutions is significantly improved, the gaps decrease by around 43% when adding feasible set inequalities. Based on these results we conclude that feasible set inequalities are a tool that helps solving mixed integer programs.

References

1. E. Balas, S. Ceria, and G. Cornuéjols. A lift-and-project cutting plane algorithm for mixed $0 - 1$ programs. *Mathematical Programming*, 58:295–324, 1993.
2. R. E. Bixby, S. Ceria, C. M. McZeal, and M. W. P. Savelsbergh. An updated mixed integer programming library: MIPLIB 3.0. 1998. Paper and problems available at http://www.caam.rice.edu/~bixby/miplib/miplib.html
3. S. Ceria, C. Cordier, H. Marchand, and L. A. Wolsey. Cutting planes for integer programs with general integer variables. Technical Report CORE DP9575, Université Catholique de Louvain, Louvain-la-Neuve, Belgium, 1997.
4. V. Chvátal. Edmonds polytopes and a hierarchy of combinatorial problems. *Discrete Mathematics*, 4:305–337, 1973.
5. H. Crowder, E. Johnson, and M. W. Padberg. Solving large-scale zero-one linear programming problems. *Operations Research*, 31:803–834, 1983.
6. R. E. Gomory. Solving linear programming problems in integers. In R. Bellman and M. Hall, editors, *Combinatorial analysis, Proceedings of Symposia in Applied Mathematics, Vol. 10*. Providence, RI, 1960.
7. H. Marchand and L. A. Wolsey. The 0–1 knapsack problem with a single continuous variable. Technical Report CORE DP9720, Université Catholique de Louvain, Louvain-la-Neuve, Belgium, 1997.
8. M. W. Padberg. On the facial structure of set packing polyhedra. *Mathematical Programming*, 5:199–215, 1973.
9. M. W. Padberg. A note on zero-one programming. *Operations Research*, 23:833–837, 1975.
10. M. W. Padberg. $(1, k)$-configurations and facets for packing problems. *Mathematical Programming*, 18:94–99, 1980.
11. M. W. Padberg, T. J. Van Roy, and L. A. Wolsey. Valid inequalities for fixed charge problems. *Operations Research*, 33:842–861, 1985.
12. A. Schrijver. On cutting planes. *Annals of Discrete Mathematics*, 9:291–296, 1980.
13. L. A. Wolsey. Faces of linear inequalities in 0-1 variables. *Mathematical Programming*, 8:165–178, 1975.

New Classes of Lower Bounds for Bin Packing Problems *

Sándor P. Fekete and Jörg Schepers

Center for Parallel Computing, Universität zu Köln
D–50923 Köln, Germany
{fekete, schepers}@@zpr.uni-koeln.de

Abstract. The bin packing problem is one of the classical NP-hard optimization problems. Even though there are many excellent theoretical results, including polynomial approximation schemes, there is still a lack of methods that are able to solve practical instances optimally. In this paper, we present a fast and simple generic approach for obtaining new lower bounds, based on dual feasible functions. Worst case analysis as well as computational results show that one of our classes clearly outperforms the currently best known "economical" lower bound for the bin packing problem by Martello and Toth, which can be understood as a special case. This indicates the usefulness of our results in a branch and bound framework.

1 Introduction

The bin packing problem (BPP) can be described as follows: Given a set of n "items" with integer size $x_1, \ldots, x_n$, and a supply of identical "containers" of capacity C, decide how many containers are necessary to pack all the items. This task is one of the classical problems of combinatorial optimization and NP-hard in the strong sense – see Garey and Johnson [9]. An excellent survey by Coffmann, Garey, and Johnson can be found as Chapter 2 in the recent book [2].

Over the years, many clever methods have been devised to deal with the resulting theoretical difficulties. Most notably, Fernandez de la Vega and Lueker [8], and Karmarkar and Karp [12] have developed polynomial time approximation schemes that allow it to approximate an optimal solution within $1 + \varepsilon$ in polynomial (even linear) time, for any fixed ε. However, these methods can be hardly called practical, due to the enormous size of the constants. On the other hand, there is still a lack of results that allow it solve even moderately sized test problems optimally – see Martello and Toth [15,16], and the ORLIB set of benchmark problems [1]. The need for better understanding is highlighted by a recent observation by Gent [10]. He showed that even though some of these benchmark

---

* This work was supported by the German Federal Ministry of Education, Science, Research and Technology (BMBF, Förderkennzeichen 01 IR 411 C7).

R. E. Bixby, E. A. Boyd, and R. Z. Ríos-Mercado (Eds.): IPCO VI
LNCS 1412, pp. 257–270, 1998. © Springer–Verlag Berlin Heidelberg 1998

problems of 120 and 250 items had defied all systematic algorithmic solution attempts, they are not exactly hard, since they can be solved by hand.

This situation emphasizes the need for ideas that are oriented towards the exact solution of problem instances, rather than results that are mainly worst case oriented. In this paper, we present a simple and fast generic approach for obtaining lower bounds, based on *dual feasible functions*. Assuming that the items are sorted by size, our bounds can be computed in linear time with small constants, a property they share with the best lower bound for the BPP by Martello and Toth [15,16]. As it turns out, one of our classes of lower bounds can be interpreted as a systematic generalization of the bound by Martello and Toth. Worst case analysis as well as computational results indicate that our generalization provides a clear improvement in performance, indicating their usefulness in the context of a branch and bound framework. Moreover, we can show that the simplicity of our systematic approach is suited to simplify and improve other hand-tailored types of bounds.

The rest of this paper is organized as follows. In Section 2, we give an introduction to dual feasible functions and show how they can be used to obtain fast and simple lower bounds for the BPP. In Section 3, we consider the worst case performance of one of our classes of lower bounds, as well as some computational results on the practical performance. Section 4 concludes with a discussion of possible extensions, including higher-dimensional packing problems.

2   Dual Feasible Functions

For the rest of this paper, we assume without loss of generality that the items have size $x_i \in [0,1]$, and the container size C is normalized to 1. Then we introduce the following:

Definition 1 (Dual feasible functions). *A function* $u : [0,1] \to [0,1]$ *is called* dual feasible, *if for any finite set* S *of nonnegative real numbers, we have the relation*

$$\sum_{x \in S} x \leq 1 \implies \sum_{x \in S} u(x) \leq 1. \tag{1}$$

Dual feasible functions have been used in the performance analysis of heuristics for the BPP, first by Johnson [11], then by Lueker [13]; see Coffman and Lueker [3] for a more detailed description. The Term (which was first introduced by Lueker [13]) refers to the fact that for any dual feasible function u and for any bin packing instance with item sizes $x_1, \ldots, x_n$, the vector $(u(x_1), \ldots, u(x_n))$ is a feasible solution for the dual of the corresponding fractional bin packing problem (see [12]). By definition, convex combination and compositions of dual feasible functions are dual feasible.

We show in this paper that dual feasible functions can be used for improving lower bounds for the one-dimensional bin packing problem. This is based on the following easy lemma.

Lemma 1. *Let $I := (x_1, \ldots, x_n)$ be a BPP instance and let u be a dual feasible function. Then any lower bound for the transformed BPP instance $u(I) := (u(x_1), \ldots, u(x_n))$ is also a lower bound for I.*

By using a set of dual feasible functions $\mathcal{U}$ and considering the maximum value over the transformed instances $u(I)$, $u \in \mathcal{U}$, we can try to obtain even better lower bounds.

In [13], a particular class of dual feasible functions is described; it relies on a special rounding technique. For a given $k \in \mathbb{N}$, consider the stair function $u^{(k)}$ that maps (for $i \in \{1, \ldots, k\}$) all values from the interval $[\frac{i}{k+1}, \frac{i+1}{k+1})$ onto the value $\frac{i}{k}$. 1 is mapped to 1. We give a slightly improved version and a simple proof that these functions are dual feasible.

Theorem 1. *Let $k \in \mathbb{N}$. Then*

$$u^{(k)} : [0,1] \to [0,1]$$
$$x \mapsto \begin{cases} x, & \text{for } x(k+1) \in \mathbb{Z} \\ \lfloor (k+1)x \rfloor \frac{1}{k}, & \text{else} \end{cases}$$

is a dual feasible function.

Proof. Let S be a finite set of nonnegative real numbers with $\sum_{x_i \in S} x_i \leq 1$. We have to show that $\sum_{x_i \in S} u^{(k)}(x_i) \leq 1$. Let $T := \{x_i \in S \mid x_i(k+1) \in \mathbb{Z}\}$. Clearly, we only need to consider the case $S \neq T$. Then

$$(k+1) \sum_{x_i \in T} u^{(k)}(x_i) + k \sum_{x_i \in S \setminus T} u^{(k)}(x_i) = (k+1) \sum_{x_i \in T} x_i + \sum_{x_i \in S \setminus T} \lfloor (k+1)x_i \rfloor$$
$$< (k+1) \sum_{x_i \in S} x_i.$$

By definition of $u^{(k)}$, the terms $(k+1) \sum_{x \in T} u^{(k)}(x)$ and $k \sum_{x \in S \setminus T} u^{(k)}(x)$ are integer, so by virtue of $\sum_{x_i \in S} x_i \leq 1$, we have the inequality $(k+1) \sum_{x \in T} u^{(k)}(x) + k \sum_{x \in S \setminus T} u^{(k)}(x) \leq k$, implying $\sum_{x \in S} u^{(k)}(x) \leq 1$. □

The rounding mechanism is visualized in Figure 1, showing the difference of the stair functions $u^{(k)}$, $k \in \{1, \ldots, 4\}$, and the identity id over the interval $[0,1]$. For the BPP, we mostly try to increase the sizes by a dual feasible function, since this allows us to obtain a tighter bound. The hope is to find a $u^{(k)}$ for which as many items as possible are in the "win zones" – the subintervals of $[0,1]$ for which the difference is positive.

The following class of dual feasible functions is the implicit basis for the bin packing bound L_2 by Martello and Toth [15,16]. This bound is obtained by neglecting all items smaller than a given value ϵ. We account for these savings by increasing all items of size larger than $1 - \epsilon$. Figure 2 shows the corresponding win and loss zones.

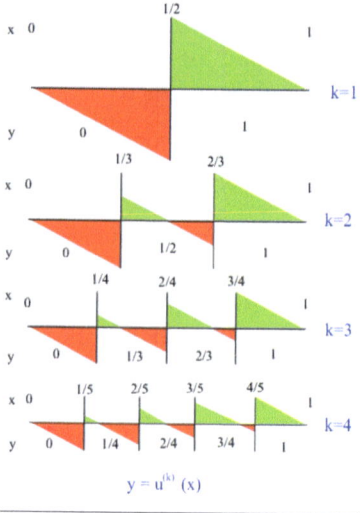

$$y = u^{(k)}(x)$$

Fig. 1. Win and loss zones for $u^{(k)}$.

Theorem 2. *Let $\epsilon \in [0, \frac{1}{2}]$. Then*

$$U^{(\epsilon)} : [0, 1] \to [0, 1]$$

$$x \mapsto \begin{cases} 1, & \text{for } x > 1 - \epsilon \\ x, & \text{for } \epsilon \le x \le 1 - \epsilon \\ 0, & \text{for } x < \epsilon \end{cases}$$

is a dual feasible function.

Proof. Let S be a finite set of nonnegative real numbers, with $\sum_{x \in S} x \le 1$. We consider two cases. If S contains an element larger than $1 - \epsilon$, then all other elements have to be smaller than ϵ. Hence we have $\sum_{x \in S} U^{(\epsilon)}(x) = 1$. If all elements of S have at most size $1 - \epsilon$, we have $\sum_{x \in S} U^{(\epsilon)}(x) \le \sum_{x \in S} x \le 1$.   □

Our third class of dual feasible functions has some similarities to some bounds that were hand-tailored for the two-dimensional and three-dimensional BPP by

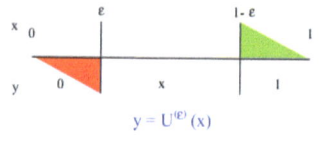

$$y = U^{(\epsilon)}(x)$$

Fig. 2. Win and loss zones for $U^{(\epsilon)}$.

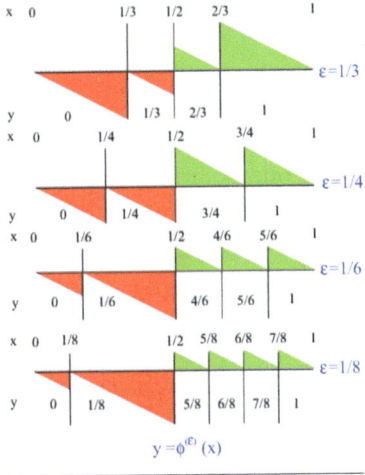

Fig. 3. Win and loss zones for $\phi^{(\epsilon)}$.

Martello and Vigo [17], and Martello, Pisinger, and Vigo [14]. However, our bounds are simpler and dominate theirs.

This third class also ignores items of size below a threshold value ϵ. For the interval $(\epsilon, \frac{1}{2}]$, these functions are constant, on $(\frac{1}{2}, 1]$ they have the form of stair functions. Figure 3 shows that for small values of ϵ, the area of loss zones for $\phi^{(\epsilon)}$ exceeds the area of win zones by a clear margin. This contrasts to the behavior of the functions $u^{(k)}$ and $U^{(\epsilon)}$, where the win and loss areas have the same size.

Theorem 3. *Let* $\epsilon \in [0, \frac{1}{2})$. *Then*

$$\phi^{(\epsilon)} : [0, 1] \to [0, 1]$$

$$x \mapsto \begin{cases} 1 - \frac{\lfloor (1-x)\epsilon^{-1} \rfloor}{\lfloor \epsilon^{-1} \rfloor}, & \text{for } x > \frac{1}{2} \\ \frac{1}{\lfloor \epsilon^{-1} \rfloor}, & \text{für } \epsilon \le x \le \frac{1}{2} \\ 0, & \text{for } x < \epsilon \end{cases}$$

is a dual feasible function.

Proof. Let S be a finite set of nonnegative real numbers, with $\sum_{x \in S} x \le 1$. Let $S' := \{x \in S \mid \epsilon \le x \le \frac{1}{2}\}$. We distinguish two cases:

If all elements of S have size at most $\frac{1}{2}$, then by definition of S',

$$1 \ge \sum_{x \in S} x \ge \sum_{x \in S'} x \ge |S'| \epsilon \tag{2}$$

holds. Since $|S'|$ is integral, it follows that $|S'| \leq \lfloor \epsilon^{-1} \rfloor$, hence

$$\sum_{x \in S} \phi^{(\epsilon)}(x) = \sum_{x \in S'} \phi^{(\epsilon)}(x) = |S'| \frac{1}{\lfloor \epsilon^{-1} \rfloor} \leq 1. \tag{3}$$

Otherwise S contains exactly one element $y > \frac{1}{2}$ and we have

$$1 \geq \sum_{x \in S} x \geq y + \sum_{x \in S'} x \geq y + |S'|\epsilon. \tag{4}$$

Therefore $|S'| \leq \lfloor (1 - y)\epsilon^{-1} \rfloor$ and hence

$$\sum_{x \in S} \phi^{(\epsilon)}(x) = \phi^{(\epsilon)}(y) + \sum_{x \in S'} \phi^{(\epsilon)}(x) = 1 - \frac{\lfloor (1 - y)\epsilon^{-1} \rfloor}{\lfloor \epsilon^{-1} \rfloor} + |S'| \frac{1}{\lfloor \epsilon^{-1} \rfloor} \leq 1. \tag{5}$$

$\square$

3   A Class of Lower Bounds for the BPP

In this section, we show how dual feasible functions can be combined by virtue of Lemma 1 in order to get good bounds for the BPP. For this purpose, we consider the lower bound L_2, suggested by Martello an Toth [15,16], which can be computed in time $O(n \log n)$. According to the numerical experiments by Martello and Toth, L_2 yields very good approximations of the optimal value. We describe L_2 with the help of dual feasible functions. Using Lemma 1, it is straightforward to see that L_2 provides a lower bound for the BPP. Our description allows an easy generalization of this bound, which can be computed in linear time. We show that this generalized bound improves the asymptotic worst-case performance from $\frac{2}{3}$ to $\frac{3}{4}$. Empirical studies provide evidence that also the practical performance is significantly improved.

3.1   The Class $L_*^{(q)}$

We give a definition of *(asymptotic) worst case performance*:

Definition 2 ((Asymptotic) worst case performance). *Let L be a lower bound for a minimization problem P. Let $opt(I)$ denote the optimal value of P for an Instance I. Then*

$$r(L) := \sup \left\{ \frac{L(I)}{opt(I)} \middle| I \text{ is an instance of } P \right\} \tag{6}$$

is called the worst case performance *and*

$$r_\infty(L) := \inf_{s \in \mathbb{R}_0^+} \sup \left\{ \frac{L(I)}{opt(I)} \middle| I \text{ is an instance of } P \text{ with } opt(I) \geq s \right\} \tag{7}$$

is called the asymptotic worst case performance *of L.*

The easiest lower bound for the BPP is the total volume of all items (the size of the container will be assumed to be 1), rounded up to the next integer. For a normalized BPP instance $I := (x_1, \ldots, x_n)$, this bound can be formulated as

$$L_1(I) := \left\lceil \sum_{i=1}^{n} x_i \right\rceil. \tag{8}$$

Theorem 4 (Martello and Toth). $r(L_1) = \frac{1}{2}$.

Martello and Toth showed that for their ILP formulation of the BPP, the bound L_1 dominates the LP relaxation, the surrogate relaxation, and the Lagrange relaxation. (See [16], pp. 224.) According to their numerical experiments, L_1 approximates the optimal value very well, as long as there are sufficiently many items of small size, meaning that the remaining capacity of bins with big items can be exploited. If this is not the case, then the ratio between L_1 and the optimal value can reach a worst case performance of $r(L_1) = \frac{1}{2}$, as shown by the class of examples with $2k$ items of size $\frac{1}{2} + \frac{1}{2k}$.

In the situation which is critical for L_1 ("not enough small items to fill up bins"), we can make use of the dual feasible functions $U^{(\epsilon)}$ from Theorem 2; these functions neglect small items in favor of big ones. This yields the bound L_2: For a BPP instance I, let

$$L_2(I) := \max_{\epsilon \in [0, \frac{1}{2}]} L_1(U^{(\epsilon)}(I)). \tag{9}$$

From Lemma 1, it follows immediately that L_2 is a lower bound for the BPP. The worst case performance is improved significantly:

Theorem 5 (Martello and Toth). $r(L_2) = \frac{2}{3}$.

This bound of $\frac{2}{3}$ is tight: consider the class of BPP instances with $6k$ items of size $\frac{1}{3} + \frac{1}{6k}$.

There is little hope that a lower bound for the BPP that can be computed in polynomial time can reach an absolute worst case performance above $\frac{2}{3}$. Otherwise, the $\mathcal{N}P$-hard problem *Partition* (see [9], p. 47) of deciding whether two sets of items can be split into two sets of equal total size could be solved in polynomial time.

L_2 can be computed in time $O(n \log n)$. The computational effort is determined by sorting the items by their size, the rest can be performed in linear time by using appropriate updates.

Lemma 2 (Martello and Toth). *Consider a BPP instance* $I := (x_1, \ldots, x_n)$. *If the sizes* x_i *are sorted, then* $L_2(I)$ *can be computed in time* $O(n)$.

Now we define for any $k \in \mathbb{N}$

$$L_2^{(k)}(I) := \max_{\epsilon \in [0, \frac{1}{2}]} L_1(u^{(k)} \circ U^{(\epsilon)}(I)) \tag{10}$$

and for $q \geq 2$ consider the following bounds:

$$L_*^{(q)}(I) := \max\{L_2(I), \max_{k=2,\ldots,q} L_2^{(k)}(I)\}. \tag{11}$$

By Lemma 1, these are valid lower bounds. As for L_2, the time needed for computing these bounds is dominated by sorting:

Lemma 3. *Let $I := (x_1, \ldots, x_n)$ be an instance of the BPP. If the items x_i are given sorted by size, then $L_*^{(q)}(I)$ can be computed in time $O(n)$ for any fixed $q \geq 2$.*

3.2   An Optimality Result for $L_*^{(q)}$

For the described class of worst case instances for L_2, the bound $L_*^{(2)}$ yields the optimal value. As a matter of fact, we have:

Theorem 6. *Let $I := (x_1, \ldots, x_n)$ be a BPP instance with all items larger than $\frac{1}{3}$. Then $L_*^{(2)}(I)$ equals the optimal value $opt(I)$.*

Proof. Without loss of generality, let $x_1 \geq x_2 \geq \ldots \geq x_n$. Consider an optimal solution, i. e., an assignment of all items to $m := opt(I)$ bins. Obviously, any bin contains one or two items. In each bin, we call the larger item the *lower* item, while (in the bins with two items) the smaller item is called the *upper* item. In the following three steps, we transform the optimal solution into normal form.

1. Sort the bins in decreasing order of the lower item.
2. Move the upper items into the bins with highest indices. The capacity constraint remains valid, since increasing bin index corresponds to decreasing size of the lower item.
3. Using appropriate swaps, make sure that in bins with two items, increasing index corresponds to increasing size of the upper item. Again, the order of the bins guarantees that no capacity constraint is violated.

Eventually, we get the normal form shown in Figure 4: Item x_i is placed in bin i. The remaining items $m+1, m+2, \ldots, n$ are placed in bins $m, m-1, \ldots, m-(n-(m+1))$. The first bin with two items has index $q := 2m+1-n$.
   For $i \in \{1, \ldots, n\}$, the assumption $x_i > \frac{1}{3}$ implies

$$u^{(2)}(x_i) = \frac{\lceil 3x_i - 1 \rceil}{2} \geq \frac{1}{2}. \tag{12}$$

For $n \geq 2m - 1$, we get

$$L_2^{(2)}(I) \geq L_1(u^{(2)}(I)) = \left\lceil \sum_{i=1}^{n} u^{(2)}(x_i) \right\rceil \geq \left\lceil \frac{(2m-1)}{2} \right\rceil = m. \tag{13}$$

Hence, the lower bound $L_2^{(2)}(I)$ yields the optimal value, leaving the case

$$n < 2m - 1. \tag{14}$$

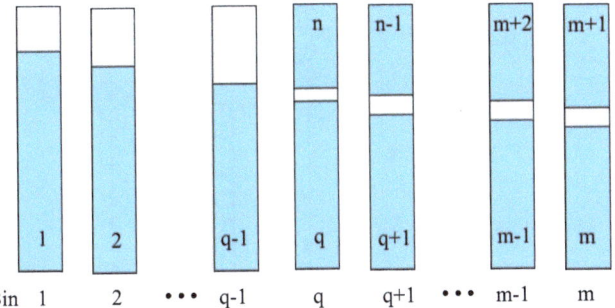

Fig. 4. Normal form of an optimal solution for a BPP instance with sizes $x_i > 1/3$.

Then $q = 2m + 1 - n > 2m + 1 - (2m - 1) = 2$, implying that at least bins 1 and 2 contain only one item.

If $x_m > \frac{1}{2}$, then $x_i > \frac{1}{2}$ holds for all $i \leq m$, as the items are sorted by decreasing size, thus $U^{(1/2)}(x_i) = 1$. Hence,

$$L_2(I) \geq L_1(U^{(1/2)}(I)) \geq \left\lceil \sum_{i=1}^{m} U^{(1/2)}(x_i) \right\rceil = m. \tag{15}$$

In this case, $L_2(I)$ equals the optimal value. Therefore, assume

$$x_m \leq \frac{1}{2} \tag{16}$$

for the rest of the proof.

If $x_{m-1} + x_m > 1$, let $\epsilon := x_m \leq 1/2$. For $i \leq m - 1$, we have $x_i \geq x_{m-1} > 1 - x_m = 1 - \epsilon$, hence

$$u^{(2)} \circ U^{(\epsilon)}(x_i) = u^{(2)}(1) = 1. \tag{17}$$

Furthermore, $1/3 < x_m \leq 1/2$ implies

$$u^{(2)} \circ U^{(\epsilon)}(x_m) = u^{(2)}(x_m) = \frac{1}{2}. \tag{18}$$

All in all, we have

$$L_2^{(2)}(I) \geq \left\lceil \sum_{i=1}^{m} u^{(2)} \circ U^{(\epsilon)}(x_i) \right\rceil = \left\lceil (m-1) + \frac{1}{2} \right\rceil = m, \tag{19}$$

i.e., the bound meets the optimum.

This leaves the case

$$x_{m-1} + x_m \leq 1. \tag{20}$$

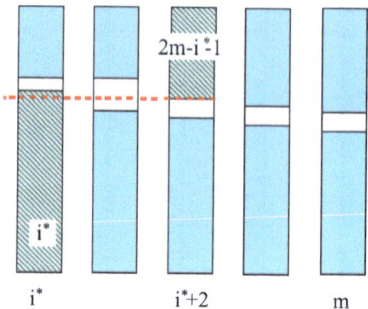

Fig. 5. Determining i^*.

Using the assumptions (14) and (20), we show that there is an $i^* \in \{2m - n, \dots, n\}$ with

$$x_{i^*} + x_{2m-i^*-1} > 1. \tag{21}$$

Figure 5 shows the meaning of this statement: At least one of the upper items cannot be combined with the lower item two bins to the left.

This is shown in the following way: If for all $i^* \in \{2m - n, \dots, n\}$, we had

$$x_i + x_{2m-i-1} \leq 1, \tag{22}$$

then all upper items could be moved two bins to the left, since the first two items do not contain more than one item by (14). This would allow it to pack item x_{m-1} with x_m, saving a bin.

Now let $\epsilon := x_{2m-i^*-1}$. By (21), we have for all $i \in \{1, \dots, i^*\}$ that

$$x_i \geq x_{i^*} > 1 - \epsilon, \tag{23}$$

hence

$$u^{(2)} \circ U^{(\epsilon)}(x_i) = u^{(2)}(1) = 1. \tag{24}$$

For $i \in \{i^* + 1, \dots, 2m - 1 - i^*\}$ we have

$$x_i \geq x_{2m-i^*-1} = \epsilon, \tag{25}$$

and therefore

$$u^{(2)} \circ U^{(\epsilon)}(x_i) \geq u^{(2)}(x_i) \geq \frac{1}{2}. \tag{26}$$

Summarizing, we get

$$L_2^{(2)}(I) \geq \left\lceil \sum_{i=1}^{i^*} u^{(2)} \circ U^{(\epsilon)}(x_i) + \sum_{i=i^*+1}^{2m-i^*-1} u^{(2)} \circ U^{(\epsilon)}(x_i) \right\rceil \tag{27}$$

$$\geq \left\lceil \sum_{i=1}^{i^*} 1 + \sum_{i=i^*+1}^{2m-i^*-1} \frac{1}{2} \right\rceil \tag{28}$$

$$= \left\lceil i^* + \frac{(2m - 2i^* - 1)}{2} \right\rceil = m. \tag{29}$$

This completes the proof.   □

3.3   Worst Case Performance of $L_*^{(q)}$

As we have stated above, we cannot hope to improve the worst case performance of L_2. However, we can show that the asymptotic worst case performance is improved by $r_\infty(L_*^{(2)})$:

Theorem 7.

$$r_\infty(L_*^{(2)}) = \frac{3}{4}. \tag{30}$$

Proof. Let $I := (x_1, \ldots, x_n)$ be a BPP instance. We start by showing

$$\max\{L_2(I), L_2^{(2)}(I)\} \geq \frac{3}{4} opt(I) - 1. \tag{31}$$

By Theorem 6, all items with $x_i > \frac{1}{3}$ fit into $m \leq L_2^{(2)}(I)$ bins. Let these bins be indexed by $1, \ldots, m$. Using the *First Fit Decreasing* heuristic, we add the remaining items, i.e., sort these items by decreasing order and put each item into the first bin with sufficient capacity; if necessary, use a new bin. Let q denote the number of bins in this solution.

Suppose we need m bins for the big items and not more than $\frac{m}{3}$ items for the rest, then Theorem 6 yields the first part of the statement:

$$opt(I) \leq q \leq \frac{4}{3}m \leq \frac{4}{3} \max\{L_2^{(2)}(I), L_2(I)\}. \tag{32}$$

Therefore, assume

$$\frac{3q}{4} > m \tag{33}$$

for the rest of the proof.

Let α denote the largest free capacity of one of the bins 1 through m, i.e., the total size of all items in this bin is at least $(1 - \alpha)$.

No item that was placed in one of the bins $m + 1$ through q can fit into the bin with free capacity α. This means that all these bins can only contain items $x_i > \alpha$. On the other hand, the size of these items does not exceed $\frac{1}{3}$. This implies that the bins $m + 1$ through $q - 1$ must contain at least three items of size $x_i > \alpha$, while bin q holds at least one item $x_i > \alpha$.

Thus, we get

$$L_2(I) \geq L_1(I) = \left\lceil \sum_{i=1}^{n} x_i \right\rceil \geq \lceil (1-\alpha)m + 3\alpha(q-1-m) + \alpha \rceil \qquad (34)$$

$$= \lceil (1-4\alpha)m + 3\alpha(q-1) + \alpha \rceil .$$

Now consider two cases.

Case 1: Assume $\alpha > \frac{1}{4}$. Since $(1-4\alpha) < 0$, we can replace the term m in (34) by $\frac{3q}{4} > m$:

$$L_2(I) > \left\lceil (1-4\alpha)\frac{3}{4}q + 3\alpha(q-1) + \alpha \right\rceil \qquad (35)$$

$$= \left\lceil \frac{3}{4}q - 2\alpha \right\rceil \qquad (36)$$

$$\geq \frac{3}{4}q - 1 \geq \frac{3}{4}opt(I) - 1. \qquad (37)$$

Case 2: For $\alpha \leq \frac{1}{4}$, neglect the term $(1-4\alpha)m \geq 0$ in (34):

$$L_2(I) \geq \lceil 3\alpha(q-1) + \alpha \rceil \qquad (38)$$

$$\geq \frac{3}{4}q - 1 \geq \frac{3}{4}opt(I) - 1. \qquad (39)$$

This proves $r_\infty(L_*^{(2)}) \geq \frac{3}{4}$. For showing that equality holds, consider the family of bin packing instances with $3k$ items of size $1/4+\delta$ with $\delta > 0$. This needs at least k bins. For sufficiently small δ, we have $L_2^{(2)} = 0$ and $L_2(I) = L_1(I) \leq \frac{3}{4}k+1$. □

3.4   Computational Performance of $L_*^{(q)}$

Generally speaking, L_2 yields results that are orders of magnitude better than the worst case performance (see [15,16]). In the following, we compare the computational performance of $L_*^{(q)}$ on the same type of benchmark problems. As it turns out, we get a clear improvement.

For our computational investigation, we generated random instances in the same way as described in [16], pp. 240. The bin size is normalized to 100. For a given number n of items, the sizes were generated randomly with even distribution on the sets $\{1, \ldots, 100\}$, $\{20, \ldots, 100\}$, and $\{35, \ldots, 100\}$. For each problem class, and $n \in \{100, 500, 1000\}$, we generated 1000 instances. In the table, we compare L_2 with the bounds $L_*^{(2)}$, $L_*^{(5)}$, and $L_*^{(10)}$. Shown is the average relative error in percent (% err) and the number of instances, for which the optimal value was reached (# opt). The optimum was computed with the routine MTP from [16] with a limit of 100000 search nodes.

Especially for large instances, we see a clear improvement by the new bounds. Compared with L_2, the number of times that the optimal value is met is increased by 50 % for $n = 1000$ and the first two problem classes. For the third problem class, we always get the optimum, as shown in Theorem 6. The differences between $L_*^{(2)}$, $L_*^{(5)}$ and $L_*^{(10)}$ are significant, but small.

Table 1. Performance of lower bounds for the BPP.

Interval	n	L_2 % err	# opt	$L_*^{(2)}$ % err	# opt	$L_*^{(5)}$ % err	# opt	$L_*^{(10)}$ % err	# opt
[1, 100]	100	0.432	774	0.324	832	0.303	843	0.272	859
	500	0.252	474	0.157	644	0.154	645	0.130	693
	1000	0.185	381	0.116	578	0.114	578	0.100	606
[20, 100]	100	0.419	732	0.297	812	0.265	832	0.232	853
	500	0.231	443	0.144	634	0.138	642	0.123	677
	1000	0.181	366	0.104	605	0.103	605	0.091	632
[35, 100]	100	0.229	827	0.000	1000	0.000	1000	0.000	1000
	500	0.160	553	0.000	1000	0.000	1000	0.000	1000
	1000	0.114	507	0.000	1000	0.000	1000	0.000	1000

4   Conclusions

We have presented a fast new method for generating lower bounds for the bin packing problem. The underlying method of dual feasible functions can also be used in the case of higher dimensions by combining our ideas the approach for modeling higher-dimensional orthogonal packings that we developed for finding exact solutions for the d-dimensional knapsack problem [4]. Details will be contained in the forthcoming papers [5,6,7].

References

1. J. E. Beasley. OR-Library: Distributing test problems by electronic mail. *Journal of Operations Research Society*, 41:1069–1072, 1990. http://mscmga.ms.ic.ac.uk/info.html

2. E. G. Coffmann, Jr., M. R. Garey, and D. S. Johnson. Approximation algorithms for bin packing: A survey. In D. S. Hochbaum, editor, *Approximation Algorithms for NP-Hard Problems*, pages 46–93. PWS Publishing, Boston, 1997.

3. E. G. Coffmann, Jr. and G. S. Lueker. *Probabilistic Analysis of Packing and Partitioning Algorithms*. Wiley, New York, 1991.

4. S. P. Fekete and J. Schepers. A new exact algorithm for general orthogonal d-dimensional knapsack problems. *Algorithms – ESA '97, LNCS, Vol. 1284*, pages 144–156. Springer, 1997.

5. S. P. Fekete and J. Schepers. On more-dimensional packing I: Modeling. ZPR Technical Report 97-288. http://www.zpr.uni-koeln.de

6. S. P. Fekete and J. Schepers. On more-dimensional packing II: Bounds. ZPR Technical Report 97-289. http://www.zpr.uni-koeln.de

7. S. P. Fekete and J. Schepers. On more-dimensional packing III: Exact Algorithms. ZPR Technical Report 97-290. http://www.zpr.uni-koeln.de

8. W. Fernandez de la Vega and G. S. Lueker. Bin packing can be solved within $1 + \varepsilon$ in linear time. *Combinatorica*, 1:349–355, 1981.

9. M. R. Garey and D. S. Johnson. *Computers and Intractability: A Guide to the Theory of NP-Completeness*. Freeman, San Francisco, 1979.

10. I. P. Gent. Heuristic solution of open bin packing problems. (To appear in *Journal of Heuristics*). http://www.cs.strath.ac.uk/~apes/papers
11. D. S. Johnson. *Near-optimal bin packing algorithms*. PhD thesis, Massachusetts Institute of Technology, Cambridge, Massachusetts, 1973.
12. N. Karmarkar and R. M. Karp. An efficient approximation scheme for the one-dimensional bin packing problem. *Proc. 23rd Annual Symp. Found. Comp. Sci. (FOCS 1982)*, pages 312–320, 1982.
13. G. S. Lueker. Bin packing with items uniformly distributed over intervals $[a, b]$. *Proc. 24th Annual Symp. Found. Comp. Sci. (FOCS 1983)*, pages 289–297, 1983.
14. S. Martello, D. Pisinger, and D. Vigo. The three-dimensional bin packing problem. Technical Report DEIS-OR-97-6, 1997. http://www.deis.unibo.it
15. S. Martello and P. Toth. Lower bounds and reduction procedures for the bin packing problem. *Discrete Applied Mathematics*, 28:59–70, 1990.
16. S. Martello and P. Toth. *Knapsack Problems*. Wiley, New York, 1990.
17. S. Martello and D. Vigo. Exact solution of the two-dimensional finite bin packing problem. Technical Report DEIS-OR-96-3, 1996. http://www. deis.unibo.it
18. J. Schepers. *Exakte Algorithmen für Orthogonale Packungsprobleme*. PhD thesis, Mathematisches Institut, Universität zu Köln, 1997.

Solving Integer and Disjunctive Programs by Lift and Project

Sebastián Ceria[1] and Gábor Pataki[2][*]

[1] Graduate School of Business, and
Computational Optimization Research Center
Columbia University, New York, NY 10027, USA
sebas@@cumparsita.gsb.columbia.edu
http://www.columbia.edu/~sc244
[2] Department of Industrial Engineering and Operations Research, and
Computational Optimization Research Center
Columbia University, New York, NY 10027, USA
gabor@@ieor.columbia.edu
http://www.ieor.columbia.edu/~gabor

Abstract. We extend the theoretical foundations of the branch-and-cut method using lift-and-project cuts for a broader class of disjunctive constraints, and also present a new, substantially improved disjunctive cut generator. Employed together with an efficient commercial MIP solver, our code is a robust, general purpose method for solving mixed integer programs. We present extensive computational experience with the most difficult problems in the MIPLIB library.

1 Introduction

Disjunctive programming is optimization over a finite union of convex sets. Its foundations were developed, and the term itself coined in the early seventies by Balas [4,5]; since then it attracted the attention of numerous researchers, including Jeroslow [18,19], Blair [12], Williams [25], Hooker [15], Beaumont [11], Sherali and Shetty [23], Meyer [21]. Besides having an elegant theory, disjunctive programming provides a way to formulate a wide variety of optimization problems, such as mixed integer programs, linear complementarity, job-shop scheduling, equilibrium problems, and so on.

There is a natural connection between disjunctive programming problems and logic. In fact, a recent paper of Hooker and Osorio [16] proposes to solve discrete optimization problems that can be formulated by using logic and linear programming. They call this area of mathematical programming MLLP (Mixed Logical Linear Programming). Even though MLLP is, in principle, more general

---

[*] Both authors were supported by NSF grant DMS 95-27-124

R. E. Bixby, E. A. Boyd, and R. Z. Ríos-Mercado (Eds.): IPCO VI
LNCS 1412, pp. 271–283, 1998. © Springer–Verlag Berlin Heidelberg 1998

than disjunctive programming, since it allows more flexibility in the representation of logical formulas, every MLLP can be represented as a general disjunctive program.

Recently, Balas, Ceria and Cornuéjols [8,9] proposed the *lift-and-project* method, which is related to the work of Balas on disjunctive programming, the *matrix-cuts* of Lovász and Schrijver [20], the hierarchy of relaxations for mixed-integer programming of Sherali and Adams [22], and the *intersection cuts* of Balas [3]. The implementation of the lift-and-project method in a branch-and-cut framework (see [9]) proved to be very effective when tackling difficult mixed 0–1 programs.

The goal of our work is twofold. First, we show that the lift-and-project method for the 0–1 case can be extended quite naturally for a large class of disjunctions, called *facial* disjunctions. Such disjunctive constraints abound in practice, and our preliminary computational experience shows that – when the disjunctions are carefully chosen – the cuts generated from them outperform 0–1 disjunctive cuts.

Second, we attempt to answer the challenge posed by a new generation of commercial MIP solvers. We implemented a new, much improved lift-and-project cut generator, and we present our computational experience with it. Combining our separator with an efficient commercial code (CPLEX 5.0) in a cut-and-branch framework yields an extremely robust general purpose MIP solver.

In the rest of this section we provide the basic notation and definitions. In Section 2 we give a brief overview of known properties of facial disjunctive programs, and describe how disjunctive cuts generated from facial disjunctions can be lifted. In Section 3 we discuss implementation issues and present new computational results on the problems of MIPLIB 3.0 with a substantially improved version of a lift-and-project cut generator. Finally, in Section 4 we give our conclusions, discuss ongoing work, and future directions.

Disjunctive Sets and Disjunctive Cuts. A *disjunctive set* is a set of points satisfying a collection of inequalities connected by the logical connectors $\wedge$ (conjunction, "AND") and $\vee$ (disjunction, "OR").

A disjunctive set is in *Disjunctive Normal Form* or *DNF*, if its terms do not contain further disjunctions. For simplicity, we shall be dealing with sets in *DNF* that contain only two terms, i.e. sets of the form

$$K_0 \cup K_1 \tag{1}$$

Moreover, we shall assume that there is a polyhedron K that contains both K_0 and K_1, and K_0 and K_1 are defined by one additional inequality, that is

$$\begin{aligned} K &= \{x \mid Ax \ge b\} \\ K_j &= \{x \mid Ax \ge b, \, d^j x \ge g_j\} \ (j = 0, 1) \end{aligned} \tag{2}$$

We shall denote

$$P = \operatorname{cl} \operatorname{conv}(K_0 \cup K_1) \tag{3}$$

A disjunctive set is in *conjunctive normal form*, or *CNF* if its conjunctions do not contain further conjunctions. E.g. if we are given the set K as above and

$$K_{ij} = \{x \mid Ax \geq b, d^{ij}x \geq g_{ij}\} \ (i = 1, \ldots, p, \ j = 0, 1)$$

then the set

$$\{x \mid \bigwedge_{i=1,\ldots,p} (x \in K_{i0} \vee x \in K_{i1})\} \tag{4}$$

or, equivalently

$$\{x \mid Ax \geq b, \bigwedge_{i=1,\ldots,p} (d^{i0}x \geq g_{i0} \vee d^{i1}x \geq g_{i1})\} \tag{5}$$

is in conjunctive normal form.

As an example, consider the feasible set of a mixed 0–1 program. In this case K is the feasible set of the linear programming relaxation, and

$$K_{i0} = \{x \mid Ax \geq b, \ x_i \leq 0\}$$
$$K_{i1} = \{x \mid Ax \geq b, \ x_i \geq 1\}$$

for $(i = 1, \ldots, p)$. Then with this definition of K_{i0} and K_{i1} (5) is the usual way of expressing the feasible set of the mixed 0–1 program. Moreover, if for some i we choose

$$K_0 = K_{i0}, \ K_1 = K_{i1} \tag{6}$$

then the set in *DNF* in the strengthening of the LP-relaxation obtained by imposing the 0–1 condition on the variable x_i.

A *disjunctive program* (DP for short) is an optimization problem with the feasible set being is a disjunctive set. Optimizing a linear function over the set in (1) can be done by optimizing over P.

If Π is a polyhedron, then we denote

$$\Pi^* = \{(\alpha, \beta) \mid \alpha x \geq \beta \text{ is a valid inequality for } \Pi\}$$

The next theorem, due to Balas, provides a representation of the set P in (1). This representation will be used in the next section as the basis for disjunctive cutting plane generation.

Theorem 1 (Balas [4]). *Assume that the sets K_0 and K_1 are nonempty. Then $(\alpha, \beta) \in P^*$ if and only if there exists u^0, v^0, u^1, v^1 such that*

$$\begin{aligned} u^j A + v^j d^j &= \alpha \ (j = 0, 1) \\ u^j b + v^j g_j &\geq \beta \ (j = 0, 1) \\ u^j, v^j &\geq 0 \ (j = 0, 1) \end{aligned} \tag{7}$$

In fact, this result holds under a more general regularity condition, than the nonemptyness of all K_i's (assuming nonemptyness, the result follows by simply using Farkas' lemma).

The lift-and-project method is based on the generation of *disjunctive or lift-and-project cuts*, $\alpha x \geq \beta$ which are valid for P, and violated by the current LP-solution $\bar{x}$, i.e. $\alpha \bar{x} < \beta$. Theorem 7 provides a way of generating disjunctive cuts for a general disjunctive program through the solution of a linear program, called the *cut-generation LP*, of the form

$$
\begin{aligned}
\max \quad & \beta - \alpha \bar{x} \\
\text{s.t.} \quad & (\alpha, \beta) \in P^* \\
& (\alpha, \beta) \in S
\end{aligned}
\tag{8}
$$

where S is a normalization set ensuring boundedness of the CLP.

There are several possible choices for the set S (see [13] for a complete description). In our current implementation we use the following normalization constraint:

$$
S = \{(u^0, v^0, u^1, v^1) : \sum_{j=0}^{1} (u^j + v^j)^T e \leq 1\}
$$

where e is a vector of all ones of appropriate dimension.

2   Facial Disjunctions

Our purpose is to use disjunctive cuts for solving general disjunctive programs. We will devote particular attention to a special class of *facial* disjunctive programs. A disjunctive set is called facial if the sets K_0 and K_1 are faces of K.

Some examples of facial and non-facial disjunctive programs include:

- **Variable upper bound constraints:** Suppose that we wish to model the following situation: if an arc i in a network is installed, we can send up to u_i units of flow on it; if it is not installed, we can send none. If we denote by y_i the amount of flow on the arc, and x_i is a 0–1 variable indicating whether the arc is installed, or not, then clearly,

$$
\begin{aligned}
& x_i = 1 \vee y_i = 0 \\
& 0 \leq y_i \leq u
\end{aligned}
\tag{9}
$$

is a correct model.
- **Linear complementarity:** The linear complementarity problem (LCP for short) is finding x, z satisfying

$$
\begin{aligned}
& x, z \geq 0 \\
& Mx + z = q \\
& x^T z = 0
\end{aligned}
\tag{10}
$$

Clearly, (10) is a facial disjunctive program in CNF, with the disjunctions being $x_i = 0 \vee z_i = 0$.

- **Ryan-Foster disjunctions:** This disjunction is used as a very successful *branching rule* in solving set-partitioning problems. Precisely, suppose that the feasible set of an SPP is given as

$$x_k \in \{0,1\} \,\forall k$$
$$Ax = e$$

with A being a matrix of zeros and ones, and e the vector of all ones. Denote by R_i the *support* of the ith row of A. Then the disjunction

$$\sum_{k \in R} x_k = 1 \vee \sum_{k \in R} x_k = 0$$

is valid for all the feasible solutions of the SPP, if R is a subset of *any* R_i. However, if R is chosen as the *intersection* of R_{i_1} and R_{i_2} for two rows i_1 and i_2, then the disjunction will perform particularly well as a branching rule, when solving the SPP by branch-and-bound.

- **Machine scheduling (see [6]):** In the machine scheduling problem we are given a number of operations $\{1,\ldots,n\}$ with processing times $p_1,\ldots,p_n$ that need to be performed on different items using a set of machines. The objective is to minimize total completion time while satisfying precedence constraints between operations and the condition that a machine can process one operation at a time, and operations cannot be interrupted. The problem can be formulated as

$$
\begin{aligned}
Min \ & t_n \\
& t_j - t_i \geq d_i, \quad (i,j) \in Z \\
& t_i \geq 0, \quad i = \{1,\ldots,n\} \\
& t_j - t_i \geq d_i \vee t_i - t_j \geq d_j \quad (i,j) \in W
\end{aligned}
\tag{11}
$$

where t_i is the starting time of job i, Z is the set of pairs of operations constrained by precedence relations and W is the set of pairs that use the same machine and therefore cannot overlap.
This disjunctive program is NOT facial.

One of the most important theoretical properties of 0–1 disjunctive cuts is the ease with which they can be *lifted* ([8,9]). For clarity, recall that

$$
\begin{aligned}
K &= \{x \mid Ax \geq b\} \\
K_j &= \{x \mid x \in K, \, d^j x \geq g_j\} \ (j = 0, 1) \\
P &= \mathrm{cl\,conv}\,(K_0 \cup K_1)
\end{aligned}
\tag{12}
$$

Let K' be a face of K, and

$$
\begin{aligned}
K'_j &= \{x \mid x \in K', \, d^j x \geq g_j\} \ (j = 0, 1) \\
P' &= \mathrm{cl\,conv}\,(K'_0 \cup K'_1)
\end{aligned}
\tag{13}
$$

Suppose that we are given a disjunctive cut (α', β') valid for P' and violated by $\bar{x} \in K'$. Is it possible to quickly compute a cut (α, β), which is valid for P, and violated by $\bar{x}$?

The answer is yes, if (α', β') is a cut obtained from a 0–1 disjunction ([8,9]), and K' is obtained from K by setting several variables to their bounds. Moreover, it is not only sufficient to solve the CLP with the constraints representing K' in place of K, its size can be reduced by removing those columns from A which correspond to the variables at bounds. In practice these variables are the ones fixed in branch-and-cut, plus the ones that happen to be at their bounds in the optimal LP-solution at the current node. Cut lifting is vital for the viability of the lift-and-project method within branch-and-cut; if the cut generation LP's are solved with putting all columns of A into the CLP, the time spent on cut generation is an order of magnitude larger, while the cuts obtained are rarely better ([9]). The main result of this section is :

Theorem 2. *Let K' be a an arbitrary face of K and K'_j and P' as above. Let $(\alpha', \beta') \in (P')^*$ with the corresponding multipliers given. Then we can compute a cut (α, β) that is valid for P and for all $x \in K'$*

$$\alpha x - \beta = \alpha' x - \beta' \tag{14}$$

Proof. Somewhat surprisingly, our proof is even simpler, than the original for the 0–1 case. Let K' be represented as

$$K' = \{x \mid A_= x = b_=, \ A_+ x \geq b_+\}$$

where the systems $A_= x \geq b_=$ and $A_+ x \geq b_+$ form a partition of $Ax \geq b$. Since $(\alpha', \beta') \in (P')^*$, we have

$$
\begin{aligned}
\alpha &= u^0_= A_= + u^0_+ A_+ + v_0 s \\
&= u^1_= A_= + u^1_+ A_+ + v_1 t \\
\beta &\leq u^0_= b_= + u^0_+ b_+ + v_0 s_0 \\
\beta &\leq u^1_= b_= + u^1_+ b_+ + v_1 t_0
\end{aligned}
\tag{15}
$$

with $u^0_+ \geq 0$, $u^1_+ \geq 0, v_0 \geq 0, v_1 \geq 0$, and $u^0_=$ and $u^1_=$ unconstrained. Then if we replace the negative components of $u^0_=$ and $u^1_=$ by 0, and compute the corresponding (α, β) it will clearly be valid for P and satisfy (14).   □

It is interesting to note that the above proof also implies that if the CLP is solved with a normalization that is imposed only on β, such as $\beta = \pm 1$, or $|\beta| \leq 1$, then the lifted cut will also be optimal. Hence the above theorem also generalizes the cut-lifting theorem in [8].

Suppose that after possibly complementing variables, multiplying rows by a scalar, adding rows, and permuting columns $A_=$ and $b_=$ can be brought into the form

$$A_= = [I, 0], \ b_= = 0$$

This condition is satisfied for most facial disjunctions of importance. Then just as in the 0–1 case, we can remove the columns from A that correspond to the columns of I, and solve the cut generation LP in the smaller space.

The consequence of these results is that facial disjunctions can be used for branching in a branch-and-cut algorithm, in place of the usual 0–1 branching. At any given node of the tree, the LP-relaxation is always a system that arises from the system defining K by imposing equality in some valid inequalities. In other words, the LP-relaxation at any given node defines a face of the LP-relaxation at the root. Therefore, if we also wish to generate disjunctice *cuts*, this can always be done using the LP-relaxation at the current node of the branch-and-cut tree, then lifting the resulting cut to be globally valid. Notice, that for this scheme to work, we only require the disjunctions for *branching* to be facial; the disjunctions for *cutting* can be arbitrary.

3   Computations

3.1   The Implementation

The computational issues that need to be addressed when generating lift-and-project cuts were thoroughly studied in [9]. Their experience can be briefly summarized as:

- It is better to generate cuts
    - in large rounds, before adding them to the linear programming relaxation, and reoptimizing.
    - in the space of the variables which are strictly between their upper and lower bounds, then to lift the cut to the full space.
- The distance of the current fractional point from the cut hyperplane is a reliable measure of cut quality.

We adopted most of their choices, and based on our own experience, we added several new features to our code, namely,

- In every round of cutting, we choose 50, (or less, if fewer are available) 0–1 disjunctions for cut generation. In an analogous way, we also choose a set of general integer disjunctions of the form $x_i \le \lfloor \bar{x}_i \rfloor \ \lor \ x_i \ge \lceil \bar{x}_i \rceil$.
- The 50 0–1 disjunctions are chosen from a candidate set of 150. It is rather conceivable that a disjunction will give rise to a strong cut if and only if it would perform well when used for branching; i.e. the improvement of the objective function on the two branches would be substantial. Therefore, we use "branching" information to choose the variables from the candidate set. In the current implementation we used the **strongbranch** routine of CPLEX 5.0, which returns an estimate of the improvements on both generated branches. Our strategy is testing the 150 candidate variables (or fewer if less are available) then picking those 50 which maximize some function (currently we use the harmonic mean) of the two estimates. We call this procedure the *strong choice* of cutting variables. We then repeat this process for the general integer variables, if any. We are in the process of testing other rules commonly used for selecting branching variables, like pseudo-costs and integer estimates.
- We pay particular attention to the *accuracy* of the cuts. If the cuts we derive are numerically unstable we resolve the CLP with a better accuracy.

- We use a normalization constraint (the set S in CLP) that bounds the sum of all multipliers (u^i, v^i).
- In joint work with Avella and Rossi [1], we have chosen to generate more than one cut from one disjunction using a simple heuristic. After solving the CLP, we fix a nonzero multiplier to zero, then resolve. We repeat this procedure several times always checking whether the consecutive cuts are close to being parallel; if so, one of them is discarded.

3.2   The Test-Bed and the Comparison

As a benchmark for comparison, we used the commercial MIP solver CPLEX 5.0, with the default parameters.

As the testbed, we used problems from MIPLIB 3.0, a collection of publicly available mixed integer programming problems. We excluded those problems which were too easy for CPLEX, namely the ones that could be solved within 100 nodes, and also the *fast0507* problem, since it is too large. We divided the remaining problems into two groups.

- "Hard" problems; the ones that cannot be solved within one thousand seconds by CPLEX with the default setting.
- "Medium" problems. All the rest.

Finally, since the enumeration code of MIPO was written 4 years ago, currently it is not competitive with the best commercial solvers. Therefore, we tested our cut-generator in a cut-and-branch framework. We generated 2 and 5 rounds of 50-100 cuts, after every round adding them to the LP formulation, reoptimizing, and dropping inactive constraints. After the fifth round we fed the strengthened formulation to the CPLEX 5.0 MIP solver with the above setting. Also, all the cut-generation LP's were solved using the CPLEX dual simplex code.

All of our tests were performed on a Sun Enterprise 4000 with 8-167MHz CPU; we set a memory limit of 200 MB, and ran all our tests using one processor only.

3.3   The Computational Results

The results for those "hard" problems which could be solved with, or without cuts, are summarized in Table 2. Their description is included in Table 1. The problems not solved by any of the two methods are: *danoint, dano3mip, noswot, set1ch, seymour*. Nevertheless, on the last two problems lift-and-project cuts perform quite well; *set1ch* can be solved with 10 rounds, and on *seymour* we were able to get the best bound known to date (see the next section).

Also, the medium problems were run, and solved by CPLEX and cut-and-branch as well. The comparisons for these problems are not presented here, but cut-and-branch was roughly twice as fast if we aggregate all the results.

Table 1. Problem description.

Problem	LP value	IP value
10teams	897.00	904.00
air04	55,264.43	55,866.00
air05	25,877.60	26,374.00
arki001	7,009,391.43	7,010,963.84
gesa2	25,476,489.68	25,781,982.72
gesa2_o	25,476,489.68	25,781,982.72
harp2	-74,325,169.35	-73,893,948.00
misc07	1415.00	2,810.00
mod011	-62,121,982.55	-54,558,535.01
modglob	20,430,947.62	20,740,51
p6000	-2,350,838.33	-2,349,787.00
pk1	0.00	11.00
pp08a	2748.35	7350.00
pp08aCUTS	5480.61	7350.00
qiu	-931.64	-132.87
rout	-1393.38	-1297.69
set1ch	30426.61	49,846.25
vpm2	9.89	13.75

The following preliminary conclusions can be drawn.

(1) In 9 problems out of the 22, either 2, or 5 rounds (in most cases 5) of lift-and-project cuts substantially improve the solution time. In 5 problems the difference is "make-or-break"; between solving, or not solving a problem, or improving the solution time by orders of magnitude.
(2) On 6 problems, our cuts do not make much difference in the solution time.
(3) On 3 problems, adding our cuts is actually detrimental. It is important to note, that the deterioration in the computing time is *not* due to the time spent on generating the cuts, rather to the fact, that they make the linear programming relaxation harder to solve. In fact, it is most likely possible to catch this effect by monitoring, e.g. the density of the cuts, and the deterioration of the LP relaxation's condition number.

Lift-and-Project Cuts on Two Very Difficult Problems. There were 5 problems that neither CPLEX alone, nor our code (with at most 5 rounds of cuts) was able to solve. These are : *danoint, dano3mip, noswot, set1ch* and *seymour*. All of them are notoriously hard, and currently unsolvable by general purpose MIP-solvers within a reasonable time. In fact, *dano3mip, noswot,* and *seymour* have never been solved to optimality (although an optimal value for *noswot* is reported in MIPLIB 3.0, we could not find anyone to confirm the existence of such a solution).

Our cuts do not perform well on the first 3 problems; *danoint* and *dano3mip* are network design problems with a combinatorial structure, already containing

Table 2. Computational results for cut-and-branch.

Problem	CPLEX 5.0		Cut–and–Branch$_2$		Cut–and–Branch$_5$	
	Time	Nodes	Time	Nodes	Time	Nodes
10teams	5404	2265	1274	306	5747	1034
air04	2401	146	1536	110	5084	120
air05	1728	326	1411	141	4099	213
arki001	6994	21814	18440	68476	13642	12536
gesa2	9919	86522	3407	22601	1721	6464
gesa2_o	12495	111264	4123	28241	668	4739
harp2	14804	57350	10686	28477	13377	31342
misc07	2950	15378	2910	12910	4133	14880
mod011	22344	18935	63481	24090	+++	+++
modglob	+++	+++	10033	267015	435	5623
p6000	1115	2911	1213	2896	805	1254
pk1	3903	130413	5094	122728	6960	150243
pp08a	+++	+++	1924	47275	178	1470
pp08aCUTS	50791	1517658	277	3801	134	607
qiu	35290	27458	15389	10280	27691	15239
rout	19467	133075	26542	155478	40902	190531
vpm2	8138	481972	1911	63282	974	18267

many special purpose cuts, and *noswot* is highly symmetric. However, disjunctive cuts perform strikingly well on the last two instances.

Until now, *set1ch* could be solved to optimality only by using special purpose *path-inequalities* [26]. After exhausting the memory limits, CPLEX could only find a solution within 15.86 % of the optimum. We ran our cutting plane generator for 10 rounds, raising the lower bound to within 1.4 % of the integer optimum. CPLEX was then able to solve the strengthened formulation in 28 seconds by enumerating 474 nodes. It is important to note that no numerical difficulties were encountered during the cutting phase, (even if we generated 15 rounds, although this proved unnecessary) and the found optimal solution precisely agrees with the one reported in MIPLIB (the objective coefficients are one-fourth integral).

The problem *seymour* is an extremely difficult setcovering instance; it was donated to MIPLIB by Paul Seymour, and its purpose is to find a minimal "irreducible configuration" in the proof of the four-colour conjecture. It has not been solved to optimality. The value of the LP-relaxation is 403.84, and an integer solution of 423.0 is known. The best previously known *lower* bound of 412.76 [2] was obtained by running CPLEX 4.0 on an HP SPP2000 with 16 processors, each processor having 180 MHz frequency, and 720 Mflops peak performance, for the total of approximately 58 hours and using approx. 1360 Mbytes of memory.

Due to the difficulty of the problem, we ran our cutting plane algorithm on this problem with a rather generous setting. We generated 10 rounds of cuts, in each round choosing the 50 cutting variables picked by our strong choice from among *all* fractional variables with the iteration limit set to 1000. The total time spent on generating the 10 rounds was approximately 10.5 hours, and the lower bound was raised to 413.16. The memory useage was below 50Mbytes. Running CPLEX 4.0 on the strengthened formulation for approximately 10 more hours raised the lower bound to 414.20 - a bound that currently seems unattainable without using our cuts.

Computational Results with Other Disjunctions. We ran our cut-generator on two of the most difficult set-partitioning problems in MIPLIB, namely *air04* and *air05*, by using the Ryan-Foster disjunctions described in the previous section. The results with two rounds of cuts are summarized in Table 3.

Table 3. Results with the Ryan-Foster disjunctions.

Problem	CPLEX 5.0 Time	CPLEX 5.0 Nodes	C&B Time	C&B Nodes
air04	2401	146	1300	115
air05	1728	326	1150	123

4   Conclusions and Future Directions

In the near future we plan to explore the following topics:

- Making our computational results with cut-and-branch more consistent. The key here is, finding the right amount of cutting, that sufficiently strengthens the LP-relaxation, but does not make it too difficult to solve.
- We are currently implementing a branch-and-cut method that uses disjunctive cuts and allows branching on facial disjunctions, using cut lifting based on Theorem 2. We will use the disjunctions which are given as part of the formulation, and in some other cases, we will use the structure of the problem to generate other valid disjunctions. Our goal is to treat disjunctions in a way similarly to inequalities (that is, to maintain a set of "active" disjunctions, and to keep the rest in a "pool"), and handle them efficiently throughout the code.
- We are in the process of testing our cutting plane generator with other commercial LP and MIP solvers (XPRESS-MP). This program allows for the generation of cutting planes within the enumeration tree without the need of programming our own enumeration, and hence improving on the efficiency.

References

1. P. Avella and F. Rossi. Private communication.
2. G. Astfalk and R. Bixby. Private communication.
3. E. Balas. Intersection cuts – A new type of cutting planes for integer programming. *Operations Research*, 19:19–39, 1971.
4. E. Balas. Disjunctive programming: Facets of the convex hull of feasible points. Technical Report No. 348, GSIA, Carnegie Mellon University, 1974.
5. E. Balas. Disjunctive programming. *Annals of Discrete Mathematics*, 5:3–51, 1979.
6. E. Balas. Disjunctive programming and a hierarchy of relaxations for discrete optimization problems. *SIAM J. Alg. Disc. Meth.*, 6:466–486, 1985.
7. E. Balas. Enhancements of lift-and-project. Technical Report, GSIA, Carnegie Mellon University, 1997.
8. E. Balas, S. Ceria, and G. Cornuéjols. A lift-and-project cutting plane algorithm for mixed 0–1 programs. *Mathematical Programming*, 58:295–324, 1993.
9. E. Balas, S. Ceria, and G. Cornuéjols. Mixed 0–1 Programming by lift-ad-project in a branch-and-cut framework. *Management Science*, 42:1229–1246, 1996.
10. E. Balas, S. Ceria, G. Cornuéjols, and G. Pataki. Polyhedral Methods for the Maximum Clique Problem. *AMS, DIMACS Series on Discrete Mathematics and Computer Science*, 26:11–28, 1996.
11. N. Beaumont. An algorithm for disjunctive programs. *European Journal of Operational Research*, 48:362–371, 1990.
12. C. Blair. Two rules for deducing valid inequalities for 0–1 problems. *SIAM Journal of Applied Mathematics*, 31:614–617, 1976.
13. S. Ceria and J. Soares. Disjunctive cuts for mixed 0–1 programming: Duality and lifting. Working paper, Graduate School of Business, Columbia University, 1997.
14. S. Ceria and J. Soares. Convex programming for disjunctive optimization. Working paper, Graduate School of Business, Columbia University, 1997.
15. J. Hooker. Logic based methods for optimization. In A. Borning, editor, *Principles and practice of constraint programming, LNCS, Vol. 626*, pages 184–200, 1992.
16. J. Hooker, M. Osorio. Mixed logical/linear programming. Technical report, GSIA, Carnegie Mellon University, 1997.
17. J. Hooker, H. Yan, I. Grossman, and R. Raman. Logic cuts for processing networks with fixed charges. *Computers and Operations Research*, 21:265–279, 1994.
18. R. Jeroslow. Representability in mixed-integer programming I: Characterization results. *Discrete Applied Mathematics*, 17:223–243, 1987.
19. R. Jeroslow. Logic based decision support: Mixed-integer model formulation. *Annals of Discrete Mathematics*, 40, 1989.
20. L. Lovász and A. Schrijver. Cones of matrices and set-functions and 0–1 optimization. *SIAM J. Optimization*, 1:166–190, 1991.
21. R. Meyer. Integer and mixed-integer programming models: General properties. *Journal of Optimization Theory and Applications*, 16:191–206, 1975.
22. H. Sherali and W. Adams. A hierarchy of relaxations between the continuous and convex hull representations for zero-one programming problems. *SIAM J. Disc. Math.*, 3:411–430, 1990.
23. H. Sherali and C. Shetty. Optimization with disjunctive constraints, In M. Beckman and H. Kunzi, editors, *Lecture notes in Economics and Mathematical Systems, Vol. 181*. Springer-Verlag, 1980.
24. J. Soares. *Disjunctive Methods for Discrete Optimization Problems*, PhD thesis, Graduate School of Business, Columbia University, 1997. In preparation.

25. H. P. Williams. An alternative explanation of disjunctive formulations. *European Journal of Operational Research*, 72:200–203, 1994.
26. L. Wolsey. Private communication.

A Class of Hard Small 0–1 Programs [*]

Gérard Cornuéjols[1] and Milind Dawande[2][**]

[1] Graduate School of Industrial Administration
Carnegie Mellon University, Pittsburgh, PA 15213, USA
[2] IBM, T. J. Watson Research Center
Yorktown Heights, NY 10598, USA

Abstract. In this paper, we consider a class of 0–1 programs which, although innocent looking, is a challenge for existing solution methods. Solving even small instances from this class is extremely difficult for conventional branch-and-bound or branch-and-cut algorithms. We also experimented with basis reduction algorithms and with dynamic programming without much success. The paper then examines the performance of two other methods: a group relaxation for 0,1 programs, and a sorting-based procedure following an idea of Wolsey. Although the results with these two methods are somewhat better than with the other four when it comes to checking feasibility, we offer this class of small 0,1 programs as a challenge to the research community. As of yet, instances from this class with as few as seven constraints and sixty 0–1 variables are unsolved.

1 Introduction

Goal programming [2] is a useful model when a decision maker wants to come "as close as possible" to satisfying a number of incompatible goals. It is frequently cited in introductory textbooks in management science and operations research. This model usually assumes that the variables are continuous but, of course, it can also arise when the decision variables must be 0,1 valued. As an example, consider the following market-sharing problem proposed by Williams [18]: A large company has two divisions D_1 and D_2. The company supplies retailers with several products. The goal is to allocate each retailer to either division D_1 or division D_2 so that D_1 controls 40% of the company's market for each product and D_2 the remaining 60% or, if such a perfect 40/60 split is not possible for all the products, to minimize the sum of percentage deviations from the 40/60 split. This problem can be modeled as the following integer program (IP):

---

[*] This work was supported in part by NSF grant DMI-9424348.
[**] Part of the work was done while this author was affiliated with Carnegie Mellon University.

R. E. Bixby, E. A. Boyd, and R. Z. Ríos-Mercado (Eds.): IPCO VI
LNCS 1412, pp. 284–293, 1998. © Springer–Verlag Berlin Heidelberg 1998

$$\text{Min} \quad \sum_{i=1}^{m} |s_i|$$

$$\text{s.t.} \sum_{j=1}^{n} a_{ij}x_j + s_i = b_i \quad i = 1, ..., m$$

$$x_j \qquad \in \{0,1\} \quad \text{for} \quad j = 1, ..., n$$

$$s_i \qquad \textit{free} \quad \text{for} \quad i = 1, ..., m.$$

where n is the number of retailers, m is the number of products, a_{ij} is the demand of retailer i for product j, and the right hand side vector b_i is determined from the desired market split among the two divisions D_1 and D_2. Note that the objective function of IP is not linear but it is straightforward to linearize it. This integer program also models the following basic *Feasibility Problem* (FP) in geometry:

Feasibility Problem: Given m hyperplanes in $\Re^n$, does there exist a point $x \in \{0, 1\}^n$ which lies on the intersection of these m hyperplanes?

If the optimum solution to IP is 0, the answer to FP is "yes", else the answer to FP is "no". Clearly, FP is NP-complete since for $m = 1$, FP is the subset-sum problem which is known to be NP-complete [6]. Problems of this form can be very difficult for existing IP solvers even for a relatively small number n of retailers and number m of products (e.g. $n = 50$, $m = 6$, and uniform integer demand between 0 and 99 for each product and retailer). More generally, with this choice of a_{ij}, asking for a 50/50 split and setting $n = 10(m - 1)$ produces a class of hard instances of 0–1 programs for existing IP solvers.

In this paper, we consider instances from the above class generated as follows: a_{ij} uniform integer between 0 and 99 ($= D - 1$), $n = 10(m - 1)$ and $b_i = \lfloor \frac{1}{2} \sum_{j=1}^{n} a_{ij} \rfloor$ or, more generally, in the range $\lfloor \frac{1}{2}(-D + \sum_{j=1}^{n} a_{ij}) \rfloor$ to $\lfloor \frac{1}{2}(-D + \sum_{j=1}^{n} a_{ij}) \rfloor + D - 1$.

2   Available Approaches for Solving IP

In this section, we report on our computational experience with four different IP solvers available in the literature.

2.1   Branch and Bound

We found that even small instances of IP are *extremely* hard to solve using the conventional branch-and-bound approach. We offer the following explanation. For instances chosen as described above, there is often no 0–1 point in the intersection of the hyperplanes, that is the optimum solution to IP is strictly greater than 0, whereas the solution to the linear programming relaxation is 0, even after fixing numerous variables to 0 or to 1. Because the lower bound stays at 0, nodes of the branch-and-bound tree are not pruned by the lower bound until very deep into the enumeration tree. We illustrate this point in Table 1. We generated 5 instances of IP (each having 30 variables and 4 constraints) using the setup described above. We indicate the number of nodes enumerated to solve IP using CPLEX 4.0.3. For each of the 5 instances, the number of nodes enumerated by the branch-and-bound tree is greater than 2^{20}. Note that, in each

Table 1. Size of the branch-and-bound tree (5 instances).

Problem size	Number of nodes enumerated	Optimal solution
4 × 30	1224450	1.00
4 × 30	1364680	2.00
4 × 30	2223845	3.00
4 × 30	1922263	1.00
4 × 30	2415059	2.00

case, the solution to IP is strictly greater than 0. Instances with 40-50 variables take several weeks before running to completion.

Note: The problem class IP is related to the knapsack problems considered by Chvátal [3]. It is shown in [3] that these knapsack problems are hard to solve using branch-and-bound algorithms. However, the coefficients of the knapsack constraint are required to be very large $(U[1, 10^{\frac{n}{2}}])$. For the instances in the class IP that we consider, the coefficients a_{ij} are relatively small $(U[0, 99])$. By combining the constraints of IP with appropriate multipliers (e.g. multiplying constraint i by $(nD)^{i-1}$) and obtaining a *surrogate* constraint, we get an equivalent problem by choosing D large enough, say $D = 100$ in our case. The resulting class of knapsack instances is similar to that considered in [3].

2.2   Branch and Cut

The idea of branch-and-cut is to enhance the basic branch-and-bound approach by adding cuts in an attempt to improve the bounds. Here, we used MIPO, a branch-and-cut algorithm which uses lift-and-project cuts [1]. The computational results were even worse than with the basic branch-and-bound approach (see Table 2). This is not surprising since, in this case, the linear programming relaxation has a huge number of basic solutions with value 0. Each cutting plane cuts off the current basic solution but tends to leave many others with value 0. As we add more cuts, the linear programs become harder to solve and, overall, time is wasted in computing bounds that remain at value 0 in much of the enumeration tree.

2.3   Dynamic Programming

Using the surrogate constraint approach described above, we get a 0–1 knapsack problem which is equivalent to the original problem. Clearly, this technique is suitable for problems with only a few constraints. Dynamic programming algorithms can be used to solve this knapsack problem. Here, we use an implementation due to Martello and Toth [13] pages 108–109. The complexity of the algorithm is $O(\min(2^{n+1}, nc))$ where c is the right-hand-side of the knapsack constraint. For the instances of size 3×20 and 4×30, we used the multiplier $(nD)^{i-1}$ with $D = 100$ for constraint i to obtain the surrogate constraint. For the larger instances, we faced space problems: We tried using smaller multipliers (e.g. $(nD')^{i-1}$ with $D' = 20$) but then the one-to-one correspondence between solutions of the original problem and that of the surrogate constraint is lost. So one has to investigate *all* the solutions to the surrogate constraint. Unfortunately, this requires that we store all the states of the dynamic program (the worst case bound for the number of states is 2^n). Hence, even when we decreased the multipliers, we faced space problems. See Table 2. We note that there are other more sophisticated dynamic programming based procedures [14] which could be used here. Hybrid techniques which use, for example, dynamic programming within branch-and-bound are also available [13].

2.4 Basis Reduction

For the basis reduction approach, we report results obtained using an implementation of the generalized basis reduction by Xin Wang [16][17]. It uses the ideas from Lenstra [11], Lovász and Scarf [12] and Cook, Rutherford, Scarf and Shallcross [4]. We consider the feasibility question FP here. Given the polytope $P = \{0 \leq x \leq 1 : Ax = b\}$, the basis reduction algorithm either finds a 0,1 point in P or generates a direction d in which P is "flat". That is, $\max\{dx - dy : x, y \in P\}$ is small. Without loss of generality, assume d has integer coordinates. For each integer t such that $\lceil \min\{dx : x \in P\} \rceil \leq t \leq \lfloor \max\{dx : x \in P\} \rfloor$ the feasibility question is recursively asked for $P \cap \{x : dx = t\}$. The dimension of each of the polytopes $P \cap \{x : dx = t\}$ is less than the dimension of P. Thus, applying the procedure to each polytope, a search tree is built which is at most n deep. A direction d in which the polytope is "flat" is found using a generalized basis reduction procedure [12][4].

2.5 Computational Experience

Table 2 contains our computational experience with the different approaches given in Section 2. We choose 4 settings for the problem size: $m \times n = 3 \times 20$, 4×30, 5×40 and 6×50. For each of these settings, we generated 5 instances as follows: a_{ij} random integers chosen uniformly in the range $[0,99]$ and $b_i = \lfloor (\sum_{j=1} a_{ij})/2 \rfloor$. For branch-and-bound, we use the CPLEX 4.0.3 optimizer. For branch-and-cut, we use MIPO [1]. For dynamic programming, we use DPS [13]. For basis reduction, we use Wang's implementation [16]. Times reported refer to seconds on an HP720 Apollo desktop workstation with 64 megabytes of memory. None of the problems with 40 or 50 variables could be solved within a time limit of 15 hours.

Table 2. Computing times for various solution procedures.

Problem size $(m \times n)$	B & B CPLEX 4.0.3	B & C MIPO	DP DPS	Basis Red. Wang
3×20	11.76	213.62	0.93	300.42
3×20	13.04	125.47	1.28	209.79
3×20	13.16	208.76	0.94	212.03
3×20	13.64	154.71	1.31	277.41
3×20	11.21	190.81	1.11	197.37
4×30	1542.76	***	20.32	***
4×30	1706.84	***	20.37	***
4×30	2722.52	***	20.31	***
4×30	2408.84	***	18.43	***
4×30	2977.28	***	18.94	***
5×40	***	***	+++	***
5×40	***	***	+++	***
5×40	***	***	+++	***
5×40	***	***	+++	***
5×40	***	***	+++	***
6×50	***	***	+++	***
6×50	***	***	+++	***
6×50	***	***	+++	***
6×50	***	***	+++	***
6×50	***	***	+++	***

*** Time limit (54000 seconds) exceeded.
+++ Space limit exceeded.

3   Two Other Approaches

In this section, we introduce two other approaches to FP.

3.1   The Group Relaxation

The feasible set of FP is

$$S = \{x \in \{0,1\}^n : Ax = b\}$$

where (A, b) is an integral $m \times (n+1)$ matrix. We relax S to the following *group problem* (GP).

$$S_\delta = \{x \in \{0,1\}^n : Ax \equiv b \pmod{\delta}\}$$

where $\delta \in Z_+^m$. GP is interesting because (i) In general, GP is easier to solve than IP [15] and (ii) Every solution of FP satisfies GP. The feasible solutions to GP can be represented as s-t paths in a directed acyclic layered network G. The digraph G has a layer corresponding to each variable $x_j, j \in N$, a source node s

and a sink node t. The layer corresponding to variable x_j has $\delta_1 \times \delta_2 ... \times \delta_m$ nodes. Node j^k where $k = (k_1, ..., k_m)$, $k_i = 0, ..., \delta_i - 1$ for $i = 1, ..., m$, can be reached from the source node s if variables $x_1, x_2, ..., x_j$ can be assigned values 0 or 1 such that $\sum_{\ell=1}^{j} a_{i\ell} x_\ell \equiv k_i \pmod{\delta_i}$, $i = 1, 2, ..., m$. When this is the case, node j^k has two outgoing arcs $(j^k, (j+1)^k)$ and $(j^k, (j+1)^{k'})$, where $k'_i \equiv k_i + a_{i,j+1} \pmod{\delta_i}$, corresponding to setting variable x_{j+1} to 0 or to 1. The only arc to t is from node $n^b \pmod{\delta}$. So, the digraph G has $N = 2 + n \times \delta_1 ... \times \delta_m$ nodes and at most twice as many arcs.

Example 2: Consider the set

$$S = \{x \in \{0,1\}^4 : 3x_1 + 2x_2 + 3x_3 + 4x_4 = 5$$
$$6x_1 + 7x_2 + 3x_3 + 3x_4 = 10\}$$

Corresponding to the choice $\delta_i = 2$, $i = 1, 2$, we have the following group relaxation

$$S_{22} = \{x \in \{0,1\}^4 : 1x_1 + 0x_2 + 1x_3 + 0x_4 \equiv 1 \pmod 2$$
$$0x_1 + 1x_2 + 1x_3 + 1x_4 \equiv 0 \pmod 2\}$$

Figure 1 gives the layered digraph G for this relaxation.

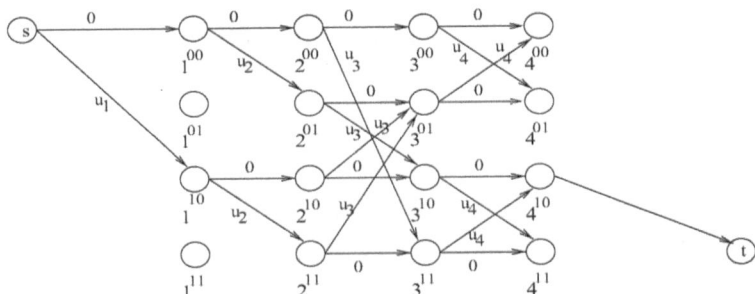

Fig. 1. The layered digraph, G, for Example 2.

Every s-t path in G represents a feasible 0–1 solution to S_{22}. What is more important is that every feasible solution to FP is also a s-t path in G. But this relationship is not reversible. That is, an s-t path in G may not be feasible for IP. Also, G may contain several arcs which do not belong to any s-t path. Such arcs can be easily identified and discarded as follows: Among the outgoing arcs from nodes in layer $n - 1$, we only keep those arcs which reach the node $n^b \pmod{\delta}$ and delete all other arcs. This may introduce paths which terminate at layer $n - 1$. Hence, going through the nodes in layer $n - 2$, we discard all outgoing arcs on paths which terminate at layer $n - 1$, and so on. It can be easily seen that performing a "backward sweep" in G in this manner, in time $O(N)$, we get a new graph G' which consists of only the arcs in solutions to the group relaxation. For the graph G in Figure 1, the graph G' is shown in Figure 2.

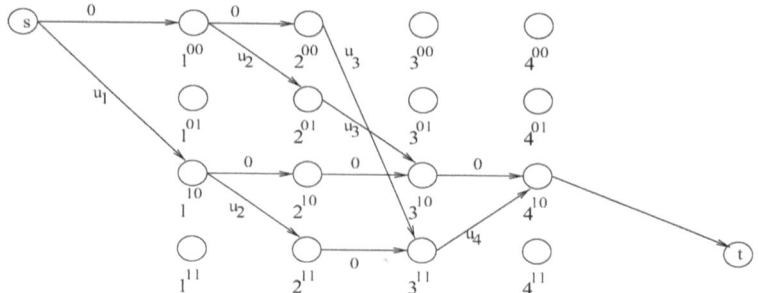

Fig. 2. The solution digraph, G', for Example 2.

For each s-t path in G', we check whether it corresponds to a feasible solution of FP. Thus, a depth-first-search backtracking algorithm solves FP in time $O(pmn)$ where p is the number of s-t paths in G'. Two issues are crucial with regard to the complexity of this approach: the size of p and the size of the digraph G'. A simple estimate shows that the expected value of p is $\frac{2^n}{\delta_1 \times ... \times \delta_m}$ when the data a_{ij} are uniformly and independently distributed between 0 and $D-1$, the δ_i's divide D and the b_i's are uniformly and independently distributed in the range $\lfloor \frac{1}{2}(-D + \sum_{j=1}^{n} a_{ij}) \rfloor$ to $\lfloor \frac{1}{2}(-D + \sum_{j=1}^{n} a_{ij}) \rfloor + D - 1$. On the other hand, the size of G' is of the order $n \times \delta_1 ... \times \delta_m$. The two issues, namely the size of the digraph G' and the number of solutions to the group relaxation, are complementary to each other. As the size of G' increases, the number of solutions to the group relaxation decreases. The best choice is when these two sizes are equal, that is $\delta_1 \times \delta_2 \times ... \delta_m \approx \frac{2^{\frac{n}{2}}}{\sqrt{n}}$. Then, under the above probabilistic assumptions, the expected complexity of the group relaxation approach is $O(\sqrt{n} 2^{\frac{n}{2}})$.

3.2   An $O(n2^{(n/2)})$ Sorting-Based Procedure

Laurence Wolsey [21] suggested a solution approach based on sorting. We first describe this procedure for the subset sum feasibility problem (SSFP).

SSFP: Given an integer n, an integral n-vector $a = (a_1, ..., a_n)$ and an integer b, is $\{x \in \{0,1\}^n : \sum_{j=1}^{n} a_j x_j = b\} \neq \phi$?

The procedure SSP(n,a,b) described below decomposes the problem into two subproblems each of size $\frac{n}{2}$. It then sorts the $2^{\frac{n}{2}}$ subset sums for both the subproblems and traverses the two lists containing these subset sums in opposite directions to find two values that add up to b. WLOG, we assume n is even.

SSP(n,a,b)

1. Let $p = \frac{n}{2}$, $v^1 = \{a_1, ..., a_p\}$ and $v^2 = \{a_{p+1}, ..., a_n\}$. Compute SS^1 and SS^2, the arrays of subset sums of the power sets of v^1 and v^2 respectively.
2. Sort SS^1 and SS^2 in ascending order.

3. Let $k = 2^p$, $i = 1$ and $j = k$.
   do while $((i \le k)$ OR $(j \ge 1))$ {
   If $(SS^1[i] + SS^2[j] = b)$ then **quit**. (Answer to SSFP is "yes")
   else if $(SS^1[i] + SS^2[j] < b)$ then set $i = i + 1$
   else set $j = j - 1$ }.
4. Answer to SSFP is "No".

The complexity of this procedure is dominated by the sorting step and therefore it is $O(n2^{\frac{n}{2}})$. This procedure can be extended to answer FP as follows: For $i = 1, ..., m$, multiply the ith constraint by $(nD)^{i-1}$ and add all the constraints to obtain a single *surrogate* constraint

$$\sum_{j=1}^{n}\sum_{i=1}^{m}(nD)^{i-1}a_{ij}x_j = \sum_{i=1}^{m}(nD)^{i-1}b_i$$

Let $a = (\sum_{i=1}^{m}(nD)^{i-1}a_{i1}, ..., \sum_{i=1}^{m}(nD)^{i-1}a_{in})$, $b = \sum_{i=1}^{m}(nD)^{i-1}b_i$. Call SSP$(n,a,b)$.

Note: As for dynamic programming, this technique is suitable only for problems with a few constraints. If $(nD)^{m-1}$ is too large, a smaller number can be used but then the one-to-one correspondence between the solutions of FP and the solutions of the surrogate constraint is lost. In this case note that, if for some i and j we have $SS^1[i] + SS^2[j] = b$, the corresponding 0–1 solution may not satisfy FP. So, in order to solve FP, we need to find all the pairs i, j such that $SS^1[i] + SS^2[j] = b$.

3.3   Computational Experience

See Table 3. For the group relaxation, we use $\delta_i = 8$, for all $i = 1, ..., m$, for the 5 instances of size 3×20 and $\delta_i = 16$, for all $i = 1, ..., m$, for the remaining instances. Times reported refer to seconds on an HP720 Apollo desktop workstation with 64 megabytes of memory. The sorting-based procedure dominates the group relaxation for instances up to 40 variables. The group relaxation could solve all the instances but is very expensive for larger problems. Within each problem setting, there is very little difference in the amount of time taken by the group relaxation. This is not surprising since, within each problem setting, the number of solutions to the group relaxation is about the same. A similar observation holds for the subset sum approach as well as dynamic programming.

4   Conclusions

In this paper, we consider a class of 0,1 programs with n variables, where n is a multiple of 10. As noted in the previous section, although the group relaxation is able to solve problem instances with up to 50 variables, its running time increases rapidly. Its space complexity and expected time complexity can be estimated to be $O(\sqrt{n}2^{\frac{n}{2}})$. It is an open question to find an algorithm with expected time

Table 3. Computing time comparisons.

Problem size ($m \times n$)	B & B	B & C	DP	Basis	Group	Subset Sort
3×20	11.76	213.62	0.93	300.42	0.13	0.01
3×20	13.04	125.47	1.28	209.79	0.11	0.01
3×20	13.16	208.76	0.94	212.03	0.11	0.01
3×20	13.64	154.71	1.31	277.41	0.12	0.03
3×20	11.21	190.81	1.11	197.37	0.12	0.03
4×30	1542.76	***	20.32	***	18.06	0.99
4×30	1706.84	***	20.37	***	17.83	1.03
4×30	2722.52	***	20.31	***	17.92	0.98
4×30	2408.84	***	18.43	***	17.93	1.00
4×30	2977.28	***	18.94	***	18.04	1.01
5×40	***	***	+++	***	1556.43	46.10
5×40	***	***	+++	***	1562.66	46.18
5×40	***	***	+++	***	1604.02	45.61
5×40	***	***	+++	***	1548.55	46.20
5×40	***	***	+++	***	1606.24	45.51
6×50	***	***	+++	***	26425.01	+++
6×50	***	***	+++	***	26591.28	+++
6×50	***	***	+++	***	26454.30	+++
6×50	***	***	+++	***	27379.04	+++
6×50	***	***	+++	***	27316.98	+++

** Time limit (54000 seconds) exceeded.
+++ Space limit exceeded.

complexity better than $O(2^{\frac{n}{2}})$. Our computational experience indicates that the standard approaches to integer programming are not well suited for this class of problems. As such, we would like to present these small-sized 0–1 integer programs as a challenge for the research community and we hope that they may lead to new algorithmic ideas.

References

1. E. Balas, S. Ceria, and G. Cornuéjols. Mixed 0–1 programming by lift-and-project in a branch-and-cut framework. *Management Science*, 42:1229–1246, 1996.
2. A. Charnes and W. W. Cooper. *Management Models and Industrial Applications of Linear Programming*. Wiley, New York, 1961.
3. V. Chvátal. Hard knapsack problems. *Operations Research*, 28:1402–1411, 1980.
4. W. Cook, T. Rutherford, H. E. Scarf, and D. Shallcross. An implementation of the generalized basis reduction algorithm for integer programming. *ORSA Journal of Computing*, 3:206–212, 1993.
5. H. P. Crowder and E. L. Johnson. Use of cyclic group methods in branch-and-bound. In T. C. Hu and S. M. Robinson, editors, *Mathematical Programming*, pages 213–216. Academic Press, 1973.

6. M. R. Garey and D. S. Johnson. *Computers and Intractability: A Guide to the Theory of NP-Completeness*. Freeman, San Francisco, 1979.
7. R. E. Gomory. On the relation between integer and non-integer solutions to linear programs. *Proceedings of the National Academy of Sciences, Vol. 53*, pages 260–265, 1965.
8. G. A. Gorry, W. D. Northup, and J. F. Shapiro. Computational experience with a group theoretic integer programming algorithm. *Mathematical Programming*, 4:171–192, 1973.
9. G. A. Gorry and J. F. Shapiro. An adaptive group theoretic algorithm for integer programming problems. *Management Science*, 7:285–306, 1971.
10. G. A. Gorry, J. F. Shapiro, and L. A. Wolsey. Relaxation methods for pure and mixed integer programming problems. *Management Science*, 18:229–239, 1972.
11. H. W. Lenstra. Integer programming with a fixed number of variables. *Mathematics of Operations Research*, 8:538–547, 1983.
12. L. Lovász and H. Scarf. The generalized basis reduction algorithm. *Mathematics of Operations Research*, 17:751–763, 1992.
13. S. Martello and P. Toth. *Knapsack Problems: Algorithms and Computer Implementations*. Wiley, Chichester, U.K, 1990.
14. S. Martello and P. Toth. A mixture of dynamic programming and branch-and-bound for the subset sum problem. *Management Science* 30:765–771, 1984.
15. M. Minoux. *Mathematical Programming: Theory and Algorithms*. Wiley, New York, 1986.
16. X. Wang. *A New Implementation of the Generalized Basis Reduction Algorithm for Convex Integer Programming*, PhD thesis, Yale University, New Haven, CT, 1997. In preparation.
17. X. Wang. Private communication, 1997.
18. H. P. Williams. *Model Building in Mathematical Programming*. Wiley, 1978.
19. L. A. Wolsey. Group-theoretic results in mixed integer programming. *Operations Research*, 19:1691–1697, 1971.
20. L. A. Wolsey. Extensions of the group theoretic approach in integer programming. *Management Science*, 18:74–83, 1971.
21. L. A. Wolsey. Private communication, 1997.

Building Chain and Cactus Representations of All Minimum Cuts from Hao-Orlin in the Same Asymptotic Run Time [⋆]

Lisa Fleischer

Department of Industrial Engineering and Operations Research
Columbia University, New York, NY 10027, USA
lisa@@ieor.columbia.edu

Abstract. A cactus tree is a simple data structure that represents all minimum cuts of a weighted graph in linear space. We describe the first algorithm that can build a cactus tree from the asymptotically fastest deterministic algorithm that finds all minimum cuts in a weighted graph — the Hao-Orlin minimum cut algorithm. This improves the time to construct the cactus in graphs with n vertices and m edges from $O(n^3)$ to $O(nm \log n^2/m)$.

1 Introduction

A minimum cut of a graph is a non-empty, proper subset of the vertices such that the sum of the weights of the edges with only one endpoint in the set is minimized. An undirected graph on n vertices can contain up to $\binom{n}{2}$ minimum cuts [7,4]. For many applications, it is useful to know many, or all minimum cuts of a graph, for instance, in separation algorithms for cutting plane approaches to solving integer programs [5,8,2], and in solving network augmentation and reliability problems [3]. Many other applications of minimum cuts are described in [19,1].

In 1976, Dinits, Karzanov, and Lomonosov [7] published a description of a very simple data structure called a cactus that represents all minimum cuts of an undirected graph in linear space. This is notable considering the number of possible minimum cuts in a graph, and the space needed to store one minimum cut. Ten years later, Karzanov and Timofeev [16] outlined the first algorithm to build such a structure for an unweighted graph. Although their outline lacks some important details, it does provide a framework for constructing correct algorithms [20]. In addition, it can be extended to weighted graphs.

[⋆] The full paper is available at http://www.ieor.columbia.edu/~lisa/papers.html. This work supported in part by ONR through an NDSEG fellowship, by AASERT through grant N00014-95-1-0985, by an American Association of University Women Educational Foundation Selected Professions Fellowship, by the NSF PYI award of Éva Tardos, and by NSF through grant DMS 9505155.

R. E. Bixby, E. A. Boyd, and R. Z. Ríos-Mercado (Eds.): IPCO VI
LNCS 1412, pp. 294–309, 1998. © Springer–Verlag Berlin Heidelberg 1998

The earliest algorithm for finding all minimum cuts in a graph uses maximum flows to compute minimum (s,t)-cuts for all pairs of vertices (s,t). Gomory and Hu [12] show how to do this with only n (s,t)-flow computations. The fastest known deterministic maximum flow algorithm, designed by Goldberg and Tarjan [11], runs in $O(nm\log(n^2/m))$ time. Hao and Orlin [13] show how a minimum cut can be computed in the same asymptotic time as one run of the Goldberg-Tarjan algorithm. Using ideas of Picard and Queyranne [18], the Hao-Orlin algorithm can be easily modified to produce all minimum cuts. Karger and Stein [15] describe a randomized algorithm that finds all minimum cuts in $O(n^2\log^3 n)$ time. Recently, Karger has devised a new, randomized algorithm that finds all minimum cuts in $O(n^2\log n)$ time.

The Karzanov-Timofeev outline breaks neatly into two parts: generating a sequence of all minimum cuts, and constructing the cactus from this sequence. This two-phase approach applies to both unweighted and weighted graphs. It is known that the second phase can be performed in $O(n^2)$ time for both weighted and unweighted graphs [16,17,20]. The bottleneck for the weighted case is the first phase. Thus the efficiency of an algorithm to build a cactus tree depends on the efficiency of the algorithm used to generate an appropriate sequence of all minimum cuts of an undirected graph. The Karzanov-Timofeev framework requires a sequence of all minimum cuts found by generating all (s,t) minimum cuts for a given s and a *specific* sequence of t's. The Hao-Orlin algorithm also generates minimum cuts by finding all (s,t) minimum cuts for a given s and a sequence of t's. However, the order of the t's cannot be predetermined in the Hao-Orlin algorithm, and it may not be an appropriate order for the Karzanov-Timofeev framework.

All minimum cuts, produced by any algorithm, can be sequenced appropriately for constructing a cactus tree in $O(n^3)$ time. This is not hard to do considering there are at most $\binom{n}{2}$ minimum cuts, and each can be stored in $O(n)$ space. The main result of this paper is an algorithm that constructs an appropriate sequence of minimum cuts within the same time as the asymptotically fastest, deterministic algorithm that finds all minimum cuts in weighted graphs, improving the deterministic time to construct the cactus in graphs with n vertices and m edges from $O(n^3)$ to $O(nm\log(n^2/m))$.

Why build a cactus tree? Any algorithm that is capable of computing all minimum cuts of a graph implicitly produces a data structure that represents all the cuts. For instance, Karger's randomized algorithm builds a data structure that represents all minimum cuts in $O(k + n\log n)$ space, where k is the number of minimum cuts. The cactus tree is special because it is simple. The size of a cactus tree is linear in the number of vertices in the original graph, and any cut can be retrieved in time linearly proportional to the size of the cut. In addition, the cactus displays explicitly all nesting and intersection relations among minimum cuts. This, as well as the compactness of a cactus, is unique among representations of minimum cuts of weighted graphs.

Karzanov and Timofeev [16] propose a method of constructing a cactus tree using the *chain representation* of minimum cuts (described in Section 2.4). Their algorithm is lacking details, some of which are provided by Naor and Vazirani [17] in a paper that describes a parallel cactus algorithm. Both of these algorithms are not correct since they construct cacti that may not contain all minimum cuts of the original graph. De Vitis [20] provides a complete and correct description of an algorithm to construct a cactus, based on the ideas in [16] and [17]. Karger and Stein [15] give a randomized algorithm for constructing the chain representation in $O(n^2 \log^3 n)$ time. Benczúr [3] outlines another approach to build a cactus without using the chain representation. However, it is not correct since it constructs cacti that may not contain all minimum cuts of the original graph. Gabow [9,10] describes a linear-sized representation of minimum cuts of an unweighted graph and gives an $O(m + \lambda^2 n \log(n/\lambda))$ time algorithm to construct this representation, where λ is the number of edges in the minimum cut.

This paper describes how to modify the output of the Hao-Orlin algorithm and rearrange the minimum cuts into an order suitable for a cactus algorithm based on Karzanov-Timofeev framework. The algorithm presented here runs in $O(nm + n^2 \log n)$ time—at least as fast as the Hao-Orlin algorithm. Since an algorithm based on the Karzanov-Timofeev outline requires $O(n^2)$ time, plus the time to find and sort all minimum cuts, this leads to the fastest known deterministic algorithm to construct a cactus of a weighted graph, and the fastest algorithm for sparse graphs.

2   Preliminaries

2.1   Definitions and Notation

We assume the reader is familiar with standard graph terminology as found in [6]. A *graph* $G = (V, E)$ is defined by a set of vertices V, with $|V| = n$ and a set of edges $E \subseteq V \mathrm{x} V$, with $|E| = m$. A *weighted graph* also has a weight function on the edges, $w : E \to \Re$. For the purposes of this paper, we assume w is nonnegative. A *cut* in a graph is a non-empty, proper subset of the vertices. The weight of a cut C is the sum of weights of edges with exactly one endpoint in the set. The weight of edges with one endpoint in each of two disjoint vertex sets S and T is denoted as $w(S, T)$. A *minimum cut* is a cut C with $w(C, \overline{C}) \leq w(C', \overline{C'})$ for all cuts C'. An (S, T)-*cut* is a cut that contains S and is disjoint from T. A (S, T) *minimum cut* is a minimum cut that separates S and T. Note that if the value of the minimum (S, T)-cut is greater than the value of the minimum cut, there are no (S, T) minimum cuts.

2.2   The Structure of Minimum Cuts of a Graph

A graph can have at most $\binom{n}{2}$ minimum cuts. This is achieved by a simple cycle on n vertices: there are $\binom{n}{2}$ choices of pairs of edges broken by a minimum cut. This is also an upper bound as shown in [7,4] or, with a simpler proof, in [14].

Let λ be the value of a minimum cut in graph $G = (V, E)$. The proof of the next lemma is in the full paper.

Lemma 1. *If S_1 and S_2 are minimum cuts such that none of $A = S_1 \cap S_2$, $B = S_1 \backslash S_2$, $C = S_2 \backslash S_1$, or $D = \overline{S_1 \cup S_2}$ is empty, then*

1. *A, B, C, and D are minimum cuts,*
2. *$w(A, D) = w(B, C) = 0$,*
3. *$w(A, B) = w(B, D) = w(D, C) = w(C, A) = \lambda/2$.*

Two cuts $(S_1, \overline{S_1})$ and $(S_2, \overline{S_2})$ that meet the conditions of the above lemma are called *crossing* cuts. A fundamental lemma further explaining the structure of minimum cuts in a graph is the Circular Partition Lemma. This lemma is proven by Bixby [4] and Dinits, et al. [7], with alternate proofs in [3,20].

Definition 2. *A circular partition is a partition of V into $k \geq 3$ disjoint subsets $V_1, V_2, \dots, V_k$, such that*

- *$w(V_i, V_j) = \lambda/2$ when $i - j = 1 \mod k$ and equals zero otherwise.*
- *For $1 \leq a < b \leq k$, $A = \cup_{i=a}^{b-1} V_i$ is a minimum cut, and if B is a minimum cut such that B or $\overline{B}$ is not of this form, then B or $\overline{B}$ is contained in some V_i.*

Lemma 3 (Bixby [4]; Dinits, Karzanov, and Lomonosov [7]). *If G contains crossing cuts, then G has a circular partition.*

The proof of this lemma is long and technical, and it is omitted here. The proof uses Lemma 1 to argue that any minimum cut not represented in a circular partition must be contained in one of the sets of the partition.

It may be that G has more than one circular partition. Let $P := \{V_1, \dots, V_k\}$ and $Q := \{U_1, \dots, U_l\}$ be two distinct circular partitions of V. These partitions are *compatible* if there is a unique i and j such that $U_r \subset V_i$ for all $r \neq j$ and $V_s \subset U_j$ for all $s \neq i$. Proofs of the next two statements are in the full paper.

Corollary 4. *Any two circular partitions of a graph G are compatible.*

A set of sets is called *laminar* if every pair sets is either disjoint or one set is contained in the other. Laminar sets that do not include the empty set can be represented by a tree where the root node corresponds to the entire underlying set, and the leaves correspond to the sets that contain no other sets. The parent of a set A is the smallest set properly containing A. Since the total number of nodes in a tree with n leaves is less than $2n$, the size of the largest set of laminar sets of n objects is at most $2n - 1$.

Lemma 5. *There are at most $n - 2$ distinct circular partitions of a graph on n vertices.*

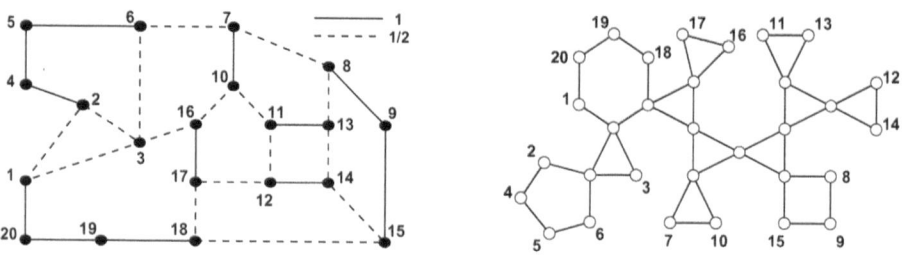

Fig. 1. Cactus of a graph.

2.3   The Cactus Representation

If G has no circular partitions, then G has no crossing cuts. In this case, by considering the smaller side of every cut, the minimum cuts of G are laminar sets. Hence we can represent the minimum cuts of G by the tree used to represent laminar sets mentioned in the previous section. There is a 1-1 correspondence between the minimum cuts of G and the minimum cuts of this tree.

In order to represent crossing minimum cuts, it is necessary to add cycles to this tree representing circular partitions of G. A cactus of a graph is a data structure that succinctly represents all minimum cuts of G by succinctly representing all circular partitions of G.

A *cactus* is a tree-like graph that may contain cycles as long as no two cycles share more than one vertex: every edge in the cactus lies on at most one cycle. Figure 1 contains an example of a cactus. A *cactus of a graph* G, denoted $\mathcal{H}(G)$, has in addition a mapping π that maps vertices of G to vertices of $\mathcal{H}(G)$, and a weight assigned to every edge. To distinguish the vertices of a graph from those of its cactus, we refer to the vertices of a cactus as *nodes*. If λ is the value of the minimum cut in G, then each cycle edge of $\mathcal{H}(G)$ has weight $\lambda/2$, and each path edge has weight λ. The mapping π is such that every minimal cut M of $\mathcal{H}(G)$ corresponds to a minimum cut $\pi^{-1}(M)$ of G, and every minimum cut in G equals $\pi^{-1}(M)$ for some minimal cut M of $\mathcal{H}(G)$. If $\pi^{-1}(i) = \emptyset$, we say that i is an *empty* node. Figure 1 contains a graph and a corresponding cactus. The proofs of the next theorem and corollary are in the full paper.

Theorem 6 (Dinits, Karzanov, and Lomonosov [7]). *Every weighted graph has a cactus.*

Corollary 7. *Every weighted graph on n vertices has a cactus with no more than $2n$ vertices.*

As defined here, and in [7], a graph does not necessarily have a unique cactus. For instance, a cycle on three nodes can also be represented by a cactus that is a star: an empty node of degree three, and three nodes of degree one each containing a distinct vertex. There are many rules that could make the definition unique [17,20]. We will follow De Vitis [20] in the definition of a canonical cactus.

Let i be a node on cycle Y of $\mathcal{H}(G)$. Let C_i^Y be the component containing i formed by removing the edges adjacent to i on Y, and let $V_i^Y = \pi^{-1}(C_i^Y)$. We call i *trivial* if $V_i^Y = \emptyset$. By removing C_i^Y and making the neighbors of i on Y neighbors in Y, we can assume $\mathcal{H}(G)$ has no trivial nodes. We also assume that $\mathcal{H}(G)$ has no empty, 3-way cut-nodes: an empty cut-node whose removal breaks $\mathcal{H}(G)$ into exactly three components can be replaced by a 3-cycle. Finally, we assume there are no empty nodes i in $\mathcal{H}(G)$ of degree three or less: either i is a 3-way cut-node, and handled as above, or it is a 2-way cut-node, and hence we can contract an incident cut-edge and still maintain a representation of all minimum cuts of G in $\mathcal{H}(G)$. A *canonical* cactus of a graph is a cactus with no trivial nodes, no empty 3-way cut-nodes, and no empty nodes with degree ≤ 3. The following theorem is not hard to prove.

Theorem 8 (De Vitis [20]). *Every weighted graph has a unique canonical cactus.*

Henceforth, cactus will be used to mean canonical cactus. In this canonical representation, every circular partition of G corresponds to a cycle of $\mathcal{H}(G)$ and every cycle of $\mathcal{H}(G)$ represents a circular partition of G. For cycle Y, the V_i^Y are the sets of the corresponding circular partition.

2.4 The Chain Representation of Minimum Cuts

Karzanov and Timofeev [16] make an important observation about the structure of minimum cuts, which they use to build the cactus: if two vertices s and t are adjacent in the graph G, then the minimum cuts that separate these vertices are nested sets. This motivates assigning vertices of G an *adjacency order* $\{v_1, \ldots, v_n\}$ so that v_{i+1} is adjacent to a vertex in $V_i := \{v_1, \ldots, v_i\}$. Let M_i be the set of minimum cuts that contain V_i but not v_{i+1}. We will refer to such minimum cuts as (V_i, v_{i+1}) *minimum cuts*. The following lemma summarizes the observation of [16].

Lemma 9. *If the vertices in G are adjacency ordered, then all cuts in M_i are non-crossing.*

This implies that the cuts in M_i form a nested chain and can be represented by a *path* P_i of sets that partition V: let $A_1 \subset A_2 \subset \cdots \subset A_l$ be the minimum cuts separating V_i and v_{i+1}. The first set X_1 of the path is A_1, and each additional set $X_r = A_r \backslash A_{r-1}$, with the final set X_{l+1} equal to $V \backslash A_l$. Each link of the path represents the minimum cut A_r that would result if the path were broken between sets X_r and X_{r+1}.

Note that, for any ordering of vertices, the M_i form a partition of the minimum cuts of G: v_i and v_{i+1} are the smallest, consecutively indexed pair of vertices that are separated by a cut, if and only if the cut is in M_i. The set of P_i for all $1 \leq i < n$ is called the *chain representation of minimum cuts*. For this reason, we will refer to these paths as *chains*.

2.5   From the Chain Representation to the Cactus

The algorithm outlined by Karzanov and Timofeev [16], refined by Naor and Vazirani [17], and corrected by De Vitis [20], builds the cactus representation from the chain representation of minimum cuts. For completeness, an outline of the algorithm is presented here. Let G_i represent the graph G with nodes in V_i contracted into one node. Let G_r be the smallest such graph that has a minimum cut of value λ (r is the largest index of such a graph). The algorithm starts with the cactus for G_r, a pair of vertices incident to one edge. It then builds the cactus for G_{i-1} from G_i using the following observation that follows from the fact that cuts in M_i are non-crossing.

Corollary 10. *Let the vertices in G have an adjacency ordering. For each non-empty M_i, there is a path in the cactus of G_i that shares at most one edge with every cycle, and such that all cuts in M_i cut through this path.*

By contracting the path described in the lemma into a single node, all cuts in M_i are removed from $\mathcal{H}(G_i)$, and no other minimum cuts are removed. Thus the resulting structure is $\mathcal{H}(G_{i+1})$.

Working in the opposite direction, in iteration i, the cactus algorithm replaces the node u in $\mathcal{H}(G_{i+1})$ that contains vertices $\{v_1, \dots, v_i\}$ with the path P_i that represents the chain of cuts C_i in M_i. Edges that were incident to u are then joined to an appropriate node in P_i. The mapping of vertices of G_i to vertices of the cactus are updated. Lastly, post-processing forms the canonical cactus of G_i. More precisely, the algorithm proceeds as follows.

(i) If $C_i = X_1, \dots, X_k$ then replace u by k new nodes $u_1, \dots, u_k$ with edges (u_j, u_{j+1}) for $1 \leq j < k$.

(ii) For any tree or cycle edge (u, w) in $\mathcal{H}(G_{i+1})$, let $W \neq \emptyset$ be the set of vertices in w, or if w is an empty node, the vertices in any non-empty node w' reachable from w by some path of edges disjoint from a cycle containing (u, w). Find the subset X_j such that $W \subset X_j$ and connect w to u_j.

(iii) Let U be the set of vertices mapped to $\mathcal{H}(G_{i+1})$. Assign to u_j the set $X_j \cap U$. All other mappings remain unchanged.

(iv) Remove all empty nodes of degree ≤ 2 and all empty 2-way cut nodes by contracting an adjacent tree edge. Replace all empty 3-way cut-nodes with 3-cycles.

The correctness of this procedure is easy to prove by induction using the previous observations and the following two lemmas.

Lemma 11. *If (u, w) is a tree edge of $\mathcal{H}(G_{i+1})$ and $T \subset V$ is the set of vertices in the subtree attached to u by (u, w), then $T \subseteq X_j$ for some j.*

Lemma 12. *Let (u, w) be a cycle edge of $\mathcal{H}(G_{i+1})$ that lies on cycle Y, and let $T_1, T_2, \dots, T_r$ be the circular partition represented by Y with $V_i \subset T_1$. Then either $\bigcup_{l=2}^r T_l \subset X_j$ for some j, or there are indices a and b such that $T_2 = X_a$, $T_3 = X_{a+1}, \dots, T_r = X_b$.*

Clearly, operations (i)-(iii) of the above procedure can be performed in linear time, thus requiring $O(n^2)$ time to build the entire cactus from the chain representation of minimum cuts. Operation (iv) requires constant time per update. Each contraction removes a node from the cactus, implying no more than $O(n)$ contractions per iteration. Each cycle created is never destroyed, and since the total number of circular partitions of a graph is at most n, there can not be more than n of these operations over the course of the algorithm.

Theorem 13. *The canonical cactus of a graph can be constructed from the chain representation of all minimum cuts of the graph in $O(n^2)$ time.*

2.6   Finding All Minimum Cuts in a Weighted Graph

To find all minimum cuts, we make minor modifications to the minimum cut algorithm of Hao and Orlin [13]. The Hao-Orlin algorithm is based on Goldberg and Tarjan's preflow-push algorithm for finding a maximum flow [11]. It starts by designating a source vertex, u_1, assigning it a label of n, and labeling all other vertices 0. For the first iteration, it selects a sink vertex, u_2, and sends as much flow as possible from the source to the sink (using the weights as arc capacities), increasing the labels on some of the vertices in the process. The algorithm then repeats this procedure $n-2$ times, starting each time by contracting the current sink into the source, and designating a node with the lowest label as the new sink. The overall minimum cut is the minimum cut found in the iteration with the smallest maximum flow value, where the *value* of a flow is the amount of flow that enters the sink.

   To find all minimum cuts with this algorithm, we use the following result. Define a *closure* of a set A of vertices in a directed graph to be the smallest set containing A that has no arcs leaving the set.

Lemma 14. (Picard and Queyranne [18]) *There is a 1-1 correspondence between the minimum (s,t)-cuts of a graph and the closed vertex sets in the residual graph of a maximum (s,t)-flow.*

Let U_i be the set of vertices in the source at iteration i of the Hao-Orlin routine, $1 \le i \le n-1$. If a complete maximum flow is computed at iteration i, Lemma 14 implies that the set of (U_i, u_{i+1}) minimum cuts equals the set of closures in the residual graph of this flow. This representation can be made more compact by contracting strongly connected components of the residual graph, creating a directed, acyclic graph (DAG). We refer to the graph on the same vertex set, but with the direction of all arcs reversed (so that they direct from U_i to u_{i+1}), as the *DAG representation of (U_i, u_{i+1}) minimum cuts.*

   The Hao-Orlin routine does not necessarily compute an entire maximum flow at iteration i. Instead it computes a flow of value $\ge \lambda$ that obeys the capacity constraint for each edge, but leaves excess flow at some nodes contained in a *dormant* region that includes the source. Let R_i denote the vertices in the dormant region at the end of the ith iteration of the Hao-Orlin algorithm. The following lemma follows easily from the results in [13]. Together with the above observations, this implies that at the end of iteration i of the Hao-Orlin algorithm, we can build a linear-sized DAG representation of all (U_i, u_{i+1}) minimum cuts.

Lemma 15. R_i *is contained in the source component of the DAG representation of* (U_i, u_{i+1}) *minimum cuts.*

The second and more serious problem with the Hao-Orlin algorithm is that the ordering of the vertices implied by the selection of sinks in the algorithm may not be an adjacency ordering as required for the construction phase of the Karzanov-Timofeev framework. This means that some of the DAGs may not correspond to directed paths. We address this problem in the next section.

3   The Algorithm

In this section, we discuss how to transform the DAG representation of all minimum cuts into the chains that can be used to construct a cactus representation of all minimum cuts. To do this, we use an intermediate structure called an (S, T)-cactus.

3.1   (S,T)-Cacti

An (S, T)-*cactus* is a cactus representation of all minimum cuts separating vertex sets S and T. Note that an (S, T)-cactus is not necessarily a structure representing minimum (S, T)-cuts, but overall minimum cuts that separate S and T. If the minimum (S, T)-cut has value greater than the minimum cut, then the (S, T)-cactus is a single node containing all vertices. The proof of the following lemma is in the full paper.

Lemma 16. *An* (S, T)-*cactus is a path of edges and cycles.*

Each cycle in an (S, T)-cactus has two nodes that are adjacent to the rest of the path, one on the source side, one on the sink side. The other nodes on the cycle form two paths between these two end nodes. We will call these *cycle-paths*. Thus each cycle in the (S, T)-cactus can be described by a source-side node, a sink-side node, and two cycle-paths of nodes connecting them. The *length* of a cycle-path refers to the number of nodes on the cycle-path. A zero length cycle-path implies that it consists of just one arc joining the source-side node to the sink-side node.

Without loss of generality, we assume that no cycle of an (S, T)-cacti has a zero length cycle-path. Any cycle of an (S, T)-cactus that contains a zero length cycle-path can be transformed into a path by deleting the arc in the zero length cycle-path. This operation does not delete any (S, T)-cut represented by the (S, T)-cactus.

Note that Lemma 9 implies that an (S, T)-cactus of a chain is a path without cycles. A first step in building these chains needed for the cactus algorithm is constructing the $n-1$ (S,T)-cacti of the $n-1$ DAG's output by the Hao-Orlin algorithm. From these (S,T)-cacti, we can then construct chains needed for the cactus algorithm.

Benczúr [3] outlines how to build an (S,T)-cactus from the DAG representation of all minimum cuts separating S and T. The algorithm is a topological sort on the DAG, with some post-processing that can also be performed in $O(m+n)$ time. For completeness, we describe the algorithm here.

For two different cycles of the (S,T)-cactus, all the nodes on the cycle closer to the source must precede all the nodes on the further cycle in any topological ordering. Thus a topological sort sequences the nodes of the DAG so that nodes on a cycle in the (S,T)-cactus appear consecutively. It remains to identify the start and end of a cycle, and to place the nodes of a cycle on the proper chain.

For a node i on cycle Y of a cactus $\mathcal{H}(G)$, define C_i^Y to be the component containing i formed by removing the edges adjacent to i on Y, and define $V_i^Y = \pi^{-1}(C_i^Y)$. If $V_i^Y = \emptyset$ for some Y, we can remove C_i^Y and make the neighbors of i on Y neighbors in Y. Thus we assume $\mathcal{H}(G)$ has no cycle nodes i with $V_i^Y = \emptyset$. The proof of the following lemma is contained in the full paper.

Lemma 17. *Let Y be a cycle on three or more nodes of cactus $\mathcal{H}(G)$ then*

1. *The sum of the costs on edges between any two vertex sets V_i^Y and V_j^Y of adjacent nodes i and j on cycle Y equals $\lambda/2$.*
2. *There are no edges between vertex sets V_i^Y and V_h^Y of nonadjacent nodes i and h on Y.*

Since we can construct an (S,T)-cactus from the cactus of the complete graph by contracting nodes, this lemma also applies to (S,T)-cacti: on a cycle of an (S,T)-cactus, only adjacent nodes share edges.

We can now walk through the topological sort of a DAG produced by the Hao-Orlin algorithm to create the corresponding (S,T)-cactus. As long as the weight of the edges between two consecutive components in the topological order equals λ, the value of a minimum cut, the (S,T)-cactus is a path. The first time we encounter a value $< \lambda$, the path divides into two cycle-paths of a cycle. We stay with one cycle-path as long as the weight of the edges between two consecutive components $= \lambda/2$. If it is zero, we jump to the other cycle-path. If it is anything in between $\lambda/2$ and zero, we end the cycle with an empty vertex, and start a new cycle. If it is λ, we end the cycle with the current vertex, and resume a path. At the end of the topologically ordered list, we end the cycle (if started) with the last component.

Lemma 18. (Benczúr [3]) *A DAG representing strongly connected components of the residual graph of a maximum (s,t)-flow can be transformed into an (s,t)-cactus in $O(m+n)$ time.* □

3.2   Constructing the Chains

In this section we describe how to efficiently build chains P_i of minimum cuts that can be used to construct a cactus representation of all minimum cuts from the (U_{j-1}, u_j)-cacti, $1 < j \leq n$. Let D_j be the (U_{j-1}, u_j)-cactus for $1 < j \leq n$. We start by fixing an adjacency ordering $\{v_1, \dots, v_n\}$ with $v_1 = u_1$ and let $\sigma : V \to V$ be the permutation $\sigma(i) = j$ if $v_i = u_j$.

The algorithm we present below builds each chain P_i, $1 < i \leq n$, one at a time by reading cuts off of the D_j. The basic idea is to use the following fact established below: If any two cuts from D_j are such that one is completely contained within the other, then the index of the chain to which the first belongs is no more than the index of the chain to which the second belongs. Thus, all the minimum cuts belonging in P_i that are contained in one D_j are consecutive; and, as we walk through the cuts from U_{j-1} to u_j in D_j, the index of the chain to which a cut belongs is non-decreasing. The validity of these statements is established in the next two subsections.

The Code

(s,t)-CactiToChains (v)

```
0. For i = 2 to n,
0.   P_i = ∅.
0.      For j = 2 to n − 1,
1.          n_ij := the node in D_j that contains v_i.
2.          If n_ij is not the source node,
3.              Identify the path of nodes from source to predecessor of n_ij in D_j.
4.              If P_i = ∅, let the source node of D_j be the source node of P_i.
5.              Else, add vertices in the source but not in P_i to the last node of P_i.
6.              Append the remaining path to P_i, skipping empty nodes;
7.              and, in D_j, contract this path into the source node.
8.          If n_ij is on a cycle in D_j,
9.              Append to P_i the chain of the cycle that does not contain n_ij.
10.             Contract n_ij into the source node of D_j.
11.             Add a component to P_i that contains only u_j.
12.     If P_i ≠ ∅, add the vertices not in P_i to the last node of P_i.
```

Correctness: In this section we prove that each P_i constructed by the above algorithm represents all cuts in M_i. Define a *phase* to be one pass through the outer while loop. P_i is *complete* when it contains all vertices. A node of P_i is *complete* when either a new node of P_i is started, or P_i is complete. For example, a node started at Step 11 is not complete until after the next execution of Step 5 or Step 12. A cut in D_j is obtained by either removing an arc on a path or a pair of arcs on two different cycle-paths of a cycle. The one or two nodes that are endpoints of these arcs on the source side are said to *represent* this cut.

Theorem 19. *Each chain P_i constructed by the algorithm contains precisely the cuts in M_i.*

The proof of this theorem establishes the correctness of the algorithm and depends upon the following lemmas.

source

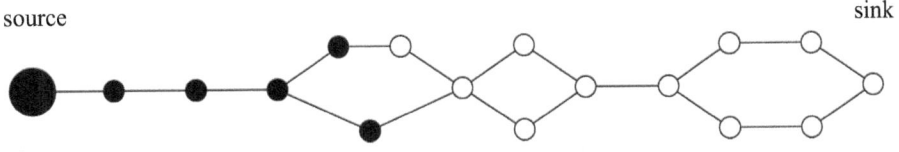

sink

Fig. 2. The (s,t)-cactus D_j, updated for iteration i. The black nodes represent the possible locations of n_{ij}.

Lemma 20. *At the end of phase i, all vertices $v_k, k \leq i$ are contained in the source node of every remaining D_j.*

Proof. Proof is by induction. At start, all sources contain vertex $v_1 = u_1$. Assume vertices v_k, $k < i$ are contained in the sources of the remaining D_j at the start of phase i. For each j, if n_{ij} is not the source node, then n_{ij} is contracted into the source in Step 10. □

Lemma 21. *In phase i, for all j, n_{ij} either lies on the path in D_j from the source to the first node of the cycle closest to the source, or is one of the two nodes on the cycle adjacent to this path, as in Figure 2.*

Proof. Figure 2 shows the possible locations of n_{ij} in D_j as specified in the lemma. We show that if there is a cycle between n_{ij} and the source S_j, then we can find crossing cuts in M_i, contradicting our choice of the ordering of the v_i. If n_{ij} is not in one of the locations stipulated in the lemma, then some cycle contains at least one node that either precedes, or is unrelated to n_{ij}, on both of its cycle-paths. (Recall that we are assuming that every cycle-path has length greater than zero.) Thus there are two crossing minimum cuts that separate the source of D_j from v_i. Lemma 20 implies that V_{i-1} is contained in the source, and hence these are cuts in M_i, contradicting Lemma 1. □

Lemma 22. *The algorithm constructs chains that contain every vertex exactly once.*

Proof. By construction, each P_i is a chain that contains every vertex. Suppose, in iteration j, we are about to add a vertex u_r to the chain P_i. If u_r is in the source component of D_j, we explicitly check before we add u_r. Otherwise, $r \geq j$ and there is some cut that separates vertices in U_{j-1} and V_{i-1} from vertices u_r and u_j. All cuts added to P_i before this iteration come from a D_k with $k < j$, and thus contain a strict subset of vertices in U_{j-1}. By Lemma 9, the cuts separating V_{i-1} from v_i are nested, and so none of the previous cuts contain u_r in the source side. Thus we add u_r to P_i exactly once. □

Lemma 23. *Each complete node added to P_i represents the same cut that is represented by the node in the current D_j.*

Proof. Proof is by induction on j. The first node added to P_i is the source node of some D_j and clearly both these cuts are the same. Now assume that up until this last node, all previous nodes added to P_i represented the corresponding cut of the relevant D_k, $k \leq j$. Adding the current node from D_j, we add or have added in previous steps to P_i all the vertices that are in the source side of this cut. We must argue that there are no additional vertices in any node preceding this one in P_i. Recall that the cuts in M_i are nested. Thus all previous nodes added to P_i represent cuts strictly contained in the current cut. Hence the node added to P_i represents the same cut this node represents in the current D_j. □

Lemma 24. *Each complete node added to P_i, $1 < i \leq n$, except the last, represents a distinct cut in M_i.*

Proof. Since each node added to P_i contains some vertex, and no vertex is repeated, each represents a distinct cut. When P_i is started, Lemma 20 implies that all vertices in V_{i-1} are contained in the source. Vertex v_i is only added in Step 11, when $j = \sigma(i)$. This starts the final node of P_i since $v_i = u_j$ is contained in the source nodes of all D_k, $k > j$. Hence all previous nodes added to P_i represent cuts that contain V_{i-1} but not v_i, and are thus cuts in M_i. □

Proof of Theorem 19. For a fixed cut C, let v_i be the vertex of least index in $\overline{C}$. We show that this cut is represented in P_i. By the end of each phase k, $k < i$, the algorithm contracts all nodes on the path from the source to the node containing v_k. Since V_{i-1} is contained in the source side of C, this cut is not contracted in D_j at the start of phase i. If the cut C is represented in D_j by one node, it lies on the path of D_j from the expanded source to n_{ij}. Thus this node is added to P_i in either Step 5 or 6. Otherwise the cut is through the same cycle that contains n_{ij}. One node representing the cycle is the node immediately preceding n_{ij}, currently source-side node of the cycle. The other node lies on the cycle-path not containing n_{ij}. This cut is added to P_i in Step 9, when this latter node is added. □

Efficiency: Some care is needed to implement the above algorithm to complete within the same asymptotic time as the Hao-Orlin algorithm. Within the inner "for" loops, operations should not take more than $O(\log n)$ time. We show, that with minor preprocessing and record keeping, all operations will take no more than logarithmic time per operation, or can be charged to some set of minimum cuts.

Before starting the algorithm, we walk through each (s, t)-cactus from t to s, labeling every node with the smallest index of a vertex v_i contained in that node, and labeling every arc with the smallest index on the sink side of the arc. This takes linear time per (s, t)-cactus, for a total of $O(n^2)$ time.

We perform one more preprocessing step. For each vertex, we create an array of length $n - 1$. In the array for vertex v_i, entry j indicates if v_i is in the source component of D_j. This takes $O(n^2)$ time to initialize; and, through the course of the algorithm, it will take an additional $O(n^2)$ time to maintain, since once a vertex is contracted into the source, it stays there.

Theorem 25. *Algorithm* (s,t)-*CactiToChains runs in* $O(n^2 \log n)$ *time.*

Proof. In Step 1, we use the labeling of the arcs to walk through D_j starting from the source until we find n_{ij}. By Lemmas 20 and 21, all these cuts are (V_{i-1}, v_i)-cuts, and hence the arcs are labeled with v_i. The time to do this depends on the out degree of each node encountered in D_j (at most 2), and is charged to the minimum cuts we encounter. Note that if none of the arcs leaving the source is labeled with the current chain index i, then n_{ij} must be the source node. This takes care of Steps 1-3, and Step 8.

Let's put aside Steps 5 and 12 and concentrate on the remaining steps. In these, we are either adding nodes and vertices to P_i, or contracting nodes of D_j. This takes constant time per addition or contraction of vertex. Since each vertex is added at most once to P_i and contracted into the source of D_j at most once, over the course of the algorithm these steps take no more than $O(n^2)$ time.

Finally, we consider Steps 5 and 12. In these, we need to determine the difference of two sets that are increasing over the course of the algorithm. To do this efficiently, we wait until the end of a phase to add the vertices that would have been added at Step 5. At the end of each phase, we make a list of the vertices that are missing from P_i. This can be done in linear time per phase by maintaining an array in each phase that indicates if a vertex has been included in P_i. For each of the vertices not in P_i, we then need to determine to which node of P_i it should be added. The nodes that are possible candidates are the incomplete nodes—any node created in Step 11.

Vertex v is added to the node containing u_k in Step 5 if v is contained in the source of D_j, $j > k$ but is not contained in the source of D_k, for iterations k and j where n_{ik} and n_{ij} are not source nodes, and k is the highest index with these properties. Consider the D_j for which n_{ij} is not the source node. The sources of these D_j are nested for any fixed i, by the observation that the cuts of M_i are nested. That is, once a vertex is contained in one of them, it is contained in the rest. During each phase, we maintain a array-list of the consecutive j for which n_{ij} is not the source node. For each vertex, we can now binary search through these nested sources using the arrays described before the statement of the theorem to find the node of P_i that should contain v in $O(\log n)$ time. The time spent on this step over the course of the algorithm is then bounded by $O(n^2 \log n)$: logarithmic time per vertex per phase.

Note that we have added vertices to the sources of these D_j since we started the phase. All the vertices we added in Step 7 have also been added to P_i, so they are not among the vertices we need to place at the end (i.e. the vertices not in P_i). We do need to worry about the vertices added in Step 10, however. To avoid this problem, simply postpone the contractions described in Step 10 until after we have assigned all vertices to a node of P_i. $\square$

Note that $mn \log n^2/m$ is never smaller than $n^2 \log n$.

Corollary 26. *The chain and the cactus representations of all minimum cuts of a weighted graph can both be constructed in* $O(mn \log n^2/m)$ *time.* $\square$

Acknowledgement

I would like to thank Éva Tardos for assistance with the proof of Theorem 25.

References

1. R. K. Ahuja, T. L. Magnanti, and J. B. Orlin. *Network Flows*. Prentice Hall, 1993.
2. D. Applegate, R. Bixby, V. Chvátal, and W. Cook. Finding cuts in the TSP (a preliminary report). Technical Report 95-05, DIMACS, 1995.
3. A. A. Benczúr. *Cut Structures and Randomized Algorithms in Edge-Connectivity Problems*. PhD thesis, Department of Mathematics, Massachusetts Institute of Technology, June 1997.
4. R. E. Bixby. The minimum number of edges and vertices in a graph with edge connectivity n and m n-bonds. *Networks*, 5:253–298, 1975.
5. A. Caprara, M. Fischetti, and A. N. Letchford. On the separation of maximally violated mod-k cuts. Unpublished manuscript, 1997.
6. T. H. Cormen, C. E. Leiserson, and R. L. Rivest. *Introduction to Algorithms*. The MIT Press / McGraw-Hill, 1990.
7. E. A. Dinits, A. V. Karzanov, and M. V. Lomonosov. On the structure of a family of minimal weighted cuts in a graph. In A. A. Fridman, editor, *Studies in Discrete Optimization*, pages 290–306. Moscow Nauka, 1976. Original article in Russian. This is an English translation obtained from ITC-International Translations Centre, Schuttersveld 2, 2611 WE Delft, The Netherlands. (ITC 85-20220); also available from NTC-National Translations Center, Library of Congress, Cataloging Distribution Service, Washington DC 20541, USA (NTC 89-20265).
8. L. Fleischer and É. Tardos. Separating maximally violated comb inequalities in planar graphs. In *Integer Programming and Combinatorial Optimization: 5th International IPCO Conference, LNCS, Vol. 1084*, pages 475–489, 1996.
9. H. N. Gabow. Applications of a poset representation to edge connectivity and graph rigidity. In *Proc. 32nd Annual Symp. on Found. of Comp. Sci.*, pages 812–821, 1991.
10. H. N. Gabow. Applications of a poset representation to edge connectivity and graph rigidity. Technical Report CU-CS-545-91, Department of Computer Science, University of Colorado at Boulder, 1991.
11. A. V. Goldberg and R. E. Tarjan. A new approach to the maximum flow problem. *Journal of ACM*, 35:921–940, 1988.
12. R. E. Gomory and T. C. Hu. Multi-terminal network flows. *J. Soc. Indust. Appl. Math*, 9(4):551–570, 1991.
13. J. Hao and J. B. Orlin. A faster algorithm for finding the minimum cut in a graph. In *Proc. of 3rd ACM-SIAM Symp. on Discrete Algorithms*, pages 165–174, 1992.
14. D. R. Karger. Random sampling in cut, flow, and network design problems. In *Proc. of 6th Annual ACM-SIAM Symposium on Discrete Algorithms*, pages 648–657, 1995.
15. D. R. Karger and C. Stein. A new approach to the minimum cut problem. *Journal of the ACM*, 43(4):601–640, 1996.
16. A. V. Karzanov and E. A. Timofeev. Efficient algorithms for finding all minimal edge cuts of a nonoriented graph. *Cybernetics*, 22:156–162, 1986. Translated from *Kibernetika*, 2:8–12, 1986.

17. D. Naor and V. V. Vazirani. Representing and enumerating edge connectivity cuts in RNC. In *Proc. Second Workshop on Algorithms and Data Structures, LNCS, Vol. 519*, pages 273–285. Springer-Verlag, 1991.
18. J.-C. Picard and M. Queyranne. On the structure of all minimum cuts in a network and applications. *Mathematical Programming Study*, 13:8–16, 1980.
19. J.-C. Picard and M. Queyranne. Selected applications of minimum cuts in networks. *INFOR*, 20(4):394–422, 1982.
20. A. De Vitis. The cactus representation of all minimum cuts in a weighted graph. Technical Report 454, IASI-CNR, 1997.

Simple Generalized Maximum Flow Algorithms

Éva Tardos[1]* and Kevin D. Wayne[2]**

[1] Computer Science Department, and
Operations Research and Industrial Engineering Department
Cornell University, Ithaca, NY 14853, USA
`eva@@cs.cornell.edu`
[2] Operations Research and Industrial Engineering Department
Cornell University, Ithaca, NY 14853, USA
`wayne@@orie.cornell.edu`

Abstract. We introduce a *gain-scaling* technique for the generalized maximum flow problem. Using this technique, we present three simple and intuitive polynomial-time combinatorial algorithms for the problem. Truemper's augmenting path algorithm is one of the simplest combinatorial algorithms for the problem, but runs in exponential-time. Our first algorithm is a polynomial-time variant of Truemper's algorithm. Our second algorithm is an adaption of Goldberg and Tarjan's preflow-push algorithm. It is the first polynomial-time preflow-push algorithm in generalized networks. Our third algorithm is a variant of the Fat-Path capacity-scaling algorithm. It is much simpler than Radzik's variant and matches the best known complexity for the problem. We discuss practical improvements in implementation.

1 Introduction

In this paper we present new algorithms for the *generalized maximum flow problem*, also known as the *generalized circulation problem*. In the traditional maximum flow problem, the objective is to send as much flow through a network from one distinguished node called the source to another called the sink, subject to capacity and flow conservation constraints. In generalized networks, a fixed percentage of the flow is lost when it is sent along arc. Specifically, each arc (v, w) has an associated *gain factor* $\gamma(v, w)$. When $g(v, w)$ units of flow enter arc (v, w) at node v then $\gamma(v, w)g(v, w)$ arrive at w. The gains factors can represent physical transformations due to evaporation, energy dissipation, breeding, theft, or interest rates. They can also represent transformations from one commodity to another as a result of manufacturing, blending, or currency exchange. They

---

* Research supported in part by an NSF PYI award DMI-9157199, by NSF through grant DMS 9505155, and by ONR through grant N00014-96-1-0050.
** Research supported in part by ONR through grant AASERT N00014-97-1-0681.

R. E. Bixby, E. A. Boyd, and R. Z. Ríos-Mercado (Eds.): IPCO VI
LNCS 1412, pp. 310–324, 1998. © Springer–Verlag Berlin Heidelberg 1998

may also represent arc failure probabilities. Many applications are described in [1,3,5].

Since the generalized maximum flow problem is a special case of linear programming, it can be solved using simplex, ellipsoid, or interior-point methods. Many general purpose linear programming algorithms can be tailored for the problem. The network simplex method can handle generalized flows. Kapoor and Vaidya [16] showed how to speed up interior-point methods on network flow problems by exploiting the structured sparsity in the underlying constraint matrix. Murray [18] and Kamath and Palmon [15] designed different interior-point algorithms for the problem. We note that these simplex and interior-point methods can also solve the generalized minimum cost flow problem.

The first *combinatorial* algorithms for the generalized maximum flow problem were the augmenting path algorithms of Jewell [14] and Onaga [19] and exponential-time variants. Truemper [22] observed that the problem is closely related to the minimum cost flow problem, and that many of the early generalized maximum flow algorithms were, in fact, analogs of pseudo-polynomial minimum cost flow algorithms. Goldberg, Plotkin and Tardos [7] designed the first two combinatorial polynomial-time algorithms for the problem: Fat-Path and MCF. The Fat-Path algorithm uses capacity-scaling and a subroutine that cancels flow-generating cycles. The MCF algorithm performs minimum cost flow computations. Radzik [20] modified the Fat-Path algorithm, by canceling only flow-generating cycle with sufficiently large gains. Goldfarb and Jin [12] modified the MCF algorithm by replacing the minimum cost flow subroutine with a simpler computation. Goldfarb and Jin [11] also presented a dual simplex variant of this algorithm. Recently, Goldfarb, Jin and Orlin [13] designed a new capacity-scaling algorithm, motivated by the Fat-Path algorithm. Tseng and Bertsekas [23] proposed an ϵ-relaxation method for solving the more general *generalized minimum cost flow problem with separable convex costs*. However, their running time may be exponential in the input size.

Researchers have also developed algorithms for the *approximate generalized maximum flow problem*. Here, the objective is to find a ξ-optimal flow, i.e., a flow that generates excess at the sink that is within a $(1 - \xi)$ factor of the optimum, where ξ is an input parameter. Cohen and Megiddo [2] showed that the approximate generalized maximum flow problem can be solved in strongly polynomial-time. Their algorithm uses a subroutine which tests feasibility of a linear system with two variables per inequality. Radzik [20] observed that the Fat-Path algorithm can be used to compute approximates flow faster than optimal flows. His Fat-Path variant, that cancels only flow-generating cycles with large gain, is the fastest algorithm for computing approximate flows. Subsequently, Radzik [21] gave a new strongly polynomial-time analysis for canceling all flow-generating cycles, implying that the original Fat-Path algorithm computes an approximate flow in strongly polynomial-time. For the linear programming algorithms, it is not known how to improve the worst-case complexity of the exact algorithms to find approximate flows.

We present a new rounding technique for generalized flows, which can be viewed as a type of *gain-scaling*. Using this technique, we propose three simple combinatorial algorithms for the generalized maximum flow problem. Our first algorithm is a polynomial-time variant of Truemper's [22] algorithm. Truemper's algorithm is a very simple maximum flow based augmenting path algorithm, analogous to Jewell's primal-dual algorithm for the minimum cost flow problem. Truemper's algorithm may require exponential-time, but by applying our new gain-scaling technique, we develop a polynomial-time variant. Our second algorithm is an adaption Goldberg and Tarjan's [10] preflow-push algorithm for the minimum cost flow problem. Using gain-scaling, we establish the first polynomial-time preflow-push algorithm for generalized flows. Our third algorithm is a simple variant of the Fat-Path capacity-scaling algorithm. By using gain-scaling, our Fat-Path variant improves the complexity of canceling flow-generating cycles, and hence of the overall algorithm. In contrast, Radzik's Fat-Path variant modifies this subroutine, canceling only flow-generating cycles with sufficiently large gain. Both Fat-Path variants have the same complexity, but Radzik's variant and proof of correctness are quite complicated.

2   Preliminaries

2.1   Generalized Networks

Since some of our algorithms are iterative and recursive, it is convenient to solve a seemingly more general version of the problem which allows for multiple sources. An instance of the generalized maximum flow problem is a *generalized network* $G = (V, E, t, u, \gamma, e)$, where V is an n-set of nodes, E is an m-set of directed arcs, $t \in V$ is a distinguished node called the *sink*, $u \colon E \to \Re_{\geq 0}$ is a *capacity function*, $\gamma \colon E \to \Re_{>0}$ is a *gain function*, and $e \colon V \to \Re_{\geq 0}$ is an *initial excess function*. A *residual arc* is an arc with positive capacity. A *lossy network* is a generalized network in which no residual arc has gain factor exceeding one. We consider only simple directed paths and cycles. The *gain of a path* P is denoted by $\gamma(P) = \prod_{e \in P} \gamma(e)$. The *gain of a cycle* is defined similarly. A *flow-generating cycle* is a cycle whose gain is more than one.

For notational convenience we assume that G has no parallel arcs. Our algorithms easily extend to allow for parallel arcs and the running times we present remain valid. Without loss of generality, we assume the network is *symmetric* and the gain function is *antisymmetric*. That is, for each arc $(v, w) \in E$ there is an arc $(w, v) \in E$ (possibly with zero capacity) and $\gamma(w, v) = 1/\gamma(v, w)$. We assume the capacities and initial excesses are given as integers between 1 and B, and the gains are given as ratios of integers which are between 1 and B. To simplify the running times we assume $B \geq m$, and use $\tilde{\mathcal{O}}(f)$ to denote $f \log^{\mathcal{O}(1)} m$.

2.2   Generalized Flows

A *generalized pseudoflow* is a function $g \colon E \to \Re$ that satisfies the *capacity constraints* $g(v, w) \leq u(v, w)$ for all $(v, w) \in E$ and the *antisymmetry constraints*

$g(v, w) = -\gamma(w, v)g(w, v)$ for all $(v, w) \in E$. The *residual excess* of g at node v is $e_g(v) = e(v) - \sum_{(v,w) \in E} g(v, w)$, i.e., the initial excess minus the the net flow out of v. If $e_g(v)$ is positive (negative) we say that g has *residual excess (deficit)* at node v. A *flow* g is a pseudoflow that has no residual deficits; it may have residual excesses. A *proper flow* is a flow which does not generate any additional residual excess, except possibly at the sink. We note that a flow can be converted into a proper flow, by removing flow on useless paths and cycles. For a flow g we denote its *value* $|g| = e_g(t)$ to be the residual excess at the sink. Let $\text{OPT}(G)$ denote the maximum possible value of any flow in network G. A flow g is *optimal* in network G if $|g| = \text{OPT}(G)$ and ξ-*optimal* if $|g| \geq (1 - \xi)\text{OPT}(G)$. The *(approximate) generalized maximum flow problem* is to find a (ξ-) optimal flow. We sometimes omit the adjective *generalized* when its meaning is clear from context.

2.3 Residual and Relabeled Networks

Let g be a generalized flow in network $G = (V, E, s, u, \gamma, e)$. With respect to the flow g, the *residual capacity function* is defined by $u_g(v, w) = u(v, w) - g(v, w)$. The *residual network* is $G_g = (V, E, s, u_g, \gamma, e_g)$. Solving the problem in the residual network is equivalent to solving it in the original network.

Our algorithms use the technique of *relabeling*, which was originally introduced by Glover and Klingman [4]. A *labeling function* is a function $\mu: V \to \Re_{>0}$ such that $\mu(t) = 1$. The *relabeled network* is $G_\mu = (V, E, t, u_\mu, \gamma_\mu, e_\mu)$, where the *relabeled capacity*, *relabeled gain* and *relabeled initial excess* functions are defined by: $u_\mu(v, w) = u(v, w)/\mu(v)$, $\gamma_\mu(v, w) = \gamma(v, w)\mu(v)/\mu(w)$, and $e_\mu(v) = e(v)/\mu(v)$. The relabeled network provides an equivalent instance of the generalized maximum flow problem. Intuitively, node label $\mu(v)$ changes the local units in which flow is measured at node v; it is the number of old units per new unit. The inverses of the node labels correspond to the linear programming dual variables, for the primal problem with decision variables $g(v, w)$. With respect to a flow g and labels μ we define the the *relabeled residual network* by $G_{g,\mu} = (V, E, t, u_{g,\mu}, \gamma_\mu, e_{g,\mu})$, where the *relabeled residual capacity* and *relabeled residual excess* functions are defined by $u_{g,\mu}(v, w) = (u(v, w) - g(v, w))/\mu(v)$ and $e_{g,\mu}(v) = e_g(v)/\mu(v)$. We define the *canonical label* of a node v in network G to be the inverse of the highest gain residual path from v to the sink. If G has no residual flow-generating cycles, then we can compute the canonical labels using a single shortest path computation with costs $c(v, w) = -\log \gamma(v, w)$.

2.4 Optimality Conditions

An *augmenting path* is a residual path from a node with residual excess to the sink. A *generalized augmenting path* (GAP) is a residual flow-generating cycle, together with a (possibly trivial) residual path from a node on this cycle to the sink. By sending flow along augmenting paths or GAPs we increase the net flow into the sink. The following theorem of Onaga [19] says that the nonexistence of augmenting paths and GAPs implies that the flow is optimal.

Theorem 1. *A flow g is optimal in network G if and only if there are no augmenting paths or GAPs in G_g.*

2.5   Finding a Good Starting Flow

Our approximation algorithms require a rough estimate of the optimum value in a network. Radzik [20] proposed a $\tilde{O}(m^2)$ time greedy augmentation algorithm that finds a flow that is within a factor n of the optimum. His greedy algorithm repeatedly sends flow along highest-gain augmenting paths, but does *not* use arcs in "backward" direction. Using this algorithm, we can determine an initial parameter Δ_0 which satisfies $\text{OPT}(G) \leq \Delta_0 \leq n\text{OPT}(G)$.

2.6   Canceling Flow-Generating Cycles

Subroutine CANCELCYCLES converts a generalized flow g into another generalized flow g' whose residual network contains no flow-generating cycles. In the process, the net flow into every node, including the sink, can only be increased. It also finds node labels μ so that $G_{g',\mu}$ is a lossy network. This subroutine is used by all of our algorithms. CANCELCYCLES was designed by Goldberg, Plotkin, and Tardos and is described in detail in [7]. It is an adaptation of Goldberg and Tarjan's [9] cancel-and-tighten algorithm for the minimum cost flow problem using costs $c(v, w) = -\log_b \gamma(v, w)$ for any base $b > 1$. Note that negative cost cycles correspond to flow-generating cycles. In Section 7 we discuss practical implementation issues.

Theorem 2. *Let $b > 1$. If all of the costs are integral and at least $-C$, then* CANCELCYCLES *runs in $\tilde{O}(mn \log C)$ time.*

In generalized flows, the costs will typically not be integral. In this case, the next theorem is useful.

Theorem 3. *If the gains are given as ratios of integers between 1 and B, then* CANCELCYCLES *requires $\tilde{O}(mn^2 \log B)$ time.*

Radzik [21] showed that CANCELCYCLES runs in strongly polynomial-time. We have a variant that limits the relabeling increases, allowing a simpler proof of the strongly polynomial running time.

2.7   Nearly Optimal Flows

The following lemma derived from [7] says that if a flow is ξ-optimal for sufficiently small ξ, then we can efficiently convert it into an optimal flow. It is used to provide termination of our exact algorithms. The conversion procedure involves one call to CANCELCYCLES and a single (nongeneralized) maximum flow computation.

Lemma 1. *Given a B^{-4m}-optimal flow, we can compute an optimal flow in $\tilde{O}(mn^2 \log B)$ time.*

3   Gain-Scaling

In this section we present a rounding and scaling framework. Together, these ideas provide a technique which can be viewed as a type of *gain-scaling*. By rounding the gains, we can improve the complexity of many generalized flow computations (e.g., canceling flow-generating cycles above). However, our approximation from rounding creates error. Using an iterative or recursive approach, we can gradually refine our approximation, until we obtain the desired level of precision.

3.1   Rounding Down the Gains

In our algorithms we round down the gains so that they are all integer powers of a base $b = (1 + \xi)^{1/n}$. Our rounding scheme applies in lossy networks, i.e., networks in which no residual arc has a gain factor above one. This implies that the network has no residual flow-generating cycles. We round the gain of each residual arc down to $\bar{\gamma}(v, w) = b^{-\bar{c}(v,w)}$ where $\bar{c}(v, w) = -\lfloor \log_b \gamma(v, w) \rfloor$, maintaining antisymmetry by setting $\bar{\gamma}(w, v) = 1/\bar{\gamma}(v, w)$. Note that if both (v, w) and (w, v) are residual arcs, then each has unit gain, ensuring that $\bar{\gamma}$ is well-defined. Let H denote the resulting *rounded network*. H is also a lossy network, sharing the same capacity function with G. Let h be a flow in network H. The *interpretation of flow* h as a flow in network G is defined by: $g(v, w) = h(v, w)$ if $g(v, w) \geq 0$ and $g(v, w) = -\gamma(w, v)h(w, v)$ if $g(v, w) < 0$. Flow interpretation in lossy networks may create additional excesses, but no deficits. We show that approximate flows in the rounded network induce approximate flows in the original network. First we show that the rounded network is close to the original network.

Theorem 4. *Let G be a lossy network and let H be the rounded network constructed as above. If $0 < \xi < 1$ then $(1 - \xi)OPT(G) \leq OPT(H) \leq OPT(G)$.*

Proof. Clearly $OPT(H) \leq OPT(G)$ since we only decrease the gain factors of residual arcs. To prove the other inequality, we consider the path formulation of the maximum flow problem in lossy networks. We include a variable x_j for each path P_j, representing the amount of flow sent along the path. Let x^* be an optimal path flow in G. Then x^* is also a feasible path flow in H. From path P_j, $\gamma(P_j)x_j^*$ units of flow arrive at the sink in network G, while only $\bar{\gamma}(P_j)x_j^*$ arrive in network H. The theorem then follows, since for each path P_j,

$$\bar{\gamma}(P_j) \geq \frac{\gamma(P_j)}{b^{|P|}} \geq \frac{\gamma(P_j)}{1 + \xi} \geq \gamma(P_j)(1 - \xi).$$

Corollary 1. *Let G be a lossy network and let H be the rounded network constructed as above. If $0 < \xi < 1$ then the interpretation of a ξ'-optimal flow in H is a $\xi + \xi'$-optimal flow in G.*

Proof. Let h be a ξ'-optimal flow in H. Let g be the interpretation of flow h in G. Then we have

$$|g| \geq |h| \geq (1 - \xi')\text{OPT}(H) \geq (1 - \xi)(1 - \xi')\text{OPT}(G) \geq (1 - \xi - \xi')\text{OPT}(G).$$

3.2   Error-Scaling and Recursion

In this section we describe an *error-scaling* technique which can be used to speed up computations for generalized flow problems. Radzik [20] proposed a recursive version of error-scaling to improve the complexity of his Fat-Path variant when finding nearly optimal and optimal flows. We use the technique in a similar manner to speed up our Fat-Path variant. We also use the idea to convert constant-factor approximation algorithms into fully polynomial-time approximation schemes.

Suppose we have a subroutine which finds a 1/2-optimal flow. Using error-scaling, we can determine a ξ-optimal flow in network G by calling this subroutine $\log_2(1/\xi)$ times. To accomplish this we first find a 1/2-optimal flow g in network G. Then we find a 1/2-optimal flow h in the residual network G_g. Now $g + h$ is a 1/4-optimal flow in network G, since each call to the subroutine captures at least half of the remaining flow. In general, we can find a ξ-optimal flow with $\log_2(1/\xi)$ calls to the subroutine.

The following lemma of Radzik [20] is a recursive version of error-scaling. It says that we can compute an ξ-optimal flow by combining two appropriate $\sqrt{\xi}$-optimal flows.

Lemma 2. *Let g be a $\sqrt{\xi}$-optimal flow in network G. Let h be a $\sqrt{\xi}$-optimal flow in network G_g. Then the flow $g + h$ is ξ-optimal in G.*

4   Truemper's Algorithm

Truemper's maximum flow based augmenting path algorithm is one of the simplest algorithms for the generalized maximum flow problem. We apply our gain-scaling techniques to Truemper's algorithm, producing perhaps the cleanest and simplest polynomial-time algorithms for the problem. In this section we first review Truemper's [22] algorithm. Our first variant runs Truemper's algorithm in a rounded network. It computes a ξ-optimal flow in polynomial-time, for any constant $\xi > 0$. However, it requires exponential-time to compute optimal flows, since we would need ξ to be very small. By incorporating error-scaling, we show that a simple variant of Truemper's algorithm computes an optimal flow in polynomial-time.

A natural and intuitive algorithm for the maximum flow problem in lossy networks is to repeatedly send flow from excess nodes to the sink along highest-gain (most-efficient) augmenting paths. Onaga observed that if the input network has no residual flow-generating cycles, then the algorithm maintains this property. Thus, we can find a highest-gain augmenting path using a single shortest path computation with costs $c(v, w) = -\log \gamma(v, w)$. By maintaining canonical labels, we can ensure that all relabeled gains are at most one, and a Dijkstra shortest path computation suffices. Unit gain paths in the canonically relabeled network correspond to highest gain paths in the original network. This is essentially Onaga's [19] algorithm. If the algorithm terminates, then the resulting flow is optimal by Theorem 1. However, this algorithm may not terminate in finite time if the capacities are irrational. Truemper's algorithm [22] uses a (nongeneralized) maximum flow computation to simultaneously augment flow along all highest-gain augmenting paths. It is the generalized flow analog of Jewell's primal-dual minimum cost flow algorithm.

Theorem 5. *In Truemper's algorithm, the number of maximum flow computations is bounded by n plus the number of different gains of paths in the original network.*

Proof. After each maximum flow computation, $\mu(v)$ strictly increases for each excess node $v \neq t$.

4.1   Rounded Truemper (RT)

Algorithm RT computes a ξ-optimal flow by running Truemper's algorithm in a rounded network. The input to Algorithm RT is a lossy network G and an error parameter ξ. Algorithm RT first rounds the gains to integer powers of $b = (1 + \xi)^{1/n}$, as described in Section 3.1. Let H denote the rounded network. Then RT computes an optimal flow in H using Truemper's algorithm. Finally the algorithm interprets the flow in the original network. Algorithm RT is described in Figure 1.

Input: lossy network G, error parameter $0 < \xi < 1$
Output: ξ-optimal flow g
  Set base $b = (1 + \xi)^{1/n}$ and round gains in G to powers of b
  Let H be resulting network
  Initialize $h \leftarrow 0$
  while $\exists$ augmenting path in H_h **do**
    $\mu \leftarrow$ canonical labels in H_h
    $f \leftarrow$ max flow from excess nodes to the sink in $H_{h,\mu}$ using only gain one
        relabeled residual arcs
    $h(v, w) \leftarrow h(v, w) + f(v, w)\mu(v)$
  end while
  $g \leftarrow$ interpretation of flow h in G

Fig. 1: $\mathrm{RT}(G, \xi)$.

The gain of a path in network G is between B^{-n} and B^n. Thus, after rounding to powers of b, there are at most $1 + \log_b B^{2n} = \mathcal{O}(n^2 \xi^{-1} \log B)$ different gains of paths in H. Using Goldberg and Tarjan's [8] preflow-push algorithm, each (nongeneralized) maximum flow computation takes $\tilde{\mathcal{O}}(mn)$ time. Thus, by

Theorem 5, RT finds an optimal flow in H in $\tilde{\mathcal{O}}(mn^3\xi^{-1}\log B)$ time. The following theorem follows using Corollary 1.

Theorem 6. *Algorithm RT computes a ξ-optimal flow in a lossy network in* $\tilde{\mathcal{O}}(mn^3\xi^{-1}\log B)$ *time.*

4.2 Iterative Rounded Truemper (IRT)

RT does not compute an optimal flow in polynomial-time, since the precision required to apply Lemma 1 is roughly $\xi = B^{-m}$. In Algorithm IRT, we apply error-scaling, as described in Section 3.2. IRT iteratively calls RT with error parameter $1/2$ and the current residual network. Since RT sends flow along highest-gain paths in the rounded network, not in the original network, it creates residual flow-generating cycles. So, before calling RT in the next iteration, we must first cancel all residual flow-generating cycles with subroutine CANCEL-CYCLES, because the input to RT is a lossy network. Intuitively, this can be interpreted as rerouting flow from its current paths to highest-gain paths, but not all of the rerouted flow reaches the sink.

Theorem 7. *Algorithm IRT computes a ξ-optimal flow in $\tilde{\mathcal{O}}(mn^3\log B\log \xi^{-1})$ time. It computes an optimal flow in $\tilde{\mathcal{O}}(m^2n^3\log^2 B)$ time.*

In the full paper we prove that Algorithm IRT actually finds an optimal flow in $\tilde{\mathcal{O}}(m^2n^3\log B + m^2n^2\log^2 B)$ time.

5 Preflow-Push

In this section we adapt Goldberg and Tarjan's [10] preflow-push algorithm to the generalized maximum flow problem. This is the first polynomial-time preflow push algorithm for generalized network flows. Tseng and Bertsekas [23] designed a preflow push-like algorithm for the generalized minimum cost flow problem, but it may require more than B^n iterations. Using our rounding technique, we present a preflow-push algorithm that computes a ξ-optimal flow in polynomial-time for any constant $\xi > 0$. Then by incorporating error-scaling, we show how to find an optimal flow in polynomial-time.

5.1 Rounded Preflow-Push (RPP)

Algorithm RPP is a generalized flow analog of Goldberg and Tarjan's preflow push algorithm for the minimum cost flow problem. Conceptually, RPP runs the minimum cost flow algorithm with costs $c(v, w) = -\log\gamma(v, w)$ and error parameter $\epsilon = \frac{1}{n}\log b$ where $b = (1 + \xi)^{1/n}$. This leads to the following natural definitions and algorithm. An *admissible arc* is a residual arc with relabeled gain above one. The *admissible graph* is the graph induced by admissible arcs. An *active node* is a node with positive residual excess and a residual path to the sink. We note that if no such residual path exists and an optimal solution sends flow through this node, then that flow does not reach the sink. So we can safely disregard this useless residual excess. (Periodically RPP determines which nodes have residual paths to the sink.) Algorithm RPP maintains a flow h and node labels μ. The algorithm repeatedly selects an active node v. If there is an admissible arc (v, w) emanating from node v, IPP pushes $\delta = \min\{e_h(v), u_h(v, w)\}$ units of flow from node v to w. If $\delta = u_h(v, w)$ the push is called *saturating*; otherwise it is *nonsaturating*. If there is no such admissible arc, RPP increases the label of node v by a factor of $2^\epsilon = b^{1/n}$; this corresponding to an additive potential increase for minimum cost flows. This process is referred to as a *relabel* operation. Relabeling node v can create new admissible arcs emanating from v. To ensure that we do not create residual flow-generating cycles, we only increase the label by a relatively small amount.

The input to Algorithm RPP is a lossy network G and error parameter ξ. Before applying the preflow-push method, IPP rounds the gains to powers of $b = (1+\xi)^{1/n}$, as described in Section 3.1. The method above is then applied to the rounded network H. Algorithm RPP is described in Figure 2.

We note that our algorithm maintains a pseudoflow with excesses, but no deficits. In contrast, the Goldberg-Tarjan algorithm allows both excesses and deficits. Also their algorithm scales ϵ. We currently do not see how to improve the worst-case complexity by a direct scaling of ϵ.

Input: lossy network G, error parameter $0 < \xi < 1$
Output: ξ-optimal flow g
  Set base $b = (1+\xi)^{1/n}$ and round gains in network G to powers of b.
  Let H be resulting network
  Initialize $h \leftarrow 0, \mu \leftarrow 1$
  while $\exists$ active node v **do**
    if $\exists$ admissible arc (v, w) **then**
      Push $\delta = \min\{e_h(v), u_h(v, w)\}$ units of flow from v to w and update h {push}
    else
      $\mu(v) \leftarrow b^{1/n}\mu(v)$ {relabel}
    end if
  end while
  $g \leftarrow$ interpretation of flow h in G

Fig. 2: RPP(G, ξ).

The bottleneck computation is performing nonsaturating pushes, just as for computing minimum cost flows with the preflow-push method. By carefully choosing the order to examine active nodes (e.g., the wave implementation), we can reduce the number of nonsaturating pushes. A dual approach is to use more clever data structures to reduce the amortized time per nonsaturating push. Using a version of dynamic trees specialized for generalized networks [6], we obtain the following theorem.

Theorem 8. *Algorithm RPP computes a ξ-optimal flow in $\tilde{\mathcal{O}}(mn^3\xi^{-1}\log B)$ time.*

5.2   Iterative Rounded Preflow-Push (IRPP)

RPP does not compute an optimal flow in polynomial time, since the precision required is roughly $\xi = B^{-m}$. Like Algorithm IRT, Algorithm IRPP adds error-scaling, resulting in the following theorem.

Theorem 9. *IRPP computes a ξ-optimal flow in $\tilde{\mathcal{O}}(mn^3\log B\log\xi^{-1})$ time. It computes an optimal flow in $\tilde{\mathcal{O}}(m^2n^3\log^2 B)$ time.*

6   Rounded Fat Path

In this section we present a simple variant of Goldberg, Plotkin, and Tardos' [7] Fat-Path algorithm which has the same complexity as Radzik's [20] Fat-Path variant. Our algorithm is intuitive and its proof of correctness is much simpler than Radzik's. The Fat-Path algorithm can be viewed as an analog of Orlin's capacity scaling algorithm for the minimum cost flow problem. The original Fat-Path algorithm computes a ξ-optimal flow in $\tilde{\mathcal{O}}(mn^2\log B\log\xi^{-1})$ time, while Radzik's and our variants require only $\tilde{\mathcal{O}}(m(m+n\log\log B)\log\xi^{-1})$ time.

The bottleneck computation in the original Fat-Path algorithm is canceling residual flow-generating cycles. Radzik's variant reduces the bottleneck by canceling only residual flow-generating cycles with big gains. The remaining flow-generating cycles are removed by decreasing the gain factors. Analyzing the precision of the resulting solution is technically complicated. Instead, our variant rounds down the gains to integer powers of a base b, which depends on the precision of the solution desired. Our rounding is done in a lossy network, which makes the quality of the resulting solution easy to analyze. Subsequent calls to CANCELCYCLES are performed in a rounded network, improving the complexity.

We first review the FATAUGMENTATIONS subroutine which finds augmenting paths with sufficiently large capacity. Then we present Algorithm RFP, which runs the Fat-Path algorithm in a rounded network. It computes approximately optimal and optimal flows in polynomial-time. We then present a recursive version of RFP, which improves the complexity when computing nearly optimal and optimal flows.

6.1   Fat Augmentations

The FATAUGMENTATIONS subroutine was originally developed by Goldberg, Plotkin, and Tardos for their Fat-Path algorithm and is described in detail in [7]. The input is a lossy network and fatness parameter δ. The subroutine repeatedly augments flow along *highest-gain δ-fat paths*, i.e. highest-gain augmenting paths among paths that have enough residual capacity to increase the excess at the sink by δ, given sufficient excess at the first node of the path. This process is repeated until no δ-fat paths remain. There are at most $n + \text{OPT}(G)/\delta$ augmentations. By maintaining appropriate labels μ, an augmentation takes $\tilde{\mathcal{O}}(m)$ time, using an algorithm based on Dijkstra's shortest path algorithm. Upon termination, the final flow has value at least $\text{OPT}(G) - m\delta$.

6.2   Rounded Fat Path (RFP)

Algorithm RFP runs the original Fat-Path algorithm in a rounded network. The idea of the original Fat-Path algorithm is to call FATAUGMENTATIONS and augment flow along δ-fat paths, until no such paths remain. At this point δ is decreased by a factor of 2 and a new phase begins. However, since FATAUGMEN-TATIONS selects only paths with large capacity, it does not necessarily send flow on overall highest-gain paths. This creates residual flow-generating cycles which must be canceled so that we can efficiently compute $\delta/2$-fat paths in the next phase.

The input to Algorithm RFP is a lossy network and an error parameter ξ. First, RFP rounds down the gains as described in Section 3.1. It maintains a flow h in the rounded network H and an upper bound Δ on the *excess discrepancy*, i.e., the difference between the value of the current flow $|h|$ and the optimum $\text{OPT}(H)$. The scaling parameter Δ is initialized using Radzik's greedy augmentation algorithm, as described in Section 2.5. In each phase, Δ is decreased by a factor of 2. To achieve this reduction, Algorithm RFP cancels all residual flow-generating cycles in H_h, using the CANCELCYCLES subroutine. By Theorem 2 this requires $\tilde{\mathcal{O}}(mn \log C)$ time where C is the biggest cost. Recall $c(v, w) = -\lfloor \log_b \gamma(v, w) \rfloor$ so $C \leq 1 + \log_b B = \mathcal{O}(n\xi^{-1} \log B)$. Then subroutine FATAUGMENTATIONS is called with fatness parameter $\delta = \Delta/(2m)$. After this call, the excess discrepancy is at most $m\delta = \Delta/2$, and Δ is decreased accordingly. Since each δ-fat augmentation either empties the residual excess of a node or increases the flow value by at least δ, there are at most $n + \Delta/\delta = n + 2m$ augmentations per Δ-phase, which requires a total of $\tilde{\mathcal{O}}(m^2)$ time. Algorithm RFP is given in Figure 3.

Theorem 10. *Algorithm RFP computes a 2ξ-optimal flow in a lossy network in $\tilde{\mathcal{O}}((m^2 + mn \log(\xi^{-1} \log B)) \log \xi^{-1})$ time.*

Proof. To bound the running time, we note that there are at most $\log_2(n/\xi)$ phases. FATAUGMENTATIONS requires $\tilde{\mathcal{O}}(m^2)$ time per phase, and CANCELCY-CLES requires $\tilde{\mathcal{O}}(mn \log C)$ time, where we bound $C = \mathcal{O}(n\xi^{-1} \log B)$ as above. The algorithm terminates when $\Delta \leq \xi \text{OPT}(H)$. At this point h is ξ-optimal in network H, since we maintain $\Delta \geq \text{OPT}(H) - |h|$. The quality of the resulting solution then follows using Theorem 4.

6.3   Recursive Rounded Fat-Path (RRFP)

Algorithm RFP computes a ξ-optimal flow in $\tilde{\mathcal{O}}(m^2 + mn \log \log B)$ time when $\xi > 0$ is inversely polynomial in m. However it may require more time to

---

Input: lossy network G, error parameter $0 < \xi < 1$
Output: 2ξ-optimal flow g
    Set base $b = (1 + \xi)^{1/n}$ and round gains in network G to powers of b
    Let H be resulting network
    Initialize $\Delta \leftarrow \Delta_0$ and $h \leftarrow 0$ $\{\mathrm{OPT}(H) \leq \Delta_0 \leq n\mathrm{OPT}(H)\}$
    repeat
        $(h', \mu) \leftarrow$ CancelCycles(H_h)
        $h \leftarrow h + h'$
        $h' \leftarrow$ FatAugmentations$(H_h, \mu, \Delta/(2m))$ $\{\mathrm{OPT}(H_h) - |h'| \leq \Delta/2\}$
        $h \leftarrow h + h'$
        $\Delta \leftarrow \Delta/2$
    until $\Delta \leq \xi\mathrm{OPT}(H)$
    $g \leftarrow$ interpretation of h in network G

Fig. 3: RFP(G, ξ).

compute optimal flows than the original Fat-Path algorithm. By using the recursive scheme from Section 3.2, we can compute nearly optimal and optimal flows faster than the original Fat-Path algorithm. In each recursive call, we reround the network. We cancel flow-generating cycles in an already (partially) rounded network. The benefit is roughly to decrease the average value of C from $\mathcal{O}(n\xi^{-1} \log B)$ to $\mathcal{O}(n \log B)$.

Theorem 11. *Algorithm RRFP computes a ξ-optimal flow in a lossy network in $\tilde{\mathcal{O}}(m(m + n \log\log B) \log \xi^{-1})$ time. If the network has residual flow-generating cycles, then an extra $\tilde{\mathcal{O}}(mn^2 \log B)$ preprocessing time is required. Algorithm RRFP computes an optimal flow in $\tilde{\mathcal{O}}(m^2(m + n \log\log B) \log B)$ time.*

7   Practical Cycle-Canceling

We implemented a version of the preflow-push algorithm, described in Section 5, in C++ using Mehlhorn and Näher's [17] Library of Efficient Data types and Algorithms (LEDA). We observed that as much as 90% of the time was spent canceling flow-generating cycles. We focused our attention on reducing this bottleneck.

Recall, CancelCycles is an adaption of Goldberg and Tarjan's cancel-and-tighten algorithm using costs $c(v, w) = -\log \gamma(v, w)$. Negative cost cycles correspond to flow-generating cycles. The underlying idea of the Goldberg-Tarjan algorithm is to cancel the most negative mean cost cycle until no negative cost cycles remain. To improve efficiency, the actual cancel-and-tighten algorithm only approximates this strategy. It maintains a flow g and node potentials (corresponding to node labels) π that satisfy ϵ-*complementary slackness.* That is, $c_\pi(v, w) = c(v, w) - \pi(v) + \pi(w) \geq -\epsilon$ for all residual arcs (v, w). The subroutine makes progress by reducing the value of ϵ, using the following two computations which comprise a phase: (i) canceling residual cycles in the subgraph induced by negative reduced cost arcs and (ii) updating node potentials so that finding new negative cost (flow-generating) cycles is efficient. The cancel-and-tighten algorithm uses the following two types of potential updates. Loose updating uses a computationally inexpensive topological sort, but may only decrease ϵ by a factor of $(1 - 1/n)$ per phase. Tight updating reduces ϵ by the maximum possible amount, but involves computing the value of the minimum mean cost cycle ϵ^*, which is relatively expensive.

We observed that the quality of the potential updates was the most important factor in the overall performance of CANCELCYCLES. So, we focused our attention on limiting the number of iterations by better node potential updates. Using only loose updates, we observed that CANCELCYCLES required a large number of phases. By using tight updates, we observed a significant reduction in the number of phases, but each phase is quite expensive. The goal is to a reach a middle ground. We introduce a *medium updating* technique which is much cheaper than tight updating, yet more effective than loose updating; it reduces the overall running time of the cycle canceling computation. Our implementation uses a combination of loose, medium, and tight potential updates.

Tight updating requires $\tilde{\mathcal{O}}(mn)$ time in the worst case, using either a dynamic programming or binary search method. We incorporated several heuristics to improve the actual performance. These heuristics are described in the full paper. However, we observed that tight relabeling was still quite expensive.

We introduce a *medium potential updating* which is a middle ground between loose and tight updating. In medium updating, we find a value ϵ' which is close to ϵ^*, without spending the time to find the actual minimum mean cost cycle. In our algorithm, we only need to estimate ϵ^* in networks where the subgraph induced by negative cost arcs is acyclic. To do this efficiently, we imagine that the in addition to the original arcs, a zero cost link exists between every pair of nodes. We can efficiently find a minimum mean cost cycle in this modified network by computing a minimum mean cost path in the acyclic network induced by only negative cost arcs, without explicitly considering the imaginary zero cost arcs. Let ϵ' denote the value of the minimum mean cost cycle in the modified network. Clearly $\epsilon' \le \epsilon^*$, and it is not hard to see that $\epsilon' \ge (1 - 1/n)\epsilon$. We can binary search for ϵ' using a shortest path computation in *acyclic* graphs. This requires only $\mathcal{O}(m)$ time per iteration. If we were to determine ϵ' exactly, in $\tilde{\mathcal{O}}(n \log B)$ iterations the search interval would be sufficiently small. If the gains in the network are rounded to powers of $b = (1 + \xi)^{1/n}$ then $\tilde{\mathcal{O}}(\log C)$ iterations suffice, where $C = \mathcal{O}(n\xi^{-1} \log B)$. In our implementation we use an approximation to ϵ'.

References

1. R. K. Ahuja, T. L. Magnanti, and J. B. Orlin. *Network Flows: Theory, Algorithms, and Applications*. Prentice Hall, Englewood Cliffs, New Jersey, 1993.
2. Edith Cohen and Nimrod Megiddo. New algorithms for generalized network flows. *Math Programming*, 64:325–336, 1994.
3. F. Glover, J. Hultz, D. Klingman, and J. Stutz. Generalized networks: A fundamental computer based planning tool. *Management Science*, 24:1209–1220, 1978.

4. F. Glover and D. Klingman. On the equivalence of some generalized network flow problems to pure network problems. *Math Programming*, 4:269–278, 1973.
5. F. Glover, D. Klingman, and N. Phillips. Netform modeling and applications. *Interfaces*, 20:7–27, 1990.
6. A. V. Goldberg, S. A. Plotkin, and É Tardos. Combinatorial algorithms for the generalized circulation problem. Technical Report STAN-CS-88-1209, Stanford University, 1988.
7. A. V. Goldberg, S. A. Plotkin, and É Tardos. Combinatorial algorithms for the generalized circulation problem. *Mathematics of Operations Research*, 16:351–379, 1991.
8. A. V. Goldberg and R. E. Tarjan. A new approach to the maximum flow problem. *Journal of the ACM*, 35:921–940, 1988.
9. A. V. Goldberg and R. E. Tarjan. Finding minimum-cost circulations by canceling negative cycles. *Journal of the ACM*, 36:388–397, 1989.
10. A. V. Goldberg and R. E. Tarjan. Solving minimum cost flow problems by successive approximation. *Mathematics of Operations Research*, 15:430–466, 1990.
11. D. Goldfarb and Z. Jin. A polynomial dual simplex algorithm for the generalized circulation problem. Technical report, Department of Industrial Engineering and Operations Research, Columbia University, 1995.
12. D. Goldfarb and Z. Jin. A faster combinatorial algorithm for the generalized circulation problem. *Mathematics of Operations Research*, 21:529–539, 1996.
13. D. Goldfarb, Z. Jin, and J. B. Orlin. Polynomial-time highest gain augmenting path algorithms for the generalized circulation problem. *Mathematics of Operations Research*. To appear.
14. W. S. Jewell. Optimal flow through networks with gains. *Operations Research*, 10:476–499, 1962.
15. Anil Kamath and Omri Palmon. Improved interior point algorithms for exact and approximate solution of multicommodity flow problems. In *Proceedings of the 6th Annual ACM-SIAM Symposium on Discrete Algorithms*, pages 502–511, 1995.
16. S. Kapoor and P. M. Vaidya. Speeding up Karmarkar's algorithm for multicommodity flows. *Math Programming*. To appear.
17. K. Mehlhorn and S. Näher. A platform for combinatorial and geometric computing. *CACM*, 38(1):96–102, 1995. http://ftp.mpi-sb.mpg.de/LEDA/leda.html
18. S. M. Murray. *An interior point approach to the generalized flow problem with costs and related problems.* PhD thesis, Stanford University, 1993.
19. K. Onaga. Dynamic programming of optimum flows in lossy communication nets. *IEEE Trans. Circuit Theory*, 13:308–327, 1966.
20. T. Radzik. Faster algorithms for the generalized network flow problem. *Mathematics of Operations Research*. To appear.
21. T. Radzik. Approximate generalized circulation. Technical Report 93-2, Cornell Computational Optimization Project, Cornell University, 1993.
22. K. Truemper. On max flows with gains and pure min-cost flows. *SIAM J. Appl. Math*, 32:450–456, 1977.
23. P. Tseng and D. P. Bertsekas. An ϵ-relaxation method for separable convex cost generalized network flow problems. In *5th International Integer Programming and Combinatorial Optimization Conference*, 1996.

The Pseudoflow Algorithm and the Pseudoflow-Based Simplex for the Maximum Flow Problem

Dorit S. Hochbaum *

Department of Industrial Engineering and Operations Research,
and Walter A. Haas School of Business
University of California, Berkeley
`dorit@@hochbaum.berkeley.edu`

Abstract. We introduce an algorithm that solves the maximum flow problem without generating flows explicitly. The algorithm solves directly a problem we call the maximum s-excess problem. That problem is equivalent to the minimum cut problem, and is a direct extension of the maximum closure problem. The concepts used also lead to a new parametric analysis algorithm generating all breakpoints in the amount of time of a single run.

The insights derived from the analysis of the new algorithm lead to a new simplex algorithm for the maximum flow problem – a pseudoflow-based simplex. We show that this simplex algorithm can perform a parametric analysis in the same amount of time as a single run. This is the first known simplex algorithm for maximum flow that generates all possible breakpoints of parameter values in the same complexity as required to solve a single maximum flow instance and the fastest one.

The complexities of our pseudoflow algorithm, the new simplex algorithm, and the parametric analysis for both algorithms are $O(mn \log n)$ on a graph with n nodes and m arcs.

1   Introduction

This extended abstract describes an efficient new approach to the maximum flow and minimum cut problems. The approach is based on a new certificate of optimality inspired by the algorithm of Lerchs and Grossmann, [LG64]. This certificate, called *normalized tree*, partitions the set of nodes into subsets some of which have excess capacity and some have capacity deficit. The nodes that belong to the subsets with excess form the source set of a candidate minimum cut. The algorithm solves, instead of the maximum flow problem, another problem which we call the maximum *s-excess problem*. That problem is defined on a directed

---

* Research supported in part by NEC, by NSF award No. DMI-9713482, and by SUN Microsystems.

R. E. Bixby, E. A. Boyd, and R. Z. Ríos-Mercado (Eds.): IPCO VI
LNCS 1412, pp. 325–337, 1998. © Springer–Verlag Berlin Heidelberg 1998

graph with arc capacities and node weights and does *not* contain distinguished source and sink nodes. The objective of the s-excess problem is to find a subset of the nodes that maximizes the sum of node weights, minus the weight of the arcs separating the set from the remainder of the nodes. The new problem is shown to be equivalent to the minimum cut problem that is traditionally solved by deriving a maximum flow first. With the new algorithm these problems can be solved without considering flows explicitly. The steps of the algorithm can be interpreted as manipulating pseudoflow – a flow that does not satisfy flow balance constraints. For this reason we choose to call the algorithm *the pseudoflow algorithm*.

The main feature that distinguishes the pseudoflow algorithm from other known algorithms for the maximum flow problem is that it does not seek to either preserve or progress towards feasibility. Instead the algorithm creates "pockets" of nodes so that at optimum there is no residual arc that can carry additional flow between an "excess pocket" and a "deficit pocket". The set of nodes in all the "excess pockets" form the source set of a minimum cut and also the maximum s-excess set.

The certificate maintained by our algorithm bears a resemblance to the basic arcs tree maintained by simplex. It is demonstrated that this certificate is analogous to the concept of a strong basis introduced by Cunningham [C76] if implemented in a certain "extended network" permitting violations of flow balance constraints. It is further shown that the algorithmic steps taken by our algorithm are substantially different from those of the simplex and lead to a different outcome in the next iteration based on the same strong basis in the given iteration.

The contributions in this paper include:

1. The introduction of the pseudoflow maximum s-excess problem that provides a new perspective on the maximum flow problem.
2. A pseudoflow algorithm for the maximum flow problem of complexity $O(mn \log n)$.
3. Parametric analysis conducted with the pseudoflow algorithm that generates all breakpoints in the same complexity as a single run.
4. A new pseudoflow-based simplex algorithm for maximum flow using the lowest label approach.
5. A parametric analysis simplex method that finds all possible parameter breakpoints in the same time as a single run, $O(mn \log n)$.

The parametric simplex method is the first known parametric implementation of simplex to date that finds all breakpoints in the same running time as a single application.

1.1 Notation

For $P, Q \subset V$, the set of arcs going from P to Q is denoted by, $(P, Q) = \{(u, v) \in A | u \in P$ and $v \in Q\}$. Let the capacity of arc (u, v) be denoted by c_{uv} or $c(u, v)$.

For $P, Q \subset V$, $P \cap Q = \emptyset$, the capacity of the cut separating P from Q is, $C(P, Q) = \sum_{(u,v) \in (P,Q)} c_{uv}$. For $S \subseteq V$, let $\bar{S} = V \setminus S$.

For a graph $G = (V, A)$ we denote the number of arcs by $m = |A|$ and the number of nodes by $n = |V|$.

An arc (u, v) of an unspecified direction is referred to as *edge* $[u, v]$. $[v_1, v_2, \ldots, v_k]$ denotes an *undirected* path from v_1 to v_k. That is, $[v_1, v_2], \ldots, [v_{k-1}, v_k] \in A$.

We use the convention that the capacity of an arc that is not in the graph is zero. Thus for $(a, b) \in A$ and $(b, a) \notin A$, $c_{b,a} = 0$.

The capacity of an edge e is denoted either by $c(e)$ or c_e. The flow on an edge e is denoted either by $f(e)$ or f_e. We use the convention that $f(a, b) = -f(b, a)$. For a given flow or pseudoflow f, the residual capacity of e is denoted by $c_f(e)$ which is $c_e - f_e$.

Given a rooted tree, T, T_v is the subtree suspended from node v that contains all the descendants of v in T. $T_{[v,p(v)]} = T_v$ is the subtree suspended from the edge $[v, p(v)]$. An immediate descendant of a node v, a *child* of v, is denoted by $ch(v)$, and the unique immediate ancestor of a node v, the *parent* of v, by $p(v)$.

2   The Maximum Flow Problem and the Maximum s-Excess Problem

The pseudoflow algorithm described here ultimately finds a maximum flow in a graph. Rather than solving the problem directly the algorithm solves instead the maximum s-excess problem.

Problem Name: *Maximum s-Excess*
Instance: *Given a directed graph $G = (V, A)$, node weights (positive or negative) w_i for all $i \in V$, and nonnegative arc weights c_{ij} for all $(i, j) \in A$.*
Optimization Problem: *Find a subset of nodes $S \subseteq V$ such that*
$\sum_{i \in S} w_i - \sum_{i \in S, j \in \bar{S}} c_{ij}$ *is maximum.*

We elaborate here further on this problem and its relationship to the maximum flow and minimum cut problems.

The maximum flow problem is defined on a directed graph with distinguished source and sink nodes and the arcs adjacent to source and sink, $A(s)$ and $A(t)$, $G_{st} = (V \cup \{s, t\}, A \cup A(s) \cup A(t))$.

The standard formulation of the maximum flow problem with zero lower bounds and x_{ij} variables indicating the amount of flow on arc (i, j) is,

$$\begin{aligned}
\text{Max} \quad & x_{ts} \\
\text{subject to} \quad & \sum_i x_{ki} - \sum_j x_{jk} = 0 \quad k \in V \\
& 0 \leq x_{ij} \leq c_{ij} \quad \forall (i, j) \in A.
\end{aligned}$$

In this formulation the first set of (equality) constraints is called the *flow balance* constraints. The second set of (inequality) constraints is called the *capacity* constraints.

Definition 1. *The s- excess capacity of a set* $S \subseteq V$ *in the graph* $G_{st} = (V \cup \{s,t\}, A \cup A(s) \cup A(t))$ *is,* $C(\{s\}, S) - C(S, \bar{S} \cup \{t\})$.

We claim that finding a subset $S \subseteq V$ that maximizes $C(\{s\}, S) - C(S, \bar{S} \cup \{t\})$ is equivalent to the maximum s-excess problem.

Lemma 2. *A subset of nodes* $S \subseteq V$ *maximizes* $C(\{s\}, S) - C(S, \bar{S} \cup \{t\})$ *in* G_{st} *if and only if it is of maximum s-excess in the graph* $G = (V, A)$.

We next prove that the s-excess problem is equivalent to the minimum cut problem,

Lemma 3. *S is the source set of a minimum cut if and only if it is a set of maximum s- excess capacity* $C(s, S) - C(S, \bar{S} \cup \{t\})$ *in the graph.*

Proof. Given an instance of the s-excess problem. Append to the graph $G = (V, A)$ the nodes s and t; Assign arcs with capacities equal to the weights of nodes from s to the nodes of positive weight; Assign arcs with capacities equal to the absolute value of the weights of nodes of negative weights, from the nodes to t.

The sum of weights of nodes in S is also the sum of capacities $C(\{s\}, S) - C(S, \{t\})$ where the first term corresponds to positive weights in S, and the second to negative weights in S:

$$
\begin{aligned}
\sum_{j \in S} w_j - \sum_{i \in S, j \in \bar{S}} c_{ij} &= C(\{s\}, S) - C(S, \{t\}) - C(S, \bar{S}) \\
&= C(\{s\}, S) - C(S, \bar{S} \cup \{t\}). \\
&= C(\{s\}, V) - C(\{s\}, \bar{S}) - C(S, \bar{S} \cup \{t\}). \\
&= C(\{s\}, V) - C(\{s\} \cup S, \bar{S} \cup \{t\}).
\end{aligned}
$$

The latter term is the capacity of the cut $(\{s\} \cup S, \bar{S} \cup \{t\})$. Hence maximizing the s-excess capacity is equivalent to minimizing the capacity of the cut separating s from t.

To see that the opposite is true, consider the network $(V \cup \{s, t\}, A)$ with arc capacities. Assign to node $v \in V$ a weight that is c_{sv} if the node is adjacent to s and $-c_{vt}$ if the node is adjacent to t. Note that it is always possible to remove paths of length 2 from s to t thus avoiding the presence of nodes that are adjacent to *both* source and sink. This is done by subtracting from the arcs' capacities c_{sv}, c_{vt} the quantity $\min\{c_{sv}, c_{vt}\}$. The capacities then translate into node weights that serve as input to the s-excess problem and satisfy the equalities above.   □

We conclude that the maximum s-excess problem is a complement of minimum cut which in turn is a dual of the maximum flow problem. As such, its solution does not contain more flow information than the solution to the minimum cut problem. As we see later, however, it is possible to derive a feasible flow of value equal to that of the cut from the certificate used in the algorithm, in $O(mn)$ time.

The reader may wonder about the arbitrary nature of the s-excess problem, at least in the sense that it has not been previously addressed in the literature. The explanation is that this problem is a relaxation of the maximum closure problem where the objective is to find, in a node weighted graph, a closed set of nodes of maximum total weight. In the maximum closure problem it is required that all successors of each node in the closure set will belong to the set. In the s-excess problem this requirement is replaced by a penalty assigned to arcs of immediate successors that are not included in the set. In that sense the s-excess problem is a *relaxation* of the maximum closure problem. The proof of Lemma 3 is effectively an extension of Picard's proof [Pic76] demonstrating that the maximum weight closed set in a graph (maximum closure) is the source set of a minimum cut.

We provide a detailed account of the use of the pseudoflow and other algorithms for the maximum closure problem in [Hoc96]. We also explain there how these algorithms have been used in the mining industry and describe the link to the algorithm of Lerchs and Grossmann, [LG64].

3   Preliminaries and Definitions

Pseudoflow is an assignment of values to arcs that satisfy the capacity constraints but not necessarily the flow balance constraints. Unlike preflow, that may violate the flow balance constraints only with inflow exceeding outflow, pseudoflow permits the inflow to be either strictly larger than outflow (excess) or outflow strictly larger than inflow (deficit). Let f be a pseudoflow vector with $0 \leq f_{ij} \leq c_{ij}$ the pseudoflow value assigned to arc (i, j). Let $inflow(D)$, $outflow(D)$ be the total amount of flow incoming and outgoing to and from the set of nodes D. For each subset of nodes $D \subset V$,

$$
\begin{aligned}
excess(D) &= inflow(D) - outflow(D) \\
&= \sum_{(u,v) \in (V \cup \{s\} \setminus D, D)} f_{u,v} - \sum_{(v,u) \in (D, V \cup \{t\} \setminus D)} f_{v,u}.
\end{aligned}
$$

The absolute value of excess that is less than zero is called *deficit*, i.e. $-excess(D) = deficit(D)$.

Given a rooted tree T. For a subtree $T_v = T_{[v,p(v)]}$, let $M_{[v,p(v)]} = M_v = excess(T_v)$ and be called the *mass* of the node v or the arc $[v, p(v)]$. That is, the mass is the amount of flow on $(v, p(v))$ directed towards the root. A flow directed in the opposite direction – from $p(v)$ to v – is interpreted as negative excess or mass.

We define the *extended network* as follows: The network G_{st} is augmented with a set of arcs – two additional arcs per node. Each node has one arc of infinite capacity directed into it from the sink, and one arc of infinite capacity directed from it to the source. This construction is shown in Figure 1. We refer to the appended arcs from sink t as the *deficit arcs* and the appended arcs to the source s as the *excess arcs*. The source and sink nodes are compressed into a 'root' node r. We refer to the extended network's set of arcs as A^{aug}. These include, in

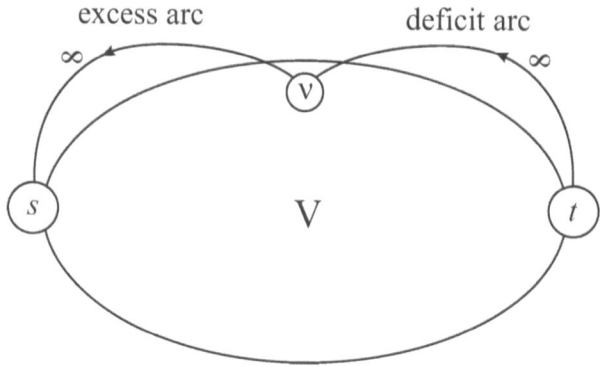

excess arc   deficit arc

Fig. 1. An extended network with excesses and deficits.

addition to the deficit and excess arcs, also the arcs adjacent to source and sink – $A(s)$ and $A(t)$. The *extended network* is the graph $G^{\mathrm{aug}} = (V \cup \{r\}, A^{\mathrm{aug}})$.

Any pseudoflow on a graph has an equivalent feasible flow on the extended network derived from the graph – a node with excess sends the excess back to the source, and a node with deficit receives a flow that balances this deficit from the sink.

Throughout our discussion of flows on extended networks, all the flows considered saturate the arcs adjacent to source and sink and thus the status of these arcs, as saturated, remains invariant. We thus omit repeated reference to the arcs $A(s)$ and $A(t)$.

4   A Normalized Tree

The algorithm maintains a construction that we call *a normalized tree* after the use of this term by Lerchs and Grossmann in [LG64] for a construction that inspired ours. Let node $r \notin V$ serve as root and represent a contraction of s and t. Let $(V \cup \{r\}, T)$ be a tree where $T \subset \bar{A}$. The children of r are called the *roots* of their respective *branches* or subtrees. The deficit and excess arcs are only used to connect r to the roots of the branches.

A normalized tree is a rooted tree in r that induces a forest in (V, A). We refer to each rooted tree in the forest as *branch*. A branch of the normalized tree rooted at a child of r, r_i, T_{r_i} is called *strong* if $excess(T_{r_i}) > 0$, and *weak* otherwise. All nodes of strong branches are considered strong, and all nodes of weak branches are considered weak.

A normalized tree is depicted in Figure 2.

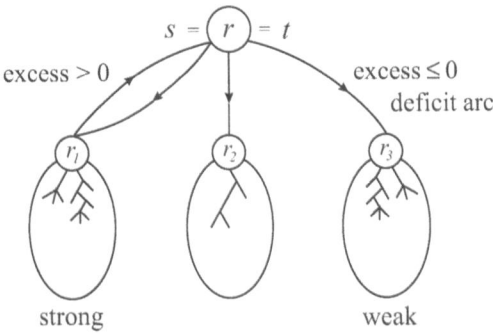

Fig. 2. A normalized tree. Each r_i is a root of a branch.

Consider a rooted forest $T \setminus \{r\}$ in $G = (V, A)$ and a pseudoflow f in G_{st} satisfying the properties:

Property 4. *The pseudoflow f saturates all source-adjacent arcs and all sink-adjacent arcs.*

Property 5. *In every branch all downwards residual capacities are strictly positive.*

Property 6. *The only nodes that do not satisfy flow balance constraints are the roots of their respective branches that are adjacent to r in the extended network.*

Definition 7. *A tree T with pseudoflow f is called* normalized *if it satisfies properties 4, 5, and 6.*

Property 6 means, in other words, that in order for T to be a normalized tree, f has to satisfy flow balance constraints in the extended network with only the roots of the branches permitted to send/receive flows along excess or deficit arcs.

We let the excess of a normalized tree T and pseudoflow f be the sum of excesses of its strong branches (or strong nodes). Property 4 implies that the excess of a set of strong nodes in a normalized tree, S, satisfies:

$$excess(S) = inflow(S) - outflow(S) = f(\{s\}, S) + f(\bar{S}, S) - f(S, \bar{S})$$
$$-f(S, \{t\}) = C(\{s\}, S) - C(S, \{t\}) + f(\bar{S}, S) - f(S, \bar{S}).$$

This equality is used to prove the superoptimality of the set of strong nodes (in the superoptimality Property).

We choose to work with normalized trees that satisfy an optional property stronger than 5:

Property 8 (Unsaturated arcs property). *The tree T has all upwards residual capacities strictly positive.*

Another optional implied property is:

Property 9. *All "free" arcs of A with flows strictly between lower and upper bound are included in T.*

With this property *all* arcs that are not adjacent to root are free. T is thus the union of all free arcs including some excess and deficit arcs. The only arcs in T that are not free are deficit arcs with 0 flow – adjacent to 0-deficit branches. All out of tree arcs are thus at their upper or at the lower bounds.

The next property – the superoptimality property – is satisfied by any normalized tree, or equivalently, by any tree-pseudoflow pair satisfying properties 4, 5, and 6. The superoptimality of a normalized tree and the conditions for optimality are stated in the next subsection.

4.1   The Superoptimality of a Normalized Tree

Property 10 (Superoptimality). *The set of strong nodes of the normalized tree T is a superoptimal solution to the s-excess problem: The sum of excesses of the strong branches is an upper bound on the maximum s-excess.*

From the proof of the superoptimality property it follows that when all arcs from S to $\bar{S}$ are saturated, then no set of nodes other than S has a larger s-excess. We thus obtain the optimality condition as a corollary to the superoptimality property.

Corollary 11 (Optimality condition). *Given a normalized tree, a pseudoflow f and the collection of strong nodes in the tree S. If f saturates all arcs in $(S, \bar{S})$ then S is a maximum s-excess set in the graph.*

The next Corollary holds only if Property 8 is satisfied. It implies minimality of the optimal solution set S.

Corollary 12 (Minimality). *Any proper subset of the strong nodes is not a maximum s-excess set in (V, A').*

On the other hand, it is possible to append to the set of strong nodes S any collection of 0-deficit branches that have no residual arcs to weak nodes without changing the value of the optimal solution. This leads to a *maximal* maximum s-excess set.

5   The Description of the Pseudoflow Algorithm

The algorithm maintains a superoptimal s-excess solution set in the form of the set of strong nodes of a normalized tree. That is, the sum of excesses of strong branches is only greater than the maximum s-excess. Each strong branch forms an "excess pocket" with the total excess of the branch assigned to its root. Within each branch the pseudoflow is feasible.

Each iteration of the algorithm consists of identifying an infeasibility in the form of a residual arc from a strong node to a weak node. The arc is then added in and the tree is updated. The update consists of pushing the *entire* excess of the strong branch along a path from the root of the strong branch to the merger arc (s', w) and progressing towards the root of the weak branch. The first arc encountered that does not have sufficient residual capacity to accommodate the pushed flow gets *split* and the subtree suspended from that arc becomes a strong branch with excess equal to the amount of flow that could not be pushed through the arc. The process continues along the path till the next bottleneck arc is encountered

Recall that $G = (V, A)$, and A_f is the set of residual arcs with respect to a given pseudoflow f.

5.1   The Pseudoflow Algorithm

begin
    <u>Initialize</u> $\forall (s, j) \in A(s)$, $f(s, j) = c(s, j)$. $\forall (j, t) \in A(t)$, $f(j, t) = c(j, t)$.
    For all arcs $(i, j) \in A$, $f(i, j) = 0$.
    $T = \cup_{j \in V} [r, j]$, the branches of the tree are $\{j\}_{j \in V}$.
    Nodes with positive excess are strong, S, and the rest are weak, W.
    while $(S, W) \cap A_f \neq \emptyset$ do
        Select $(s', w) \in (S, W)$
        <u>Merge</u> $T \leftarrow T \setminus [r, r_{s'}] \cup (s', w)$.
        <u>Renormalize</u>
        Push $\delta = M_{[r, r_{s'}]}$ units of flow along the path $[r_{s'}, \ldots, s', w, \ldots, r_w, r]$:
        begin
        Let $[v_i, v_{i+1}]$ be the next edge on the path.
        If $c_f(v_i, v_{i+1}) > \delta$ augment flow by δ, $f(v_i, v_{i+1}) \leftarrow f(v_i, v_{i+1}) + \delta$.
        Else, split $\{(v_i, v_{i+1}), \delta - c_f(v_i, v_{i+1})\}$.
        Set $\delta \leftarrow c_f(v_i, v_{i+1})$.
        Set $f(v_i, v_{i+1}) \leftarrow c(v_i, v_{i+1})$; $i \leftarrow i + 1$
        end
    end
end
procedure split $\{(a, b), M\}$
$T \leftarrow T \setminus (a, b) \cup (r, a)$; $M_{(r, a)} = M$.
The branch T_a is strong or 0-deficit with excess M.
$A_f \leftarrow A_f \cup \{(b, a)\} \setminus \{(a, b)\}$.
end

The push step, in which we augment flow by δ if $c_f(v_i, v_{i+1}) > \delta$, can be replaced by augmentation if $c_f(v_i, v_{i+1}) \geq \delta$. The algorithm remains correct, but the set of strong nodes is no longer minimal among maximum s-excess sets, and will not satisfy Property 8. We prove in the expanded version of this paper that the tree maintained is indeed normalized, which establishes the algorithm's correctness.

5.2   Initialization

We choose an initial normalized tree with each node as a separate branch for which the node serves as root. The corresponding pseudoflow saturates all arcs adjacent to source and to sink. Thus all nodes adjacent to source are strong nodes, and all those adjacent to sink are weak nodes. All the remaining nodes have zero inflow and outflow, and are thus of 0 deficit and set as weak. If a node is adjacent to both source and sink, then the lower capacity arc among the two is removed, and the other has that value subtracted from it. Therefore each node is uniquely identified with being adjacent to either source or sink or to neither.

5.3   Termination

The algorithm terminates when there is no residual arc between any strong and weak nodes.

From Corollary 12 we conclude that at termination the set of strong nodes is a *minimal* source set of a minimum cut. In other words, any proper subset of the set of strong nodes cannot be a source set of a minimum cut. It will be necessary to identify additional optimal solutions, and in particular a minimum cut with *maximal* source set for the parametric analysis.

To that end, we identify the 0-deficit branches among the weak branches. The set of strong branches can be appended with any collection of 0-deficit branches without residual arcs to weak nodes for an alternative optimal solution. The collection of *all* such 0-deficit branches with the strong nodes forms the sink set of a minimum cut that is *maximal*. To see that, consider an analogue of Corollary 12 that demonstrates that no proper subset of a weak branch (of negative deficit) can be in a source set of a minimum cut.

5.4   About Normalized Trees and Feasible Flows

Given any tree in the extended network, and a specification of pseudoflow values on the out of tree arcs, it is possible to determine in linear time $O(n)$ whether the tree is normalized. If the tree is not normalized, then the process is used to derive a normalized tree which consists of a subset of the arcs of the given tree. Given a normalized tree it is possible to derive in $O(n)$ time the values of the pseudoflow on the tree arcs, and in time $O(mn)$ to derive an associated feasible flow. At termination, that feasible flow is a maximum flow.

5.5   Variants of Pseudoflow Algorithm and Their Complexity

Implementing the pseudoflow algorithm in its generic form results in complexity of $O(nM^+)$ iterations, for $M^+ = C(\{s\}, V)$ – the sum of all capacities of arcs adjacent to source. This running time is not polynomial.

A natural variant to apply is capacity scaling. Capacity scaling is implemented in a straightforward way with running time of $O(mn \log M)$, where M is the largest capacity of source-adjacent or sink-adjacent arcs. This running time is polynomial but not strongly polynomial.

Our strongly polynomial variant relies on the lowest label selection rule of a merger arc. With this selection rule, our algorithm runs in strongly polynomial time, $O(mn \log n)$.

The lowest label selection rule is described recursively. Initially all nodes are assigned the label 1, $\ell_v = 1$ for all $v \in V$. The arc (s', w) selected is such that w is a lowest label weak node among all possible active arcs.

Upon a merger using the arc (s', w) the label of the strong node s' becomes the label of w plus 1 and all nodes of the strong branch with labels smaller than that of s' are updated to be equal to the label of s': formally, $\ell_{s'} \leftarrow \ell_w + 1$ and for all nodes v in the same branch with s', $\ell_v \leftarrow \max\{\ell_v, \ell_{s'}\}$.

The lowest label rule guarantees a bound of mn on the total number of iterations. In the *phase implementation* all weak nodes of the same label are processed in one phase. The phase implementation runs in time $O(mn \log n)$.

The lowest label implementation of the algorithm is particularly suitable for parametric implementation. The algorithm has features that make it especially easy to adjust to changes in capacities. The common type of analysis finding all breakpoints in parametric capacities of arcs adjacent to source and sink that are linear functions of the parameter λ can also be implemented in $O(mn \log n)$.

6   Pseudoflow-Based Simplex

The simplex algorithm adapted to the s-excess problem maintains a pseudoflow with all source and sink adjacent arcs saturated. At termination, the optimal solution delivered by this s-excess version of simplex identifies a minimum cut. The solution is optimal when only source adjacent nodes have positive excess and only sink adjacent nodes have deficits. Additional running time is required to reconstruct the feasible flows on the given tree, that constitute maximum flow.

We show in the full version of the paper that the concept of a strong basis, introduced by Cunningham [C76], is a tree in the extended network satisfying Properties 5, 6 and 9.

6.1   Pseudoflow-Based Simplex Iteration

An entering arc is a merger arc. It completes a cycle in the residual graph. It is thus an arc between two branches. We include an auxiliary arc from sink to source, thus the merger arc completes a cycle. Alternatively we shrink source and sink into a single node r as before.

Nodes that are on the source side of the tree are referred to as *strong*, and those that are on the sink side, as *weak*, with the notation of S and W respectively. Let an *entering arc* with positive residual capacity be (s', w). The cycle created is $[r, r_{s'}, \ldots, s', w, \ldots, r_w, r]$.

The largest amount of the flow that can be augmented along the cycle is the bottleneck residual capacity along the cycle. The first arc attaining this bottleneck capacity is the *leaving arc*.

In the simplex the amount of flow pushed is determined by the bottleneck capacity. In the pseudoflow algorithm the entire excess is pushed even though it may be blocked by one or more arcs that have insufficient residual capacity.

The use of the lowest label selection rule in the pseudoflow-based simplex algorithm for the choice of an entering arc leads to precisely the same complexity as that of our pseudoflow algorithm, $O(mn \log n)$.

6.2 Parametric Analysis for Pseudoflow-Based Simplex

Given a series of ℓ parameter values for λ, $\{\lambda_1, \ldots, \lambda_\ell\}$. Let the source adjacent arcs capacities and the sink adjacent arc capacities be a linear function of λ with the source adjacent capacities monotone nondecreasing with λ and the sink adjacent capacities monotone nonincreasing with λ. Recently Goldfarb and Chen [GC96] presented a dual simplex method with running time $O(mn^2)$. This method is adaptable to use for sensitivity analysis for such a sequence of ℓ parameter values, in the same amount of time as a single run, $O(mn^2 + n\ell)$. The algorithm, however, does not generate all parameter breakpoints in a complete parametric analysis. Our algorithm is the first simplex algorithm that does generate all parameter breakpoints.

The parametric analysis process is not described here. We only mention that in order to implement the complete parametric analysis we must recover from the simplex solution the minimal and maximal source sets minimum cuts. To do that, we scan the tree at the end of each computation for one parameter value, and separate 0-deficit branches. This process is equivalent to the normalization of a tree. It adds only linear time to the running time and may be viewed as *basis adjustment*. The running time is linear in the number of nodes in the currently computed graph.

The overall running time of the procedure is identical to that of the pseudoflow algorithm with lowest label, $O(mn \log n + n\ell)$.

6.3 Comparing Simplex to Pseudoflow

Although the simplex implementation in the extended network has the same complexity as that of the pseudoflow algorithm, the two algorithms are not the same. Starting from the same normalized tree a simplex iteration will produce different trees in the following iteration. The pseudoflow algorithm tends to produce trees that are shallower than those produced by simplex thus reducing the average work per iteration (which depends on the length of the path from the root to the merger node). Other details on the similarities and differences between simplex and the pseudoflow algorithm are provided in the full version of the paper.

References

C76.   W. H. Cunningham. A network simplex method. *Mathematical Programming*, 1:105–116, 1976.

GGT89.   G. Gallo, M. D. Grigoriadis and R. E. Tarjan. A fast parametric maximum flow algorithm and applications. *SIAM Journal of Computing*, 18(1):30–55, 1989.

GT86.   A. V. Goldberg and R. E. Tarjan. A new approach to the maximum flow problem. *J. Assoc. Comput. Mach.*, 35:921–940, 1988.

GC96.   D. Goldfarb and W. Chen. On strongly polynomial dual algorithms for the maximum flow problem. Special issue of *Mathematical Programming B*, 1996. To appear.

GH90.   D. Goldfarb and J. Hao. A primal simplex method that solves the Maximum flow problem in at most nm pivots and $O(n^2 m)$ time. *Mathematical Programming*, 47:353–365, 1990.

Hoc96.   D. S. Hochbaum. A new – old algorithm for minimum cut on closure graphs. Manuscript, June 1996.

LG64.   H. Lerchs, I. F. Grossmann. Optimum design of open-pit mines. *Transactions, C.I.M.*, LXVIII:17–24, 1965.

Pic76.   J. C. Picard. Maximal closure of a graph and applications to combinatorial problems. *Management Science*, 22:1268–1272, 1976.

An Implementation of a Combinatorial Approximation Algorithm for Minimum-Cost Multicommodity Flow

Andrew V. Goldberg[1], Jeffrey D. Oldham[2*], Serge Plotkin[2**], and Cliff Stein[3***]

[1] NEC Research Institute, Inc., Princeton, NJ 08540, USA
avg@research.nj.nec.com
[2] Department of Computer Science, Stanford University
Stanford, CA 94305–9045, USA
{oldham, plotkin}@cs.stanford.edu
http://theory.stanford.edu/~oldham
[3] Department of Computer Science, Dartmouth College
Hanover, NH 03755, USA
cliff@cs.dartmouth.edu

Abstract. The *minimum-cost multicommodity flow problem* involves simultaneously shipping multiple commodities through a single network so that the total flow obeys arc capacity constraints and has minimum cost. Multicommodity flow problems can be expressed as linear programs, and most theoretical and practical algorithms use linear-programming algorithms specialized for the problems' structures. Combinatorial approximation algorithms in [GK96,KP95b,PST95] yield flows with costs slightly larger than the minimum cost and use capacities slightly larger than the given capacities. Theoretically, the running times of these algorithms are much less than that of linear-programming-based algorithms. We combine and modify the theoretical ideas in these approximation algorithms to yield a fast, practical implementation solving the minimum-cost multicommodity flow problem. Experimentally, the algorithm solved our problem instances (to 1% accuracy) two to three orders of magnitude faster than the linear-programming package CPLEX [CPL95] and the linear-programming based multicommodity flow program PPRN [CN96].

* Research partially supported by an NSF Graduate Research Fellowship, ARO Grant DAAH04-95-1-0121, and NSF Grants CCR-9304971 and CCR-9307045.
** Research supported by ARO Grant DAAH04-95-1-0121, NSF Grants CCR-9304971 and CCR-9307045, and a Terman Fellowship.
*** Research partly supported by NSF Award CCR-9308701 and NSF Career Award CCR-9624828. Some of this work was done while this author was visiting Stanford University.

R. E. Bixby, E. A. Boyd, and R. Z. Ríos-Mercado (Eds.): IPCO VI
LNCS 1412, pp. 338–352, 1998. © Springer–Verlag Berlin Heidelberg 1998

1   Introduction

The *minimum-cost multicommodity flow problem* involves simultaneously shipping multiple commodities through a single network so the total flow obeys the arc capacity constraints and has minimum cost. The problem occurs in many contexts where different items share the same resource, e.g., communication networks, transportation, and scheduling problems [AMO93,HL96,HO96].

Traditional methods for solving minimum-cost and no-cost multicommodity flow problems are linear-programming based [AMO93,Ass78,CN96,KH80]. Using the ellipsoid [Kha80] or the interior-point [Kar84] methods, linear-programming problems can be solved in polynomial time. Theoretically, the fastest algorithms for solving the minimum-cost multicommodity flow problem exactly use the problem structure to speed up the interior-point method [KP95a,KV86,Vai89].

In practice, solutions to within, say 1%, often suffice. More precisely, we say that a flow is ϵ-optimal if it overflows the capacities by at most $1 + \epsilon$ factor and has cost that is within $1 + \epsilon$ of the optimum. Algorithms for computing approximate solutions to the multicommodity flow problem were developed in [LMP+95] (no-cost case) and [GK96,KP95b,PST95] (minimum-cost case). Theoretically, these algorithms are much faster than interior-point method based algorithms for constant ϵ. The algorithm in [LMP+95] was implemented [LSS93] and was shown that indeed it often outperforms the more traditional approaches. Prior to our work, it was not known whether the combinatorial approximation algorithms for the minimum-cost case could be implemented to run quickly.

In this paper we describe MCMCF, our implementation of a combinatorial approximation algorithm for the minimum-cost multicommodity flow problem. A direct implementation of [KP95b] yielded a correct but practically slow implementation. Much experimentation helped us select among the different theoretical insights of [KP95b,LMP+95,LSS93,PST95,Rad97] to achieve good practical performance.

We compare our implementation with CPLEX [CPL95] and PPRN [CN96]. (Several other efficient minimum-cost multicommodity flow implementations, e.g., [ARVK89], are proprietary so we were unable to use these programs in our study.) Both are based on the simplex method [Dan63] and both find exact solutions. CPLEX is a state-of-the-art commercial linear programming package, and PPRN uses a primal partitioning technique to take advantage of the multicommodity flow problem structure.

Our results indicate that the theoretical advantages of approximation algorithms over linear-programming-based algorithms can be translated into practice. On the examples we studied, MCMCF was several orders of magnitude faster than CPLEX and PPRN. For example, for 1% accuracy, it was up to three orders of magnitude faster. Our implementation's dependence on the number of commodities and the network size is also smaller, and hence we are able to solve larger problems.

We would like to compare MCMCF's running times with modified CPLEX and PPRN programs that yield approximate solutions, but it is not clear how to make the modifications. Even if we could make the modifications, we would probably need to use CPLEX's primal simplex to obtain a feasible flow before an exact solution is found. Since its primal simplex is an order of magnitude slower than its dual simplex for the problem instances we tested, the approximate code would probably not be any faster than computing an exact solution using dual simplex.

To find an ϵ-optimal multicommodity flow, MCMCF repeatedly chooses a commodity and then computes a single-commodity minimum-cost flow in an auxiliary graph. This graph's arc costs are exponential functions of the current flow. The base of the exponent depends on a parameter α, which our implementation chooses. A fraction σ of the commodity's flow is then rerouted to the corresponding minimum-cost flow. Each rerouting decreases a certain potential function. The algorithm iterates this process until it finds an ϵ-optimal flow.

As we have mentioned above, a direct implementation of [KP95b], while theoretically fast, is very slow in practice. Several issues are crucial for achieving an efficient implementation:

Exponential Costs: The value of the parameter α, which defines the base of the exponent, must be chosen carefully: Using a value that is too small will not guarantee any progress, and using a value that is too large will lead to very slow progress. Our adaptive scheme for choosing α leads to significantly better performance than using the theoretical value. Importantly, this heuristic does not invalidate the worst-case performance guarantees proved for algorithms using fixed α.

Stopping Condition: Theoretically, the algorithm yields an ϵ-optimal flow when the potential function becomes sufficiently small [KP95b]. Alternative algorithms, e.g., [PST95], explicitly compute lower bounds. Although these stopping conditions lead to the same asymptotic running time, the latter one leads to much better performance in our experiments.

Step Size: Theory specifies the rerouting fraction σ as a fixed function of α. Computing σ that maximizes the exponential potential function reduction experimentally decreases the running time. We show that is it possible to use the Newton-Raphson method [Rap90] to quickly find a near-optimal value of σ for every rerouting. Additionally, a commodity's flow usually differs from its minimum-cost flow on only a few arcs. We use this fact to speed up these computations.

Minimum-Cost Flow Subroutine: Minimum-cost flow computations dominate the algorithm's running time both in theory and in practice. The arc costs and capacities do not change much between consecutive minimum-cost flow computations for a particular commodity. Furthermore, the problem size is moderate by minimum-cost flow standards. This led us to decide to use the primal network simplex method. We use the current flow and a basis from a previous minimum-cost flow to "warm-start" each minimum-cost flow computation. Excepting the warm-start idea, our primal simplex code is similar to that of Grigoriadis [Gri86].

In the rest of this paper, we first introduce the theoretical ideas behind the implementation. After discussing the various choices in translating the theoretical ideas into practical performance, we present experimental data showing that

MCMCF's running time's dependence on the accuracy ϵ is smaller than theoretically predicted and its dependence on the number k of commodities is close to what is predicted. We conclude by showing that the combinatorial-based implementation solves our instances two to three orders of magnitude faster than two simplex-based implementations. In the longer version of this paper, we will also show that a slightly modified MCMCF solves the *concurrent flow problem*, i.e., the optimization version of the no-cost multicommodity flow problem, two to twenty times faster than Leong et al.'s approximation implementation [LSS93].

2   Theoretical Background

2.1   Definitions

The *minimum-cost multicommodity flow problem* consists of a directed network $G = (V, A)$, a positive arc capacity function u, a nonnegative arc cost function c, and a specification (s_i, t_i, d_i) for each commodity i, $i \in \{1, 2, \ldots, k_0\}$. Nodes s_i and t_i are the *source* and the *sink* of commodity i, and a positive number d_i is its *demand*.

A *flow* is a nonnegative arc function f. A *flow* f_i *of commodity* i is a flow obeying conservation constraints and satisfying its demand d_i. We define the total flow $f(a)$ on arc a by $f(a) = \sum_{i=1}^{k_0} f_i(a)$. Depending on context, the symbol f represents both the (multi)flow $(f_1, f_2, \ldots, f_k)$ and the total flow $f_1 + f_2 + \cdots + f_k$, summed arc-wise. The cost of a flow f is the dot product $c \cdot f = \sum_{a \in A} c(a) f(a)$.

Given a problem and a flow f, the *congestion* of arc a is $\lambda_a = f(a)/u(a)$, and the *congestion* of the flow is $\lambda_A = \max_a \lambda_a$. Given a budget B, the *cost congestion* is $\lambda_c = c \cdot f / B$, and the *total congestion* is $\lambda = \max\{\lambda_A, \lambda_c\}$. A *feasible problem instance* has a flow f with $\lambda_A \leq 1$.

Our implementation approximately solves the minimum-cost multicommodity flow problem. Given an accuracy $\epsilon > 0$ and a feasible multicommodity flow problem instance, the algorithm finds an ϵ-*optimal flow* f with ϵ-*optimal congestion*, i.e., $\lambda_A \leq (1 + \epsilon)$, and ϵ-*optimal cost*, i.e., if B^* is the minimum cost of any feasible flow, f's cost is at most $(1 + \epsilon) B^*$. Because we can choose ϵ arbitrarily small, we can find a solution arbitrarily close to the optimal.

We combine commodities with the same source nodes to form commodities with one source and (possibly) many sinks (see [LSS93,Sch91]). Thus, the number k of commodity groups may be smaller than the number k_0 of simple commodities in the input.

2.2   The Algorithmic Framework

Our algorithm is mostly based on [KP95b]. Roughly speaking, the approach in that paper is as follows. The algorithm first finds an initial flow satisfying demands but which may violate capacities and may be too expensive. The algorithm repeatedly modifies the flow until it becomes $O(\epsilon)$-optimal. Each iteration, the algorithm first computes the theoretical values for the constant α

and the step size σ. It then computes the dual variables $y_r = e^{\alpha(\lambda_r - 1)}$, where r ranges over the arcs A and the arc cost function c, and a potential function $\phi(f) = \sum_r y_r$. The algorithm chooses a commodity i to reroute in a round robin order, as in [Rad97]. It computes, for that commodity, a minimum-cost flow f_i^* in a graph with arc costs related to the gradient $\nabla\phi(f)$ of the potential function and arc capacities $\lambda_A u$. The commodity's flow f_i is changed to the convex combination $(1 - \sigma)f_i + \sigma f_i^*$. An appropriate choice of values for α and σ lead to $\tilde{O}(\epsilon^{-3}nmk)$ running time (suppressing logarithmic terms). Grigoriadis and Khachiyan [GK96] decreased the dependence on ϵ to ϵ^{-2}.

Since these minimum-cost algorithms compute a multiflow having arc cost at most a budget bound B, we use binary search on B to determine an ϵ-optimal cost. The arc cost of the initial flow gives the initial lower bound because the flow is the union of minimum-cost single-commodity flows with respect to the arc cost function c and arc capacities u. Lower bound computations (see Sect. 3.1) increase the lower bound and the algorithm decreases the congestion and cost until an ϵ-optimal flow is found.

3   Translating Theory into Practice

The algorithmic framework described in the previous section is theoretically efficient, but a direct implementation requires orders of magnitude larger running time than commercial linear-programming packages [CPL95]. Guided by the theoretical ideas of [KP95b,LMP+95,PST95], we converted the theoretically correct but practically slow implementation to a theoretically correct and practically fast implementation. In some cases, we differentiated between theoretically equivalent implementation choices that differ in practicality, e.g, see Sect. 3.1. In other cases, we used the theory to create heuristics that, in practice, reduce the running time, but, in the worst case, do not have an effect on the theoretical running time, e.g., see Sect. 3.3.

To test our implementation, we produced several families of random problem instances using three generators, MULTIGRID, RMFGEN, and TRIPARTITE. Our implementation, like most combinatorial algorithms, is sensitive to graph structure. MCMCF solves the MULTIGRID problem instances (based on [LO91]) very quickly, while RMFGEN instances (based on [Bad91]) are more difficult. We wrote the TRIPARTITE generator to produce instances that are especially difficult for our implementation to solve. More data will appear in the full paper. Brief descriptions of our problem generators and families will appear in [GOPS97].

3.1   The Termination Condition

Theoretically, a small potential function value and a sufficiently large value of the constant α indicates the flow is ϵ-optimal [KP95b], but this pessimistic indicator leads to poor performance. Instead, we periodically compute the lower bound on the optimal congestion λ^* as found in [LMP+95,PST95]. Since the problem instance is assumed to be feasible, the computation indicates when the current guess for the minimum flow cost is too low.

The weak duality inequalities yield a lower bound. Using the notation from [PST95],

$$\lambda \sum_r y_r \geq \sum_{\text{comm. } i} C_i(\lambda_A) \geq \sum_{\text{comm. } i} C_i^*(\lambda_A) \ . \tag{1}$$

For commodity i, $C_i(\lambda_A)$ represents the cost of the current flow f_i with respect to arc capacities $\lambda_A u$ and the cost function $y^t A$, where A is the $km \times m$ arc adjacency matrix together with the arc cost function c. $C_i^*(\lambda_A)$ is the minimum-cost flow. For all choices of dual variables and $\lambda_A \geq 1$, $\lambda^* \geq \sum_i C_i^*(1)/\sum_r y_r \geq \sum_i C_i^*(\lambda_A)/\sum_r y_r$. Thus, this ratio serves as a lower bound on the optimal congestion λ^*.

3.2   Computing the Step Size σ

While, as suggested by the theory, using a fixed step size σ to form the convex combination $(1 - \sigma)f_i + \sigma f_i^*$ suffices to reduce the potential function, our algorithm computes σ to maximize the potential function reduction. Brent's method and similar strategies, e.g., see [LSS93], are natural strategies to maximize the function's reduction. We implemented Brent's method [PFTV88], but the special structure of the potential function allows us to compute the function's first and second derivatives. Thus, we can use the Newton-Raphson method [PFTV88,Rap90], which is faster.

Given the current flow f and the minimum-cost flow f_i^* for commodity i, the potential function $\phi(\sigma)$ is a convex function (with positive second derivative) of the step size σ. Over the range of possible choices for σ, the potential function's minimum occurs either at the endpoints or at one interior point. Since the function is a sum of exponentials, the first and second derivatives $\phi'(\sigma)$ and $\phi''(\sigma)$ are easy to compute.

Using the Newton-Raphson method reduces the running time by two orders of magnitude compared with using a fixed step size. (See Table 1.) As the accuracy increases, the reduction in running time for the Newton-Raphson method increases. As expected, the decrease in the number of minimum-cost flow computations was even greater.

3.3   Choosing α

The algorithm's performance depends on the value of α. The larger its value, the more running time the algorithm requires. Unfortunately, α must be large enough to produce an ϵ-optimal flow. Thus, we developed heuristics for slowly increasing its value. There are two different theoretical explanations for α than can be used to develop two different heuristics.

Karger and Plotkin [KP95b] choose α so that, when the potential function is less than a constant factor of its minimum, the flow is ϵ-optimal. The heuristic

Table 1. Computing an (almost) optimal step size reduces the running time and number of minimum-cost flow (MCF) computations by two orders of magnitude.

problem	ϵ	time (seconds)			number of MCFs		
		Newton	fixed	ratio	Newton	fixed	ratio
rmfgen1	0.12	830	17220	20.7	399	8764	22.0
rmfgen1	0.06	1810	83120	45.9	912	45023	49.4
rmfgen1	0.03	5240	279530	53.3	2907	156216	53.7
rmfgen1	0.01	22930	2346800	102.3	13642	1361900	99.8
rmfgen2	0.01	3380	650480	192.4	3577	686427	191.9
rmfgen3	0.01	86790	9290550	107.0	17544	1665483	94.9
multigrid1	0.01	980	57800	59.0	1375	75516	54.9

of starting with a small α and increasing it when the potential function's value became too small experimentally failed to decrease significantly the running time.

Plotkin, Shmoys, and Tardos [PST95] use the weak duality inequalities (1) upon which we base a different heuristic. The product of the gaps bounds the distance between the potential function and the optimal flow. The algorithm's improvement is proportional to the size of the right gap, and increasing α decreases the left gap's size. Choosing α too large, however, can impede progress because progress is proportional to the step size σ which itself depends on how closely the potential function's linearization approximates its value. Thus, larger α reduces the step size.

Our heuristic attempts to balance the left and right gaps. More precisely, it chooses α dynamically to ensure the ratio of inequalities

$$\frac{(\lambda \sum_r y_r / \sum_{\text{comm. } i} C_i(\lambda_A)) - 1}{(\sum_{\text{comm. } i} C_i(\lambda_A) / \sum_{\text{comm. } i} C_i^*(\lambda_A)) - 1} \tag{2}$$

remains balanced. We increase α by factor β if the ratio is larger than 0.5 and otherwise decrease it by γ. After limited experimentation, we decided to use the golden ratio for both β and γ. The α values are frequently much lower than those from [PST95]. Using this heuristic rather than using the theoretical value of $\ln(3m)/\epsilon$ [KP95b] usually decreases the running time by a factor of between two and six. See Table 2.

3.4   Choosing a Commodity to Reroute

Several strategies for selecting the next commodity to reroute, proposed for concurrent flows, also apply to the minimum-cost multicommodity flow problem. These strategies include weighted [KPST94] and uniform [Gol92] randomization, and round-robin [Rad97] selection. Our experiments, which also included adaptive strategies, suggest that round robin works best.

Table 2. Adaptively choosing α requires fewer minimum-cost flow (MCF) computations than using the theoretical, fixed value of α.

problem	ϵ	number of MCFs			time (seconds)		
		adaptive	fixed	ratio	adaptive	fixed	ratio
GTE	0.01	2256	15723	6.97	1.17	7.60	6.49
rmfgen3	0.03	3394	10271	3.03	19.34	68.30	3.53
rmfgen4	0.10	4575	6139	1.34	24.65	33.00	1.34
rmfgen4	0.05	6966	22579	3.24	36.19	117.12	3.24
rmfgen4	0.03	16287	53006	3.25	80.10	265.21	3.31
rmfgen5	0.03	18221	64842	3.56	140.85	530.03	3.76
multigrid2	0.01	1659	1277	0.77	20.82	23.56	1.13

3.5   "Restarting" the Minimum-Cost Flow Subroutine

Theoretically, MCMCF can use any minimum-cost flow subroutine. In practice, the repeated evaluation of single-commodity problems with similar arc costs and capacities favor an implementation that can take advantage of *restarting* from a previous solution. We show that using a primal network simplex implementation allows restarting and thereby reduces the running time by one-third to one-half.

To solve a single-commodity problem, the primal simplex algorithm repeatedly *pivots* arcs into and out of a spanning tree until the tree has minimum cost. Each pivot maintains the flow's feasibility and can decrease its cost. The simplex algorithm can start with any feasible flow and any spanning tree. Since the cost and capacity functions do not vary much between MCF calls for the same commodity, we can speed up the computation, using the previously-computed spanning tree. Using the previously-found minimum-cost flow requires $O(km)$ additional storage. Moreover, it is more frequently unusable because it is infeasible with respect to the capacity constraints than using the current flow. In contrast, using the current flow requires no additional storage, this flow is known to be feasible, and starting from this flow experimentally requires a very small number of pivots.

Fairly quickly, the number of pivots per MCF iteration becomes very small. See Fig. 1. For the 2121-arc rmfgen-d-7-10-040 instance, the average number of pivots are 27, 13, and 7 for the three commodities shown. Less than two percent of arcs served as pivots. For the 260-arc GTE problem, the average numbers are 8, 3, and 1, i.e., at most three percent of the arcs.

Instead of using a commodity's current flow and its previous spanning tree, a minimum-cost flow computation could start from an arbitrary spanning tree and flow. On the problem instances we tried, restarting reduces the running time by a factor of about 1.3 to 2. See Table 3. Because optimal flows are not unique, the number of MCF computations differ, but the difference of usually less than five percent.

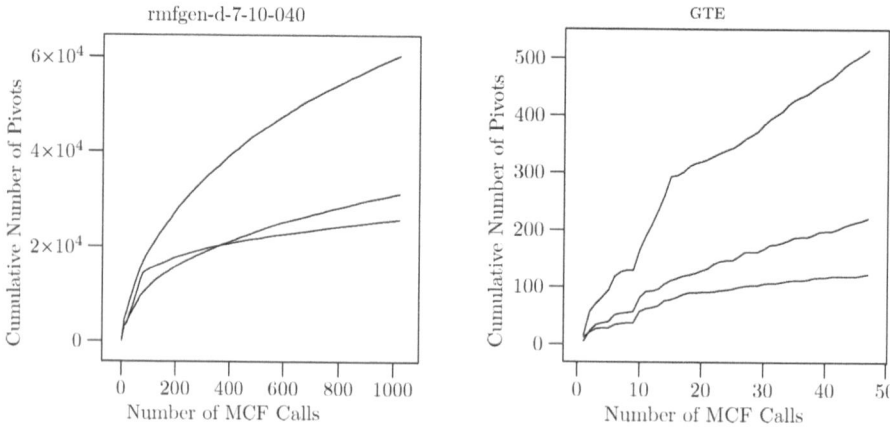

Fig. 1. The cumulative number of pivots as a function of the number of minimum-cost flow (MCF) calls for three different commodities in two problem instances

4   Experimental Results

4.1   Dependence on the Approximation Factor ϵ

The approximation algorithm MCMCF yields an ϵ-optimal flow. Plotkin, Shmoys, and Tardos [PST95] solve the minimum-cost multicommodity flow problem using shortest-paths as a basic subroutine. Karger and Plotkin [KP95b] decreased the running time by m/n using minimum-cost flow subroutines and adding a linear-cost term to the gradient to ensure each flow's arc cost is bounded. This change increases the ϵ-dependence of [PST95] by $1/\epsilon$ to ϵ^{-3}. Grigoriadis and Khachiyan [GK96] improved the [KP95b] technique, reducing the ϵ-dependence back to ϵ^{-2}. MCMCF implements the linear-cost term, but experimentation showed the minimum-cost flows' arc costs were bounded even without using the linear-cost term. Furthermore, running times usually decrease when omitting the term.

Table 3. Restarting the minimum-cost flow computations using the current flow and the previous spanning tree reduces the running time by at least 35%.

problem instance $\quad\epsilon$	restarting time (seconds)	no restarting time (seconds)	ratio
rmfgen-d-7-10-020 0.01	88	180	2.06
rmfgen-d-7-10-240 0.01	437	825	1.89
rmfgen-d-7-12-240 0.01	613	993	1.62
rmfgen-d-7-14-240 0.01	837	1396	1.67
rmfgen-d-7-16-240 0.01	1207	2014	1.67
multigrid-032-032-128-0080 0.01	21	37	1.77
multigrid-064-064-128-0160 0.01	275	801	2.92

The implementation exhibits smaller dependence than the worst-case no-cost multicommodity flow dependence of $O(\epsilon^{-2})$. We believe the implementation's searching for an almost-optimal step size and its regularly computing lower bounds decreases the dependence. Figure 2 shows the number of minimum-cost flow computations as a function of the desired accuracy ϵ. Each line represents a problem instance solved with various accuracies. On the log-log scale, a line's slope represents the power of $1/\epsilon$. For the RMFGEN problem instances, the dependence is about $O(\epsilon^{-1.5})$. For most MULTIGRID instances, we solved to a maximum accuracy of 1% but for five instances, we solved to an accuracy of 0.2%. These instances depend very little on the accuracy; MCMCF yields the same flows for several different accuracies. Intuitively, the grid networks permit so many different routes to satisfy a commodity that very few commodities need to share the same arcs. MCMCF is able to take advantage of these many different routes, while, as we will see in Sect. 5, some linear-programming based implementations have more difficulty.

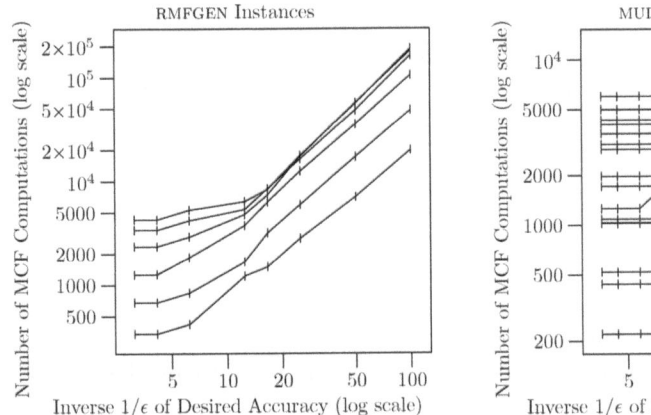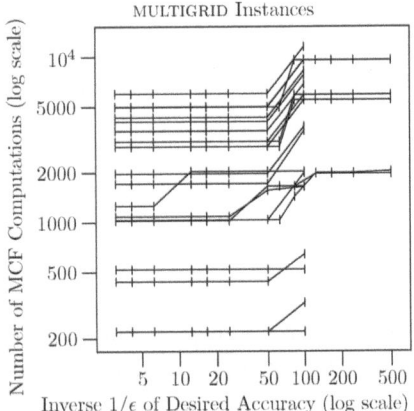

Fig. 2. The number of minimum-cost flow (MCF) computations as a function of $1/\epsilon$ for RMFGEN instances is $O(\epsilon^{-1.5})$ and for MULTIGRID instances

4.2 Dependence on the Number k of Commodity Groups

The experimental number of minimum-cost flow computations and the running time of the implementation match the theoretical upper bounds. Theoretically, the algorithm performs $\tilde{O}(\epsilon^{-3}k)$ minimum-cost flow computations, as described in Sect. 2.2. These upper bounds (ignoring the ϵ dependence and logarithmic dependences) match the natural lower bound where the joint capacity constraints are ignored and the problem can be solved using k single-commodity minimum-cost flow problems. In practice, the implementation requires at most a linear (in k) number of minimum-cost flows.

Figure 3 shows the number of minimum-cost flow computations as a function of the number k of commodity groups. Each line represents a fixed network with various numbers of commodity groups. The MULTIGRID figure shows a dependence of approximately $25k$ for two networks. For the RMFGEN instances, the dependence is initially linear but flattens and even decreases. As the number of commodity groups increases, the average demand per commodity decreases because the demands are scaled so the instances are feasible in a graph with 60% of the arc capacities. Furthermore, the randomly distributed sources and sinks are more distributed throughout the graph reducing contention for the most congested arcs. The number of minimum-cost flows depends more on the network's congestion than on the instance's size so the lines flatten.

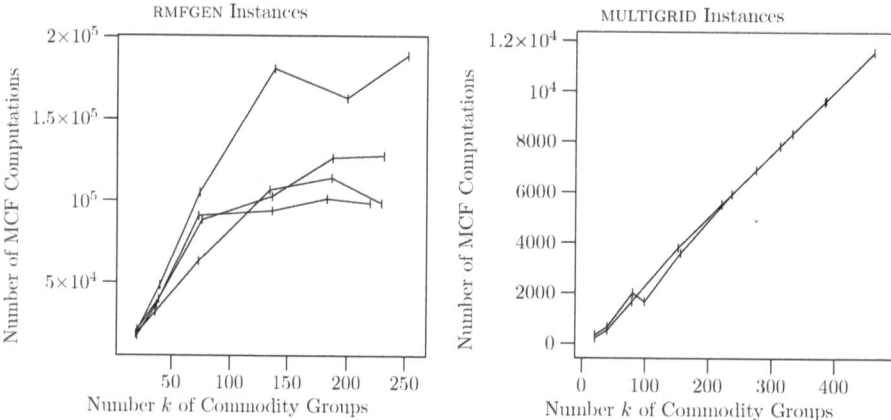

Fig. 3. The number of minimum-cost flow (MCF) computations as a function of the number k of commodity groups for RMFGEN and MULTIGRID instances

5   Comparisons with Other Implementations

5.1   The Other Implementations: CPLEX and PPRN

We compared MCMCF (solving to 1% accuracy) with a commercial linear-programming package CPLEX [CPL95] and the primal partitioning multicommodity flow implementation PPRN [CN96].

CPLEX (version 4.0.9) yields exact solutions to multicommodity flow linear programs. When forming the linear programs, we group the commodities since MCMCF computes these groups at run-time. CPLEX's dual simplex method yields a feasible solution only upon completion, while the primal method, in principle, could be stopped to yield an approximation. Despite this fact, we compared MCMCF with CPLEX's dual simplex method because it is an order of magnitude faster than its primal simplex for the problems we tested.

PPRN [CN96] specializes the primal partitioning linear programming technique to solve multicommodity problems. The primal partitioning method splits the instance's basis into bases for the commodities and another basis for the joint capacity constraints. Network simplex methods then solve each commodity's subproblem. More general linear-programming matrix computations applied to the joint capacity basis combine these subproblems' solutions to solve the problem.

5.2   Dependence on the Number k of Commodity Groups

The combinatorial algorithm MCMCF solves our problem instances two to three orders of magnitude faster than the linear-programming-based implementations CPLEX and PPRN. Furthermore, its running time depends mostly on the network structure and much less on the arc costs' magnitude.

We solved several different RMFGEN networks (see Fig. 4) with various numbers of commodities and two different arc cost schemes. Even for instances having as few as fifty commodities, MCMCF required less running time. Furthermore, its dependence on the number k of commodities was much smaller. For the left side of Fig. 4, the arc costs were randomly chosen from the range $[1, 100]$. For these problems, CPLEX's running time is roughly quadratic in k, while MCMCF's is roughly linear. Although for problems with few commodities, CPLEX is somewhat faster, for larger problems MCMCF is faster by an order of magnitude. PPRN is about five times slower than CPLEX for these problems. Changing the cost of interframe arcs significantly changes CPLEX's running time. (See the right side of Fig. 4.) Both MCMCF's and PPRN's running times decrease slightly. The running times' dependences on k do not change appreciably.

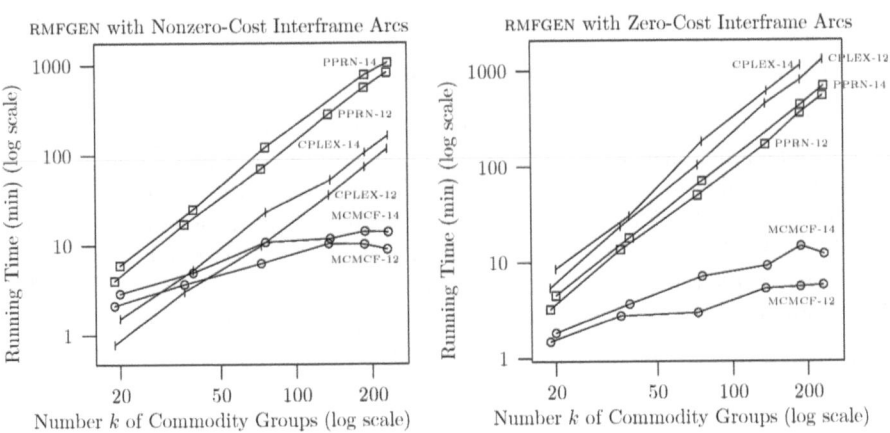

Fig. 4. The running time in minutes as a function of the number k of commodity groups for two different RMFGEN networks with twelve and fourteen frames. CPLEX's and PPRN's dependences are larger than MCMCF's

MCMCF solves MULTIGRID networks two to three orders of magnitude faster than CPLEX and PPRN. The left side of Fig. 5 shows MCMCF's running time using a log-log scale for two different networks: the smaller one having 1025 nodes and 3072 arcs and the larger one having 4097 nodes and 9152 arcs. CPLEX and PPRN required several days to solve the smaller network instances so we omitted solving the larger instances. Even for the smallest problem instance, MCMCF is eighty times faster than CPLEX, and its dependence on the number of commodities is much smaller. PPRN is two to three times slower than CPLEX so we solved only very small problem instances using PPRN.

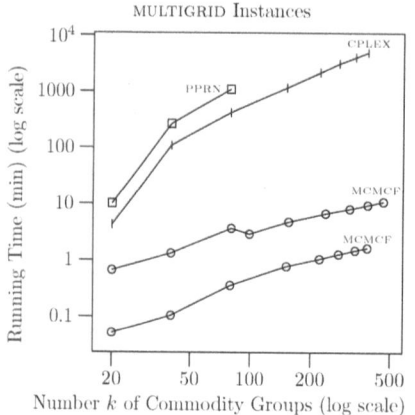
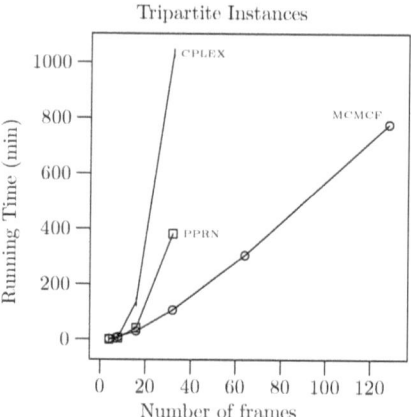

Fig. 5. The running time in minutes as a function of the number k of commodity groups (left figure) and the number of frames (right figure)

5.3   Dependence on the Network Size

To test the implementations' dependences on the problem size, we used TRIPAR-TITE problem instances with increasing numbers of frames. Each frame has fixed size so the number of nodes and arcs is linearly related to the number of frames. For these instances, MCMCF's almost linear dependence on problem size is much less than CPLEX's and PPRN's dependences. See the right side of Fig. 5. (MCMCF solved the problem instances to two-percent accuracy.) As described in Sect. 3.5, the minimum-cost flow routine needs only a few pivots before a solution is found. (For the sixty-four frame problem, PPRN required 2890 minutes so it was omitted from the figure.)

6   Concluding Remarks

For the problem classes we studied, MCMCF solved minimum-cost multicommodity flow problems significantly faster than state-of-the-art linear-programming-based programs. This is strong evidence the approximate problem is simpler, and that combinatorial-based methods, appropriately implemented, should be considered for this problem. We believe many of these techniques can be extended to other problems solved using the fractional packing and covering framework of [PST95].

We conclude with two unanswered questions. Since our implementation never needs to use the linear-cost term [KP95b], it is interesting to prove whether the term is indeed unnecessary. Also, it is interesting to try to prove the experimental $O(\epsilon^{-1.5})$ dependence of Sect. 4.1.

References

AMO93.   R. K. Ahuja, T. L. Magnanti, and J. B. Orlin. *Network Flows: Theory, Algorithms, and Applications*. Prentice Hall, Englewood Cliffs, NJ, 1993.

ARVK89.   I. Adler, M. G. C. Resende, G. Veiga, and N. Karmarkar. An implementation of Karmarkar's algorithm for linear programming. *Mathematical Programming A*, 44(3):297–335, 1989.

Ass78.   A. A. Assad. Multicommodity network flows – A survey. *Networks*, 8(1):37–91, Spring 1978.

Bad91.   T. Badics. GENRMF. 1991. ftp://dimacs.rutgers.edu/pub/netflow/generators/network/genrmf/

CN96.   J. Castro and N. Nabona. An implementation of linear and nonlinear multicommodity network flows. *European Journal of Operational Research*, 92(1):37–53, 1996.

CPL95.   CPLEX Optimization, Inc., Incline Village, NV. *Using the CPLEX Callable Library*, 4.0 edition, 1995.

Dan63.   G. B. Dantzig. *Linear Programming and Extensions*. Princeton University Press, Princeton, NJ, 1963.

GK96.   M. D. Grigoriadis and L. G. Khachiyan. Approximate minimum-cost multicommodity flows in $\tilde{O}(\epsilon^{-2}knm)$ time. *Mathematical Programming*, 75(3):477–482, 1996.

Gol92.   A. V. Goldberg. A natural randomization strategy for multicommodity flow and related algorithms. *Information Processing Letters*, 42(5):249–256, 1992.

GOPS97.   A. Goldberg, J. D. Oldham, S. Plotkin, and C. Stein. An implementation of a combinatorial approximation algorithm for minimum-cost multicommodity flow. Technical Report CS-TR-97-1600, Stanford University, December 1997.

Gri86.   M. D. Grigoriadis. An efficient implementation of the network simplex method. *Mathematical Programming Study*, 26:83–111, 1986.

HL96.   R. W. Hall and D. Lotspeich. Optimized lane assignment on an automated highway. *Transportation Research—C*, 4C(4):211–229, 1996.

HO96.   A. Haghani and S.-C. Oh. Formulation and solution of a multi-commodity, multi-modal network flow model for disaster relief operations. *Transportation Research—A*, 30A(3):231–250, 1996.

Kar84.   N. Karmarkar. A new polynomial-time algorithm for linear programming. *Combinatorica*, 4(4):373–395, 1984.

KH80.   J. L. Kennington and R. V. Helgason. *Algorithms for Network Programming*. John Wiley & Sons, New York, 1980.

Kha80.   L. G. Khachiyan. Polynomial algorithms in linear programming. *Zhurnal Vychislitel'noi Matematiki i Matematicheskoi Fiziki (Journal of Computational Mathematics and Mathematical Physics)*, 20(1):51–68, 1980.

KP95a.   A. Kamath and O. Palmon. Improved interior point algorithms for exact and approximate solution of multicommodity flow problems. In *Proceedings of the 6th Annual ACM-SIAM Symposium on Discrete Algorithms, Vol. 6*, pages 502–511. Association for Computing Machinery, January 1995.

KP95b.   D. Karger and S. Plotkin. Adding multiple cost constraints to combinatorial optimization problems, with applications to multicommodity flows. In *Symposium on the Theory of Computing, Vol. 27*, pages 18–25. Association for Computing Machinery, ACM Press, May 1995.

KPST94.   P. Klein, S. Plotkin, C. Stein, and É. Tardos. Faster approximation algorithms for the unit capacity concurrent flow problem with applications to routing and finding sparse cuts. *SIAM Journal on Computing*, 23(3):466–487, 1994.

KV86.   S. Kapoor and P. M. Vaidya. Fast algorithms for convex quadratic programming and multicommodity flows. In *Proceedings of the 18th Annual ACM Symposium on Theory of Computing, Vol. 18*, pages 147–159. Association for Computing Machinery, 1986.

LMP+95.   T. Leighton, F. Makedon, S. Plotkin, C. Stein, É. Tardos, and S. Tragoudas. Fast approximation algorithms for multicommodity flow problems. *Journal of Computer and System Sciences*, 50(2):228–243, 1995.

LO91.   Y. Lee and J. Orlin. GRIDGEN. 1991. ftp://dimacs.rutgers.edu/pub/netflow/generators/network/gridgen/

LSS93.   T. Leong, P. Shor, and C. Stein. Implementation of a combinatorial multicommodity flow algorithm. In David S. Johnson and Catherine C. McGeoch, editors, *Network Flows and Matching, Series in Discrete Mathematics and Theoretical Computer Science, Vol. 12*, pages 387–405. American Mathematical Society, 1993.

PFTV88.   W. H. Press, B. P. Flannery, S. A. Teukolsky, and W. T. Vetterling. *Numerical Recipes in C*. Cambridge University Press, Cambridge, 1988.

PST95.   S. A. Plotkin, D. B. Shmoys, and É. Tardos. Fast approximation algorithms for fractional packing and covering problems. *Mathematics of Operations Research*, 20(2):257–301, 1995.

Rad97.   T. Radzik. Fast deterministic approximation for the multicommodity flow problem. *Mathematical Programming*, 78(1):43–58, 1997.

Rap90.   J. Raphson. *Analysis Æquationum Universalis, seu, Ad Æquationes Algebraicas Resolvendas Methodus Generalis, et Expedita*. Prostant venales apud Abelem Swalle, London, 1690.

Sch91.   R. Schneur. *Scaling Algorithms for Multicommodity Flow Problems and Network Flow Problems with Side Constraints*. PhD thesis, MIT, Cambridge, MA, February 1991.

Vai89.   P. M. Vaidya. Speeding up linear programming using fast matrix multiplication. In *Proceedings of the 30th Annual Symposium on Foundations of Computer Science, Vol. 30*, pages 332–337. IEEE Computer Society Press, 1989.

Non-approximability Results for Scheduling Problems with Minsum Criteria

Han Hoogeveen[1], Petra Schuurman[1], and Gerhard J. Woeginger[2*]

[1] Department of Mathematics and Computing Science
Eindhoven University of Technology
P.O. Box 513, 5600 MB Eindhoven, The Netherlands
{slam, petra}@@win.tue.nl
[2] Institut für Mathematik, Technische Universität Graz
Steyrergasse 30, A-8010 Graz, Austria
gwoegi@@opt.math.tu-graz.ac.at

Abstract. We provide several non-approximability results for deterministic scheduling problems whose objective is to minimize the total job completion time. Unless $\mathcal{P} = \mathcal{NP}$, none of the problems under consideration can be approximated in polynomial time within arbitrarily good precision. Most of our results are derived by MAX SNP hardness proofs. Among the investigated problems are: scheduling unrelated machines with some additional features like job release dates, deadlines and weights, scheduling flow shops, and scheduling open shops.

1 Introduction

Since the early 1970s, the algorithms and optimization community has put lots of efforts into identifying the computational complexity of various combinatorial optimization problems. Nowadays it is common knowledge that, when dealing with an $\mathcal{NP}$-hard optimization problem, one should not expect to find a polynomial-time solution algorithm. This insight motivates the search for approximation algorithms that output provably good solutions in polynomial time. It also immediately raises the question of how well we can approximate a specific optimization problem in polynomial time.

We say that a polynomial-time approximation algorithm for some optimization problem has a *performance guarantee* or *worst-case ratio* ρ, if it outputs a feasible solution with cost at most ρ times the optimum value for all instances of the problem; such an algorithm is also called a *polynomial-time ρ-approximation algorithm*. Now, given an $\mathcal{NP}$-hard optimization problem, for which values of ρ does there exist and for which values of ρ does there not exist a polynomial-time ρ-approximation algorithm? In this paper we focus on 'negative' results for scheduling problems in this area, i.e., we demonstrate for several scheduling

---

* Supported by the START program Y43–MAT of the Austrian Ministry of Science.

R. E. Bixby, E. A. Boyd, and R. Z. Ríos-Mercado (Eds.): IPCO VI
LNCS 1412, pp. 353–366, 1998. © Springer–Verlag Berlin Heidelberg 1998

problems that they do not have polynomial-time ρ-approximation algorithms with ρ arbitrarily close to 1, unless $\mathcal{P} = \mathcal{NP}$.

Until now, the literature only contains a small number of non-approximability results for scheduling problems. Most of the known non-approximability results have been derived for the objective of minimizing the makespan. As far as we know, the only non-approximability results for scheduling problems with minsum objective are presented in the papers of Kellerer, Tautenhahn & Woeginger [7] and Leonardi & Raz [12]. In the first paper, the authors prove that the problem of minimizing total flow time on a single machine subject to release dates cannot be approximated in polynomial time within any constant factor. In the second one, the authors derive a similar result for the total flow time problem subject to release dates on parallel machines. Below, we give a short list with the most important results for makespan minimization, all under the assumption that $\mathcal{P} = \mathcal{NP}$; for more detailed information on the scheduling problems mentioned, see the first paragraphs of Sections 3 and 4, or see the survey article by Lawler, Lenstra, Rinnooy Kan & Shmoys [8]. We refer the reader interested in more results on approximability and non-approximability in scheduling to Lenstra & Shmoys [10]. Throughout this paper, we use the notation introduced by Graham, Lawler, Lenstra & Rinnooy Kan [4] to denote the scheduling problems.

- Lenstra & Rinnooy Kan [9] prove that $P \,|\, prec, p_j = 1 \,|\, C_{\max}$ (makespan minimization on parallel machines with precedence constraints and unit processing times) does not have a polynomial-time approximation algorithm with performance guarantee strictly better than 4/3.
- Lenstra, Shmoys & Tardos [11] show that $R \,|\,|\, C_{\max}$ (makespan minimization on unrelated machines) does not have a polynomial-time approximation algorithm with performance guarantee strictly better than 3/2.
- Hoogeveen, Lenstra & Veltman [5] prove that $P \,|\, prec, c = 1, p_j = 1 \,|\, C_{\max}$ and $P\infty \,|\, prec, c = 1, p_j = 1 \,|\, C_{\max}$ (two variants of makespan minimization on parallel machines with unit processing times and unit communication delays) cannot be approximated with worst-case ratios better than 5/4 and 7/6, respectively.
- Williamson et al. [16] prove that $O \,|\,|\, C_{\max}$ and $F \,|\,|\, C_{\max}$ (makespan minimization in open shops and flow shops, respectively) cannot be approximated in polynomial time with performance guarantees better than 5/4.

All the listed non-approximability results have been derived via the so-called *gap technique*, i.e., via $\mathcal{NP}$-hardness reductions that create gaps in the cost function of the constructed instances. More precisely, such a reduction transforms the YES-instances of some $\mathcal{NP}$-hard problem into scheduling instances with objective value at most c^*, and it transforms the NO-instances into scheduling instances with objective value at least $g \cdot c^*$, where $g > 1$ is some fixed real number. Then a polynomial-time approximation algorithm for the scheduling problem with performance guarantee strictly better than g (i.e., with guarantee $g - \varepsilon$ where $\varepsilon > 0$) would be able to separate the YES-instances from the NO-instances, thus yielding a polynomial-time solution algorithm for an $\mathcal{NP}$-

complete problem. Consequently, unless $\mathcal{P} = \mathcal{NP}$, the scheduling problem cannot have a polynomial-time ρ-approximation algorithm with $\rho < g$.

In this paper, we present non-approximability results for the corresponding minsum versions of the makespan minimization problems listed above. We prove that none of the scheduling problems

- $R \,|\, r_j \,|\, \sum C_j$, $R \,|\, \bar{d}_j \,|\, \sum C_j$, and $R \,|\,\,|\, \sum w_j C_j$;
- $F \,|\,\,|\, \sum C_j$ and $O \,|\,\,|\, \sum C_j$;
- $P \,|\, prec, p_j = 1 \,|\, \sum C_j$;
- $P \,|\, prec, c = 1, p_j = 1 \,|\, \sum C_j$ and $P\infty \,|\, prec, c = 1, p_j = 1 \,|\, \sum C_j$

can be approximated in polynomial time within arbitrarily good precision, unless $\mathcal{P} = \mathcal{NP}$. Our main contribution is the non-approximability result for the problem $R \,|\, r_j \,|\, \sum C_j$, which answers an open problem posed by Skutella [15]. Interestingly, we do not prove this non-approximability result by applying the standard gap technique that we sketched above, but by establishing MAX SNP-hardness of this problem (see Section 2 for some information on MAX SNP-hardness). In the MAX SNP-hardness proof for $R \,|\, r_j \,|\, \sum C_j$ we are recycling some of the ideas that Lenstra, Shmoys & Tardos [11] used in their gap reduction for $R \,|\,\,|\, C_{\max}$. Also the non-approximability results for $F \,|\,\,|\, \sum C_j$ and $O \,|\,\,|\, \sum C_j$ are established via MAX SNP-hardness proofs; part of these arguments are based on combinatorial considerations that first have been used by Williamson et al. [16]. The non-approximability results for $P \,|\, prec, p_j = 1 \,|\, \sum C_j$, for $P \,|\, prec, c = 1, p_j = 1 \,|\, \sum C_j$, and for $P\infty \,|\, prec, c = 1, p_j = 1 \,|\, \sum C_j$ follow from the well-known gap reductions for the corresponding makespan minimization problems $P \,|\, prec, p_j = 1 \,|\, C_{\max}$, $P \,|\, prec, c = 1, p_j = 1 \,|\, C_{\max}$, and $P\infty \,|\, prec, c = 1, p_j = 1 \,|\, C_{\max}$ in a straightforward way, and therefore they are not described in this paper.

The paper is organized as follows. In Section 2, we summarize some useful information on approximation schemes, non-approximability, MAX SNP-hardness, and L-reductions, as we need it in the remainder of the paper. The non-approximability results are presented in Sections 3 and 4: Section 3 deals with scheduling unrelated machines and Section 4 considers flow shops and open shops. The paper is concluded with a short discussion in Section 5.

2   Preliminaries on Non-approximability

This section gives some information on approximation algorithms, and it summarizes some basic facts on MAX SNP-hardness. For a more extensive explanation we refer the reader to Papadimitriou & Yannakakis [14]. For a compendium of publications on non-approximability results we refer to Crescenzi & Kann [2].

A *polynomial-time approximation scheme* for an optimization problem, PTAS for short, is a family of polynomial-time $(1 + \varepsilon)$-approximation algorithms for all $\varepsilon > 0$. Essentially, a polynomial-time approximation scheme is the strongest possible approximability result for an $\mathcal{NP}$-hard problem. For $\mathcal{NP}$-hard problems an important question is whether such a scheme exists. The main tool for dealing with this question is the *L-reduction* as introduced by Papadimitriou & Yannakakis [14]:

Definition 1 (Papadimitriou and Yannakakis [14]). *Let A and B be two optimization problems. An L-reduction from A to B is a pair of functions R and S, both computable in polynomial time, with the following two additional properties:*

- *For any instance I of A with optimum cost* $\text{OPT}(I)$*, $R(I)$ is an instance of B with optimum cost* $\text{OPT}(R(I))$*, such that*

$$\text{OPT}(R(I)) \leq \alpha \cdot \text{OPT}(I), \tag{1}$$

  for some positive constant α*.*
- *For any feasible solution s of $R(I)$, $S(s)$ is a feasible solution of I such that*

$$|\text{OPT}(I) - c(S(s))| \leq \beta \cdot |\text{OPT}(R(I)) - c(s)|, \tag{2}$$

  for some positive constant β*, where $c(S(s))$ and $c(s)$ represent the costs of $S(s)$ and s, respectively.* □

Papadimitriou & Yannakakis [14] prove that *L*-reductions in fact are *approximation preserving* reductions. If there is an *L*-reduction from the optimization problem A to problem B with parameters α and β, and if there exists a polynomial-time approximation algorithm for B with performance guarantee $1 + \varepsilon$, then there exists a polynomial-time approximation algorithm for A with performance guarantee $1 + \alpha\beta\varepsilon$. Consequently, if there exists a PTAS for B, then there also exists a PTAS for A.

Papadimitriou & Yannakakis [14] also define a class of optimization problems called MAX SNP, and they prove that every problem in this class is approximable in polynomial time within some constant factor. The class MAX SNP is closed under *L*-reductions, and the hardest problems in this class (with respect to *L*-reductions) are the MAX SNP-*complete* ones. A problem that is at least as hard (with respect to *L*-reductions) as a MAX SNP-complete problem is called MAX SNP-*hard*. For none of these MAX SNP-hard problems, a PTAS has been constructed. Moreover, if there does exist a PTAS for *one* MAX SNP-hard problem, then *all* MAX SNP-hard problems have a PTAS, and

Proposition 2 (Arora, Lund, Motwani, Sudan, and Szegedy [1]). *If there exists a PTAS for some MAX SNP-hard problem, then $\mathcal{P} = \mathcal{NP}$.* □

Proposition 2 provides a strong tool for proving the non-approximability of an optimization problem X. Just provide an *L*-reduction from a MAX SNP-hard problem to X. Then unless $\mathcal{P} = \mathcal{NP}$, problem X cannot have a PTAS.

3   Unrelated Parallel Machine Scheduling

The *unrelated parallel machine scheduling problem* considered in this section is defined as follows. There are n independent jobs $1, 2, \ldots, n$ that have to be scheduled on m machines $M_1, M_2, \ldots, M_m$. Preemptions are not allowed. Every machine can only process one job at a time and every job has to be processed on exactly one machine. If job j is scheduled on machine M_i, then the processing time required is p_{ij}. Every job j has a release date r_j on which it becomes available for processing. The objective is to minimize the total job completion time. In the standard scheduling notation this problem is denoted by $R \mid r_j \mid \sum C_j$.

We prove the non-approximability of $R \mid r_j \mid \sum C_j$ by presenting an L-reduction from a MAX SNP-hard 3-dimensional matching problem to $R \mid r_j \mid \sum C_j$. This L-reduction draws some ideas from the gap reduction of Lenstra, Shmoys & Tardos [11] for $R \mid \mid C_{\max}$. Consider the following *Maximum Bounded 3-Dimensional Matching* problem (MAX-3DM-B), which has been proven to be MAX SNP-hard by Kann [6].

MAXIMUM BOUNDED 3-DIMENSIONAL MATCHING (MAX-3DM-B)

Input: Three sets $A = \{a_1, \ldots, a_q\}$, $B = \{b_1, \ldots, b_q\}$ and $C = \{c_1, \ldots, c_q\}$. A subset T of $A \times B \times C$ of cardinality s, such that any element of A, B and C occurs in exactly one, two, or three triples in T. Note that this implies that $q \leq s \leq 3q$.
Goal: Find a subset T' of T of maximum cardinality such that no two triples of T' agree in any coordinate.
Measure: The cardinality of T'.

The following simple observation will be useful.

Lemma 3. *For any instance I of MAX-3DM-B, we have* $\mathrm{OPT}(I) \geq \frac{1}{7} s$.   □

Now let $I = (q, T)$ be an instance of MAX-3DM-B. We construct an instance $R(I)$ of $R \mid r_j \mid \sum C_j$ with $3q + s$ jobs and $s + q$ machines. The first s machines correspond to the triples in T, and hence are called the *triple machines*. The remaining q machines are the *dummy machines*. The $3q + s$ jobs are divided into $3q$ *element jobs* and into s *dummy jobs*. The $3q$ element jobs correspond to the elements of A, B, and C, and are called A-jobs, B-jobs, and C-jobs, respectively.

Let $T_l = (a_i, b_j, c_k)$ be the lth triple in T. The processing times on the lth triple machine are now defined as follows. The processing time of the three element jobs that correspond to the elements a_i, b_j, and c_k is 1, whereas all the other element jobs have processing time infinity. The dummy jobs have processing time 3 on any triple machine and also on any dummy machine. All A-jobs and B-jobs have processing time 1 on the dummy machines, and all C-jobs have infinite processing time on the dummy machines. The release dates of all A-jobs and all dummy jobs are 0, all B-jobs have release date 1, and all C-jobs have release date 2.

In the following we are mainly interested in schedules of the following structure. Among the triple machines, there are k machines that process the three element jobs that belong to the triple corresponding to the machine (we call such machines *good* machines), there are $q - k$ machines that process a dummy job together with a C-job, and there are $s - q$ machines that process a single dummy

job. The q dummy machines are split into two groups: $q - k$ of them process an
A-job and a B-job, and k of them process a single dummy job. An illustration for
schedules of this structure is given in Figure 1. All jobs are scheduled as early as
possible. The cost Z_k of a schedule σ with the special structure described above

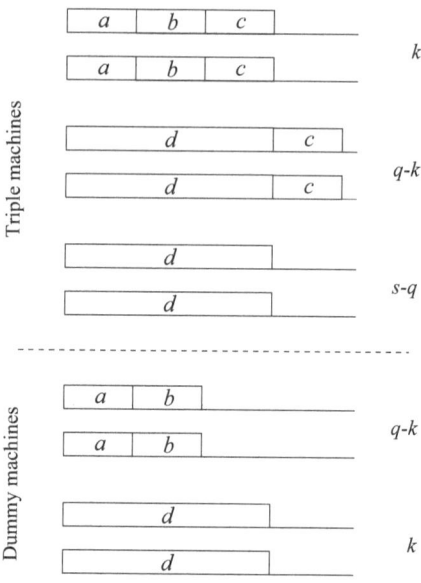

Fig. 1. A schedule for $R(I)$ with a special structure.

only depends on the number k of good machines, and it is equal to

$$Z_k = \sum_{A-jobs\ a} C_a + \sum_{B-jobs\ b} C_b + \sum_{C-jobs\ c} C_c + \sum_{dummy\ jobs\ d} C_d$$
$$= q + 2q + 3k + 4(q - k) + 3s$$
$$= 7q - k + 3s.$$

Lemma 4. *Let σ be any feasible schedule for $R(I)$ with k good machines. Then
the objective value of σ is at least equal to Z_k, and there exists a feasible schedule σ' satisfying the structure described above with the same number k of good
machines.*

Proof. We argue that the objective value of schedule σ is at least Z_k; then σ
can be replaced by a schedule σ' with the desired structure, which is readily
determined given the k good machines. Note that a schedule in which every job
starts at its release date has objective value equal to $6q + 3s$. We prove that the
extra cost caused by jobs that do not start at their release date is at least $q - k$.

Let ℓ^A, ℓ^B, and ℓ^C be the number of A-jobs, B-jobs, and C-jobs, respectively, that do not start at their release dates. Let ℓ_1^D be the number of dummy jobs that start at time 1, and let ℓ_2^D be the number of dummy jobs that start at or after time 2. Finally, let k_1 and k_2 be the number of machines that are idle during the time intervals $[0,1]$ and $[1,2]$, respectively. There are $q - \ell^C$ of the C-jobs whose processing starts at their release date. All these C-jobs are processed on triple machines, and only k triple machines are good machines. Hence, the remaining number of at least $q - \ell^C - k$ machines that process such a C-job must be idle during the time interval $[0,1]$, or during $[1,2]$, or during both intervals. This yields that

$$k_1 + k_2 \geq q - \ell^C - k. \tag{3}$$

The number of jobs that have release date zero equals the number of machines. Hence, the processing of at least k_1 of the A-jobs and dummy jobs does not start at time zero, and we conclude that $\ell^A + \ell_1^D + \ell_2^D \geq k_1$. Analogously at least k_2 of the B-jobs and dummy jobs are not processed during the time interval $[1,2]$, and therefore $\ell^B + \ell_2^D \geq k_2$ holds. Summarizing, one gets that the additional cost caused by jobs in σ that do not start at their release date is at least

$$\ell^A + \ell^B + \ell^C + \ell_1^D + 2\ell_2^D = (\ell^A + \ell_1^D + \ell_2^D) + (\ell^B + \ell_2^D) + \ell^C$$
$$\geq k_1 + k_2 + \ell^C \geq q - \ell^C - k + \ell^C = q - k.$$

Hereby the proof of the lemma is complete. $\quad\square$

Lemma 5. *For any instance I of* MAX-3DM-B, *we have that* $\text{OPT}(R(I)) \leq 69\,\text{OPT}(I)$. *Hence, the polynomial-time transformation R fulfills condition (1) of Definition 1.*

Proof. Let $k = \text{OPT}(I)$. The statement of Lemma 4 yields that $\text{OPT}(R(I)) = 7q - k + 3s$, and Lemma 3 yields that $k \geq \frac{1}{7}s \geq \frac{1}{7}q$. Hence, $\text{OPT}(R(I)) \leq 49k - k + 21k = 69\,\text{OPT}(I)$. $\quad\square$

Next, we define a polynomial-time transformation S that maps feasible solutions of $R(I)$ to feasible solutions of I. Let σ be a feasible schedule for $R(I)$. As described in the proof of Lemma 4, we find a corresponding schedule σ' for σ that has the special structure. Then the feasible solution $S(\sigma)$ for the instance I of MAX-3DM-B consists of the triples in T that correspond to the good machines in σ'.

Lemma 6. *For any feasible schedule σ of $R(I)$, the feasible solution $S(\sigma)$ of instance I fulfills the inequality $|\text{OPT}(I) - c(S(\sigma))| \leq |\text{OPT}(R(I)) - c(\sigma)|$. Hence, the polynomial-time transformation S fulfills condition (2) of Definition 1.*

Proof. The statement of Lemma 4 implies that if there exists a schedule of cost $7q + 3s - k'$ for $R(I)$, then there exists a solution T' for I with cardinality $|T'| = k'$. Let $k = \text{OPT}(I)$. Then $|\text{OPT}(I) - c(S(\sigma))| = k - k'$ and $|\text{OPT}(R(I)) - c(\sigma)| = |7q + 3s - k - (7q + 3s - k')|$. $\quad\square$

Summarizing, Lemma 5 and Lemma 6 state that the transformations R and S satisfy both conditions in Definition 1, and hence constitute a valid L-reduction. We formulate the following theorem.

Theorem 7. *The scheduling problem $R \mid r_j \mid \sum C_j$ is* MAX SNP-*hard and thus does not has a PTAS, unless $\mathcal{P} = \mathcal{NP}$.* $\qquad\square$

The same idea works to prove MAX SNP-hardness of $R \mid \bar{d}_j \mid \sum C_j$ (scheduling unrelated machines with job deadlines) and $R \mid \mid \sum w_j C_j$ (minimizing the total weighted completion time).

4 Shop Scheduling Problems

The *shop scheduling problems* considered in this section are defined as follows. There are n jobs $1, 2, \ldots, n$ and m machines $M_1, M_2, \ldots, M_m$. Each job j consists of m operations $O_{1j}, O_{2j}, \ldots, O_{mj}$. Operation O_{ij} has to be processed on M_i for a period of p_{ij} units. No machine may process more than one job at a time, and no two operations of the same job may be processed at the same time. Preemption is not allowed.

There are three types of shop models. First, there is the *open shop* in which it is immaterial in what order the operations are executed. Secondly, there is the *job shop* in which the processing order of the operations is prespecified for each job; different jobs may have different processing orders. Finally, there is the *flow shop*, which is a special case of the job shop: here, the prespecified processing order is the same for all jobs. In the standard scheduling notation, the open shop, job shop, and flow shop with total job completion time objective are denoted by $O \mid \mid \sum C_j$, $J \mid \mid \sum C_j$, and $F \mid \mid \sum C_j$, respectively.

In this section, we provide a proof for the following theorem.

Theorem 8. *The scheduling problems $O \mid \mid \sum C_j$, $J \mid \mid \sum C_j$, and $F \mid \mid \sum C_j$ are* MAX SNP-*hard. Unless $\mathcal{P} = \mathcal{NP}$, they do not have a PTAS.*

Williamson et al. [16] show by a gap reduction from a variant of 3-satisfiability that the makespan minimization shop problems $J \mid \mid C_{\max}$ and $F \mid \mid C_{\max}$ cannot be approximated in polynomial time within a factor better than 5/4 (unless $\mathcal{P} = \mathcal{NP}$). In proving Theorem 8, we use an L-reduction that is based on this gap reduction. Our L-reduction is from the following version MAX-2SAT-B of maximum 2-satisfiability.

MAXIMUM BOUNDED 2-SATISFIABILITY (MAX-2SAT-B)

Input: A set $U = \{x_1, \ldots, x_q\}$ of variables and a collection $C = \{c_1, \ldots, c_s\}$ of clauses over U. Each clause consists of exactly two (possibly identical) literals. Each variable occurs either once or twice in negated form in C, and either once or twice in unnegated form in C.
Goal: Find a truth assignment for U such that a maximum number of clauses is satisfied.
Measure: The number of satisfied clauses.

Proposition 9. *The above defined problem* MAX-2SAT-B *is* MAX SNP-*hard.*

□

In analogy to Lemma 3, we argue that for any instance I of MAX-2SAT-B, the optimum value is bounded away from zero.

Lemma 10. *For any instance I of* MAX-2SAT-B, *we have* $\mathrm{OPT}(I) \geq \frac{1}{2}s$.   □

4.1 Flow Shop Scheduling

Let $I = (U, C)$ with $U = \{x_1, \ldots, x_q\}$ and $C = \{c_1, \ldots, c_s\}$ be an instance of MAX-2SAT-B. We define a flow shop instance $R(I)$ as follows.

In constructing this instance $R(I)$, we distinguish between the first and the second unnegated (respectively, negated) occurrence of each literal. For $j = 1, 2$, we refer to the jth occurrence of the literal x_i as x_{ij}, and to the jth occurrence of $\overline{x}_i$ as $\overline{x}_{ij}$. With each variable x_i we associate four corresponding so-called *variable jobs*: the two jobs x_{i1} and x_{i2} correspond to the first and second occurrence of literal x_i, and the two jobs $\overline{x}_{i1}$ and $\overline{x}_{i2}$ correspond to the first and second occurrence of the literal $\overline{x}_i$. With each literal $x \in \{x_{i2}, \overline{x}_{i2}\}$ that does *not* occur in the formula, we further associate three so-called *dummy jobs* $d_1(x)$, $d_2(x)$, and $d_3(x)$. The processing time of every operation is either 0 or 1. Every variable job x has exactly three operations with processing time 1 and $m - 3$ operations of processing time 0. The three length 1 operations are a beginning operation $B(x)$, a middle operation $M(x)$ and an ending operation $E(x)$. A dummy job $d_i(x)$ only has a single operation $D_i(x)$ with processing time 1, and $m - 1$ operations of processing time 0. There are four classes of machines.

- For each variable x_j, there are two *assignment machines*: the first one processes the operations $B(x_{j1})$ and $B(\overline{x}_{j1})$, whereas the second one processes the operations $B(x_{j2})$ and $B(\overline{x}_{j2})$.
- For each variable x_j, there are two *consistency machines*: the first one processes the operations $M(x_{j1})$ and $M(\overline{x}_{j2})$, whereas the second one processes the operations $M(x_{j2})$ and $M(\overline{x}_{j1})$.
- For each clause $a \vee b$, there is a *clause machine* that processes $E(a)$ and $E(b)$.
- For each literal $x \in \{x_{i2}, \overline{x}_{i2}\}$ that does not occur in the formula, there is a *garbage machine* that processes the operation $E(x)$ and the dummy operations $D_1(x)$, $D_2(x)$, and $D_3(x)$.

The processing of every job first goes through the assignment machines, then through the consistency machines, then through the clause machines, and finally through the garbage machines. Since most operations have length 0, the precise processing order within every machine class is not essential; we only note that for every variable job, processing on the first assignment (first consistency) machine always precedes processing on the second assignment (second consistency) machine. Similarly as in Section 3, we are mainly interested in schedules for $R(I)$ with a special combinatorial structure. In a so-called *consistent schedule*, for every variable x either both operations $B(x_{i1})$ and $B(x_{i2})$ are processed in

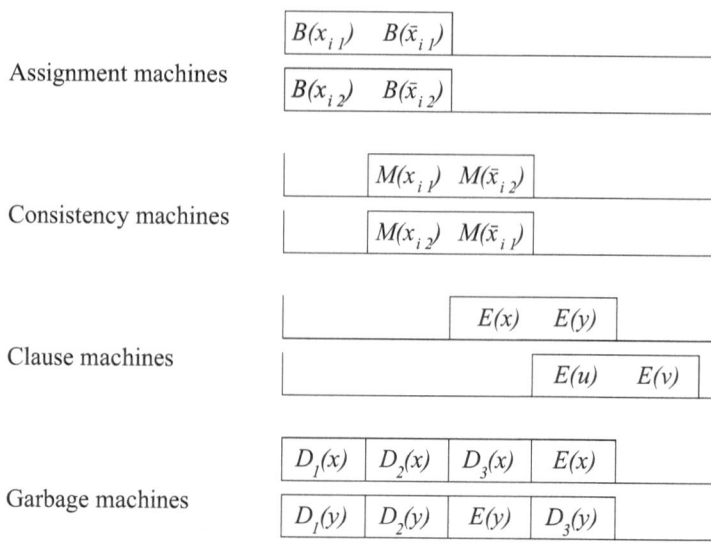

Fig. 2. A consistent schedule for the flow shop instance.

the interval $[0, 1]$, or both operations $B(\overline{x}_{i1})$ and $B(\overline{x}_{i2})$ are processed during $[0, 1]$. Moreover, in a consistent schedule the machines process the operations of length 1 during the following intervals. The assignment machines are only processing length 1 operations during $[0, 2]$, and the consistency machines are only processing such operations during $[1, 3]$. On every clause machine, the operations of length 1 are either processed during $[2, 3]$ and $[3, 4]$ or during $[3, 4]$ and $[4, 5]$. On the garbage machines, all four operations of length 1 are processed during $[0, 4]$; there are no restrictions on the processing order on the garbage machines. Figure 2 gives an illustration of a consistent schedule.

Lemma 11. *Let σ be any feasible schedule for $R(I)$. Then there exists a feasible consistent schedule σ' whose objective value is at most the objective value of σ.*

Proof. Let σ be an arbitrary feasible schedule for $R(I)$. Throughout the proof, we only deal with the placement of length 1 operations (as placing the length 0 operations is straightforward); a machine is *busy* if it is processing a length 1 operation. Our first goal is to transform σ into a schedule in which the assignment machines are only busy during $[0, 2]$ and the consistency machines are only busy during $[1, 3]$. We start by shifting all operations on the assignment and consistency machines as far to the left as possible without violating feasibility. Clearly, in the resulting schedule all operations on the assignment machines are processed during $[0, 2]$.

Now suppose that on some consistency machine some operation, say operation $M(x_{i1})$, only completes at time four. Since all operations were shifted to the left, this yields that both operations $B(x_{i1})$ and $B(\overline{x}_{i2})$ are scheduled during the time interval $[1, 2]$. Moreover, operations $B(\overline{x}_{i1})$ and $B(x_{i2})$ are both processed during $[0, 1]$. We proceed as follows: If operation $M(\overline{x}_{i1})$ is processed before $M(x_{i2})$, then we switch the order of the operations $B(x_{i2})$ and $B(\overline{x}_{i2})$ on their assignment machine. Moreover, we reschedule operation $M(\overline{x}_{i2})$ during $[1, 2]$ and operation $M(x_{i1})$ during $[2, 3]$. If operation $M(\overline{x}_{i1})$ is processed after $M(x_{i2})$, then we perform a symmetric switching and rescheduling step. Note that after these switches, the schedule is still feasible and that on the consistency machines no operation has been shifted to the right.

By performing such switches, we eventually get a schedule in which all assignment machines are busy during $[0, 2]$ and all consistency machines are busy during $[1, 3]$. Finally, by shifting all operations on the clause and garbage machines as far to the left as possible, we obtain a schedule σ'. It is routine to check that in schedule σ' every clause machine is either busy during $[2, 4]$ or during $[3, 5]$, and every garbage machine is busy during $[0, 4]$. Since in schedule σ' no job finishes later than in schedule σ, the objective value has not been increased.

It remains to be proven that the constructed schedule σ' indeed is a consistent schedule, i.e., to prove that for every variable x_i, the operations $B(x_{i1})$ and $B(x_{i2})$ are processed simultaneously. First, assume that operation $B(x_{i1})$ starts at time 0. This implies that $B(\overline{x}_{i1})$ is processed during $[1, 2]$, and that $M(\overline{x}_{i1})$ is processed during $[2, 3]$. This in turn implies that $M(x_{i2})$ starts at time 1, and thus $B(x_{i2})$ is processed during $[0, 1]$, and we are done. In case operation $B(x_{i1})$ starts at time 1, a symmetric argument works.   □

Next, we define a polynomial-time transformation S that maps feasible solutions of $R(I)$ to feasible solutions of the MAX-2SAT-B instance I. Let σ be a feasible schedule for $R(I)$. As described in the proof of Lemma 11, we find a corresponding consistent schedule σ'. We define the truth setting $S(\sigma)$ for I as follows. If in σ' operations $B(x_{i1})$ and $B(x_{i2})$ are processed during $[0, 1]$, then variable x_i is set to TRUE, and if $B(x_{i1})$ and $B(x_{i2})$ are processed during $[1, 2]$, then x_i is set to FALSE.

It can be verified that if a clause is satisfied under the truth setting $S(\sigma)$, then the length 1 operations on the corresponding clause machine are processed during $[2, 3]$ and $[3, 4]$. Conversely if a clause is not satisfied, then these operations occupy the intervals $[3, 4]$ and $[4, 5]$.

Lemma 12. *For any instance I of MAX-2SAT-B, we have that $\mathrm{OPT}(R(I)) \leq 58\,\mathrm{OPT}(I)$. Hence, the polynomial-time transformation R fulfills condition (1) of Definition 1.*

Proof. There are s clause machines and at most $2q$ garbage machines. Since the total completion time of two jobs on the same clause machine is at most $4 + 5 = 9$, the total completion time of the variable jobs that have non-zero processing requirement on some clause machine is at most $9s$. Moreover, the total completion time of the remaining jobs (i.e., the jobs with non-zero processing time on some garbage machine) is at most $(1 + 2 + 3 + 4)2q \leq 20s$. Hence $\mathrm{OPT}(R(I)) \leq 29s$, and Lemma 10 completes the proof.   □

Lemma 13. *For any feasible schedule σ of $R(I)$, the feasible solution $S(\sigma)$ of instance I fulfills the inequality $|\mathrm{OPT}(I) - c(S(\sigma))| \leq \frac{1}{2}|\mathrm{OPT}(R(I)) - c(\sigma)|$. Hence, the polynomial-time transformation S fulfills condition (2) of Definition 1.*

Proof. As has been argued before, the length 1 operations on a clause machine have completion times 3 and 4, if the corresponding clause is satisfied, and they have completion times 4 and 5, if the clause is not satisfied. In other words, every unsatisfied clause induces an extra cost of 2 in the objective function. The claim follows.   □

Lemma 12 and 13 state that the transformations R and S satisfy the conditions in Definition 1. Since both transformations are computable in polynomial time, the problem $F \mid \mid \sum C_j$ is MAX SNP-hard. This completes the proof of the flow shop and of the job shop part of Theorem 8.

4.2 Open Shop Scheduling

The essential difference between the flow shop problem and the open shop problem is that the order in which the operations belonging to the same job are processed is no longer given; therefore, we must look for a different way to enforce that the beginning operation indeed precedes the middle operation, which in turn must precede the ending operation. To this end, we introduce a number of additional jobs, which are used to fill the interval $[0, 1]$ on the consistency machines and the interval $[0, 2]$ on the clause machines; these additional jobs can be forced to go there, because our objective is to minimize the total completion time, which favors small jobs. We further need some more jobs, which are used to remove unnecessary idle time. This can be worked out as follows.

Similar to the flow shop we start from an instance $I = (U, C)$ of MAX-2SAT-B. We introduce the same set of variable jobs, dummy jobs, assignment machines, consistency machines, clause machines, and garbage machines. Additionally, the instance $R(I)$ contains $26q + 6s$ so-called *structure jobs*. Every structure job consists of $m - 1$ operations of length 0 and of a single operation of non-zero length; this operation is called the *structure operation* corresponding to the structure job.

- On each of the $2q$ assignment machines, we introduce five structure operations of length 3. Because of their large processing time, any reasonable schedule processes them during the interval $[2, 17]$.
- On each of the $2q$ consistency machines, we introduce three structure operations of length $\frac{1}{3}$ and five structure operations of length 3. Any reasonable schedule processes the operations of length $\frac{1}{3}$ during $[0, 1]$ and the operations of length 3 during $[3, 18]$.
- On each of the s clause machines, we introduce six structure operations of length $\frac{1}{3}$. Any reasonable schedule processes them during $[0, 2]$.

It can be shown that for any feasible schedule for $R(I)$, there exists a reasonable consistent schedule with non-larger objective value. With this, one can define a truth setting $S(\sigma)$ like in Subsection 4.1. Again, the constructed transformations are polynomial time computable and fulfill the conditions in Definition 1. Hence, problem $O \mid \mid \sum C_j$ is MAX SNP-hard. This completes the proof of Theorem 8.

5 Conclusions

In this paper, we have derived a number of non-approximability results for scheduling problems with total job completion time objective. The approximability status of most scheduling problems with this objective function or its weighted counterpart remains amazingly unclear: until today, all that we know is that some of these problems can be solved in polynomial time by straightforward algorithms (like the problems $1 \mid \mid \sum w_j C_j$ and $P \mid \mid \sum C_j$) and that some of these problems do not have a PTAS (like the problems investigated in this paper). However, there is not a single strongly $\mathcal{NP}$-hard scheduling problem with minsum objective for which a PTAS has been constructed. We state the following conjectures.

Conjecture 14. *The problems $1 \mid r_j \mid \sum C_j$ (scheduling a single machine with job release dates) and $P \mid \mid \sum w_j C_j$ (scheduling parallel identical machines with the objective of minimizing the total weighted job completion time) both do have a PTAS.*

Conjecture 15. *Neither of the problems $P \mid r_j \mid \sum C_j$ (scheduling parallel identical machines with job release dates) and $1 \mid prec \mid \sum C_j$ (scheduling a single machine with precedence constraints) has a PTAS.*

References

1. S. Arora, C. Lund, R. Motwani, M. Sudan, and M. Szegedy. Proof verification and hardness of approximation problems. *Proceedings of the 33rd IEEE Symposium on the Foundations of Computer Science*, pages 14–23, 1992.
2. P. Crescenzi and V. Kann. A compendium of NP-optimization problems. 1997. http://www.nada.kth.se/nada/theory/problemlist/html.
3. M. R. Garey and D. S. Johnson. *Computers and Intractability: A Guide to the Theory of $\mathcal{NP}$-Completeness.* Freeman, San Francisco, 1979.
4. R. L. Graham, E. L. Lawler, J. K. Lenstra, and A. H. G. Rinnooy Kan. Optimization and approximation in deterministic sequencing and scheduling: A survey. *Annals of Discrete Mathematics*, 5:287–326, 1979.
5. J. A. Hoogeveen, J. K. Lenstra, and B. Veltman. Three, four, five, six, or the complexity of scheduling with communication delays. *Operations Research Letters*, 16:129–137, 1994.
6. V. Kann. Maximum bounded 3-dimensional matching is MAX SNP-complete. *Information Processing Letters*, 37:27–35, 1991.
7. H. Kellerer, T. Tautenhahn, and G. J. Woeginger. Approximability and nonapproximability results for minimizing total flow time on a single machine. *Proceedings of the 28th Annual ACM Symposium on the Theory of Computing*, pages 418–426, 1996.

8. E. L. Lawler, J. K. Lenstra, A. H. G. Rinnooy Kan, and D. B. Shmoys. Sequencing and scheduling: Algorithms and complexity. In S. C. Graves, A. H. G. Rinnooy Kan, and P. H. Zipkin, editors, *Logistics of Production and Inventory. Handbooks in Operations Research and Management Science, Vol. 4*, pages 445–522. North-Holland, Amsterdam, 1993.

9. J. K. Lenstra and A. H. G. Rinnooy Kan. Complexity of scheduling under precedence constraints. *Operations Research*, 26:22–35, 1978.

10. J. K. Lenstra and D. B. Shmoys. Computing near-optimal schedules. In P. Chrétienne, E. G. Coffman Jr., J. K. Lenstra, and Z. Liu, editors, *Scheduling Theory and its Applications*, pages 1–14. Wiley, Chichester, 1995.

11. J. K. Lenstra, D. B. Shmoys, and É. Tardos. Approximation algorithms for scheduling unrelated parallel machines. *Math. Programming*, 46:259–271, 1990.

12. S. Leonardi and D. Raz. Approximating total flow time on parallel machines. *Proceedings of the 29th Annual ACM Symposium on the Theory of Computing*, pages 110–119, 1997.

13. C. H. Papadimitriou. *Computational Complexity*. Addison-Wesley, 1994.

14. C. H. Papadimitriou and M. Yannakakis. Optimization, approximation, and complexity classes. *Journal of Computer and System Sciences*, 43:425–440, 1991.

15. M. Skutella. Problem posed at the open problem session of the Dagstuhl Meeting on "Parallel Machine Scheduling". Schloß Dagstuhl, Germany, July 14–18, 1997.

16. D. P. Williamson, L. A. Hall, J. A. Hoogeveen, C. A. J. Hurkens, J. K. Lenstra, S. V. Sevastianov, and D. B. Shmoys. Short shop schedules. *Operations Research*, 45:288–294, 1997.

Approximation Bounds for a General Class of Precedence Constrained Parallel Machine Scheduling Problems

Alix Munier[1], Maurice Queyranne[2], and Andreas S. Schulz[3]

[1] Université Pierre et Marie Curie, Laboratoire LIP6
4 place Jussieu, 75 252 Paris, cedex 05, France
Alix.Munier@@lip6.fr
[2] Faculty of Commerce and Business Administration, University of British Columbia
Vancouver, B.C., Canada V6T 1Z2, and
Università di Bologna – Sede di Rimini, via Angherà 22, 47037 Rimini, Italy
queyranne@@ecosta.unibo.it
[3] Technische Universität Berlin, Fachbereich Mathematik
MA 6–1, Straße des 17. Juni 136, 10623 Berlin, Germany
schulz@@math.tu-berlin.de

Abstract. A well studied and difficult class of scheduling problems concerns parallel machines and precedence constraints. In order to model more realistic situations, we consider precedence delays, associating with each precedence constraint a certain amount of time which must elapse between the completion and start times of the corresponding jobs. Release dates, among others, may be modeled in this fashion. We provide the first constant-factor approximation algorithms for the makespan and the total weighted completion time objectives in this general class of problems. These algorithms are rather simple and practical forms of list scheduling. Our analysis also unifies and simplifies that of a number of special cases heretofore separately studied, while actually improving some of the former approximation results.

1 Introduction

Scheduling problems involving precedence constraints are among the most difficult problems in the area of machine scheduling, in particular for the design of good approximation algorithms. Our understanding of the structure of these problems and our ability to generate near-optimal solutions remain limited. The following examples illustrate this point. (i) The first approximation algorithm for $P|prec|C_{\max}$ by Graham [14] is not only more than thirty years old, but it is also still essentially the best one available for this problem. On the other hand, it is only known that no polynomial-time algorithm can have a better approximation ratio than $4/3$ unless P = NP [23]. (ii) The computational complexity of the problem $Pm|p_j = 1, prec|C_{\max}$, open problem 8 from the original list of

R. E. Bixby, E. A. Boyd, and R. Z. Ríos-Mercado (Eds.): IPCO VI
LNCS 1412, pp. 367–382, 1998. © Springer–Verlag Berlin Heidelberg 1998

Garey and Johnson [11] is still open. (iii) The situation is also unsatisfactory with machines running at different speed, for which no constant-factor approximation algorithms are known. For the makespan objective, Chudak and Shmoys [10] only recently improved to $O(\log m)$ an almost twenty year old approximation ratio of $O(\sqrt{m})$ due to Jaffe [20]. They obtained the same approximation ratio for the total weighted completion time objective. (iv) Progress is also quite recent for the latter objective on a single machine or identical parallel machines. Until recently, no constant-factor approximation algorithms were known. Lately, a better understanding of linear programming relaxations and their use to guide solution strategies led to a 2– and a 2.719–approximation algorithm for $1|\text{prec}| \sum w_j C_j$ and $1|r_j, \text{prec}| \sum w_j C_j$, respectively [16,34,35], and to a 5.328–approximation algorithm for $P|r_j, \text{prec}| \sum w_j C_j$ [7]. Few deep negative results are known for these problems (see [18] for the total completion times objective).

In this paper, we consider (a generalization of) the scheduling problem $P|r_j, \text{prec}| \sum w_j C_j$ and answer a question of Hall et al. [16, Page 530]:

> "Unfortunately, we do not know how to prove a good performance guarantee for this model by using a simple list-scheduling variant."

Indeed their algorithm, as well as its improvement by Chakrabarti et al. [7], is rather elaborate and its performance ratio does not match the quality of the lower bound it uses. We show that using the same LP relaxation in a different way (reading the list order from the LP midpoints instead of LP completion times) yields a simple 4–approximation algorithm for $P|r_j, \text{prec}| \sum w_j C_j$. We actually obtain this result in the more general framework of precedence delays.

We consider a general class of precedence-constrained scheduling problems on identical parallel machines. We have a set N of n jobs and m identical parallel machines. Each job j has a nonnegative processing requirement (size) p_j and must be processed for that amount of time on any one of the machines. A job must be processed in an uninterrupted fashion, and a machine can process only one job at a time. We are interested in constrained scheduling problems in which each job j may have a release date r_j before which it cannot be processed, and there may be a partial order A on the jobs. We associate with each precedence-constrained job pair $(i, j) \in A$ a nonnegative *precedence delay* d_{ij}, with the following meaning: in every feasible schedule, job j cannot start until d_{ij} time units after job i is completed. Special cases include ordinary precedence constraints ($d_{ij} = 0$); and release dates $r_j \geq 0$ (which may be modeled by adding a dummy job 0 with zero processing time and precedence delays $d_{0j} = r_j$ for all other jobs). Delivery times (or lags), which must elapse between the end of a job's processing and its actual completion time, may also be modeled by adding one or several dummy jobs and the corresponding precedence delays.

We denote the completion time of a job j in a schedule S as C_j^S and will drop the superscript S when it is clear to which schedule we refer. We consider the usual objectives of minimizing the makespan $C_{\max} = \max_j C_j$ and, for given nonnegative weights $w_j \geq 0$, a weighted sum $\sum_j w_j C_j$ of completion times. In an extension of the common notation introduced in [15], we may denote these problems as $P|\text{prec. delays } d_{ij}|C_{\max}$ and $P|\text{prec. delays } d_{ij}| \sum w_j C_j$, respectively. These problems are NP-hard (see, e.g., Lawler et al. [22]), and we discuss here the quality of relaxations and approximation algorithms. An α–*approximation algorithm* is a polynomial-time algorithm that delivers a solution with objective value at most α times the optimal value. Sometimes α is called the (worst-case) *performance guarantee* or *ratio* of the algorithm.

Precedence delays were considered for resource-constrained project scheduling under the name "finish-to-start lags", e.g., by Bartusch, Möhring, and Radermacher [4] and Herroelen and Demeulemeester [17], for one-machine scheduling by Wikum, Llewellyn, and Nemhauser [37] under the name "generalized precedence constraints", and by Balas, Lenstra, and Vazacopoulos [3] under that of "delayed precedence constraints"; the latter authors use the $L_{\max}$ minimization problem as a key relaxation in a modified version of the shifting bottleneck procedure for the classic job-shop scheduling problem. Most of the theoretical studies concerning this class of precedence constraints consider the one machine problem $1|$prec. delays $d_{ij} = k, p_j = 1|C_{\max}$ which corresponds to a basic pipeline scheduling problem (see [21] for a survey). Leung, Vornberger, and Witthoff [24] showed that this problem is NP-complete. Several other authors (e.g., [6,5,28]) obtained polynomial-time algorithms for particular instances by utilizing well-known algorithms for special cases of the classical m–machine problem. In the context of approximation algorithms, the main result is that Graham's list scheduling algorithm [14] was extended to $P|$prec. delays $d_{ij} = k, p_j = 1|C_{\max}$ to give a worst-case performance ratio of $2 - 1/(m(k+1))$ [21,28]. We extend this result, in Section 3, to nonidentical precedence delays and processing times.

List scheduling algorithms, first analyzed by Graham [14] are among the simplest and most commonly used approximate solution methods for parallel machine scheduling problems. These algorithms use priority rules, or job rankings, which are often derived from solutions to relaxed versions of the problems. For example, several algorithms of Hall, Schulz, Shmoys, and Wein [16] use the job completion times obtained from linear programming relaxations. In Section 2 we show that a modified, job-driven version of list scheduling based on job completion times can, in the presence of precedence constraints, lead to solutions that are about as bad as m times the optimum, for both the $C_{\max}$ and $\sum_j w_j C_j$ objectives; this behavior may also occur when using actual *optimum* completion times.

Graham's original list scheduling, however, works well for minimizing the makespan as we show in Section 3 by extending it to the case of precedence delays. For minimizing a weighted sum $\sum_j w_j C_j$ of completion times, we present in Section 4 a new algorithm with approximation ratio bounded by 4 for the general problem with precedence delays. This algorithm is based on an LP relaxation of this problem, which is a straightforward extension of earlier LP relaxations proposed by Hall et al. [16]. The decision variables in this relaxation are the completion times C_j of every job j, so we choose to ignore the machine assignments in these relaxations. There are two sets of linear constraints, one representing the precedence delays (and, through the use of a dummy job, the release dates) in a straightforward fashion; the other set of constraints is a relatively simple way of enforcing the total capacity of the m machines. Although the machine assignments are ignored and the machine capacities are modeled in a simplistic way, this is sufficient to obtain the best relaxation and approximation bounds known so far for these problems and several special cases thereof. We show that using *midpoints* derived from the LP relaxation leads to a performance ratio bounded by 4 for the general problem described above. Recall that in a given schedule the midpoint of a job is the earliest point in time at which half of its processing has been performed; that is, if the schedule is (or may be considered as) nonpreemptive then the midpoint of job j is simply $C_j^R - p_j/2$ where C_j^R is its completion time in the relaxation R. The advantage of using midpoints in the analysis of approximation algorithms was first observed by Goemans [12] and has since then been used by several authors (e.g., [35,36,13]). Our result seems to be the first, however, where midpoints are really needed within the algorithm itself. We also show how the analysis yields tighter bounds for some special cases, and then conclude with some additional remarks in Section 5. We believe that the approach of applying a list-scheduling rule in which the jobs are ordered based on their midpoints in an LP solution will have further consequences for the design of approximation algorithms.

In summary, the main contributions of this paper are as follows.

1. We clarify the relationship between two forms of List Scheduling Algorithms (LSAs): Graham's non-idling LSAs and job-based LSAs. In particular, it is shown that the former are appropriate for optimizing objectives, such as the makespan $C_{\max}$, that are related to maximizing machine utilization, whereas they are inappropriate (leading to unbounded performance ratio) for job oriented objectives, such as the weighted sum of completion times $\sum_j w_j C_j$. In contrast, we present job-based LSAs with bounded performance ratio for the latter objective.

2. We show that using *job completion times* as a basis for job-based list scheduling may yield very poor schedules for problems with parallel machines, precedence constraints and weighted sum of completion times objective. This may happen even if the completion times are those of an *optimal* schedule.

3. In contrast, we show that job-based list scheduling according to *job midpoints* from an appropriate LP relaxation leads to job-by-job error ratios of at most 4 for a broad class of problems.

4. We present a general model of scheduling with precedence delays. This also allows us to treat in a unified framework ordinary precedence constraints, release dates and delivery times. In particular, this simplifies and unifies the analysis and proof techniques.

5. Finally, we present the best polynomial-time approximation bounds known so far for a broad class of parallel machine scheduling problems with precedence constraints or delays (including release dates and delivery times) and either a makespan or total weighted completion time objective. These bounds are obtained by using relatively simple LSAs which should be of practical as well as theoretical interest. We also present the best polynomially solvable relaxations known so far for such problems with the latter objective.

The approximation results are summarized in Table 1 where the parameter ρ is defined as $\rho = \max_{(j,k) \in A} d_{jk} / \min_{i \in N} p_i$, and m denotes the number of identical parallel machines.

Table 1. Summary of results.

Problem	New bound	Best earlier bound
$1|$prec. delays $d_{ij}|C_{\max}$	$2 - \frac{1}{1+\rho}$	?
$P|$prec. delays $d_{ij}|C_{\max}$	$2 - \frac{1}{m(1+\rho)}$	?
$P|$prec. delays $d_{ij}|\sum w_j C_j$	4	?
$P|r_j, \text{prec}|\sum w_j C_j$	4	5.328 Chakrabarti et al. [7]
$P|\text{prec}|\sum w_j C_j$	$4 - \frac{2}{m}$	5.328 Chakrabarti et al. [7]
$1|$prec. delays $d_{ij}|\sum w_j C_j$	3	?

Due to space limitations some details are omitted from this paper. They can be found in the complete version, see [27].

2 List Scheduling Algorithms

In his seminal paper, Graham (1966) showed that a simple list-scheduling rule is a $(2 - \frac{1}{m})$-approximation algorithm for $P|\text{prec}|C_{\max}$. In this algorithm, the jobs are ordered in some list, and whenever one of the m machines becomes idle, the next available job on the list is started on that machine, where a job is available if all of its predecessors have completed processing. By their non-idling property, Graham's List Scheduling Algorithms (GLSAs) are well suited when machine utilization is an important consideration. Indeed, it is shown in Section 3 that, for the makespan minimization problem $P|\text{prec. delays } d_{ij}|C_{\max}$, any GLSA (i.e., no matter which list of jobs is used) produces a schedule with objective function value within a factor 2 of the optimum. In this case, a job is available if all its predecessors are completed and the corresponding precedence delays have elapsed.

In contrast, the elementary Example 1 shows that the non-idling property may lead to an arbitrarily poor performance ratio for a weighted sum of completion times objective $\sum_{j \in N} w_j C_j$.

Example 1. Consider the following two-job instance of the single machine non-preemptive scheduling problem $1|r_j|\sum w_j C_j$ (a special case of a precedence delay problem, as discussed in the introduction). For a parameter $q \geq 2$, job 1 has $p_1 = q$, $r_1 = 0$ and $w_1 = 1$, whereas job 2 has $p_2 = 1$, $r_2 = 1$ and $w_2 = q^2$. The optimum schedule is to leave the machine idle during the time interval $[0, 1)$ so as to process job 2 first. The optimum objective value is $2q^2 + (q + 2)$. Any non-idling heuristic starts processing job 1 at time 0, leading to an objective value at least $q^3 + q^2 + q$, and its performance ratio is unbounded as q may be arbitrarily large.   □

A different example of the same type but using ordinary precedence constraints rather than release dates can be found in [33, Page 82, Ex. 2.20]. Thus to obtain a bounded performance for the weighted sum of completion times objective $\sum_j w_j C_j$, we must relax the non-idleness property. One strategy, leading to *job-based* nonpreemptive list scheduling algorithms, is to consider the jobs one by one, in the given list order, starting from an empty schedule. Each job is nonpreemptively inserted in the current schedule without altering the jobs already scheduled. Specific list scheduling algorithms differ in how this principle is implemented, in particular, for parallel machines, regarding the assignment of the jobs to the machines. For definiteness, consider the following version, whereby every job is considered in the list order and is scheduled at the earliest feasible time at the end of the current schedule on a machine. Notice that the given list is assumed to be a linear extension of the poset defined by the precedence constraints.

Job-Based List Scheduling Algorithm for P|prec. delays d_{ij}|.

1. The list $L = (\ell(1), \ell(2), \ldots, \ell(n))$ is given.
2. Initially all machines are empty, with machine completion times $\Gamma_h = 0$ for all $h = 1, \ldots, m$.
3. For $k = 1$ to n do
    3.1 Let job $j = \ell(k)$. Its start time is
    $$S_j = \max\left(\max\{C_i + d_{ij} : (i, j) \in A\}, \min\{\Gamma_h : h = 1, \ldots, m\}\right)$$
    and its completion time is $C_j = S_j + p_j$.
    3.2 Assign job j to a machine h such that $\Gamma_h \leq S_j$.
        Update $\Gamma_h = C_j$.

Various rules could be used in Step 3.2 for the choice of the assigned machine h, for example one with largest completion time Γ_h (so as to reduce the idle time between Γ_h and S_j). Note also that the above algorithm can be modified to allow insertion of a job in an idle period before Γ_h on a machine h. In effect, the observations below also apply to all these variants.

One method (e.g., Phillips et al. [29] as well as Hall et al. [16]) for defining the list L consists in sorting the jobs in nondecreasing order of their completion times in a relaxation of the scheduling problem under consideration. In the presence of ordinary precedence constraints, this works well for the case of a single machine (Hall et al., ibid., see also [34]), but Example 2 shows that this may produce very poor schedules for the case of identical parallel machines. This example uses the list which is produced by an *optimal schedule*, the tightest kind of relaxation that can be defined; note that this optimal schedule defines the same completion time order as the relaxation in Hall et al. and its extension in Section 4 below.

Example 2. For a fixed number $m \geq 2$ of identical parallel machines and a positive number ϵ, let the job set be $N = \{1, \ldots, n\}$ with $n = m(m+1) + 1$.

The ordinary precedence constraints (j, k) (with $d_{jk} = 0$) are defined as follows: (i) $j = 1 + h(m+1)$ and $k = j + g$, for all $h = 0, \ldots, m - 1$ and all $g = 1, \ldots, m$; and (ii) for all $j < n$ and $k = n$. The processing times are $p_j = 1 + h(m+1)\epsilon$ for $j = 1 + h(m+1)$ and $h = 0, \ldots, m - 1$; and $p_j = \epsilon$ otherwise. The objective is either to minimize the makespan, or a weighted sum $\sum_j w_j C_j$ of job completion times with weights $w_j = 0$ for all $j < n$ and $w_n = 1$; note that, due to the precedence constraints (ii) above, these two objectives coincide for any feasible schedule.

An optimal solution has, for $h = 0, \ldots, m - 1$, job $j = 1 + h(m+1)$ starting at time $S_j^* = 0$ on machine $h + 1$, immediately followed by jobs $j + 1, \ldots, j + m$ assigned as uniformly as possible to machines $1, \ldots, h+1$. Job n is then processed last on machine m, so that the optimal objective value is $C_{\max}^* = C_n^* = 1 + (m^2 + 1)\epsilon$. A corresponding list is $L = (1, 2, \ldots, n)$. Any version of the list scheduling algorithm described above produces the following schedule from this list: job 1 is scheduled with start time $S_1^L = 0$ and completion time $C_1^L = 1$; the m jobs $k = 2, \ldots, m+1$ are then scheduled, each with $S_k^L = 1$ and $C_k^L = 1 + \epsilon$ on a different machine; this will force all subsequent jobs to be scheduled no earlier than time $1 + \epsilon$. As a result, for $h = 0, \ldots, m-1$, job $j = 1 + h(m+1)$ is scheduled with start time $S_j^L = h + (\frac{1}{2}(h-1)h(m+1) + h)\epsilon$, followed by jobs $k = j+1, \ldots, j+(m+1)$ each with $S_k^L = h + 1 + (\frac{1}{2}h(h+1)(m+1) + h)\epsilon$ on a different machine. Finally, job n is scheduled last with $S_n^L = m + (\frac{1}{2}(m-1)m(m+1) + m)\epsilon$ and thus the objective value is $C_{\max}^L = C_n^L = m + o(\epsilon)$, or arbitrarily close to m times the optimal value $C_{\max}^*$ for $\epsilon > 0$ small enough. □

The example shows that list scheduling according to completion times can lead to poor schedules on identical parallel machines with job precedence constraints. In contrast, we will present in Section 4 a linear programming relaxation of the general problem with precedence delays and show that job-based list scheduling according to job midpoints leads to a bounded performance ratio.

3 The Performance of Graham's List Scheduling

In this section we show that Graham's (non-idling) List Scheduling Algorithms generate feasible schedules with makespan less than twice the optimum for instances of P|prec. delays d_{ij}|$C_{\max}$. As discussed in the introduction, this result and proof extend and unify earlier work.

Let $S^H = (S_1^H, \ldots, S_n^H)$ denote the vector of start times of a schedule constructed by GLSA, as described in Section 2. Let $C_{\max}^H = \max_{i \in N}(S_i^H + p_i)$ denote the makespan of this schedule. For any pair (t, t') of dates such that $0 \le t \le t' \le C_{\max}^H$, let $\mathcal{I}[t, t']$ denote the total machine idle time during the interval $[t, t')$. (Thus, for example, if all m machines are idle during the whole interval, then $\mathcal{I}[t, t'] = m(t' - t)$.) Let N^+ denote the set of jobs in N that have at least one predecessor.

Lemma 1. *Let j be a job in N^+ and let i be an immediate predecessor of j with largest value $S_i^H + p_i + d_{ij}$. Then $\mathcal{I}[S_i^H, S_j^H] \le m\, d_{ij} + (m - 1)p_i$.*

Proof. Let i be an immediate predecessor of j with largest value of $S_i^H + p_i + d_{ij}$. Since H is a GLSA, job j may be processed by any available processor from date $S_i^H + p_i + d_{ij}$ on. Therefore $\mathcal{I}[S_i^H + p_i + d_{ij}, S_j^H) = 0$ and hence $\mathcal{I}[S_i^H, S_j^H) = \mathcal{I}[S_i^H, S_i^H + p_i + d_{ij})$. Since job i is processed during the time interval $[S_i^H, S_i^H + p_i)$, at most $m - 1$ machines are idle during this interval, and we obtain the requisite inequality.   □

The *precedence network* (N^0, A^0, ℓ) is defined by $N_0 = N \cup \{0\}$; $A_0 = A \cup \{(i, 0) : i \in N\}$; and arc lengths $\ell(i, j) = p_i + d_{ij}$ for all $(i, j) \in A$, and $\ell(i, 0) = p_i$ for all $i \in N$. Job 0 is a dummy job succeeding all other jobs with zero delays. It is added so that any maximal path in the precedence network is of the form $i \ldots j0$ and its length is the minimum time required to process all the jobs $i, \ldots, j$. Recall from the introduction that $\rho = \max_{(j,k) \in A} d_{jk} / \min_{i \in N} p_i$. We assume that $\max_{(j,k) \in A} d_{jk} + \min_{i \in N} p_i > 0$; otherwise we have ordinary precedence constraints and we know that, for the makespan objective (and since we only consider nonpreemptive schedules), jobs with zero processing time can then be simply eliminated. If some $p_i = 0$ (with all $d_{jk} > 0$) then we may use $\rho = +\infty$ and the ratio $1/(1 + \rho) = 0$.

Lemma 2. *Let L denote the length of a longest path in the precedence network. Then $\mathcal{I}[0, C_{\max}^H) \leq (m - 1/(1 + \rho))L$.*

Proof. Let $i_0 \in N$ such that $S_{i_0}^H + p_{i_0} = C_{\max}^H$. Starting with $q = 0$, we construct as follows a path $(i_k, i_{k-1}, \ldots, i_0)$ in the precedence network:

1. While $i_q \in N^+$ do: choose as i_{q+1} an immediate predecessor of i_q with largest value $S_{i_{q+1}}^H + p_{i_{q+1}} + d_{i_{q+1}i_q}$.
2. Set $k = q$ and stop.

Since i_k has no predecessor, $\mathcal{I}[0, S_{i_k}^H) = 0$. By repeated application of Lemma 1, we have $\mathcal{I}[S_{i_k}^H, S_{i_0}^H) \leq m \sum_{q=0}^{k-1} d_{i_{q+1}i_q} + (m - 1) \sum_{q=1}^{k} p_{i_q}$. In addition, $\mathcal{I}[S_{i_0}^H, C_{\max}^H) \leq (m - 1)p_{i_0}$, and therefore $\mathcal{I}[0, C_{\max}^H) \leq m \sum_{q=0}^{k-1} d_{i_{q+1}i_q} + (m - 1) \sum_{q=0}^{k} p_{i_q}$. The total length of the path satisfies $\sum_{q=0}^{k-1} d_{i_{q+1}i_q} + \sum_{q=0}^{k} p_{i_q} \leq L$. If $\min_{i \in N} p_i = 0$ then we immediately have $\mathcal{I}[0, C_{\max}^H) \leq mL$ and the proof is complete. Otherwise, $\sum_{q=0}^{k-1} d_{i_{q+1}i_q} \leq \rho \sum_{q=0}^{k-1} \min_{i \in N} p_i \leq \rho \sum_{q=0}^{k-1} p_{i_q}$. Therefore, $\sum_{q=0}^{k-1} d_{i_{q+1}i_q} \leq \frac{\rho}{1+\rho} L$ and the lemma follows.   □

We are now in the position to prove the main result of this section.

Theorem 1. *For the scheduling problem P|prec. delays d_{ij}|$C_{\max}$, the performance ratio of Graham's List Scheduling Algorithm is $2 - 1/(m(1 + \rho))$, no matter which priority list is used.*

Proof. We have $m C_{\max}^H = \sum_{i \in N} p_i + \mathcal{I}[0, C_{\max}^H)$, and thus, by Lemma 2,

$$C_{\max}^H \leq \frac{\sum_{i \in N} p_i}{m} + \left(1 - \frac{1}{m(1 + \rho)}\right) L \ .$$

Since $\frac{1}{m} \sum_{i \in N} p_i$ and L are two lower bounds on the optimum makespan, the result follows.   □

There exist instances showing that the bound in Theorem 1 is (asymptotically) best possible, even for $m = 1$ [27].

4   The Performance of Job-Based List Scheduling

In this section we first present a linear programming relaxation of the problem of minimizing a weighted sum $\sum w_j C_j$ of job completion times subject to precedence delays, which is then used in the design of a 4–approximation algorithm for this problem. This formulation is a direct extension of a formulation given in [16], see also [33]. The decision variables are the job completion times C_j for all jobs $j \in N$. Note that this formulation does not take into account the assignment of jobs to machines. The set of constraints is:

$$C_j \geq \qquad C_i + d_{ij} + p_j \qquad \text{all } (i,j) \in A, \qquad (1)$$

$$\sum_{j \in F} p_j C_j \geq \frac{1}{2m} \left(\sum_{j \in F} p_j \right)^2 + \frac{1}{2} \sum_{j \in F} p_j^2 \quad \text{all } F \subseteq N. \qquad (2)$$

Constraints (1) are the precedence delay constraints. Constraints (2) are a relatively weak way of expressing the requirement that each machine can process at most one job at a time; for the single-machine case ($m = 1$) they were introduced by Wolsey [38] and Queyranne [30], and studied by Queyranne and Wang [31], von Arnim, Schrader, and Wang [1], von Arnim and Schulz [2], and Schulz [33] in the presence of ordinary precedence constraints; they were extended to $m \geq 2$ parallel machines by Hall et al. [16]. Note that these constraints, for $F = \{j\}$, imply $C_j \geq 0$; these and, as indicated above, the use of a dummy job 0 allow the formulation of release date constraints.

For a weighted sum of completion times objective, the LP formulation is simply:

$$\text{minimize } \sum_{j \in N} w_j C_j \quad \text{subject to (1) - (2).} \qquad (3)$$

Let C^{LP} denote any feasible solution to the constraint set (1)–(2) of this LP; we will call C_j^{LP} the LP completion time of job j. We now use this LP solution to define a feasible schedule with completion time vector C^H and analyze the job-by-job relationship between C_j^H and C_j^{LP} for every job $j \in N$.

We define the LP start time S_j^{LP} and LP midpoint M_j^{LP} as $S_j^{LP} = C_j^{LP} - p_j$ and $M_j^{LP} = C_j^{LP} - p_j/2$, respectively. We now use the List Scheduling Algorithm of Section 2 with the LP midpoint list L defined by sorting the jobs in nondecreasing order of their midpoints M_j^{LP}. The next theorem contains our main result.

Theorem 2. *Let C^{LP} denote any feasible solution to the constraint set (1)–(2) and let M^{LP} and S^{LP} denote the associated LP midpoints and start times, respectively. Let S^H be the vector of start times of the feasible schedule constructed by the List Scheduling Algorithm using the LP midpoint list. Then*

$$S_j^H \leq 2\, M_j^{LP} + 2\, S_j^{LP} \qquad \text{for all jobs } j \in N. \tag{4}$$

Proof. Assume for simplicity that the jobs are indexed in the order of their LP midpoints, that is, $M_1^{LP} \leq M_2^{LP} \leq \cdots \leq M_n^{LP}$. We fix job $j \in N$ and consider the schedule constructed by the List Scheduling heuristic using the LP midpoint list $L = (1, 2, \ldots, n)$ up to and including the scheduling of job j, that is, up to the completion of Step 3 with $k = j$. Let $[j] = \{1, \ldots, j\}$.

Let μ denote the total time between 0 and start time S_j^H of job j when all m machines are busy at this stage of the algorithm. Since only jobs in $[j-1]$ have been scheduled so far, we have $\mu \leq \frac{1}{m} \sum_{i=1}^{j-1} p_i$. Let $\lambda = S_j^H - \mu$. To prove (4), we only need to show that

$$\text{(i)} \quad \frac{1}{m} \sum_{i=1}^{j-1} p_i \leq 2M_j^{LP} \qquad \text{and} \qquad \text{(ii)} \quad \lambda \leq 2S_j^{LP}.$$

Inequality (i) follows from a straightforward variant of Lemma 3.2 in [16]. We omit the details here and refer the reader to the full version of this paper, see [27]. To show (ii), let q denote the number of time intervals between dates 0 and S_j^H when at least one machine is idle (i.e., not processing a job in $[j-1]$) in the schedule C^H. Denote these idle intervals as (a_h, b_h) for $h = 1, \ldots, q$, so that $0 \leq a_1$; that $b_{h-1} < a_h < b_h$ for all $h = 2, \ldots, q$; and that $b_q \leq S_j^H$. We have $\lambda = \sum_{h=1}^{q} (b_h - a_h)$ and all machines are busy during the complementary intervals $[b_h, a_{h+1}]$, including intervals $[0, a_1]$ and $[b_q, S_j^H]$ if nonempty.

Consider the digraph $G^{[j]} = ([j], A^{[j]})$ where

$$A^{[j]} = \{(k, \ell) \in A : k, \ell \in [j] \text{ and } C_\ell^H = C_k^H + d_{k\ell} + p_\ell\},$$

that is, $A^{[j]}$ is the set of precedence pairs in $[j]$ for which the precedence delay constraints (1) are tight for C^H. If $b_q > 0$ then a machine becomes busy at date b_q (or starts processing job j if $b_q = S_j^H$) and thus there exists a job $x(q) \in [j]$ with start time $S_{x(q)}^H = b_q$. We repeat the following process for decreasing values of the interval index h, starting with $h = q$, until we reach the date 0 or the busy interval $[0, a_1]$. Let $(v(1), \ldots, v(s))$ denote a maximal path in $G^{[j]}$ with last node (job) $v(s) = x(h)$. Note that we must have $b_g < S_{v(1)}^H \leq a_{g+1}$ for some busy interval $[b_g, a_{g+1}]$ with $a_{g+1} < b_h$, for otherwise some machine is idle immediately before the start time $S_{v(1)}^H$ of job $v(1)$ and this job, not being constrained by any tight precedence delay constraint, should have started earlier than that date. This implies in particular that $s \geq 2$. We have

$$b_h - a_{g+1} \leq S_{v(s)}^H - S_{v(1)}^H = \sum_{i=1}^{s-1} \left(S_{v(i+1)}^H - S_{v(i)}^H \right) = \sum_{i=1}^{s-1} \left(p_{v(i)} + d_{v(i)v(i+1)} \right). \tag{5}$$

On the other hand, the precedence delay constraints (1) imply

$$M_{v(i+1)}^{LP} \geq M_{v(i)}^{LP} + \frac{1}{2}p_{v(i)} + d_{v(i)v(i+1)} + \frac{1}{2}p_{v(i+1)}$$

for all $i = 1, \ldots, s - 1$. Therefore

$$M_{x(h)}^{LP} - M_{v(1)}^{LP} \geq \frac{1}{2}p_{x(h)} + \frac{1}{2}\sum_{i=1}^{s-1} \left(p_{v(i)} + d_{v(i)v(i+1)} \right)$$

and thus

$$S_{x(h)}^{LP} - M_{v(1)}^{LP} \geq \frac{1}{2}(b_h - a_{g+1}) \ .$$

If $b_g > 0$, then let $x(g)$ denote a job with start time $S_{x(g)}^{H}$ satisfying $b_g \leq S_{x(g)}^{H} \leq a_{g+1}$ and with minimum value of $M_{x(g)}^{LP}$ under this condition. Therefore $M_{x(g)}^{LP} \leq M_{v(1)}^{LP}$. We also have $(k, x(g)) \in A^{[j]}$ for some $k \in [j]$ with $S_k^{H} < b_g \leq S_{x(g)}^{H}$, for otherwise job $x(g)$ should have started processing on some idle machine before date b_g. This implies

$$S_{x(h)}^{LP} - M_{x(g)}^{LP} \geq \frac{1}{2}(b_h - a_{g+1}) \tag{6}$$

and we may repeat the above process with $h = g$ and job $x(h) = x(g)$. Since $g < h$ at each step, this whole process must terminate, generating a decreasing sequence of indices $q = h(1) > \ldots > h(q') = 0$ such that every idle interval is contained in some interval $[a_{h(i+1)+1}, b_{h(i)}]$. Adding the inequalities (6) and using $S_j^{LP} \leq M_j^{LP}$ for all $j = h(i)$, we obtain

$$\lambda \leq \sum_{i=1}^{q'} (b_{h(i)} - a_{h(i+1)+1}) \leq 2(S_{x(h(1))}^{LP} - M_{x(h(q'))}^{LP}) \leq 2(S_j^{LP} - 0) \ . \tag{7}$$

This establishes (ii). The proof of Theorem 2 is complete. $\qquad\square$

There exist instances showing that the factors 2 in inequality (4) are (asymptotically) best possible for any number of machines, see [27].

Using for C^{LP} an optimal LP solution, Theorem 2 implies performance ratios of $1/4$ and 4 for the LP relaxation and the heuristic solution, respectively, for the $\sum w_j C_j$ objective.

Corollary 1. *Let C^{LP} denote an optimal solution to the LP defined in (3) for the problem* P|prec. delays d_{ij}| $\sum w_j C_j$. *Let C^H denote the solution constructed from C^{LP} by the List Scheduling Algorithm using the LP midpoint list, and let C^* denote an optimum schedule. Then*

$$\sum_{j \in N} w_j C_j^{LP} \geq \frac{1}{4} \sum_{j \in N} w_j C_j^* \quad and \quad \sum_{j \in N} w_j C_j^{H} \leq 4 \sum_{j \in N} w_j C_j^* \ . \tag{8}$$

Examples show that the latter bound is (asymptotically) tight for an arbitrary number of machines. We suspect that the first inequality in (8), bounding the performance ratio of the LP relaxation, is not tight. The worst instances we know have a performance ratio of $1/3$ for the LP relaxation, see [27] for details.

The analysis in Theorem 2 may be refined for some special cases, yielding tighter performance ratios. For the problem $\mathrm{P}|\mathrm{prec}|\sum w_j C_j$, observe that the list scheduling algorithm will not allow all machines to be simultaneously idle at any date before the start time of any job $i \in N$. Therefore, in the proof of Theorem 2, all the idle intervals, with total length λ, contain some processing of some job(s) $i < j$; as a result the total work during the busy intervals is at most $\sum_{i=1}^{j-1} p_i - \lambda$. Hence, we obtain the following result.

Corollary 2. *List scheduling by LP midpoints is a $(4 - 2/m)$–approximation algorithm for $\mathrm{P}|\mathrm{prec}|\sum w_j C_j$.*

Note that for $m = 1$ we recover the performance ratio of 2 for $1|\mathrm{prec}|\sum w_j C_j$ in [16,34], which is known to be tight for that special case.

In the case of a single machine, the idle intervals that add up to λ time units cannot contain any processing. Therefore, in the proof of Theorem 2 replace inequality (5) with $b_h - a_{g+1} \leq S_{v(s)}^H - C_{v(1)}^H = \sum_{i=1}^{s-1} d_{v(i)v(i+1)}$. Adding up the precedence delay constraints for all $i = 1, \ldots, s-1$ and omitting some processing times yield $M_{x(h)}^{LP} - M_{v(1)}^{LP} \geq \frac{1}{2}p_{x(h)} + \sum_{i=1}^{s-1} d_{v(i)v(i+1)}$ and thus $S_{x(h)}^{LP} - M_{v(1)}^{LP} \geq b_h - a_{g+1}$. Therefore we may replace (7) with $\lambda \leq \sum_{i=1}^{q'}(b_{h(i)} - a_{h(i+1)+1}) \leq S_{x(h(1))}^{LP} - M_{x(h(q'))}^{LP} \leq S_j^{LP}$ and thus inequality (ii) with $\lambda \leq S_j^{LP}$. This implies $S_j^H \leq 3M_j^{LP}$.

Corollary 3. *List scheduling by LP midpoints is a 3–approximation algorithm for $1|\mathrm{prec.\ delays}\ d_{ij}|\sum w_j C_j$.*

Note that for the special case $1|r_j, \mathrm{prec}|\sum w_j C_j$ we recover the performance ratio of 3 in [16,34]. The best known approximation algorithm for this problem, however, has a performance guarantee of $e \approx 2.719$ [35].

5   Concluding Remarks

The appropriate introduction of idle time is a rather important part in the design of approximation algorithms to minimize the weighted sum of completion times subject to precedence delays. As Example 1 illustrates, idle time is needed to avoid that profitable jobs which become available soon are delayed by other, less important jobs. On the other hand, too much idle time is undesired as well. The necessity to balance these two effects contributes to the difficulty of this problem. Interestingly, all former approximation algorithms for $\mathrm{P}|r_j, \mathrm{prec}|\sum w_j C_j$ with constant-factor performance ratios are based on variants of Graham's original list scheduling which actually tries to avoid machine idle time. In fact, Hall et al. [16] partition jobs into groups that are individually scheduled according to a GLSA, and then these schedules are concatenated to obtain a solution of the original problem. To find a good partition, this scheme was enriched with randomness by Chakrabarti et al. [7]. Chekuri et al. [8] presented a different variant of a GLSA by artificially introducing idle time whenever it seems that a further delay of the next available job in the list (if it is not the first) can be afforded. Hence, the algorithm analyzed in Section 4 seems the first within this context that does not take the machine-based point of view of GLSAs.

However, an advantage of the rather simple scheme of Chekuri et al. is its small running time (though the performance ratios obtained are considerably worse). In fact, so far we have not even explained that the algorithms presented above are indeed polynomial-time algorithms. Whereas this is obvious for the GLSA variant for makespan minimization, we have to argue that in case of the total weighted completion time the linear programming relaxation (3) behaves well. In fact, it can be solved in polynomial time since the corresponding separation problem is polynomially solvable [33].

It seems worth to note that Theorem 2 actually implies stronger results than stated in Corollaries 1 and 3. The performance ratios given there not only hold for the weighted sum of completion times, but even for the weighted sum of third-points and midpoints, respectively. Recall that, in a given schedule and for a given value $0 < \alpha \leq 1$, the α–point of job j is the earliest time at which a fraction α of its processing has been performed; thus if the schedule is (or may be considered as) nonpreemptive, then the α–point of job j is simply $C_j - (1-\alpha)p_j$. Talking about α–points, it is interesting to note that in the proof of Theorem 2 other α–points could be used in this analysis, provided that $\frac{1}{2} \leq \alpha < 1$, but the midpoint ($\alpha = 1/2$) leads to the best bound.

Finally, let us relate precedence delays to another kind of restrictions that has been given quite some attention recently. Assume that precedence constraints (i, j) are caused by some technological requirements. Then it seems reasonable that the information or object produced during the processing of job i needs to be transferred to job j before the processing of j can start. Hence, precedence delays might be interpreted as kind of *communication delays*. This term is coined with a slightly different meaning, however (see, e.g., [32,19,9,25,26]). In fact, in this context one usually assumes that a delay only occurs if i and j are assigned to different machines. Our results can be extended to the model with precedence and communication delays. For details, we again refer the reader to the full version of this paper [27].

Acknowledgments

This work was initiated during a workshop on "Parallel Scheduling" held at Schloß Dagstuhl, Germany, July 14–18, 1997. The authors are grateful to the organizers for providing a stimulating atmosphere. The research of the second author is supported in part by a research grant from NSERC, (the Natural Sciences and Research Council of Canada) and by the UNI.TU.RIM. S.p.a. (Società per l'Università nel riminese), whose support is gratefully acknowledged.

References

1. A. von Arnim, R. Schrader, and Y. Wang. The permutahedron of N-sparse posets. *Mathematical Programming*, 75:1–18, 1996.
2. A. von Arnim and A. S. Schulz. Facets of the generalized permutahedron of a poset. *Discrete Applied Mathematics*, 72:179–192, 1997.
3. E. Balas, J. K. Lenstra, and A. Vazacopoulos. The one machine problem with delayed precedence constraints and its use in job-shop scheduling. *Management Science*, 41:94–109, 1995.
4. M. Bartusch, R. H. Möhring, and F. J. Radermacher. Scheduling project networks with resource constraints and time windows. *Annals of Operations Research*, 16:201–240, 1988.
5. D. Bernstein and I. Gertner. Scheduling expressions on a pipelined processor with a maximum delay of one cycle. *ACM Transactions on Programming Languages and Systems*, 11:57 – 66, 1989.
6. J. Bruno, J. W. Jones, and K. So. Deterministic scheduling with pipelined processors. *IEEE Transactions on Computers*, C-29:308–316, 1980.
7. S. Chakrabarti, C. A. Phillips, A. S. Schulz, D. B. Shmoys, C. Stein, and J. Wein. Improved scheduling algorithms for minsum criteria. In F. Meyer auf der Heide and B. Monien, editors, *Automata, Languages and Programming, LNCS, Vol. 1099*, pages 646–657. Springer, Berlin, 1996.
8. C. S. Chekuri, R. Motwani, B. Natarajan, and C. Stein. Approximation techniques for average completion time scheduling. In *Proceedings of the 8th ACM–SIAM Symposium on Discrete Algorithms*, pages 609–618, 1997.
9. P. Chrétienne and C. Picouleau. Scheduling with communication delays: A survey. In P. Chrétienne, E. G. Coffman Jr., J. K. Lenstra, and Z. Liu, editors, *Scheduling Theory and its Applications*, chapter 4, pages 65–90. John Wiley & Sons, 1995.
10. F. A. Chudak and D. B. Shmoys. Approximation algorithms for precedence–constrained scheduling problems on parallel machines that run at different speeds. In *Proceedings of the 8th ACM–SIAM Symposium on Discrete Algorithms*, pages 581–590, 1997.
11. M. R. Garey and D. S. Johnson. *Computers and Intractability: A Guide to the Theory of NP–Completeness*. Freeman, San Francisco, 1979.
12. M. X. Goemans. Improved approximation algorithms for scheduling with release dates. In *Proceedings of the 8th ACM–SIAM Symposium on Discrete Algorithms*, pages 591–598, 1997.
13. M. X. Goemans, M. Queyranne, A. S. Schulz, M. Skutella, and Y. Wang. Single machine scheduling with release dates. Working paper, 1997.
14. R. L. Graham. Bounds for certain multiprocessing anomalies. *Bell System Tech. J.*, 45:1563–1581, 1966.
15. R. L. Graham, E. L. Lawler, J. K. Lenstra, and A. H. G. Rinnooy Kan. Optimization and approximation in deterministic sequencing and scheduling: A survey. *Annals of Discrete Mathematics*, 5:287–326, 1979.
16. L. A. Hall, A. S. Schulz, D. B. Shmoys, and J. Wein. Scheduling to minimize average completion time: Off–line and on–line approximation algorithms. *Mathematics of Operations Research*, 22:513–544, 1997.
17. W. Herroelen and E. Demeulemeester. Recent advances in branch-and-bound procedures for resource-constrained project scheduling problems. In P. Chrétienne, E. G. Coffman Jr., J. K. Lenstra, and Z. Liu, editors, *Scheduling Theory and its Applications*, chapter 12, pages 259–276. John Wiley & Sons, 1995.

18. J. A. Hoogeveen, P. Schuurman, and G. J. Woeginger. Non-approximability results for scheduling problems with minsum criteria. In R. E. Bixby, E. A. Boyd, and R. Z. Ríos-Mercado, editors, *Proceedings of the 6th International IPCO Conference, LNCS, Vol. 1412*, pages 344–357. Springer, 1998. This volume.

19. J. A. Hoogeveen, B. Veltman, and J. K. Lenstra. Three, four, five, six, or the complexity of scheduling with communication delays. *Operations Research Letters*, 3:129–137, 1994.

20. J. M. Jaffe. Efficient scheduling of tasks without full use of processor resources. *Theoretical Computer Science*, 12:1–17, 1980.

21. E. Lawler, J. K. Lenstra, C. Martel, B. Simons, and L. Stockmeyer. Pipeline scheduling: A survey. Technical Report RJ 5738 (57717), IBM Research Division, San Jose, California, 1987.

22. E. L. Lawler, J. K. Lenstra, A. H. G. Rinnooy Kan, and D. B. Shmoys. Sequencing and scheduling: Algorithms and complexity. In S. C. Graves, A. H. G. Rinnooy Kan, and P. H. Zipkin, editors, *Logistics of Production and Inventory, Handbooks in Operations Research and Management Science, Vol. 4*, chapter 9, pages 445–522. North–Holland, Amsterdam, The Netherlands, 1993.

23. J. K. Lenstra and A. H. G. Rinnooy Kan. Complexity of scheduling under precedence constraints. *Operations Research*, 26:22–35, 1978.

24. J. Y.–T. Leung, O. Vornberger, and J. Witthoff. On some variants of the bandwidth minimization problem. *SIAM J. Computing*, 13:650–667, 1984.

25. R. H. Möhring, M. W. Schäffter, and A. S. Schulz. Scheduling jobs with communication delays: Using infeasible solutions for approximation. In J. Diaz and M. Serna, editors, *Algorithms – ESA'96, LNCS, Vol. 1136*, pages 76–90. Springer, Berlin, 1996.

26. A. Munier and J.–C. König. A heuristic for a scheduling problem with communication delays. *Operations Research*, 45:145–147, 1997.

27. A. Munier, M. Queyranne, and A. S. Schulz. Approximation bounds for a general class of precedence constrained parallel machine scheduling problems. Preprint 584/1998, Department of Mathematics, Technical University of Berlin, Berlin, Germany, 1998. `ftp://ftp.math.tu-berlin.de/pub/Preprints/combi/Report-584-1998.ps.Z`

28. K. W. Palem and B. Simons. Scheduling time critical instructions on RISC machines. In *Proceedings of the 17th Annual Symposium on Principles of Programming Languages*, pages 270–280, 1990.

29. C. Phillips, C. Stein, and J. Wein. Scheduling jobs that arrive over time. In *Proceedings of the Fourth Workshop on Algorithms and Data Structures, LNCS, Vol. 955*, pages 86–97. Springer, Berlin, 1995.

30. M. Queyranne. Structure of a simple scheduling polyhedron. *Mathematical Programming*, 58:263–285, 1993.

31. M. Queyranne and Y. Wang. Single–machine scheduling polyhedra with precedence constraints. *Mathematics of Operations Research*, 16:1–20, 1991.

32. V. J. Rayward-Smith. UET scheduling with unit interprocessor communication delays. *Discrete Applied Mathematics*, 18:55–71, 1987.

33. A. S. Schulz. *Polytopes and Scheduling*. PhD thesis, Technical University of Berlin, Berlin, Germany, 1996.

34. A. S. Schulz. Scheduling to minimize total weighted completion time: Performance guarantees of LP–based heuristics and lower bounds. In W. H. Cunningham, S. T. McCormick, and M. Queyranne, editors, *Integer Programming and Combinatorial Optimization, LNCS, Vol. 1084*, pages 301–315. Springer, Berlin, 1996.

35. A. S. Schulz and M. Skutella. Random–based scheduling: New approximations and LP lower bounds. In J. Rolim, editor, *Randomization and Approximation Techniques in Computer Science, LNCS, Vol. 1269*, pages 119–133. Springer, Berlin, 1997.

36. A. S. Schulz and M. Skutella. Scheduling–LPs bear probabilities: Randomized approximations for min–sum criteria. In R. Burkard and G. Woeginger, editors, *Algorithms – ESA'97, LNCS, Vol. 1284*, pages 416–429. Springer, Berlin, 1997.

37. E. D. Wikum, D. C. Llewellyn, and G. L. Nemhauser. One–machine generalized precedence constrained scheduling problems. *Operations Research Letters*, 16:87–89, 1994.

38. L. A. Wolsey, August 1985. Invited Talk at the 12th International Symposium on Mathematical Programming, MIT, Cambridge.

An Efficient Approximation Algorithm for Minimizing Makespan on Uniformly Related Machines

Chandra Chekuri[1]* and Michael Bender[2]

[1] Computer Science Department
Stanford University
chekuri@@cs.stanford.edu
[2] Division of Applied Sciences
Harvard University
bender@@deas.harvard.edu

Abstract. We give a new efficient approximation algorithm for scheduling precedence constrained jobs on machines with different speeds. The setting is as follows. There are n jobs $1, \ldots, n$ where job j requires p_j units of processing. The jobs are to be scheduled on a set of m machines. Machine i has a speed s_i; it takes p_j/s_i units of time for machine i to process job j. The precedence constraints on the jobs are given in the form of a partial order. If $j \prec k$, processing of k cannot start until j's execution if finished. Let C_j denote the completion time of job j. The objective is to find a schedule to minimize $C_{\max} = \max_j C_j$, conventionally called the *makespan* of the schedule. We consider non-preemptive schedules where each job is processed on a single machine with no preemptions. Recently Chudak and Shmoys [1] gave an algorithm with an approximation ratio of $O(\log m)$ significantly improving the earlier ratio of $O(\sqrt{m})$ due to Jaffe [7]. Their algorithm is based on solving a linear programming relaxation of the problem. Building on some of their ideas, we present a combinatorial algorithm that achieves a similar approximation ratio but runs in $O(n^3)$ time. In the process we also obtain a constant factor approximation algorithm for the special case of precedence constraints induced by a collection of chains. Our algorithm is based on a new lower bound which we believe is of independent interest. By a general result of Shmoys, Wein, and Williamson [10] our algorithm can be extended to obtain an $O(\log m)$ approximation ratio even if jobs have release dates.

* Supported primarily by an IBM Cooperative Fellowship. Remaining support was provided by an ARO MURI Grant DAAH04-96-1-0007 and NSF Award CCR-9357849, with matching funds from IBM, Schlumberger Foundation, Shell Foundation, and Xerox Corporation.

R. E. Bixby, E. A. Boyd, and R. Z. Ríos-Mercado (Eds.): IPCO VI
LNCS 1412, pp. 383–393, 1998. © Springer–Verlag Berlin Heidelberg 1998

1   Introduction

The problem of scheduling precedence constrained jobs on a set of identical parallel machines to minimize makespan is one of the oldest problems for which approximation algorithms have been devised. Graham [4] showed that a simple list scheduling gives a ratio of 2 and it is the best known algorithm till date. We consider a generalization of this model in which machines have different speeds. In the scheduling literature such machines are called *uniformly related*. We formalize the problem below. There are n jobs $1, \ldots, n$, with job j requiring processing of p_j units. The jobs are to be scheduled on a set of m machines. Machine i has a speed factor s_i. Job j with a processing requirement p_j takes p_j/s_i time units to run on machine i. Let C_j denote the completion time of job j. The objective is to find a schedule to minimize $C_{\max} = \max_j C_j$ called the makespan of the schedule. We restrict ourselves to non-preemptive schedules where a job once started on a machine has to run to completion on the same machine. Our results carry over to the preemptive case as well. In the scheduling literature [5] where problems are classified in the $\alpha|\beta|\gamma$ notation, this problem is referred to as $Q|prec|C_{\max}$.

Liu and Liu [9] analyzed the performance of Graham's list scheduling algorithm for the case of different speeds and showed that the approximation guarantee depends on the ratio of the largest to the smallest speed. This ratio could be arbitrarily large even for a small number of machines. The first algorithm to have a bound independent of the speeds was given by Jaffe [7]. He showed that list scheduling restricted to the set of machines with speeds that are within a factor of $1/\sqrt{m}$ of the fastest machine speed results in an $O(\sqrt{m})$ bound. More recently, Chudak and Shmoys [1] improved the ratio considerably and gave an algorithm which has a guarantee of $O(\log m)$. At a more basic level their algorithm has a guarantee of $O(K)$ where K is the number of *distinct* speeds. The above mentioned algorithm relies on solving a linear programming relaxation and uses the information obtained from the solution to allocate jobs to processors. We present a new algorithm which finds an allocation without solving a linear program. The ratio guaranteed by our algorithm is also $O(\log m)$ but is advantageous for the following reason. Our algorithm runs in $O(n^3)$ time and is combinatorial, hence is more efficient than the algorithm in [1]. Further, the analysis of our algorithm relies on a new lower bound which is very natural, and might be useful in other contexts. In addition we show that our algorithm achieves a constant factor approximation when the precedence constraints are induced by a collection of chains. We remark here that our work was inspired by, and builds upon the ideas in [1].

The rest of the paper is organized as follows. Section 2 contains some of the ideas from the paper of Chudak and Shmoys [1] that are useful to us. We present our lower bound in Section 3, and give the approximation algorithm and the analysis in Section 4.

2   Preliminaries

We summarize below the basic ideas in the work of Chudak and Shmoys [1]. Their main result is an algorithm which gives a ratio of $O(K)$ for the problem of $Q|prec|C_{\max}$ where K is the number of distinct speeds. They also show how to reduce the general case with arbitrary speeds to one in which there are only $O(\log m)$ distinct speeds as follows.

- Ignore all machines with speed less than $1/m$ times the speed of the fastest machine.
- Round down all speeds to the nearest power of 2.

They observe that the above transformation can be done while losing only a constant factor in the approximation ratio. Using this observation, we will restrict ourselves to the case where we have K distinct speeds.

When all machines have the same speed $(K = 1)$, Graham [4] showed that list scheduling gives a 2 approximation. His analysis shows that in any schedule produced by list scheduling, we can identify a chain of jobs $j_1 \prec j_2 \ldots \prec j_r$ such that a machine is idle only when one of the jobs in the above chain is being processed. The time spent processing the above chain is a lower bound on the optimal makespan. In addition, the measure of the time instants during which all machines are busy is also a lower bound by arguments about the average load. These two bounds provide an upper bound of 2 on the approximation ratio of list scheduling. One can apply a similar analysis for the multiple speed case. As observed in [1], the difficulty is that the time spent in processing the chain identified from the list scheduling analysis, is not a lower bound. The only claim that can be made is that the processing time of any chain on the fastest machine is a lower bound. However the jobs in the chain guaranteed by the list scheduling analysis do not necessarily run on the fastest machine. Based on this observation, the algorithm in [1] tries to find an assignment of jobs to speeds (machines) that ensures that the processing time of any chain is bounded by some factor of the optimal.

We will follow the notation of [1] for sake of continuity and convenience. Recall that we have K distinct speeds. Let m_k be the number of machines with speed $\bar{s}_k$, $k = 1, \ldots, K$, where $\bar{s}_1 > \ldots > \bar{s}_K$. Let M_u^v denote the sum $\sum_{k=u}^{v} m_k$. In the sequel we will be interested in assigning jobs to speeds. For a given assignment, let $k(j)$ denote the speed at which job j is assigned to be processed. The average processing allocated to a machine of a specific speed k, denoted by D_k, is the following.

$$D_k = \frac{1}{m_k \bar{s}_k} \sum_{j:k(j)=k} p_j.$$

A chain is simply a subset of jobs which are totally ordered by the precedence constraints. Let $\mathcal{P}$ be the set of all chains induced by the precedence constraints. We compute a quantity C defined by the following equation.

$$C = \max_{P \in \mathcal{P}} \sum_{j \in P} \frac{p_j}{\bar{s}_{k(j)}}$$

A natural variant of list scheduling called speed based list scheduling is developed in [1] which is constrained to schedule according to the speed assignments of the jobs. In classical list scheduling, the first available job from the list is scheduled as soon as a machine is free. In speed based list scheduling, an available job is scheduled on a free machine provided the speed of free machine matches the speed assignment of the job. The following theorem is from [1]. The analysis is a simple generalization of Graham's analysis of list scheduling.

Theorem 1 (Chudak and Shmoys). *For any job assignment $k(j)$, $j = 1, \ldots,$ n, the speed-based list scheduling algorithm produces a schedule of length*

$$C_{\max} \leq C + \sum_{k=1}^{K} D_k.$$

The authors of [1] use a linear programming relaxation of the problem to obtain a job assignment that satisfies the two conditions: $\sum_{k=1}^{K} D_k \leq (K + \sqrt{K})C_{\max}^*$, and $C \leq (\sqrt{K}+1)C_{\max}^*$, where $C_{\max}^*$ is the optimal makespan. Combining these with Theorem 1 gives them an $O(K)$ approximation. We will show how to use an alternative method based on chain decompositions to obtain an assignment satisfying similar properties.

3   A New Lower Bound

In this section we develop a simple and natural lower bound that will be used in the analysis of our algorithm. Before formally stating the lower bound we provide some intuition. The two lower bounds used in Graham's analysis for identical parallel machines are the maximum chain length and the average load. As discussed in the previous section, a naive generalization of the first lower bound implies that the maximum chain length (a chain's length is sum of processing times of jobs in it) divided by the fastest speed is a lower bound. However it is easy to generate examples where the maximum of this bound and the average load is $O(1/m)$ times the optimal. We describe the general nature of such examples to motivate our new bound. Suppose we have two speeds with $\bar{s}_1 = D$ and $\bar{s}_2 = 1$. The precedence constraints between the jobs are induced by a collection of $\ell > 1$ chains, each of the same length D. Suppose $m_1 = 1$, and $m_2 = \ell \cdot D$. The average load can be seen to be upper bounded by 1. In addition the time to process any chain on the fastest processor is 1. However if $D \gg \ell$ it is easy to observe that the optimal is $\Omega(\ell)$ since *only* ℓ machines can be busy at any time instant. The key insight we obtain from the above example is that the amount of parallelism in an instance restricts the number of machines that can be used. We capture this insight in our lower bound in a simple way. We need a few definitions to formalize the intuition. We view the precedence relations between the jobs as a weighted poset where each element of the poset has a weight associated with it that is the same as the processing time of the associated job. We will further assume that we have the transitive closure of the poset.

Definition 1. *A chain P is a set of jobs $j_1, \ldots, j_r$ such that for all $1 \leq i < r$, $j_i \prec j_{i+1}$. The length of a chain P, denoted by $|P|$, is the sum of the processing times of the jobs in P.*

Definition 2. *A chain decomposition $\mathcal{P}$ of a set of precedence constrained jobs is an partition of the poset in to a collection of chains $\{P_1, P_2, \ldots, P_r\}$. A maximal chain decomposition is one in which P_1 is a longest chain and $\{P_2, \ldots, P_r\}$ is a maximal chain decomposition of the poset with elements of P_1 removed.*

Though we define a maximal chain decomposition as a set of chains, we will implicitly assume that it is an *ordered* set, that is $|P_1| \geq |P_2| \geq \ldots |P_r|$.

Definition 3. *Let $\mathcal{P} = \{P_1, P_2, \ldots, P_r\}$ be any maximal chain decomposition of the precedence graph of the jobs. We define a quantity called $L_{\mathcal{P}}$ associated with $\mathcal{P}$ as follows.*

$$L_{\mathcal{P}} = \max_{1 \leq j \leq \min\{r, m\}} \frac{\sum_{i=1}^{j} |P_i|}{\sum_{i=1}^{j} s_i}$$

With the above definitions in place we are ready to state and prove the new lower bound.

Theorem 2. *Let $\mathcal{P} = \{P_1, P_2, \ldots, P_r\}$ be any maximal chain decomposition of the precedence graph of the jobs. Let $AL = (\sum_{j=1}^{n} p_i)/(\sum_{i=1}^{m} s_i)$ which represents the average load. Then*

$$C_{\max}^* \geq \max\{AL, L_{\mathcal{P}}\}.$$

Moreover the lower bound is valid for the preemptive case as well.

Proof. It is easy to observe that $C_{\max}^* \geq AL$. We will show the following for $1 \leq j \leq m$

$$C_{\max}^* \geq \frac{\sum_{i=1}^{j} |P_i|}{\sum_{i=1}^{j} s_i}$$

which will prove the theorem. Consider the first j chains. Suppose our input instance was modified to have only the jobs in the first j chains. It is easy to see that a lower bound for this modified instance is a lower bound for the original instance. Since it is possible to execute only one job from each chain at any time instant, only the fastest j machines are relevant for this modified instance. The expression $(\sum_{i=1}^{j} |P_i|)/(\sum_{i=1}^{j} s_i)$ is simply the average load for the modified instance, which as we observed before, is a lower bound. Since the average load is also a lower bound for the preemptive case, the claimed lower bound applies even if preemptions are allowed.

Horvath, Lam, and Sethi [6] proved that the above lower bound gives the optimal schedule length for preemptive scheduling of chains on uniformly related machines. The idea of extending their lower bound to general precedences using maximal chain decompositions is natural but does not appear to have been effectively used before.

Theorem 3. *A maximal chain decomposition can be computed in $O(n^3)$ time. If all p_j are the same, the running time can be improved to $O(n^2\sqrt{n})$.*

Proof. It is necessary to find the transitive closure of the given graph of precedence constraints. This can be done in $O(n^3)$ time using a BFS from each vertex. From a theoretical point of view this can be improved to $O(n^\omega)$ where $\omega \leq 2.376$ using fast matrix multiplication [2]. A longest chain in a weighted DAG can be found in $O(n^2)$ time using standard algorithms. Using this at most n times, a maximal chain decomposition can be obtained in $O(n^3)$ time. If all p_j are the same (without loss of generality we can assume they are all 1), the length of a chain is the same as the number of vertices in the chain. It is possible to use this additional structure to obtain a maximal chain decomposition in $O(n^2\sqrt{n})$ time. We omit the details.

4   The Approximation Algorithm

The approximation algorithm we develop in this section will be based on the maximal chain decompositions defined in the previous section. As mentioned in Section 2, we will describe an algorithm to produces an assignment of jobs to speeds at which they will be processed. Then we use the speed based list scheduling of [1] with the job assignment produced by our algorithm.

---

1. **compute** a maximal chain decomposition of the jobs $\mathcal{P} = \{P_1, \ldots, P_r\}$.
2. **set** $\ell = 1$. **set** $B = \max\{AL, L_\mathcal{P}\}$.
3. **foreach** speed $1 \leq i \leq K$ **do**
      (a) **let** $\ell \leq t \leq r$ be maximum index such that $\sum_{\ell \leq j \leq t} |P_j|/(m_i s_i) \leq 4B$.
      (b) **assign** jobs in chains $P_\ell, \ldots, P_t$ to speed i.
      (c) **set** $\ell = t + 1$. If $\ell > r$ **return**.
4. **return**.

Fig. 1. Algorithm Chain-Alloc

The algorithm in Figure 1 computes a lower bound B on the optimal using Theorem 2. It then orders the chains in non-increasing lengths and allocates the chains to speeds such that no speed is loaded by more than four times the lower bound. We now prove several properties of the above described allocation which leads to the approximation guarantee of the algorithm.

Lemma 1. *Let $P_{\ell(u)}, \ldots, P_r$ be the chains remaining when Chain-Alloc considers speed u in step 3 of the algorithm. Then*

1. $|P_{\ell(u)}|/\bar{s}_u \leq 2B$ and
2. Either $P_{\ell(u)}, \ldots, P_r$ are allocated to speed u or $D_u > 2B$.

Proof. We prove the above assertions by induction on u. Consider the base case when $u = 1$ and $\ell(1) = 1$. From the definition of $L_\mathcal{P}$ it follows that $|P_1|/\bar{s}_1 \leq B$. Since P_1 is the longest chain, it follows that $|P_j|/\bar{s}_1 \leq B$ for $1 \leq j \leq r$. Let t be the last chain allocated to $\bar{s}_1$. If $t = r$ we are done. If $t < r$, it must be the case that adding $P_{(t+1)}$ increases the average load on $\bar{s}_1$ to more than $4B$. Since $P_{(t+1)}/\bar{s}_1 \leq B$, we conclude that $D_1 = \sum_{j=1}^{t} |P_j|/m_1\bar{s}_1 > 3B > 2B$.

Assume that the conditions of the lemma are satisfied for speeds s_1 to s_{u-1} and consider speed s_u. We will assume that $\ell(u) < r$ for otherwise there is nothing to prove. We observe that the second condition follows from the first using an argument similar to the one used above for the base case. Therefore it is sufficient to prove the first condition. Suppose $|P_{\ell(u)}|/\bar{s}_u > 2B$. We will derive a contradiction later. Let $j = \ell(u)$ and let v be the index such that $M_1^{v-1} < j \leq M_1^v$ (recall that $M_1^v = \sum_{k=1}^{v} m_k$). If $j > m$, no such index exists and we set v to K, the slowest speed. If $j \leq m$, for convenience of notation we assume that $j = M_1^v$ simply by ignoring other machines of speed $\bar{s}_v$. It is easy to see that $v \geq u$ and $j > M_1^{u-1}$. From the definition of $L_\mathcal{P}$, AL, and B, we get the following. If $j \leq m$ then $L_\mathcal{P} \geq (\sum_{i=1}^{j} |P_i|)/(\sum_{k=1}^{v} m_k\bar{s}_k)$. If $j > m$ then $AL \geq (\sum_{i=1}^{j} |P_i|)/(\sum_{k=1}^{K} m_k\bar{s}_k)$. In either case we obtain the fact that

$$\frac{\sum_{i=1}^{j} |P_i|}{\sum_{k=1}^{v} m_k\bar{s}_k} \leq \max\{L_\mathcal{P}, AL\} = B \tag{1}$$

Since $|P_j|/\bar{s}_u > 2B$, it must be the case that $|P_i|/\bar{s}_u > 2B$ for all $M_1^{u-1} < i \leq j$. This implies that

$$\sum_{M_1^{u-1} < i}^{j} |P_i| > 2B(j - M_1^{u-1})\bar{s}_u$$

$$\geq 2B \sum_{k=u}^{v} m_k\bar{s}_k$$

$$\Rightarrow \sum_{i=1}^{j} |P_i| > 2B \sum_{k=u}^{v} m_k\bar{s}_k \tag{2}$$

The last inequality follows since we are summing up more terms on the left hand side. From the induction hypothesis it follows that speeds $\bar{s}_1$ to $\bar{s}_{u-1}$ have an average load greater than $2B$. From this we obtain

$$\sum_{i=1}^{j-1} |P_i| > 2B \sum_{k=1}^{u-1} m_k\bar{s}_k \tag{3}$$

$$\Rightarrow \sum_{i=1}^{j} |P_i| > 2B \sum_{k=1}^{u-1} m_k\bar{s}_k \tag{4}$$

Combining Equations 2 and 4 we obtain the following.

$$2\sum_{i=1}^{j}|P_i| > 2B\sum_{k=1}^{u-1}m_k\bar{s}_k + 2B\sum_{k=u}^{v}m_k\bar{s}_k$$

$$> 2B\sum_{k=1}^{v}m_k\bar{s}_k$$

$$\Rightarrow \sum_{i=1}^{j}|P_i| > B\sum_{k=1}^{v}m_k\bar{s}_k \tag{5}$$

Equation 5 contradicts Equation 1.

Corollary 1. *If chain P_j is assigned to speed i, then $\frac{|P_j|}{\bar{s}_i} \leq 2B$.*

Corollary 2. *Algorithm Chain-Alloc allocates all chains.*

Lemma 2. *For $1 \leq k \leq K$, $D_k \leq 4C_{\max}^*$.*

Proof. Since $B \leq C_{\max}^*$ and the algorithm never loads a speed by more than an average load of $4B$, the bound follows.

Lemma 3. *For the job assignment produced by Chain-Alloc $C \leq 2KC_{\max}^*$.*

Proof. Let P be any chain. We will show that $\sum_{j\in P} p_j/\bar{s}_{k(j)} \leq 2KC_{\max}^*$ where $k(j)$ is the speed to which job j is assigned. Let A_i be the set of jobs in P which are assigned to speed i. Let P_ℓ be the longest chain assigned to speed i by the algorithm. We claim that $|P_\ell| \geq \sum_{j\in A_i} p_i$. This is because the jobs in A_i form a chain when we picked P_ℓ to be the longest chain in the max chain decomposition. From Corollary 1 we know that $|P_\ell|/\bar{s}_i \leq 2B \leq 2C_{\max}^*$. Therefore it follows that

$$\sum_{j\in P}\frac{p_j}{\bar{s}_{k(j)}} = \sum_{i=1}^{K}\frac{|A_i|}{\bar{s}_i} \leq 2KC_{\max}^*.$$

Theorem 4. *Using speed based list scheduling on the job assignment produced by Algorithm Chain-Alloc gives a $6K$ approximation where K is the number of distinct speeds. Furthermore the algorithm runs in $O(n^3)$ time. The running time can be improved to $O(n^2\sqrt{n})$ if all p_j are the same.*

Proof. From Lemma 2 we have $D_k \leq 4C_{\max}^*$ for $1 \leq k \leq K$ and from Lemma 3 we have $C \leq 2KC_{\max}^*$. Putting these two facts together, for the job assignment produced by the algorithm Chain-Alloc, speed based list scheduling gives the following upper bound by Theorem 1.

$$C_{\max} \leq C + \sum_{k=1}^{K} D_k \leq 2KC_{\max}^* + 4KC_{\max}^* \leq 6KC_{\max}^*.$$

It is easy to see that the speed based list scheduling can be implemented in $O(n^2)$ time. The running time is dominated by the time to do the maximum chain decomposition. Theorem 3 gives the desired bounds.

Corollary 3. *There is an algorithm which runs in $O(n^3)$ time and gives an $O(\log m)$ approximation ratio to the problem of scheduling precedence constrained jobs on uniformly related machines.*

We remark here that the leading constant in the LP based algorithm in [1] is better. We also observe that the above bound is based on our lower bound which is valid for preemptive schedules as well. Hence our approximation ratio is also valid for preemptive schedules. In [1] it is shown that the lower bound provided by the LP relaxation is a factor of $\Omega(\log m/\log\log m)$ away from the optimal. Surprisingly it is easy to show using the same example as in [1] that our lower bound from Section 3 is also a factor of $\Omega(\log m/\log\log m)$ away from the optimal.

Theorem 5. *There are instances where the lower bound given in Theorem 2 is a factor of $\Omega(\log m/\log\log m)$ away from the optimal.*

Proof. The proof of Theorem 3.3 in [1] provides the instance and it is easily verified that *any* maximum chain decomposition of that instance is a factor of $\Omega(\log m/\log\log m)$ away from the optimal.

4.1   Release Dates

Now consider the scenario where each job j has a release date r_j before which it cannot be processed. By a general result of Shmoys, Wein, and Williamson [10] an approximation algorithm for the problem without release dates can be transformed to one with release dates losing only a factor of 2 in the process. Therefore we obtain the following.

Theorem 6. *There is an $O(\log m)$ approximation algorithm for the problem $Q|prec, r_j|C_{\max}$ that runs in time $O(n^3)$.*

4.2   Scheduling Chains

In this subsection we show that Chain-Alloc followed by speed based list scheduling gives a constant factor approximation if the precedence constraints are induced by a collection of chains. We give an informal proof in this version of the paper. We first observe that any maximal chain decomposition of a collection of chains is simply the collection itself. The crucial observation is that the algorithm Chain-Alloc allocates all jobs of any chain to the same speed class. The two observation together imply that there are no precedence relations between jobs allocated to different speeds. This allows us to obtain a stronger version of Theorem 1 where we can upper bound the makespan obtained by speed

392 Chandra Chekuri and Michael Bender

based list scheduling as $C_{\max} \leq C + \max_{1 \leq k \leq K} D_k$. Further we can bound C as $\max_{1 \leq i \leq r} |P_i|/\overline{s}_{k(i)}$ where chain P_i is allocated to speed $k(i)$. From Corollary 1 it follows that $C \leq 2B$. Lemma 2 implies that $\max_{1 \leq k \leq K} D_k \leq 4B$. Combining the above observations yields a 6 approximation.

Theorem 7. *There is a 6 approximation for the problem $Q|chains|C_{\max}$ and a 12 approximation for the problem $Q|chains, r_j|C_{\max}$.*

Computing the maximal chain decomposition of a collection of chains is trivial and the above algorithm can be implemented in $O(n \log n)$ time.

5 Conclusions

The main contribution of this paper is a simple and efficient $O(\log m)$ approximation to the scheduling problem $Q|prec|C_{\max}$. Chudak and Shmoys [1] provide similar approximations for the more general case when the objective function is the average weighted completion time ($Q|prec| \sum w_j C_j$) using linear programming relaxations. We believe that the techniques of this paper can be extended to obtain a simpler and combinatorial algorithm for that case as well. It is known that the problem of minimizing makespan is hard to approximate to within a factor of 4/3 even if all machines have the same speed [8]. However, for the single speed case a 2 approximation is known, while the current ratio for the multiple speed case is only $O(\log m)$. Obtaining a constant factor approximation, or improving the hardness are interesting open problems.

Acknowledgments

We thank Monika Henzinger for simplifying the proof of Lemma 1.

References

1. F. Chudak and D. Shmoys. Approximation algorithms for precedence-constrained scheduling problems on parallel machines that run at different speeds. *Proceedings of the Eighth Annual ACM-SIAM Symposium on Discrete Algorithms (SODA),* 1997.
2. D. Coppersmith and S. Winograd. Matrix multiplication via arithmetic progression. *Proceedings of the 19th ACM Symposium on Theory of Computing,* pages 1–6, 1987.
3. M. R. Garey and D. S. Johnson. *Computers and Intractability: A Guide to the Theory of NP-Completeness.* Freeman, San Francisco, 1979.
4. R.L. Graham. Bounds for certain multiprocessor anomalies. *Bell System Tech. J.,* 45:1563–81, 1966.
5. R. L. Graham, E. L. Lawler, J. K. Lenstra, and A. H. G. Rinnooy Kan. Optimization and approximation in deterministic sequencing and scheduling: A survey. *Ann. Discrete Math.,* 5:287–326, 1979.
6. E. Horvath, S. Lam, and R. Sethi. A level algorithm for preemptive scheduling. *Journal of the ACM,* 24(1):32–43, 1977.

7. J. Jaffe. Efficient scheduling of tasks without full use of processor resources. *Theoretical Computer Science*, 26:1–17, 1980.
8. J. K. Lenstra and A. H. G. Rinnooy Kan. Complexity of scheduling under precedence constraints. *Operations Research*, 26:22–35, 1978.
9. J. W. S. Lui and C. L. Lui. Bounds on scheduling algorithms for heterogeneous computing systems. In J. L. Rosenfeld, editor, *Information Processing 74*, pages 349–353. North-Holland, 1974.
10. D. Shmoys, J. Wein, and D. Williamson. Scheduling parallel machines on-line. *SIAM Journal on Computing*, 24:1313–31, 1995.

On the Relationship Between Combinatorial and LP-Based Approaches to NP-Hard Scheduling Problems

R. N. Uma [*] and Joel Wein [**]

Department of Computer Science, Polytechnic University
Brooklyn, NY 11201, USA
ruma@@tiger.poly.edu, wein@@mem.poly.edu

Abstract. Enumerative approaches, such as branch-and-bound, to solving optimization problems require a subroutine that produces a lower bound on the value of the optimal solution. In the domain of scheduling problems the requisite lower bound has typically been derived from either the solution to a linear-programming relaxation of the problem or the solution of a combinatorial relaxation. In this paper we investigate, from both a theoretical and practical perspective, the relationship between several linear-programming based lower bounds and combinatorial lower bounds for two scheduling problems in which the goal is to minimize the average weighted completion time of the jobs scheduled.

We establish a number of facts about the relationship between these different sorts of lower bounds, including the equivalence of certain linear-programming-based lower bounds for both of these problems to combinatorial lower bounds used in successful branch-and-bound algorithms. As a result we obtain the first worst-case analysis of the quality of the lower bound delivered by these combinatorial relaxations.

We then give an empirical evaluation of the strength of the various lower bounds and heuristics. This extends and puts in a broader context a recent experimental evaluation by Savelsbergh and the authors of the empirical strength of both heuristics and lower bounds based on different LP-relaxations of a single-machine scheduling problem. We observe that on most kinds of synthetic data used in experimental studies a simple heuristic, used in successful combinatorial branch-and-bound algorithms for the problem, outperforms on average all of the LP-based heuristics. However, we identify other classes of problems on which the LP-based heuristics are superior, and report on experiments that give a qualitative sense of the range of dominance of each. Finally, we consider the impact of local improvement on the solutions.

---

[*] Research partially supported by NSF Grant CCR-9626831.
[**] Research partially supported by NSF Grant CCR-9626831 and a grant from the New York State Science and Technology Foundation, through its Center for Advanced Technology in Telecommunications.

R. E. Bixby, E. A. Boyd, and R. Z. Ríos-Mercado (Eds.): IPCO VI
LNCS 1412, pp. 394–408, 1998. © Springer–Verlag Berlin Heidelberg 1998

1   Introduction

A well-studied approach to the exact solution of NP-hard scheduling problems may be called *enumerative methods*, in which (implicitly) every possible solution to an instance is considered in an ordered fashion. An example of these methods is branch-and-bound, which uses upper and lower bounds on the value of the optimal solution to cut down the search space to a (potentially) computationally tractable size. Such methods are typically most effective when the subroutines used to calculate both the upper and lower bounds are fast and yield strong bounds, hence quickly eliminating much of the search space from consideration.

Although there are a wealth of approaches to designing the lower-bounding subroutines, we can identify two that have been particularly prominent. The first relies on a linear-programming relaxation of the problem, which itself is often derived from an integer linear-programming formulation by relaxing the integrality constraints; Queyranne and Schulz give an extensive survey of this approach [15]. The second relies on what we will call a *combinatorial relaxation* of the problem and yields what we will call a *combinatorial lower bound*. By this we simply mean that the lower bound is produced by exploiting some understanding of the structure of the problem as opposed to by solving a mathematical program. For example, in this paper we focus on combinatorial lower bounds that are obtained by relaxing the constraint (in a nonpreemptive scheduling problem) that the entire job must be processed in an uninterrupted fashion.

Another approach to an NP-hard scheduling problem is to develop an *approximation algorithm*. Here the goal is to design an algorithm that runs in polynomial time and produces a near-optimal solution of some guaranteed quality. Specifically, we define a ρ-approximation algorithm to be an algorithm that runs in polynomial time and delivers a solution of value at most ρ times optimal; see [9] for a survey. In contrast, an enumerative approach attempts to solve a (usually small) problem to optimality, with no guarantee that the solution will be obtained in time polynomial in the size of the input.

Recently various researchers have been successful in creating new connections between linear-programming relaxations used to give lower bounds for certain scheduling problems and the design of approximation algorithms. Specifically, they have used these relaxations to develop approximation algorithms with small-constant-factor worst-case performance guarantees; as a by-product one obtains worst-case bounds on the quality of the lower bound delivered by these relaxations [14,10,7,18,17]. We define a ρ-*relaxation* of a problem to be a relaxation that yields a lower bound that is always within a factor of ρ of the optimal solution.

In this paper we establish additional connections between different approaches to these problems. We consider two NP-hard scheduling problems in which the goal is to minimize the average weighted completion time of the jobs scheduled: $1|r_j|\sum w_j C_j$, the problem of scheduling n jobs with release dates on a single machine, and $P||\sum w_j C_j$, the problem of scheduling n jobs on identical parallel processors. For each problem we show that a combinatorial lower bound that was used successfully in a branch-and-bound code for the problem is equivalent to the solution of a linear-programming relaxation that had been used in the design of approximation algorithms. As a consequence we give the first worst-case analysis of these sorts of combinatorial lower bounds. We also consider several related lower bounds and establish a number of facts about their relative strength. Finally, for $1|r_j| \sum w_j C_j$, we give an empirical evaluation of the relative performance of these different lower bounds, and compare the performance of the approximation algorithms based on the LP-relaxations with the heuristics used in the successful branch-and-bound code of Belouadah, Posner and Potts [2].

Brief Discussion of Previous Related Work: We begin with $1|r_j| \sum w_j C_j$. Dyer and Wolsey considered several linear-programming relaxations of this problem as a tool for producing strong lower bounds [5]. Among those considered were two time-indexed linear programming relaxations, in which the linear program contains a variable for every job at every point in time. In the first relaxation $\{0,1\}$-variables y_{jt} determine whether job j is processed during time t, whereas in a second stronger relaxation $\{0,1\}$-variables x_{jt} determine whether job j completes at time t.

Although both linear programs are of exponential size, Dyer and Wolsey showed that the y_{jt}-LP is a transportation problem with a very special structure and thus can be solved in $O(n \log n)$ time. The x_{jt}-LP, which has been observed empirically to give strong lower bounds [19,22], is very difficult to solve due to its size. Van den Akker, Hurkens and Savelsbergh [21] developed a column-generation approach to solving these linear programs that made feasible the solution of instances with up to 50 jobs with processing times in the range of [0..30].

Inspired by the empirical strength of this relaxation, Hall, Schulz, Shmoys and Wein [10] gave a 3-approximation algorithm for $1|r_j| \sum w_j C_j$ based on time-indexed linear programs. Their approximation algorithm in fact relies only on the weaker y_{jt}-relaxation and simultaneously proves that the y_{jt}-LP (and hence the stronger x_{jt}-LP) are 3-relaxations of the problem. Subsequent papers gave improved techniques with better constant performance guarantees [7,8]. In a recent empirical study, Savelsbergh, Uma and Wein [16] demonstrated that the y_{jt} bound on many instances comes within a few percent of the x_{jt} bound at a greatly reduced computational cost. They also showed that the ideas that led to improved approximation algorithms also yielded improved empirical performance; the best heuristics gave rather good results on most sorts of randomly generated synthetic data.

In parallel with work on linear-programming lower bounds for $1|r_j| \sum w_j C_j$ there has been significant work on branch-and-bound algorithms for $1|r_j| \sum w_j C_j$ based on combinatorial lower bounds [2,3,4,12]. The most successful of these is due to Belouadah, Posner and Potts [2] who made use of two combinatorial lower bounds based on *job splitting*, and an upper bound based on a simple greedy heuristic.

Although it is difficult to compare the efficacy of the branch-and-bound code of Belouadah, Posner and Potts with the branch-and-cut code due to Van den

Akker et. al. based on x_{jt}-relaxations [20] (since they were developed several years apart in different programming languages on different architectures, etc.) the evidence seems to be that neither much dominates the other; however, that of Belouadah, Posner and Potts seems to have been somewhat stronger, as they were able to solve to optimality problems of size 50 whereas van den Akker et. al solved to optimality problems of size 30. The enhanced strength of the lower bounds due to the x_{jt}-relaxations does not make up for the amount of time it takes to solve them.

Discussion of Results: This paper was born out of an interest to make more precise, from both an analytical and empirical perspective, the comparison between the LP-based techniques and the techniques associated with the best combinatorial branch-and-bound algorithm. In this process several potentially surprising relationships between these two approaches arose. Specifically, we show that the solution delivered by the y_{jt}-based relaxation for $1|r_j|\sum w_j C_j$ is identical to that used to deliver the weaker of the two lower bounds used by Belouadah, Posner and Potts. We also show that the stronger of their two lower bounds, while empirically usually weaker than the x_{jt}-based relaxation, neither always dominates that lower bound nor is it dominated by it. A corollary of this observation is that the optimal preemptive schedule for an instance of $1|r_j|\sum w_j C_j$ neither always dominates nor is dominated by the solution to the x_{jt}-relaxation.

We then establish a similar relationship for a different problem. Webster [23] gave a series of lower bounds for $P||\sum w_j C_j$ that are based on a similar notion to the job-splitting approach of Belouadah, Posner and Potts. We show that the weakest of his lower bounds (which in fact was originally proposed by Eastman, Even and Isaacs in 1964 [6]) is equivalent to a generalization of the y_{jt} relaxation to parallel machines that was also used to give approximation algorithms for $P|r_j|\sum w_j C_j$ by Schulz and Skutella [18,17].

We then give an empirical evaluation of the quality of the different lower bounds and associated heuristics for $1|r_j|\sum w_j C_j$; this extends and puts in a broader context a recent experimental study of Savelsbergh and the authors [16]. We demonstrate that the stronger lower bound considered by Belouadah, Posner and Potts on synthetic data sets (which can be computed in $O(n^2)$ time) on average improves on the y_{jt} lower bound by a few percent, and that heuristics based on this relaxation improve by a few percent as well. However, we demonstrate that on most of the synthetic data sets we consider, the simple greedy heuristic used by Belouadah, Posner and Potts is superior to all of the heuristics based on the approximation algorithms associated with the different LP relaxations. It is only on data sets that were specifically designed to be difficult that the LP-based heuristics outperform the simple greedy approach.

Finally, we note that simple local-improvement techniques are often very successful in giving good solutions to scheduling problems [1]; we therefore consider the impact of some simple local improvement techniques when applied both "from scratch" and to the solutions yielded by the various heuristics that we consider.

2   Background

In this section we briefly review the relevant lower bounds and algorithms. In both problems we consider we have n jobs j, $j = 1, \ldots, n$, each with positive processing time p_j and nonnegative weight w_j. For $1|r_j| \sum w_j C_j$ with each job is associated a *release date* r_j before which it is not available for processing.

LP-Relaxations: We begin with the two relevant linear-programming relaxations of $1|r_j| \sum w_j C_j$. As mentioned earlier, Dyer and Wolsey introduced several integer linear programming formulations of the problem. We focus on two. In the first, with variables y_{jt} and completion-time variables C_j, $y_{jt} = 1$ if job j is being processed in the time period $[t, t + 1]$ and $y_{jt} = 0$ otherwise.

$$\text{minimize } \sum_{j=1}^{n} w_j C_j$$

$$
\begin{aligned}
\text{subject to} \quad & \sum_{j=1}^{n} y_{jt} \leq 1, & t = 1, \ldots, T; \\
& \sum_{t=1}^{T} y_{jt} = p_j, & j = 1, \ldots, n; \\
C_j = \tfrac{p_j}{2} + & \tfrac{1}{p_j} \sum_{t=1}^{T} (t + \tfrac{1}{2}) y_{jt}, & j = 1, \ldots, n; \\
& y_{jt} \geq 0, & j = 1, \ldots, n, \ t = r_j, \ldots, T.
\end{aligned}
$$

As noted earlier, this linear program is a valid relaxation of the optimal preemptive schedule as well [10], and can be solved in $O(n \log n)$ time [5]. The structure of the solution is in fact quite simple: at any point in time, schedule the available unfinished job with maximum w_j/p_j (this may involve preemption).

In the second linear-program, which is much harder to solve, the binary variable x_{jt} for each job j ($j = 1, \ldots, n$) and time period t ($t = p_j, \ldots, T$), where T is an upper bound on schedule makespan, indicates whether job j *completes* in period t ($x_{jt} = 1$) or not ($x_{jt} = 0$). This relaxation is stronger than the y_{jt} relaxation; in particular it is *not* a valid relaxation of the optimal preemptive schedule, and its integer solutions yield only nonpreemptive schedules.

$$\text{minimize } \sum_{j=1}^{n} \sum_{t=p_j}^{T} c_{jt} x_{jt}$$

$$
\begin{aligned}
\text{subject to} \quad & \sum_{t=p_j}^{T} x_{jt} = 1, & j = 1, \ldots, n; & \quad (1) \\
& \sum_{j=1}^{n} \sum_{s=t}^{t+p_j-1} x_{js} \leq 1, & t = 1, \ldots, T; & \quad (2) \\
& x_{jt} \geq 0, & j = 1, \ldots, n, \ t = p_j, \ldots, T.
\end{aligned}
$$

The assignment constraints (1) state that each job has to be completed exactly once, and the capacity constraints (2) state that the machine can process at most one job during any time period.

Job Splitting: The lower bounds of Belouadah, Posner and Potts are based on *job splitting*. This technique is based on the idea that a relaxation of a nonpreemptive scheduling problem may be obtained by splitting each job into smaller pieces that can be scheduled individually. If the objective function is $\sum w_j C_j$, when we split job j into pieces we also must split its weight w_j among the pieces as well. In essence, we create a number of smaller jobs. If we split the jobs in such a way that we can solve the resulting relaxed problem in polynomial time, we obtain a polynomial-time computable lower bound on the optimal solution to the original problem.

Note that an inherent upper bound on the quality of any lower bound achieved by this process is the value of the optimal preemptive schedule, as preemptive schedules correspond to splits in which all of the weight of the job j is assigned to the last piece to be scheduled, and such a weight assignment gives the strongest possible lower bound for a specified split of the job into pieces. Note however, that the preemptive version of $1|r_j| \sum w_j C_j$ is also NP-hard and we must settle for solving something weaker that can be computed in polynomial time.

Belouadah, Posner and Potts give two lower bounds (BPP1 and BPP2) based on job splitting. In BPP1, the pieces of job j are *exactly* those that arise in the optimal preemptive solution to the y_{jt} relaxation, and the optimal solution to the resulting split problem has the same structure as that of optimal preemptive solution to the y_{jt}-relaxation. In this lower bound each piece of job j receives a fraction of weight w_j in exact proportion to the fraction of the size of p_j that its size is. In the BPP2 lower bound as much weight as possible is shifted to later scheduled pieces of the job while maintaining the property that the structure of the optimal solution can be computed in polynomial time.

Approximation Algorithms: Recent progress on approximation algorithms for $1|r_j| \sum w_j C_j$ is based on solving the linear relaxation of either of the aforementioned LPs and then inferring an ordering from the solution [14,10,7,8]. In this abstract we focus only on a few of these heuristics that were demonstrated in [16] to be the most useful computationally.

In constructing an ordering we make use of the notion of an α-point [14,11]. We define the α-point of job j, $0 \le \alpha \le 1$, to be the first point in time, in the solution to a time-indexed relaxation, at which an α fraction of job j has been completed. We define the algorithm Schedule-by-Fixed-α, that can be applied to the solution of either LP relaxation, as ordering the jobs by their α-points and scheduling in that order. Goemans [7] has shown that for appropriate choice of α this is a $(\sqrt{2}+1)$-approximation algorithm. Goemans also showed that by choosing α randomly according to a uniform distribution and then scheduling in this order, one obtains a randomized 2-approximation algorithm [7] and if one chooses using a different distribution, a 1.7451-approximation algorithm [8]. Either randomized algorithm can be derandomized by considering n different values of α, scheduling according to each of them, and then choosing the best. We call this algorithm Best-α, or, if only consider k equally-spaced values of α, we call the algorithm k-Best-α. We note that it is simple to adapt these algorithms to be applied to combinatorial relaxations of $1|r_j| \sum w_j C_j$ as well.

3   Analytical Evaluation of Strength of Different Bounds

One Machine: We first introduce some notation to describe the two BPP lower bounds. We say job l is "*better*" than job j if $\frac{p_l}{w_l} < \frac{p_j}{w_j}$, or, equivalently, if $\frac{w_l}{p_l} > \frac{w_j}{p_j}$. For both bounds jobs are split as follows. When a *better* job arrives, we split the currently executing job into two pieces such that one piece completes at the arrival time of the new job and the the second piece is considered for scheduling later. When a job is split into pieces, its weight is also split. So if job j is split into k *pieces*, then each piece i has a processing time p_j^i, a weight w_j^i and release date r_j, such that $\sum_{i=1}^{k} p_j^i = p_j$ and $\sum_{i=1}^{k} w_j^i = w_j$.

For the BPP1 bound, the weights are assigned to the pieces of a job such that $\frac{w_j^i}{p_j^i} = \frac{w_j}{p_j}$ for all $i = 1, \ldots k$, whereas for BPP2 the weights are assigned to the pieces in a greedy fashion with the aim of maximizing the lower bound; in [2] they have shown that BPP2 is always greater than or equal to BPP1.

Let us denote the lower bound given by BPP1 as LB^{BPP1} and the lower bound given by the y_{jt} LP relaxation as $LB^{y_{jt}}$.

Theorem 1. $LB^{y_{jt}} = LB^{BPP1}$.

Proof. As a preliminary we establish a bit more understanding of BPP1. Let P be the original instance with no split jobs. Let P_1 denote the corresponding problem where one job, say j (with processing time p_j, weight w_j and release date r_j), is split into k pieces. Say each piece i is of length p_j^i and is assigned a weight w_j^i for $i = 1, \ldots, k$ such that $\sum_{i=1}^{k} p_j^i = p_j$ and $\sum_{i=1}^{k} w_j^i = w_j$. These k pieces are constrained to be scheduled contiguously in P_1. The set of jobs in P_1 are the k pieces of this split job j plus the remaining jobs in P. Obviously there is a one-to-one correspondence between the feasible schedules for P and P_1. Note that in P_1, the pieces of job j will be scheduled exactly during the interval P schedules job j. Therefore, it is sufficient to consider the contribution of this one job to the weighted completion time in both the schedules. Let the pieces of job j start at times $t_j^1, \ldots, t_j^k$ respectively. So in P, this job is scheduled during $[t_j^1, t_j^1 + p_j]$ and it contributes $w_j * (t_j^1 + p_j)$.

The following equality was shown by Belouadah, Posner and Potts in [2].

$$w_j * (t_j^1 + p_j) = \sum_{i=1}^{k} w_j^i C_j^i + CBRK_j \tag{3}$$

where C_j^i denotes the completion time of piece i of job j and

$$CBRK_j = \sum_{i=1}^{k-1} w_j^i \sum_{h=i+1}^{k} p_j^h$$

can be thought of as the *cost of breaking* job j into k pieces.

If we remove the constraint from P_1 that the pieces of the job have to be scheduled contiguously, then the resulting problem P_2 gives a lower bound on the cost of P_1. This is true even if more than one job is split. Therefore, the idea is to split the jobs so that the optimal schedule for the resulting problem P_2 can be computed easily. So $BPP1$ (and $BPP2$) split the jobs accordingly.

We now turn to proving our theorem. Consider the (weighted) contribution of just one job, say j, to the respective lower bounds (denoted LB_j^{BPP1} and

$LB_j^{y_{jt}}$). Let job j be released at r_j with a processing time requirement of p_j and weight w_j. Let this job be split into k pieces of lengths $p_j^1, \ldots, p_j^k$ starting at times $t_j^1, \ldots, t_j^k$, respectively. So we have $p_j^1 + p_j^2 + \cdots + p_j^k = p_j$. BPP1 would assign weights $w_j^i = \frac{p_j^i}{p_j} w_j$ for $i = 1, \ldots, k$. The cost of breaking job j is given by,

$$CBRK_j = \sum_{i=1}^{k-1} w_j^i \sum_{h=i+1}^{k} p_j^h$$

$$= \sum_{i=1}^{k-1} \frac{w_j}{p_j} \cdot p_j^i \sum_{h=i+1}^{k} p_j^h$$

$$= \frac{w_j}{p_j} \cdot \frac{1}{2} ((\sum_{i=1}^{k} p_j^i)^2 - \sum_{i=1}^{k} (p_j^i)^2)$$

$$= \frac{1}{2} w_j p_j - \frac{1}{2} \sum_{i=1}^{k} w_j^i p_j^i$$

Now the contribution of job j to the $BPP1$ lower bound is,

$$LB_j^{BPP1} = \sum_{i=1}^{k} w_j^i (t_j^i + p_j^i) + CBRK_j$$

$$= \sum_{i=1}^{k} w_j^i (t_j^i + p_j^i) + \frac{1}{2} w_j p_j - \frac{1}{2} \sum_{i=1}^{k} w_j^i p_j^i$$

and the contribution of job j to the y_{jt} lower bound is,

$$LB_j^{y_{jt}} = \frac{w_j p_j}{2} + \frac{w_j}{p_j} \sum_t y_{jt} (t + \frac{1}{2})$$

$$= \frac{w_j p_j}{2} + \frac{w_j}{p_j} \sum_{i=1}^{k} ((t_j^i + \frac{1}{2}) (t_j^i + 1 + \frac{1}{2}) + \cdots + (t_j^i + (p_j^i - 1) + \frac{1}{2}))$$

$$= \frac{w_j p_j}{2} + \frac{w_j}{p_j} \sum_{i=1}^{k} (p_j^i t_j^i + (p_j^i)^2 - \frac{(p_j^i)^2}{2})$$

$$= \sum_{i=1}^{k} (\frac{p_j^i}{p_j} w_j) (t_j^i + p_j^i) + \frac{w_j p_j}{2} - \frac{w_j}{2 p_j} \sum_{i=1}^{k} (p_j^i)^2$$

$$= LB_j^{BPP1}$$

Summing over all jobs j we have the required result.   □

As an immediate corollary we obtain an upper bound on the quality of lower bound provided by both BPP1 and BPP2. Goemans et al. [8] proved that the y_{jt}-relaxation is a 1.685-relaxation of $1|r_j| \sum w_j C_j$; thus we see that BPP1 and BPP2 are as well. We now turn to the relationship with the x_{jt} relaxation; it is known that this is stronger than the y_{jt} [5].

Theorem 2. *The lower bound given by BPP2 neither always dominates nor is dominated by the x_{jt}-lower bound.*

The proof is by exhibiting two instances; one on which BPP2 is better and one on which x_{jt} is better. Due to space constraints the instances are omitted in this extended abstract.

Parallel Identical Machines: In [23], Webster considers the problem $P||\sum w_j C_j$ He gives a series of progressively stronger lower bounds for this problem all of which are based on ideas similar to job-splitting; these lower bounds lead to a successful branch-and-bound algorithm. The weakest of these bounds is actually due to a 1964 paper by Eastman, Even and Isaacs [6]. By an argument of a similar flavor to that of the one machine problem we can show that in fact the weakest bound is equivalent to an extension of the y_{jt}-relaxation to parallel machines. In this extension the m machines are conceptualized as one machine of speed m. Schulz and Skutella show that this formulation is a 2-relaxation of $P|r_j|\sum w_j C_j$ and a 3/2-relaxation of $P||\sum w_j C_j$. Williamson [24] had also, independently, observed this equivalence between the lower bounds due to Eastman, Even and Isaacs and the y_{jt}-relaxation of $P||\sum w_j C_j$. By establishing this equivalence we obtain a worst-case upper bound on the performance of all of Webster's lower bounds. In fact, he considers a more general problem in which each processor can have non-trivial ready times – our results extend to this case as well. The details are omitted due to the space constraints of the extended abstract.

4   Empirical Evaluation

4.1   Experimental Design

In [16] a comprehensive experimental evaluation is given of the relative strength of the y_{jt} (BPP1) and x_{jt} LP relaxations and approximation algorithms based on them. Our goals in this section are threefold. (1) To quantify experimentally the relative strength of the BPP2 lower bound, which is no weaker than the BPP1 bound but, in contrast to the x_{jt}-lower bound, can be computed efficiently, and of the approximation algorithms based on this stronger lower bound. (2) To compare the performance of the simple WSPT heuristic used by Belouadah et al. in their branch-and-bound algorithm with the LP-based approximation algorithms. (3) To understand the impact of local improvement on solutions to the problem.

We sought to generate a rich enough set of instances to observe the fullest possible range of algorithm and lower bound performance. Our experimental design follows that both of Hariri and Potts [12] and of [16]. We worked with three sets of data. The first, which we call **OPT**, was a set of 60 instances with $n = 30$ jobs, with $w_j \in [1..10]$ and $p_j \in [1..5]$ in 20 instances and $p_j \in [1..10]$ in 40 instances. For these instances we knew the optimal solutions from the branch-and-cut code of [20].

We then generated a large **Synthetic** data set according to a number of parameters: problem size (n), distribution for random generation of the weights w_j, distribution for the random generation of the p_j, and arrival process. The problem size n was chosen from $\{50, 100, 200, 500\}$. The release dates were generated either by a Poisson process to model that, on average 2, 5, 10 or 20 jobs arrived every 10 units of time, (where 10 is the maximum job size) or uniformly in $[0, \sum p_j]$. Three distributions were used for the generation of each of the w_j and p_j: (i) uniform in $[1, 10]$ (ii) normal with a mean of 5 and a standard deviation of 2.5, (iii) a two hump distribution, where, with probability 0.5 the number is chosen from a normal distribution with mean 2.5 and standard deviation 0.5, and with probability 0.5 the number is chosen from a normal distribution with mean 7.5 and standard deviation 0.5. Ten instances were generated randomly for each combination of parameters, for a total of 1800 instances. For all of these we computed the BPP1 and BPP2 relaxations, and ran on all of these relaxations the heuristics discussed in Section 2. Solutions to the x_{jt} relaxation were available for some of the instances from our work in [16].

Finally, as in [16] we considered the **Hard** data sets that were designed to provoke poor performance from the LP-based heuristics. All of these instances attempt to exploit the fact that the y_{jt} relaxation is also a valid relaxation of the optimal preemptive schedule [10], and that therefore on instances for which the optimal preemptive schedule is much better than the optimal nonpreemptive schedule, the y_{jt}-based relaxation should perform poorly. Therefore these instances have one or several very large jobs, and a large number of tiny jobs that are released regularly at small intervals. In the **Hard1** data set we created instances with processing times 1 and 20 or 30. The size 1 jobs were generated on average 9 times more frequently than the size 20 or size 30 jobs. In the **Hard2** data set we generated job sizes in a two-hump distribution, with the humps at 5 and 85; the size 5 jobs were generated on average 9 times more frequently.

We note that all problem instances are available at http://ebbets.poly.edu/SCHED/onerj.html.

Due to space considerations in this extended abstract we focus on a few data sets which turn out to be among the most difficult in their categories for the heuristics and that are at the same time representative of the overall behavior of that set. Specifically, we focus on the **OPT**, **Hard1** and **Hard2**, **Synthetic1** and **Synthetic2**. The **Synthetic1** set corresponds to the **Synthetic** set where release dates are generated by a Poisson process with arrival rates 2 and 5 ($n = 50, 100$) and the **Synthetic2** set corresponds to the release dates being generated uniformly ($n = 50, 100, 200, 500$). We also report results with respect to the average weighted flow time $= \sum w_j(C_j - r_j)$; while equivalent to average weighted completion time at optimality this criterion is provably much harder to approximate [13] and thus more interesting.

4.2   Lower Bounds

Table 1 reports on the relative performance of the different lower bounds. We note that the BPP2 lower bound does provide some improvement over the BPP1 bound at modest additional computational cost ($O(n \log n)$ to $O(n^2)$), but since both are relaxations of the optimal preemptive schedule, on the **Hard1** data set the BPP2 bound is still far from the x_{jt}-lower bound. (Do not be misled by the small numbers recorded for **Hard2**. They are with respect to the best lower bound we know for that instance, which is BPP2.) Furthermore, note that on the **Synthetic1** data set sometimes BPP2 is better than the x_{jt}-relaxation, and this explains why none of the numbers in that column is 0 – the comparison is always with respect to the best lower bound for that instance. We note that the maximum improvement observed by BPP2 over BPP1 on any instance was 9.492%, on an instance in the **Hard1** data set.

Table 1. Quality of the lower bounds with respect to the weighted flow time given by y_{jt}, $BPP2$ and x_{jt} relaxations. We report on $\frac{(BEST-LB)}{LB} \times 100$, where $BEST$ is the best available lower bound and LB is the corresponding lower bound. The values reported are averaged over all the instances in each case. Results for the y_{jt} and x_{jt} relaxations are from the experimental work of Savelsbergh et al.

	OPT	Hard1	Hard2	Synthetic1	Synthetic2
y_{jt}(BPP1)	4.506	19.169	0.848	1.729	1.848
BPP2	2.969	15.539	0.000	0.941	0.000
x_{jt}	1.322	0.000	N/A	0.013	N/A

4.3 Upper Bounds

In [16] we demonstrated that the various heuristics based on the LP-relaxations in general yield improved performance with improved quality of relaxation, although there are many instances for which this is not the case. Therefore it is of interest to study the impact of the BPP2 lower bounds on heuristic performance.

Of greater interest is to compare the performance of all this LP "stuff" to the simple and naive heuristic used by Belouadah et al. in their combinatorial branch and bound code. This heuristic, WSPT, simply, when idle, selects for processing the unprocessed available job with the largest w_j/p_j ratio and schedules it nonpreemptively. It is trivial to see that the worst-case performance of this heuristic is unbounded. Table 2 provides a summary of the performance of the various heuristics.

The Table demonstrates that the heuristics based on the BPP2 bound do in fact do better by up to a few percent on the hard data sets. Most striking, however, is that except on the **Hard1** data set the WSPT heuristic is superior to all the LP relaxation-based approaches. Across the entire **Synthetic** data sets WSPT yields a better solution than the LP-approaches 70 − 80% of the time; on specific instances of the **Hard** data set it is up to 27% better than the best performance of the other heuristics.

Table 2. Performance of the algorithms based on the three $(y_{jt}, BPP2, x_{jt})$ formulations. We report the mean, standard deviation and maximum of the ratio of the performance of the algorithm to the best lower bound. The first three rows of the table report results from the work of Savelsbergh et al.

	Hard1			Synthetic1			Synthetic2		
	Mean	SDev	Max	Mean	SDev	Max	Mean	SDev	Max
Schedule-by-fixed-α (y_{jt})	2.207	0.623	3.906	1.126	0.093	1.686	1.334	0.126	1.892
Best-α (y_{jt})	1.683	0.263	2.164	1.042	0.042	1.282	1.155	0.067	1.371
5-Best-α (y_{jt})	1.775	0.301	2.244	1.048	0.046	1.358	1.171	0.079	1.419
Schedule-by-fixed-α $(BPP2)$	1.985	0.507	3.902	1.116	0.084	1.643	1.292	0.109	1.615
Best-α $(BPP2)$	1.628	0.245	2.122	1.040	0.039	1.233	1.144	0.059	1.311
5-Best-α $(BPP2)$	1.710	0.274	2.296	1.046	0.043	1.233	1.157	0.067	1.356
Best-α (x_{jt})	1.854	0.320	2.428	1.048	0.045	1.303	-	-	-
WSPT	1.874	0.372	3.015	1.033	0.032	1.148	1.121	0.048	1.307

In some sense this performance is quite surprising, but in another it is not very surprising at all. WSPT, while lacking any worst-case performance guarantee, is a natural choice that should work well "most of the time". It is likely that when one generates synthetic data one is generating instances that one would think of "much of the time". It is on the harder sorts of instances that the potential problems with such a heuristic arise.

Although it would not be difficult to generate instances that would yield horrendous performance for WSPT the purpose of an experimental study is to yield insight into algorithm performance on instances that may be somewhat natural. Towards this end we generated a spectrum of problems in the framework of "a few jobs of size P and lots of jobs of size 1", with P ranging from 1 to 1000. The results of the experiment, plotted in Figure 1, gives a qualitative sense of the range in which WSPT is competitive with algorithms with worst-case performance guarantee, which is essentially up to $P = 100$ or so.

Finally, we experimented with the impact of local improvement, which can be a very powerful algorithmic technique for certain scheduling problems. Specifically, given a solution we considered all pairs and all triples of jobs, switching any that led to an improved solution and iterating until no improving switch existed. We applied these to the solutions of all the heuristics. For the heuristics that considered several possible schedules we applied local improvement to all of them and took the best resulting schedule. The results, reported in Table 3, indicate that the combination of local improvement applied to a collection of high-quality schedules yield the best results. We note further that applying this local improvement from scratch (random orderings) was not competitive with its application to a good initial schedule; the latter approach on average yielded solutions of quality 40-50% better. The importance of the quality of the initial schedule for local improvement was also observed by Savelsbergh et al [16] in the case of a real resource constrained scheduling problem.

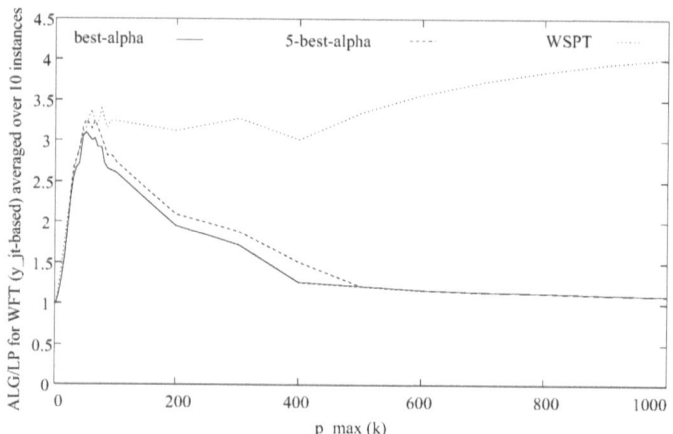

Fig. 1. Plot of $\frac{ALG}{LP}$ ratios for $\sum w_j F_j$ for Best-α, 5-Best-α and WSPT. Each data point is an average of 10 instances. On the x-axis we plot the maximum job size p_{max}. $p_{max} = k$ means that the processing time was generated so that $p = 1$ with a probability of 0.9 and $p = k$ with a probability of 0.1. The arrival times were generated uniformly in the range $[0..500]$

5   Conclusions

Enthusiasts of worst-case analysis may be disappointed by the relatively poor showing of the approximation algorithms with respect to the simple WSPT heuristic. They may be consoled by their advantages on certain classes of difficult problems and their supremacy when coupled with local improvement. We suggest, however, that although it is interesting to compare these approaches on simple models, perhaps the most important role for worst-case analysis is as a tool that forces the algorithm designer to have new ideas that may also be useful in other more complex problems[1]. Savelsbergh et al. showed that the idea of using many α values led to improved solutions for actual complex manufacturing scheduling problems; for these problems it is unlikely that there exist simple heuristics that will perform as well in such settings.

Finally, in this study we have touched on several approaches to scheduling problems: linear programming, approximation, combinatorial relaxations and branch-and-bound, and local improvement, and made a modest contribution to our further understanding of their relationship. We feel that it is an important direction to continue to try to understand in a unified fashion the different roles that these elements can play in the theory and practice of scheduling.

Acknowledgements. We are grateful to Jan Karel Lenstra, Maurice Queyranne, David Shmoys and Chris Potts for helpful discussions. The second author would

---

[1] We are grateful to David Shmoys for suggesting this idea.

Table 3. Performance of the algorithms based on the three (y_{jt}, $BPP2$, x_{jt}) formulations after some local improvement. We report the mean, standard deviation and maximum of the ratio of the performance of the algorithm to the best lower bound.

	Hard1			Synthetic1			Synthetic2		
	Mean	SDev	Max	Mean	SDev	Max	Mean	SDev	Max
Schedule-by-fixed-α (y_{jt})	1.594	0.264	2.137	1.088	0.061	1.330	1.200	0.073	1.438
20-Best-α (y_{jt})	1.402	0.148	1.679	1.022	0.022	1.138	1.096	0.035	1.196
5-Best-α (y_{jt})	1.420	0.164	1.770	1.024	0.023	1.140	1.100	0.037	1.213
Schedule-by-fixed-α ($BPP2$)	1.591	0.270	2.141	1.081	0.058	1.339	1.175	0.066	1.371
20-Best-α ($BPP2$)	1.398	0.133	1.624	1.022	0.022	1.143	1.096	0.035	1.191
5-Best-α ($BPP2$)	1.413	0.142	1.652	1.024	0.023	1.143	1.100	0.036	1.213
20-Best-α (x_{jt})	1.429	0.197	1.871	1.021	0.022	1.125	-	-	-
WSPT	1.746	0.284	2.321	1.031	0.030	1.147	1.110	0.042	1.241

also like to thank Andreas Schulz for inviting him to ISMP '97, where the idea for this paper arose in discussions with Chris Potts. Some of the results in Section 3 were independently obtained by David Williamson and a joint journal paper is forthcoming.

References

1. E. J. Anderson, C. A. Glass, and C. N. Potts. Machine scheduling. In E. Aarts and J. K. Lenstra, editors, *Local Search in Combinatorial Optimization*. Wiley Press, 1997.
2. H. Belouadah, M. E. Posner, and C. N. Potts. Scheduling with release dates on a single machine to minimize total weighted completion time. *Discrete Applied Mathematics*, 36:213–231, 1992.
3. L. Bianco and S. Ricciardelli. Scheduling of a single machine to minimize total weighted completion time subject to release dates. *Naval Research Logistics Quarterly*, 29:151–167, 1982.
4. M. I. Dessouky and J. S. Deogun. Sequencing jobs with unequal ready times to minimize mean flow time. *SIAM Journal on Computing*, 10:192–202, 1981.
5. M. E. Dyer and L. A. Wolsey. Formulating the single machine sequencing problem with release dates as a mixed integer program. *Discrete Applied Mathematics*, 26:255–270, 1990.
6. W. L. Eastman, S. Even, and I. M. Isaacs. Bounds for the optimal scheduling of n jobs on m processors. *Management Science*, 11(2):268–279, 1964.
7. M. Goemans. Improved approximation algorithms for scheduling with release dates. In *Proceedings of the 8th ACM-SIAM Symposium on Discrete Algorithms*, pages 591–598, 1997.
8. M. Goemans, M. Queyranne, A. Schulz, M. Skutella, and Y. Wang. Single machine scheduling with release dates. Preprint, 1997.
9. L. A. Hall. Approximation algorithms for scheduling. In D. S. Hochbaum, editor, *Approximation Algorithms for NP-hard Problems*, pages 1–43. PWS Publishing Company, 1997.

10. L. A. Hall, A. S. Schulz, D. B. Shmoys, and J. Wein. Scheduling to minimize average completion time: Off-line and on-line approximation algorithms. *Mathematics of Operations Research*, (3):513–544, August 1997.

11. L. A. Hall, D. B. Shmoys, and J. Wein. Scheduling to minimize average completion time: Off-line and on-line algorithms. In *Proceedings of the 7th ACM-SIAM Symposium on Discrete Algorithms*, pages 142–151, January 1996.

12. A. M. A. Hariri and C. N. Potts. An algorithm for single machine sequencing with release dates to minimize total weighted completion time. *Discrete Applied Mathematics*, 5:99–109, 1983.

13. H. Kellerer, T. Tautenhahn, and G. J. Woeginger. Approximability and nonapproximability results for minimizing total flow time on a single machine. In *Proceedings of the 28th Annual ACM Symposium on Theory of Computing*, May 1995.

14. C. Phillips, C. Stein, and J. Wein. Scheduling jobs that arrive over time. In *Proceedings of Fourth Workshop on Algorithms and Data Structures, LNCS, Vol. 955*, pages 86–97. Springer-Verlag, Berlin, 1995.

15. M. Queyranne and A. S. Schulz. Polyhedral approaches to machine scheduling. Technical Report 408/1994, Technical University of Berlin, 1994.

16. M. W. P. Savelsbergh, R. N. Uma, and J. Wein. An experimental study of LP-based approximation algorithms for scheduling problems. In *Proceedings of the 9th ACM-SIAM Symposium on Discrete Algorithms*, 1998.

17. A. S. Schulz and M. Skutella. Random-based scheduling: New approximations and LP lower bounds. In J. Rolim, editor, *Randomization and Approximation Techniques in Computer Science – Proceedings of the International Workshop RANDOM'97, LNCS, Vol. 1269*, pages 119–133. Springer, Berlin, 1997.

18. A. S. Schulz and M. Skutella. Scheduling–LPs bear probabilities: Randomized approximations for min–sum criteria. In R. Burkard and G. Woeginger, editors, *Algorithms – Proceedings of the 5th Annual European Symposium on Algorithms (ESA'97), LNCS, Vol. 1284*, pages 416–429. Springer, Berlin, 1997.

19. J. P. De Sousa and L. A. Wolsey. A time-indexed formulation of non-preemptive single-machine scheduling problems. *Mathematical Programming*, 54:353–367, 1992.

20. M. Van den Akker, C. P. M. Van Hoesel, and M. W. P. Savelsbergh. A polyhedral approach to single machine scheduling. *Mathematical Programming*, 1997. To appear.

21. M. Van den Akker, C. A. J. Hurkens, and M. W. P. Savelsbergh. A time-indexed formulation for single-machine scheduling problems: Column generation. 1996. Submitted for publication.

22. M. Van den Akker. *LP-Based Solution Methods for Single-Machine Scheduling Problems*. PhD thesis, Eindhoven University of Technology, Eindhoven, The Netherlands, 1994.

23. S. Webster. New bounds for the identical paralled processor weighted flow time problem. *Management Science*, 38(1):124–136, 1992.

24. D. P. Williamson. Personal communication, 1997.

Polyhedral Combinatorics of Quadratic Assignment Problems with Less Objects than Locations

Volker Kaibel

Institut für Informatik
Universität zu Köln
Pohligstr. 1
50969 Köln, Germany
kaibel@@informatik.uni-koeln.de
http://www.informatik.uni-koeln.de/ls_juenger/kaibel.html

Abstract. For the classical quadratic assignment problem (QAP) that requires n objects to be assigned to n locations (the $n \times n$-case), polyhedral studies have been started in the very recent years by several authors. In this paper, we investigate the variant of the QAP, where the number of locations may exceed the number of objects (the $m \times n$-case). It turns out that one can obtain structural results on the $m \times n$-polytopes by exploiting knowledge on the $n \times n$-case, since the first ones are certain projections of the latter ones. Besides answering the basic questions for the affine hulls, the dimensions, and the trivial facets of the $m \times n$-polytopes, we present a large class of facet defining inequalities. Employed into a cutting plane procedure, these polyhedral results enable us to compute optimal solutions for some hard instances from the QAPLIB for the first time without using branch-and-bound. Moreover, we can calculate for several yet unsolved instances significantly improved lower bounds.

1 Introduction

Let a set of m objects and a set of n locations be given, where $m \leq n$. We will be concerned with the following problem. Given *linear costs* $c_{(i,j)}$ for assigning object i to location j and *quadratic costs* $q_{\{(i,j),(k,l)\}}$ for assigning object i to location j and object k to location l, the task is (in the *non-symmetric case*) to find an assignment, i.e., an injective map $\varphi : \{1, \ldots, m\} \longrightarrow \{1, \ldots, n\}$, that minimizes

$$\sum_{i=1}^{m} \sum_{k=i+1}^{m} q_{\{(i,\varphi(i)),(k,\varphi(k))\}} + \sum_{i=1}^{m} c_{(i,\varphi(i))} \; .$$

R. E. Bixby, E. A. Boyd, and R. Z. Ríos-Mercado (Eds.): IPCO VI
LNCS 1412, pp. 409–422, 1998. © Springer–Verlag Berlin Heidelberg 1998

In the *symmetric case* we have quadratic costs $\hat{q}_{(\{i,k\},\{j,l\})}$ for assigning the two objects i and k anyhow to the two locations j and l, and we have to find an assignment $\varphi : \{1, \ldots, m\} \longrightarrow \{1, \ldots, n\}$ that minimizes

$$\sum_{i=1}^{m} \sum_{k=i+1}^{m} \hat{q}_{(\{i,k\},\{\varphi(i),\varphi(k)\})} + \sum_{i=1}^{m} c_{(i,\varphi(i))} \ .$$

The classical *quadratic assignment problem* (introduced by Koopmans and Beckmann [10]) is a special case of the non-symmetric formulation, where $m = n$ and $d_{\{(i,j),(k,l)\}} = f_{ik}d_{jl} + f_{ki}d_{lj}$ holds for some *flow-matrix* (f_{ik}) and some *distance-matrix* (d_{jl}). If one of the flow- or distance-matrix is symmetric then the problem is also a special case of the symmetric formulation given above.

The QAP is not only from the theoretical point of view a hard one among the classical combinatorial optimization problems (Sahni and Gonzales [14] showed that even ϵ-approximation is $\mathcal{NP}$-hard), but it has also resisted quite well most practical attacks to solve it for larger instances.

Polyhedral approaches to the classical case with $m = n$ (the $n \times n$-case) have been started during the recent years by Padberg and Rijal ([13,11]) as well as Jünger and Kaibel ([4,6,5,8]). In [5] the first large class of facet defining inequalities for the associated polytopes is presented. These inequalities turned out to yield very effective cutting planes that allowed to solve for the first time several instances from the QAPLIB (the commonly used set of test instances compiled by Burkarhd, Karisch, and Rendl [2]) to optimality without using branch-and-bound.

In this paper, we describe a polyhedral approach to the case where the number of objects m might be less than the number of locations n (the $m \times n$-case). We restrict our presentation to the symmetric version. In Sect. 2 a problem formulation in terms of certain hypergraphs is introduced and the associated polytopes are defined. The trivial facets as well as the affine hulls of these polytopes are considered in Sect. 3. In Sect. 4 rather tight relaxation polytopes are presented that are projections of certain relaxation polytopes for the $n \times n$-case. In Sect. 5 we describe a large class of facet defining inequalities for the polytopes that are associated with the (symmetric) $m \times n$-case. Strengthening the relaxations of Sect. 4 by some of these inequalities in a cutting plane procedure, we can improve for some $m \times n$-instances in the QAPLIB the lower bounds significantly. Moreover, we can solve several $m \times n$-instances by a pure cutting plane procedure. Results of these experiments are given in Sect. 6. We conclude with some remarks on promising further directions of polyhedral investigations for the $m \times n$-QAP in Sect. 7.

2   QAP-Polytopes

As indicated in the introduction, we restrict to the symmetric QAP in this paper. Since it provides convenient ways to talk about the problem, we first formulate the symmetric QAP as a problem defined on a certain hypergraph.

Throughout this paper, let $m \leq n$, $\mathcal{M} := \{1, \ldots, m\}$, and $\mathcal{N} := \{1, \ldots, n\}$. We define a hypergraph $\hat{\mathcal{G}}_{m,n} := (\mathcal{V}_{m,n}, \hat{\mathcal{E}}_{m,n})$ on the nodes $\mathcal{V}_{m,n} := \mathcal{M} \times \mathcal{N}$ with hyperedges

$$\hat{\mathcal{E}}_{m,n} := \left\{ \{(i,j), (k,l), (i,l), (k,j)\} \in \binom{\mathcal{V}_{m,n}}{4} : i \neq k, j \neq l \right\}.$$

A hyperedge $\{(i,j), (k,l), (i,l), (k,j)\}$ is denoted by $\langle i, j, k, l \rangle$. The sets $\text{row}_i := \{(i,j) \in \mathcal{V}_{m,n} : j \in \mathcal{N}\}$ and $\text{col}_j := \{(i,j) \in \mathcal{V}_{m,n} : i \in \mathcal{M}\}$ are called the i-th row and the j-th column of $\mathcal{V}_{m,n}$, respectively.

We call a subset $C \subset \mathcal{V}_{m,n}$ of nodes a *clique* of the hypergraph $\hat{\mathcal{G}}_{m,n}$ if it intersects neither any row nor any column more than once. The maximal cliques of $\hat{\mathcal{G}}_{m,n}$ are the m-cliques. The set of hyperedges that is associated with an m-clique $C \subset \mathcal{V}_{m,n}$ of $\hat{\mathcal{G}}_{m,n}$ consists of all hyperedges that share two nodes with C. This set is denoted by $\hat{\mathcal{E}}_{m,n}(C)$. Solving symmetric QAPs then is equivalent to finding minimally node- and hyperedge-weighted m-cliques in $\hat{\mathcal{G}}_{m,n}$.

We denote by $x^{(\cdots)} \in \mathbb{R}^{\mathcal{V}_{m,n}}$ and $z^{(\cdots)} \in \mathbb{R}^{\hat{\mathcal{E}}_{m,n}}$ the characteristic vectors of subsets of $\mathcal{V}_{m,n}$ and $\hat{\mathcal{E}}_{m,n}$, respectively. Thus the following polytope encodes the structure of the symmetric QAP in an adequate fashion (where we simplify $(x^C, z^C) := (x^C, z^{\hat{\mathcal{E}}_{m,n}(C)})$):

$$\mathcal{SQAP}_{m,n} := \text{conv} \left\{ (x^C, z^C) : C \text{ is an } m\text{-clique of } \hat{\mathcal{G}}_{m,n} \right\}$$

The (mixed) integer linear programming formulation in Theorem 1 is quite basic for the polyhedral approach. Let $\Delta_{(k,l)}^{(i,j)}$ be the set of all hyperedges that contain both nodes (i,j) and (k,l). As usual, for any vector $u \in \mathbb{R}^L$ and any subset $L' \subset L$ of indices we denote $u(L') := \sum_{\lambda \in L'} u_\lambda$.

Theorem 1. *Let $1 \leq m \leq n$. A vector $(x,z) \in \mathbb{R}^{\mathcal{V}_{m,n}} \times \mathbb{R}^{\hat{\mathcal{E}}_{m,n}}$ is a vertex of $\mathcal{SQAP}_{m,n}$, i.e., the characteristic vector of an m-clique of $\hat{\mathcal{G}}_{m,n}$, if and only if it satisfies the following conditions:*

$$
\begin{align}
x(\text{row}_i) &= 1 & (i \in \mathcal{M}) \tag{1} \\
x(\text{col}_j) &\leq 1 & (j \in \mathcal{N}) \tag{2} \\
-x_{(i,j)} - x_{(k,j)} + z\left(\Delta_{(k,j)}^{(i,j)} \right) &= 0 & (i, k \in \mathcal{M}, i < k, j \in \mathcal{N}) \tag{3} \\
z_h &\geq 0 & (h \in \hat{\mathcal{E}}_{m,n}) \tag{4} \\
x_v &\in \{0, 1\} & (v \in \mathcal{V}_{m,n}) \tag{5}
\end{align}
$$

Proof. The "only if part" is obvious. To prove the "if part", let $(x, z) \in \mathbb{R}^{\mathcal{V}_{m,n}} \times \mathbb{R}^{\hat{\mathcal{E}}_{m,n}}$ satisfy (1), ... ,(5). Since x is a $0/1$-vector that satisfies (1) and (2) it must be the characteristic vector of some m-clique $C \subset \mathcal{V}_{m,n}$ of $\hat{\mathcal{G}}_{m,n}$. Considering two appropriate equations from (3), one obtains (by the nonnegativity of z) that $z_{\langle i,j,k,l \rangle} > 0$ implies $x_{(i,j)} = x_{(k,l)} = 1$ or $x_{(i,l)} = x_{(k,j)} = 1$. But then, in each of the equations (3) there is at most one hyperedge involved that corresponds to a non-zero component of z. This leads to the fact that $z_{\langle i,j,k,l \rangle} > 0$ implies

$z_{\langle i,j,k,l\rangle} = 1$, and that $x_{(i,j)} = x_{(k,l)} = 1$ implies $z_{\langle i,j,k,l\rangle} = 1$. Hence, z must be the characteristic vector of $\hat{\mathcal{E}}_{m,n}(C)$ □

How is the $n \times n$-case related to the $m \times n$-case? Obviously, $\mathcal{SQAP}_{m,n}$ arises from $\mathcal{SQAP}_{n,n}$ by the canonical orthogonal projection $\hat{\sigma}^{(m,n)} : \mathbb{R}^{\mathcal{V}_{n,n}} \times \mathbb{R}^{\hat{\mathcal{E}}_{n,n}} \longrightarrow \mathbb{R}^{\mathcal{V}_{m,n}} \times \mathbb{R}^{\hat{\mathcal{E}}_{m,n}}$. Let $W_{m,n} = \mathcal{V}_{n,n} \setminus \mathcal{V}_{m,n}$ and $F_{m,n} = \{\langle i,j,k,l\rangle \in \hat{\mathcal{E}}_{n,n} : \langle i,j,k,l\rangle \cap W_{m,n} \neq \emptyset\}$ be the sets of nodes and hyperedges that are "projected out" this way. The following connection is very useful.

Remark 1. If an inequality $(a,b)^T(x,z) \leq \alpha$ defines a facet of $\mathcal{SQAP}_{n,n}$ and $a_v = 0$ holds for all $v \in W_{m,n}$ as well as $b_h = 0$ for all $h \in F_{m,n}$, then the "projected inequality" $(a',b')^T(x',z') \leq \alpha$ with $(a',b') = \hat{\sigma}^{(m,n)}(a,b)$ defines a facet of $\mathcal{SQAP}_{m,n}$.

The following result shows that in order to investigate the $m \times n$-case with $m < n$ it suffices to restrict even to $m \leq n - 2$. In fact, it turns out later that the structures of the polytopes for $m \leq n - 2$ differ a lot from those for $m = n$ or $m = n - 1$.

Theorem 2. *For $n \geq 2$ the canonical orthogonal projection $\hat{\sigma}^{(n,n-1)} : \mathbb{R}^{\mathcal{V}_{n,n}} \times \mathbb{R}^{\hat{\mathcal{E}}_{n,n}} \longrightarrow \mathbb{R}^{\mathcal{V}_{n-1,n}} \times \mathbb{R}^{\hat{\mathcal{E}}_{n-1,n}}$ induces an isomorphism between the polytopes $\mathcal{SQAP}_{n,n}$ and $\mathcal{SQAP}_{n-1,n}$.*

Proof. In the $n \times n$-case, in addition to (1) and (3) the equations

$$x(\text{col}_j) = 1 \quad (j \in \mathcal{N}) \tag{6}$$

$$-x_{(i,j)} - x_{(i,l)} + z\left(\Delta_{(i,l)}^{(i,j)}\right) = 0 \quad (i,j,l \in \mathcal{N}, j < l) \tag{7}$$

are valid for $\mathcal{SQAP}_{n,n}$. Obviously, the columns of these equations that correspond to nodes in $W_{n-1,n}$ or to hyperedges in $F_{n-1,n}$ are linearly independent. This implies the theorem. □

3   The Basic Facial Structures of $\mathcal{SQAP}_{m,n}$

The questions for the affine hull, the dimension, and the trivial facets are answered by the following theorem.

Theorem 3. *Let $3 \leq m \leq n - 2$.*

(i)   The affine hull of $\mathcal{SQAP}_{m,n}$ is precisely the solution space of (1) and (3).

(ii) $\mathcal{SQAP}_{m,n}$ has dimension $\dim(\mathbb{R}^{\mathcal{V}_{m,n}} \times \mathbb{R}^{\hat{\mathcal{E}}_{m,n}}) - (m + mn(m-1)/2)$.
(iii) The nonnegativity constraints $(x,z) \geq 0$ define facets of $\mathcal{SQAP}_{m,n}$.
(iv) The inequalities $(x,z) \leq 1$ are implied by the (1) and (3) together with the nonnegativity constraints $(x,z) \geq 0$.

Proof. Part (iv) is a straightforward calculation. While (iii) follows immediately from Remark 1 and the fact that the nonnegativity constraints define facets in the $n \times n$-case (see [4]), part (i) (which implies (ii)) needs some more techniques, which are not introduced here due to the restricted space. They can be found in detail in [8]. The key step is to project the polytope isomorphically into a lower dimensional vector space, where the vertices have a more convenient coordinate structure. $\qquad\square$

4 Projecting a Certain Relaxation Polytope

In [4] it is proved that the affine hull of the polytope $\mathcal{SQAP}_{n,n}$ (the $n \times n$-case) is described by (1), (3), (6), and (7), i.e., aff$(\mathcal{SQAP}_{n,n})$ is the solution space of the following system (where we use capital letters (X, Z) in order to avoid confusion between the $n \times n$- and the $m \times n$-case):

$$X(\mathrm{row}_i) = 1 \quad (i \in \mathcal{N}) \tag{8}$$
$$X(\mathrm{col}_j) = 1 \quad (j \in \mathcal{N}) \tag{9}$$
$$-X_{(i,j)} - X_{(k,j)} + Z\left(\Delta_{(k,j)}^{(i,j)}\right) = 0 \quad (i,j,k \in \mathcal{N}, i < k) \tag{10}$$
$$-X_{(i,j)} - X_{(i,l)} + Z\left(\Delta_{(i,l)}^{(i,j)}\right) = 0 \quad (i,j,l \in \mathcal{N}, j < l) \tag{11}$$

The fact that (9) and (11) are needed additionally to describe the affine hull of the polytope in the $n \times n$-case is the most important difference to the $m \times n$-case with $m \leq n - 2$.

It turned out ([12,8]) that minimizing over the intersection $\mathcal{SEQP}_{n,n}$ of aff$(\mathcal{SQAP}_{n,n})$ and the nonnegative orthant empirically yields a very strong lower bound for the symmetric $n \times n$-QAP. In contrast to that, minimizing over the intersection $\mathcal{SEQP}_{m,n}$ of aff$(\mathcal{SQAP}_{m,n})$ and the nonnegative orthant usually gives rather poor lower bounds (for $m \leq n - 2$). However, solving the corresponding linear programs is much faster in the $m \times n$ case (as long as m is much smaller than n).

In order to obtain a good lower bound also in the $m \times n$-case, one could add $n - m$ dummy objects to the instance and after that calculate the bound in the $n \times n$-model. Clearly it would be desirable to be able to compute that bound without "blowing up" the model by adding dummies. The following result provides a possibility to do so, and hence, enables us to compute good lower bounds fast in case of considerably less objects than locations. One more notational convention is needed. For two disjoint columns col$_j$ and col$_l$ of $\hat{\mathcal{G}}_{m,n}$ we denote by $\langle$col$_j$: col$_l\rangle$ the set of all hyperedges that share two nodes with col$_j$ and two nodes with col$_l$.

Theorem 4. *Let* $3 \leq m \leq n - 2$. *A point* $(x, z) \in \mathbb{R}^{\mathcal{V}_{m,n}} \times \mathbb{R}^{\hat{\mathcal{E}}_{m,n}}$ *is contained in* $\hat{\sigma}^{(m,n)}(\mathcal{SEQP}_{n,n})$ *if and only if it satisfies the following linear system:*

$$x(\mathrm{row}_i) = 1 \quad (i \in \mathcal{M}) \tag{12}$$
$$x(\mathrm{col}_j) \leq 1 \quad (j \in \mathcal{N}) \tag{13}$$
$$-x_{(i,j)} - x_{(k,j)} + z\left(\Delta_{(k,j)}^{(i,j)}\right) = 0 \quad (i,k \in \mathcal{M}, i < k, j \in \mathcal{N}) \tag{14}$$

$$- x_{(i,j)} - x_{(i,l)} + z\left(\Delta_{(i,l)}^{(i,j)}\right) \le 0 \quad (j,l \in \mathcal{N}, j < l, i \in \mathcal{M}) \tag{15}$$

$$x(\mathrm{col}_j \cup \mathrm{col}_l) - z(\langle \mathrm{col}_j : \mathrm{col}_l \rangle) \le 1 \quad (j,l \in \mathcal{N}, j < l) \tag{16}$$

$$x_v \ge 0 \quad (v \in \mathcal{V}_{m,n}) \tag{17}$$

$$z_h \ge 0 \quad (h \in \hat{\mathcal{E}}_{m,n}) \tag{18}$$

Proof. It should be always clear from the context whether a symbol like row_i is meant to be the i-th row of $\mathcal{V}_{n,n}$ or of $\mathcal{V}_{m,n}$. The rule is that in connection with variables denoted by lower-case letters always $\mathcal{V}_{m,n}$ is the reference set, while variables denoted by upper-case letters refer to $\mathcal{V}_{n,n}$.

We shall first consider the "only if claim" of the theorem. Let $(X, Z) \in \mathcal{SEQP}_{n,n}$, and let $(x, z) = \hat{\sigma}^{(m,n)}(X, Z)$ be the projections of (X, Z). Obviously, (12), (14), and the nonnegativity constraints hold for (x, z). The inequalities (13) and (15) follow from (9), (11) and the the nonnegativity of (X, Z).

It remains to show that the inequalities (16) are satisfied by (x, z). The equations

$$X(\mathrm{col}_j \cup \mathrm{col}_l) - Z(\langle \mathrm{col}_j : \mathrm{col}_l \rangle) = 1 \quad (j,l \in \mathcal{N}, j < l) \tag{19}$$

hold, since $(X, Z) \in \mathrm{aff}(\mathcal{SQAP}_{n,n})$ and they are easily seen to be valid for $\mathcal{SQAP}_{n,n}$. For $j \in \mathcal{N}$ we denote

$$Q_j := \{(i,j) : 1 \le i \le m\} \quad \text{and} \quad \bar{Q}_j := \{(i,j) : m < i \le n\} .$$

Adding some equations of (11) (the ones with $i > m$) to (19) yields equations

$$X(Q_j) + X(Q_l) - Z(\langle Q_j : Q_l \rangle) + Z(\langle \bar{Q}_j : \bar{Q}_l \rangle) = 1 \quad (j,l \in \mathcal{N}, j < l) . \tag{20}$$

Thus, by the nonnegativity of Z, the inequalities (16) hold for the projected vector (x, z).

We come to the more interesting "if claim". In order to show that the given system (12), ... ,(18) of linear constraints forces the point (x, z) to be contained in the projected polytope $\hat{\sigma}^{(m,n)}(\mathcal{SEQP}_{n,n})$, we shall exhibit a map $\phi : \mathbb{R}^{\mathcal{V}_{m,n}} \times \mathbb{R}^{\hat{\mathcal{E}}_{m,n}} \longrightarrow \mathbb{R}^{\mathcal{V}_{n,n}} \times \mathbb{R}^{\hat{\mathcal{E}}_{n,n}}$ that maps such a point (x, z) satisfying (12), ... ,(18) to a point $(X, Z) = \phi(x, z) \in \mathcal{SEQP}_{n,n}$ which coincides with (x, z) on the components belonging to $\hat{\mathcal{G}}_{m,n}$ (as a subgraph of $\hat{\mathcal{G}}_{n,n}$). Hence, the first step is to define $(X, Z) = \phi(x, z)$ as an extension of (x, z), and the second step is to prove that this (X, Z) indeed satisfies (8), ... ,(11), as well as $(X, Z) \ge 0$. The following extension turns out to be a suitable choice (recall that $m \le n-2$):

$$X_{(i,j)} := \frac{1 - x(\mathrm{col}_j)}{n - m} \quad (i > m) \tag{21}$$

$$Z_{\langle i,j,k,l \rangle} := \frac{x_{(i,j)} + x_{(i,l)} - z\left(\Delta_{(i,l)}^{(i,j)}\right)}{n - m} \quad (i \le m, k > m) \tag{22}$$

$$Z_{\langle i,j,k,l \rangle} := \frac{2\left(1 - x(\mathrm{col}_j \cup \mathrm{col}_l) + z\left(\langle \mathrm{col}_j : \mathrm{col}_l \rangle\right)\right)}{(n - m - 1)(n - m)} \quad (i, k > m) \tag{23}$$

Let $(x, z) \in \mathbb{R}^{\mathcal{V}_{m,n}} \times \mathbb{R}^{\hat{\mathcal{E}}_{m,n}}$ satisfy (12), ... ,(18), and let $(X, Z) = \phi(x, z)$ be the extension defined by (21), ... ,(23). Clearly, X is nonnegative (by (13)) and Z is nonnegative (by (15) for $i \leq m$, $k > m$ and by (16) for $i, k > m$).

The validity of (8), ... ,(11) for (X, Z) is shown by the following series of calculations. We use the notation

$$\Delta(i, j) := \{h \in \hat{\mathcal{E}}_{m,n} : (i, j) \in h\}$$

for the set of all hyperedges of $\hat{\mathcal{G}}_{m,n}$ that contain the node (i, j). Note that by (14) we have

$$z(\Delta(i, j)) = (m - 2)x_{(i,j)} + x(\mathrm{col}_j) . \tag{24}$$

Equations (8): For $i \leq m$ this is clear, and for $i > m$ we have

$$X(\mathrm{row}_i) \stackrel{(21)}{=} \sum_{j \in \mathcal{N}} \frac{1 - x(\mathrm{col}_j)}{n - m}$$

$$= \tfrac{1}{n-m} \left(n - \sum_{j \in \mathcal{N}} x(\mathrm{col}_j)\right)$$

$$= \tfrac{1}{n-m} \left(n - \sum_{i \in \mathcal{M}} x(\mathrm{row}_i)\right)$$

$$\stackrel{(12)}{=} 1 .$$

Equations (9): For $j \in \mathcal{N}$ we have

$$X(\mathrm{col}_j) \stackrel{(21)}{=} x(\mathrm{col}_j) + (n - m)\frac{1 - x(\mathrm{col}_j)}{n - m}$$

$$= 1 .$$

Equations (10): For $i, k \leq m$ these equations clearly hold. In case of $i \leq m$ and $k > m$ we have

$$-X_{(i,j)} - X_{(k,j)} + Z\left(\Delta_{(k,j)}^{(i,j)}\right)$$

$$\stackrel{(21),(22)}{=} -x_{(i,j)} - \frac{1 - x(\mathrm{col}_j)}{n - m} + \sum_{l \in \mathcal{N} \setminus j} \frac{x_{(i,j)} + x_{(i,l)} - z\left(\Delta_{(i,l)}^{(i,j)}\right)}{n - m}$$

$$= -x_{(i,j)} + \frac{1}{n - m}\left(x(\mathrm{col}_j) - 1 + (n - 2)x_{(i,j)} + x(\mathrm{row}_i) - z(\Delta(i, j))\right)$$

$$\stackrel{(12),(24)}{=} -x_{(i,j)} + \frac{1}{n - m}(n - m)x_{(i,j)}$$

$$= 0 .$$

It remains to consider the case $i, k > m$. Here, we get (using $\alpha := \frac{1}{(n-m-1)(n-m)}$ in order to increase readability)

$$-X_{(i,j)} - X_{(k,j)} + Z\left(\Delta_{(k,j)}^{(i,j)}\right)$$

$$\overset{(23)}{=} \quad -X_{(i,j)} - X_{(k,j)} + 2\alpha \sum_{l \in \mathcal{N} \backslash j} (1 - x(\mathrm{col}_j \cup \mathrm{col}_l) + z(\langle \mathrm{col}_j : \mathrm{col}_l \rangle))$$

$$= \quad -X_{(i,j)} - X_{(k,j)}$$

$$+2\alpha\left(n - 1 - (n-2)x(\mathrm{col}_j) - x(\mathcal{V}_{m,n}) + \sum_{i \in M} \frac{1}{2} z(\Delta(i,j))\right)$$

$$\overset{(12),(24)}{=} \quad -X_{(i,j)} - X_{(k,j)} + 2\alpha\left(n - m - 1 + (m - n + 1)x(\mathrm{col}_j)\right)$$

$$= \quad -X_{(i,j)} - X_{(k,j)} + \frac{2}{n-m}(1 - x(\mathrm{col}_j))$$

$$\overset{(21)}{=} \quad 0 \ .$$

Equations (11): If $i \leq m$ holds, then we have

$$-X_{(i,j)} - X_{(i,l)} + Z\left(\Delta_{(i,l)}^{(i,j)}\right)$$

$$= \quad -x_{(i,j)} - x_{(i,l)} + \sum_{k \in \mathcal{M} \backslash i} Z_{\langle i,j,k,l \rangle} + \sum_{k \in \mathcal{N} \backslash \mathcal{M}} Z_{\langle i,j,k,l \rangle}$$

$$\overset{(22)}{=} \quad -x_{(i,j)} - x_{(i,l)} + z\left(\Delta_{(i,l)}^{(i,j)}\right) + \sum_{k \in \mathcal{N} \backslash \mathcal{M}} \frac{x_{(i,j)} + x_{(i,l)} - z\left(\Delta_{(i,l)}^{(i,j)}\right)}{n - m}$$

$$= \quad 0 \ .$$

For $i > m$ we have

$$-X_{(i,j)} - X_{(i,l)} + Z\left(\Delta_{(i,l)}^{(i,j)}\right)$$

$$= \quad -X_{(i,j)} - X_{(i,l)} + \sum_{k \in \mathcal{M}} Z_{\langle i,j,k,l \rangle} + \sum_{k \in \mathcal{N} \backslash \mathcal{M} \backslash i} Z_{\langle i,j,k,l \rangle}$$

$$\overset{(22),(23)}{=} \quad -X_{(i,j)} - X_{(i,l)} + \sum_{k \in \mathcal{M}} \frac{x_{(k,l)} + x_{(k,j)} - z\left(\Delta_{(k,j)}^{(k,l)}\right)}{n - m}$$

$$+2 \sum_{k \in (\mathcal{N} \backslash \mathcal{M}) \backslash i} \frac{1 - x(\mathrm{col}_j \cup \mathrm{col}_l) + z(\langle \mathrm{col}_j : \mathrm{col}_l \rangle)}{(n - m - 1)(n - m)}$$

$$= -X_{(i,j)} - X_{(i,l)} + \frac{1}{n-m}\Big(x(\mathrm{col}_l + x(\mathrm{col}_j) - 2z(\langle \mathrm{col}_l : \mathrm{col}_j \rangle) \Big)$$

$$+ \frac{2}{n-m}\Big(1 - x(\mathrm{col}_j \cup \mathrm{col}_l) + z(\langle \mathrm{col}_j : \mathrm{col}_l \rangle) \Big)$$

$$= -X_{(i,j)} - X_{(i,l)} + \frac{1}{n-m}\Big(2 - x(\mathrm{col}_j) - x(\mathrm{col}_l) \Big)$$

$$\overset{(21)}{=} 0 \ .$$

$\square$

Finally, we investigate the system (12), ... ,(18) with respect to the question of redundancies. From Theorem 3 we know already that the nonnegativity constraints (17) and (18) define facets of $\mathcal{SQAP}_{m,n}$ as well as that (12) and (14) are needed in the linear description of the affine hull of the polytope. Thus it remains to investigate (13), (15), and (16). And in fact, it turns out that one of these classes is redundant.

Theorem 5. *Let* $4 \le m \le n-2$.

(i) *The inequalities* (13) *are redundant in* (12), ... ,(18).
(ii) *The inequalities* (15) *and* (16) *define facets of* $\mathcal{SQAP}_{m,n}$.

Proof. To prove part (i) observe that the equations

$$(m-1)x(\mathrm{col}_j) - z(\langle \mathrm{col}_j : \mathcal{V}_{m,n} \setminus \mathrm{col}_j \rangle) = 0 \quad (j \in \mathcal{N}) \tag{25}$$

and

$$x(\mathcal{V}_{m,n}) = m \tag{26}$$

hold for $\mathcal{SQAP}_{m,n}$. Thus, they are implied by the linear system (12), (14) due to Theorem 3. Adding up all inequalities (16) for a fixed $j \in \mathcal{N}$ and all $l \in \mathcal{N} \setminus j$ and subtracting (25) (for that j) and (26) yields

$$n-m-1 \ge \sum_{l \in \mathcal{N} \setminus j} \Big(x(\mathrm{col}_j \cup \mathrm{col}_l) - z(\langle \mathrm{col}_j : \mathrm{col}_l \rangle) \Big)$$
$$-(m-1)x(\mathrm{col}_j) + z(\langle \mathrm{col}_j : \mathcal{V}_{m,n} \setminus \mathrm{col}_j \rangle) - x(\mathcal{V}_{m,n})$$
$$= (n-1)x(\mathrm{col}_j) + \sum_{l \in \mathcal{N} \setminus j} x(\mathrm{col}_l) - z(\langle \mathrm{col}_j : \mathcal{V}_{m,n} \setminus \mathrm{col}_j \rangle)$$
$$-(m-1)x(\mathrm{col}_j) + z(\langle \mathrm{col}_j : \mathcal{V}_{m,n} \setminus \mathrm{col}_j \rangle) - x(\mathcal{V}_{m,n})$$
$$= (n-m-1)x(\mathrm{col}_j) \ ,$$

what proves part (i) due to $n-m-1 \ge 1$.

The proof of (ii) needs the same techniques as mentioned in the proof of Theorem 3, and is omitted here as well. It can be found in [8]. $\square$

5   A Large Class of Facets

In [5] a large class of facet defining inequalities for the $n \times n$-case is investigated. Many of them satisfy the requirements stated in Remark 1. We briefly introduce these inequalities here, and demonstrate in Sect. 6 how valuable they are for computing good lower bounds or even optimal solutions for QAPs with less objects than locations.

Let $P_1, P_2 \subseteq \mathcal{M}$ and $Q_1, Q_2 \subseteq \mathcal{N}$ be two sets of row respectively column indices with $P_1 \cap P_2 = \emptyset$ and $Q_1 \cap Q_2 = \emptyset$. Define $\mathcal{S} := (P_1 \times Q_1) \cup (P_2 \times Q_2)$ and $\mathcal{T} := (P_1 \times Q_2) \cup (P_2 \times Q_1)$. In [5] it is shown that the following inequality is valid for $\mathcal{SQAP}_{n,n}$ for every $\beta \in \mathbb{Z}$ (where $z(\mathcal{S})$ and $z(\mathcal{T})$ are the sums over all components of z that belong to hyperedges with all four endnodes in $\mathcal{S}$ respectively in $\mathcal{T}$, and $\langle \mathcal{S} : \mathcal{T} \rangle$ is the set of all hyperedges with two endnodes in $\mathcal{S}$ and the other two endnodes in $\mathcal{T}$):

$$- \beta x(\mathcal{S}) + (\beta - 1)x(\mathcal{T}) - z(\mathcal{S}) - z(\mathcal{T}) + z(\langle \mathcal{S} : \mathcal{T} \rangle) \le \frac{\beta(\beta - 1)}{2} \qquad (27)$$

Clearly the validity carries over to the $m \times n$-case. The inequality (27) is called the *4-box inequality* determined by the triple $(\mathcal{S}, \mathcal{T}, \beta)$.

In this paper, we concentrate on 4-box inequalities that are generated by a triple $(\emptyset, \mathcal{T}, \beta)$ (i.e., we have $P_1 = Q_2 = \emptyset$ or $P_2 = Q_1 = \emptyset$), which we call *1-box inequalities*. Empirically, they have turned out to be the most valuable ones within cutting plane procedures among the whole set of 4-box inequalities.

In [5], part (i) of the following theorem is proved, from which part (ii) follows immediately by Remark 1.

Theorem 6. *Let $n \ge 7$.*

(i) *Let $P, Q \subseteq \mathcal{N}$ generate $\mathcal{T} = P \times Q \subseteq \mathcal{V}_{n,n}$, and let $\beta \in \mathbb{Z}$ be an integer number such that*
- $\beta \ge 2$,
- $|P|, |Q| \ge \beta + 2$,
- $|P|, |Q| \le n - 3$, and
- $|P| + |Q| \le n + \beta - 5$

hold. Then the 1-box inequality

$$(\beta - 1)x(\mathcal{T}) - z(\mathcal{T}) \le \frac{\beta(\beta - 1)}{2} \qquad (28)$$

defined by the triple $(\emptyset, \mathcal{T}, \beta)$ defines a facet of $\mathcal{SQAP}_{n,n}$.

(ii) *For $m \le n$ and $P \subseteq \mathcal{M}$ the 1-box inequality (28) defines a facet of $\mathcal{SQAP}_{m,n}$ as well.*

6   Computational Results

Using the ABACUS framework (Jünger and Thienel [7]) we have implemented a simple cutting plane algorithm for (symmetric) $m \times n$-instances (with $m \le n-2$)

that uses (12), (14), ... ,(18) as the initial set of constraints. Thus, by Theorem 4 (and Theorem 5 (i)), the first bound that is computed is the *symmetric equation bound (SEQB)*, which is obtained by optimizing over the intersection of aff($SQAP_{n,n}$) with the nonnegative orthant.

The separation algorithm that we use is a simple 2-opt based heuristic for finding violated 1-box inequalities with $\beta = 2$. We limited the experiments to this small subclass of box inequalities since on the one hand they emerged as the most valuable ones from initial tests, and on the other hand even our simple heuristic usually finds many violated inequalities among the 1-box inequalities with $\beta = 2$ (if it is called several times with different randomly chosen initial boxes T).

The experiments were carried out on a Silicon Graphics Power Challenge computer. For solving the linear programs we used the barrier code of CPLEX 4.0, which was run in its parallel version on four processors.

We tested our code on the esc instances of the QAPLIB, which are the only ones in that problem library that have much less objects than locations. Note that all these instances have both a symmetric (integral) flow as well as a symmetric (integral) distance matrix, yielding that only even numbers occur as objective function values of feasible solutions. Thus, every lower bound can be rounded up to the next even integer number greater than or equal to it. Our tests are restricted to those ones among these instances that have 16 or 32 locations. All the esc16 instances (with 16 locations) were solved for the first time to optimality by Clausen and Perreghard [3]. The esc32 (with 32 locations) instances are still unsolved, up to three easy ones among them.

Table 1 shows the results for the esc16 instances. The instances esc16b, esc16c, and esc16h are omitted since they do not satisfy $m \leq n - 2$ (esc16f was removed from the QAPLIB since it has an all-zero flow matrix). The bounds

Table 1. The column *objects* contains the number of objects in the respective instance, *opt* is the optimal solution value, *SEQB* is the symmetric equation bound (i.e., the bound after the first LP), *box* is the bound obtained after some cutting plane iterations, *LPs* shows the number of linear programs being solved, *time* is the CPU time in seconds, and *speed up* is the quotient of the running times for working in the $n \times n$- and in the $m \times n$-model. The last column gives the running times Clausen and Perregard needed to solve the instances on a parallel machine with 16 i860 processors.

name	objects	opt	SEQB	box	LPs	time (s)	speed up	CP (s)
esc16a	10	68	48	64	3	522	4.87	65
esc16d	14	16	4	16	2	269	2.74	492
esc16e	9	28	14	28	4	588	3.37	66
esc16g	8	26	14	26	3	58	14.62	7
esc16i	9	14	0	14	4	106	28.18	84
esc16j	7	8	2	8	2	25	32.96	14

produced by our cutting plane code match the optimal solution values for all instances but esc16a (see also Fig. 1). Working in the $m \times n$-model speeds up the cutting plane code quite much for some instances. The running times in the $m \times n$-model are comparable with the ones of the branch-and-bound code of [3].

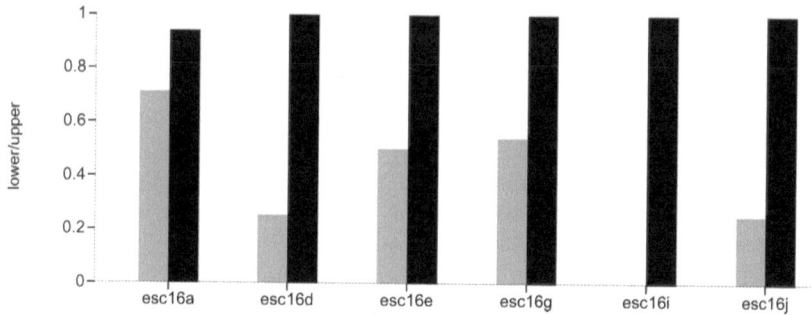

Fig. 1. The bars (gray for SEQB and black for the bound obtained by the box inequalities) show the ratios of the lower and upper bounds, where the upper bounds are always the optimal solution values here.

For the esc32 instances, our cutting plane algorithm computes always the best known lower bounds (see Tab. 2 and Fig. 2). The three instances esc32e, esc32f, and esc32g were solved to optimality for the first time by Brüngger, Marzetta, Clausen, and Perregard [1]. Our cutting plane code is able to solve these instances to optimality within a few hundred seconds of CPU time (on

Table 2. The column labels have the same meanings as in Tab. 1. Additionally, *upper* gives the objective function value of the best known feasible solution and *prev lb* denotes the best previously known lower bound. (Running times with a ⋆ are only approximately measured due to problems with the queuing system of the machine).

name	objects	upper	prev lb	SEQB	box	LPs	time (s)
esc32a	25	130	36	40	88	3	62988
esc32b	24	168	96	96	100	4	⋆60000
esc32c	19	642	506	382	506	8	⋆140000
esc32d	18	200	132	112	152	8	⋆80000
esc32e	9	2	2	0	2	2	576
esc32f	9	2	2	0	2	2	554
esc32g	7	6	6	0	6	2	277
esc32h	19	438	315	290	352	6	119974

four processors). These are about the same running times as needed by [1] with a branch-and-bound code on a 32 processor NEC Cenju-3 machine.

The formerly best known lower bounds for the other `esc32` instances were calculated by the *triangle decomposition* bounding procedure of Karisch and Rendl [9]. The bounds obtained by the cutting plane code improve (or match, in case of `esc32c`) all these bounds. The most impressive gain is the improvement of the bound quality from 0.28 to 0.68 for `esc32a`. While for the `esc16` instances switching from the $n \times n$- to the $m \times n$-model yields a significant speed up, in case of the `esc32` instances to solve the linear programs even became only possible in the $m \times n$-model.

Nevertheless, for the hard ones among the `esc32` instances the running times of the cutting plane code are rather large. Here, a more sophisticated cutting plane algorithm is required in order to succeed in solving these instances to optimality. This concerns the separation algorithms and strategies, the treatment of the linear programs, as well as the exploitation of sparsity of the objective functions, which will be briefly addressed in the following section.

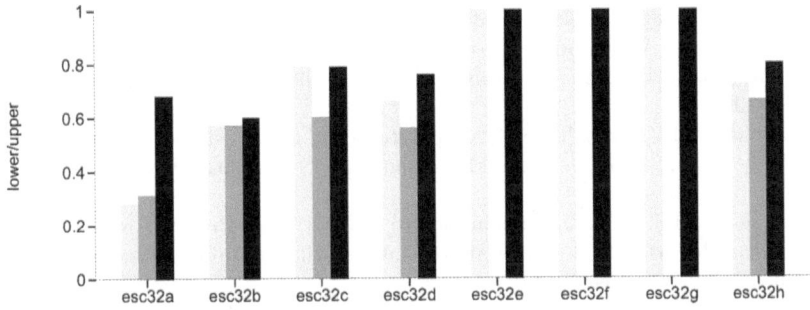

Fig. 2. The dark gray and the black bars have the same meaning as in Fig. 1. Additionally, the light gray bars show the qualities of the previously best known lower bounds.

7 Conclusion

The polyhedral studies reported in this paper have enabled us to build for the first time a cutting plane code for QAPs with less objects than locations that has a similar performance as current parallel branch-and-bound codes for smaller instances and gives new lower bounds for the larger ones. More elaborated separation procedures (including parallelization) and a more sophisticated handling of the linear programs will surely increase the performance of the cutting plane algorithm still further.

At the moment, the limiting factor for the cutting plane approach is the size (and the hardness) of the linear programs. But if one considers the instances in the QAPLIB more closely, it turns out that the flow matrices very often are extremely sparse. If one exploits this sparsity, one can "project out" even more variables than we did by passing from the $n \times n$- to the $m \times n$-model. In our opinion, investigations of the associated projected polytopes will eventually lead to cutting plane algorithms in much smaller models, which perhaps will push the limits for exact solutions of quadratic assignment problems far beyond the current ones.

References

1. A. Brüngger, A. Marzetta, J. Clausen, and M. Perregaard. Joining forces in solving large-scale quadratic assignment problems. In *Proceedings of the 11th International Parallel Processing Symposium IPPS*, pages 418–427, 1997.
2. R. E. Burkard, S. E. Karisch, and F. Rendl. QAPLIB - A quadratic assignment problem library. *Journal of Global Optimization*, 10:391–403, 1997.
3. J. Clausen and M. Perregaard. Solving large quadratic assignment problems in parallel. *Computational Optimization and Applications*, 8(2):111–127, 1997.
4. M. Jünger and V. Kaibel. On the SQAP-polytope. Technical Report 96.241, Angewandte Mathematik und Informatik, Universität zu Köln, 1996.
5. M. Jünger and V. Kaibel. Box-inequalities for quadratic assignment polytopes. Technical Report 97.285, Angewandte Mathematik und Informatik, Universität zu Köln, 1997.
6. M. Jünger and V. Kaibel. The QAP-polytope and the star-transformation. Technical Report 97.284, Angewandte Mathematik und Informatik, Universität zu Köln, 1997.
7. M. Jünger and S. Thienel. Introduction to ABACUS – A Branch-And-CUt System. Technical Report 97.263, Angewandte Mathematik und Informatik, Universität zu Köln, 1997. (To appear in *OR Letters*).
8. V. Kaibel. *Polyhedral Combinatorics of the Quadratic Assignment Problem*. PhD thesis, Universität zu Köln, 1997. http://www.informatik.uni-koeln.de/ls_juenger/staff/kaibel/diss.html.
9. S. E. Karisch and F. Rendl. Lower bounds for the quadratic assignment problem via triangle decompositions. *Mathematical Programming*, 71(2):137–152, 1995.
10. T. C. Koopmans and M. J. Beckmann. Assignment problems and the location of economic activities. *Econometrica*, 25:53–76, 1957.
11. M. Padberg and M. P. Rijal. *Location, Scheduling, Design and Integer Programming*. Kluwer Academic Publishers, 1996.
12. M. G. C. Resende, K. G. Ramakrishnan, and Z. Drezner. Computing lower bounds for the quadratic assignment problem with an interior point solver for linear programming. *Operations Research*, 43:781–791, 1995.
13. M. P. Rijal. *Scheduling, Design and Assignment Problems with Quadratic Costs*. PhD thesis, New York University, 1995.
14. S. Sahni and T. Gonzales. P-complete approximation problems. *Journal of the Association for Computing Machinery*, 1976.

Incorporating Inequality Constraints in the Spectral Bundle Method

Christoph Helmberg[1], Krzysztof C. Kiwiel[2], and Franz Rendl[3]

[1] Konrad-Zuse-Zentrum für Informationstechnik Berlin
Takustraße 7, 14195 Berlin, Germany
helmberg@@zib.de, http://www.zib.de/helmberg
[2] Systems Research Institute, Polish Academy of Sciences
Newelska 6, 01-447 Warsaw, Poland
kiwiel@@ibspan.waw.pl
[3] Technische Universität Graz, Institut für Mathematik
Steyrergasse 30, A-8010 Graz, Austria
rendl@@opt.math.tu-graz.ac.at

Abstract. Semidefinite relaxations of quadratic 0-1 programming or graph partitioning problems are well known to be of high quality. However, solving them by primal-dual interior point methods can take much time even for problems of moderate size. Recently we proposed a spectral bundle method that allows to compute, within reasonable time, approximate solutions to structured, large equality constrained semidefinite programs if the trace of the primal matrix variable is fixed. The latter property holds for the aforementioned applications. We extend the spectral bundle method so that it can handle inequality constraints without seriously increasing computation time. This makes it possible to apply cutting plane algorithms to semidefinite relaxations of real world sized instances. We illustrate the efficacy of the approach by giving some preliminary computational results.

1 Introduction

Since the landmark papers [9,4,10,2] it is well known that semidefinite programming allows to design powerful relaxations for constrained quadratic 0-1 programming and graph partitioning problems. The most commonly used algorithms for solving these relaxations, primal-dual interior point methods, offer little possibilities to exploit problem structure. Typically their runtime is governed by the factorization of a dense symmetric positive definite matrix in the number of constraints and by the line search that ensures the positive definiteness of the matrix variables. Computation times for problems with more than 3000 constraints or matrix variables of order 500, say, are prohibitive. Very recently, a pure dual approach has been proposed in [1] that is able to exploit the sparsity of the cost matrix in the case of the max-cut relaxation with diagonal

R. E. Bixby, E. A. Boyd, and R. Z. Ríos-Mercado (Eds.): IPCO VI
LNCS 1412, pp. 423–435, 1998. © Springer–Verlag Berlin Heidelberg 1998

constraints. It is not yet clear whether these results extend to problems with a huge number of less structured constraints.

The spectral bundle method [6] works on a reformulation of semidefinite relaxations as eigenvalue optimization problems and was developed to provide approximate solutions to structured problems fast. In contrast to standard bundle methods [7,11], a non-polyhedral semidefinite cutting plane model is constructed from the subgradients. Reinterpreted in terms of the original semidefinite program, the semidefinite model ensures positive semidefiniteness of the dual matrix variable on a subspace only. Subgradients correspond to eigenvectors to negative eigenvalues and are used to correct the subspace. By means of an aggregate subgradient the dimension of the subspace can be kept small, thus ensuring efficient solvability of the subproblems. Lanczos methods (see e.g. [3]) allow to compute a few extremal eigenvalues and their eigenvectors efficiently by a series of matrix-vector multiplications which do not require the matrix in explicit form. Thus structural properties of cost and coefficient matrices can be exploited.

Like most first order methods, the spectral bundle method exhibits fast progress in the beginning, but shows a strong tailing off effect as the optimal solution is approached. Fortunately many semidefinite relaxations do not have to be solved exactly. Rather an approximate solution is used to improve, e.g., by cutting planes, the current relaxation, which is then resolved.

In its original form the spectral bundle method is designed for equality constraints only because sign constraints on the dual variables may increase computation times for the semidefinite subproblem significantly. In this paper we employ Lagrangian relaxation to approximate the solution of a sign constrained semidefinite subproblem. Surprisingly just one update of Lagrange multipliers per function evaluation suffices to ensure convergence. The semidefinite subproblem can be solved as efficiently as in the unconstrained case, thus rendering this method an attractive choice for large scale semidefinite cutting plane algorithms.

Section 2 introduces some notation and explains the connection to semidefinite programming. This is followed by a very brief review of important properties of the maximal eigenvalue function. Section 4 explains the extension of the spectral bundle method to inequality constraints. In Section 5 we discuss efficiency aspects of the subproblem solution. Section 6 gives computational results.

2 Semidefinite Programs

Let $\mathcal{S}_n$ denote the space of symmetric matrices of order n. The inner product in this space is the usual matrix inner product, $\langle A, B \rangle = \mathrm{tr}(B^T A)$ for $A, B \in \mathbb{R}^{m \times n}$. Let $\mathcal{S}_n^+$ denote the set of symmetric positive semidefinite matrices. $\mathcal{S}_n^+$ is a pointed closed convex cone. Except for its apex $\{0\}$, a face F of this cone can be described as

$$F = \left\{ P V P^T : V \in \mathcal{S}_r^+ \right\},$$

where $P \in \mathbb{R}^{n \times r}$ is some fixed matrix with orthonormal columns (w.l.o.g.). The dimension of such a face F is $\binom{r+1}{2}$. For $A, B \in \mathcal{S}_n$, $A \succeq B$ refers to the Löwner partial order induced by the cone $\mathcal{S}_n^+$ ($A \succeq B \iff A - B \in \mathcal{S}_n^+$).

In the following $\mathcal{A} : \mathcal{S}_n \rightarrow \mathrm{I\!R}^m$ is a linear operator and $\mathcal{A}^T(\cdot)$ is the corresponding adjoint operator satisfying $\langle \mathcal{A}(X), y \rangle = \langle X, \mathcal{A}^T(y) \rangle$ for all $X \in \mathcal{S}_n$ and $y \in \mathrm{I\!R}^m$. The operators are of the form

$$
\mathcal{A}(X) = \begin{bmatrix} \langle A_1, X \rangle \\ \vdots \\ \langle A_m, X \rangle \end{bmatrix} \qquad \text{and} \qquad \mathcal{A}^T(y) = \sum_{i=1}^{m} y_i A_i.
$$

We will denote a subset $J \subset \{1, \ldots, m\}$ of the rows of $\mathcal{A}$ by $\mathcal{A}_J$. Likewise we will speak of $\mathcal{A}_J^T$ and y_J for some vector $y \in \mathrm{I\!R}^m$.

We consider semidefinite programs of the following form with $J \subset \{1, \ldots, m\}$ and $\bar{J} = \{1, \ldots, m\} \setminus J$,

$$
\text{(P)} \quad \begin{aligned} \max \ & \langle C, X \rangle \\ \text{s.t.} \ & \mathcal{A}_{\bar{J}}(X) = b_{\bar{J}} \\ & \mathcal{A}_J(X) \le b_J \\ & X \succeq 0. \end{aligned} \qquad \text{(D)} \quad \begin{aligned} \min \ & \langle b, y \rangle \\ \text{s.t.} \ & Z = \mathcal{A}^T(y) - C \\ & Z \succeq 0, y_J \ge 0. \end{aligned}
$$

Under a constraint qualification (which we tacitly assume to hold), any optimal solution X^* of (P) and any optimal solution (y^*, Z^*) of (D) satisfy $X^* Z^* = 0$.

We assume that there exists $\hat{y} \in \mathrm{I\!R}^m$ with $I = \mathcal{A}_{\bar{J}}^T(\hat{y})$, which is equivalent to the requirement that $\mathrm{tr}(X) = a = b_{\bar{J}}^T \hat{y}_{\bar{J}}$ for all feasible X. Most semidefinite relaxations of combinatorial optimization problems satisfy this requirement or can be scaled appropriately. To avoid the trivial case of $X = 0$, assume $a > 0$.

Under these assumptions it is well known that (D) is equivalent to the eigenvalue optimization problem (see e.g. [8])

$$
\min_{y \in \mathrm{I\!R}^m, y_J \ge 0} a\lambda_{\max}(C - \mathcal{A}^T(y)) + \langle b, y \rangle.
$$

To simplify notation we drop, without loss of generality, the coefficient a and assume that $|J| = m$, i.e., there are inequality constraints only. Thus we concentrate on the following problem

$$
\min_{y \ge 0} f(y) \qquad \text{with} \qquad f(y) = \lambda_{\max}(C - \mathcal{A}^T(y)) + \langle b, y \rangle, \tag{1}
$$

or equivalently $\min f_S := f + \iota_S$ with ι_S being the indicator function of $S = \mathrm{I\!R}_+^m$ ($\iota_S(y) = 0$ if $y \in S$, ∞ otherwise).

3   The Function $\lambda_{\max}(X)$

The maximal eigenvalue $\lambda_{\max}(X)$ of a matrix is a nonsmooth convex function. The kinks of $\lambda_{\max}(X)$ appear at points where the maximal eigenvalue has multiplicity at least two. If, for some $\hat{X} \in \mathcal{S}_n$, the columns of $P \in \mathrm{I\!R}^{n \times r}$ form an orthonormal basis of the eigenspace of the maximal eigenvalue of $\hat{X}$, then the subdifferential (the set of subgradients) of $\lambda_{\max}(\cdot)$ at $\hat{X}$ is

$$\partial \lambda_{\max}(\hat{X}) = \left\{ PVP^T : \mathrm{tr}(V) = 1, V \in \mathcal{S}_r^+ \right\} \tag{2}$$
$$= \mathrm{conv} \left\{ vv^T : v^T \hat{X} v = \lambda_{\max}(\hat{X}), \|v\| = 1 \right\}.$$

It is a well known fact that

$$\lambda_{\max}(X) = \max \left\{ \langle W, X \rangle : \mathrm{tr}(W) = 1, W \succeq 0 \right\}.$$

Any subset $\hat{\mathcal{W}} \subset \{ W \succeq 0 : \mathrm{tr}(W) = 1 \}$ gives rise to a non-polyhedral cutting plane model minorizing $\lambda_{\max}(X)$,

$$\lambda_{\max}(X) \geq \max \left\{ \langle W, X \rangle : W \in \hat{\mathcal{W}} \right\}. \tag{3}$$

4   The Spectral Bundle Method with Inequality Constraints

It follows from the properties of $\lambda_{\max}(\cdot)$ that (1) is a nonsmooth convex minimization problem. A standard approach to solve such problems is the proximal bundle method [7]. We first sketch the method in general and give the corresponding symbols that we will use for the spectral bundle method. Afterwards we specialize the algorithm to the eigenvalue problem (1).

At iteration k the algorithm generates a test point $y^{k+1} = \arg\min_S \hat{f}^k + \frac{u}{2}\|\cdot - x^k\|^2$, where $\hat{f}^k$ is an accumulated cutting plane model of f and the weight $u > 0$ keeps y^{k+1} near the current iterate x^k. A *serious step* $x^{k+1} = y^{k+1}$ occurs if $f(x^k) - f(y^{k+1}) \geq m_L[f(x^k) - \hat{f}^k(y^{k+1})]$, where $m_L \in (0, \frac{1}{2})$. Otherwise a *null step* $x^{k+1} = x^k$ is made but the new subgradient computed at y^{k+1} improves the next model.

The spectral bundle method is tailored for the eigenvalue problem (1) in that, instead of the usual polyhedral $\hat{f}^k$, it uses a semidefinite cutting plane model of the form (cf. (3))

$$\hat{f}^k(y) = \max_{W \in \hat{\mathcal{W}}^k} \langle W, C - \mathcal{A}^T(y) \rangle + \langle b, y \rangle. \tag{4}$$

We will specify the choice of $\hat{\mathcal{W}}^k$ and discuss the computation of $\hat{f}^k$ in detail in Sect. 5. The usual bundle subproblem of finding

$$y_*^{k+1} = \arg\min_{y \geq 0} \hat{f}^k(y) + \frac{u}{2}\|y - x^k\|^2 \tag{5}$$

turns out to be rather expensive computationally because of the sign constraints on y. Instead we compute, by means of Lagrangian relaxation, an approximation y^{k+1} to y_*^{k+1}. To this end we define the Lagrangian of (4)–(5)

$$L(y; W, \eta) = \langle C, W \rangle + \langle b - \mathcal{A}(W), y \rangle - \langle \eta, y \rangle + \frac{u}{2}\|y - x^k\|^2$$

for $W \in \hat{\mathcal{W}}^k$ and $\eta \in \mathbb{R}^m_+$. It has the form $L(\cdot; W, \eta) = \bar{f}_{W,\eta}(\cdot) + \frac{u}{2}\| \cdot -x^k\|^2$, where

$$\bar{f}_{W,\eta}(\cdot) = \langle C, W \rangle + \langle b - \eta - \mathcal{A}(W), \cdot \rangle$$

is an affine function minorizing $\hat{f}^k_S = \hat{f}^k + \imath_S$ and hence f_S. The dual function $\varphi(W, \eta) = \min L(\cdot; W, \eta)$, computed by finding

$$y_{W,\eta} = \arg\min_y \bar{f}_{W,\eta}(y) + \frac{u}{2}\|y - x^k\|^2 = x^k - \frac{b - \eta - \mathcal{A}(W)}{u},$$

has the form

$$\varphi(W, \eta) = \langle C, W \rangle + \langle b - \eta - \mathcal{A}(W), x^k \rangle - \frac{1}{2u}\|b - \eta - \mathcal{A}(W)\|^2.$$

Denote by η^k the multipliers of the previous iteration. We find an approximate maximizer (W^{k+1}, η^{k+1}) to φ over $\hat{\mathcal{W}}^k \times \mathbb{R}^m_+$ by first computing

$$W^{k+1} \in \text{Arg}\max_{W \in \hat{\mathcal{W}}^k} \varphi(W, \eta^k) \tag{6}$$

via an interior point algorithm, and then setting

$$\eta^{k+1} = \arg\max_{\eta \geq 0} \varphi(W^{k+1}, \eta) = -u \min\left\{0, x^k - \frac{b - \mathcal{A}(W)}{u}\right\}. \tag{7}$$

It is not difficult to check that

$$y^{k+1} := y_{W^{k+1}, \eta^{k+1}} \geq 0 \qquad \text{and} \qquad \langle \eta^{k+1}, y^{k+1} \rangle = 0.$$

At the test point y^{k+1} the function is evaluated and a new subgradient is computed.

By the observations above, we have

$$\bar{f}^{k+1} = \bar{f}_{W^{k+1}, \eta^{k+1}} \leq \hat{f}^k_S \leq f_S.$$

This and the following lemma motivate a stopping criterion of the form

$$f(x^k) - \bar{f}^{k+1}(y^{k+1}) \leq \varepsilon_{\text{opt}}(|f(x^k)| + 1).$$

Lemma 1. $f(x^k) \geq \bar{f}^{k+1}(y^{k+1})$, and if $f(x^k) = \bar{f}^{k+1}(y^{k+1})$ then x^k is optimal.

Proof. Since $f(x^k) \geq \bar{f}^{k+1}(x^k)$ and

$$\bar{f}^{k+1}(x^k) - \bar{f}^{k+1}(y^{k+1}) = \langle b - \eta^{k+1} - \mathcal{A}(W^{k+1}), x^k - y^{k+1} \rangle$$
$$= \frac{1}{u}\|b - \eta^{k+1} - \mathcal{A}(W^{k+1})\|^2,$$

the inequality follows. If equality holds then $f(x^k) = \bar{f}^{k+1}(x^k)$ and $0 = b - \eta^{k+1} - \mathcal{A}(W^{k+1}) = \nabla \bar{f}^{k+1} \in \partial f_S(x^k)$. $\qquad\square$

For practical reasons we state the algorithm with an inner loop allowing for several repetitions of the two "coordinatewise" maximization steps (6) and (7).

Algorithm 1.
Input: $y^0 \in \mathbb{R}_+^m$, $\varepsilon_{\mathrm{opt}} \geq 0$, $\varepsilon_{\mathrm{M}} \in (0, \infty]$, *an improvement parameter* $m_L \in (0, \frac{1}{2})$, *a weight* $u > 0$.

1. *(Initialization)* Set $k = 0$, $x^0 = y^0$, $\eta^0 = 0$, $f(x^0)$ *and* $\hat{\mathcal{W}}^0$ *(cf. Sect. 5)*.
2. *(Direction finding)* Set $\eta^+ = \eta^k$.
   (a) Find $W^+ \in \mathrm{Arg\,max}_{W \in \hat{\mathcal{W}}^k} \varphi(W, \eta^+)$.
   (b) Compute $\eta^+ = \arg\max_{\eta \geq 0} \varphi(W^+, \eta)$.
   (c) Compute $y^+ = y_{W^+, \eta^+}$.
   (d) *(Termination)* If $f(x^k) - \bar{f}_{W^+, \eta^+}(y^+) \leq \varepsilon_{\mathrm{opt}}(|f(x^k)| + 1)$ then STOP.
   (e) If $\hat{f}^k(y^+) - \bar{f}_{W^+, \eta^+}(y^+) > \varepsilon_{\mathrm{M}}[f(x^k) - \bar{f}_{W^+, \eta^+}(y^+)]$ then go to (a).
   Set $\eta^{k+1} = \eta^+$, $W^{k+1} = W^+$, $y^{k+1} = y^+$.
3. *(Evaluation)* Compute $f(y^{k+1})$,
   find $W_S^{k+1} \in \mathrm{Arg\,max}\{\langle C - \mathcal{A}^T(y^{k+1}), W\rangle : \mathrm{tr}(W) = 1, W \succeq 0\}$.
4. *(Model updating)* Determine $\hat{\mathcal{W}}^{k+1} \supseteq \mathrm{conv}\{W^{k+1}, W_S^{k+1}\}$ *(cf. Sect. 5)*.
5. *(Descent test)* If $f(x^k) - f(y^{k+1}) \geq m_L[f(x^k) - \bar{f}^{k+1}(y^{k+1})]$ then set $x^{k+1} = y^{k+1}$ *(serious step)*; otherwise set $x^{k+1} = x^k$ *(null step)*.
6. *Increase* k *by 1 and go to Step 2.*

For $\varepsilon_{\mathrm{M}} = \infty$ the algorithm performs exactly one inner iteration as described before. This single inner iteration suffices to guarantee convergence.

Theorem 1. *Let* $\varepsilon_{\mathrm{opt}} = 0$ *and* $\varepsilon_{\mathrm{M}} = \infty$. *If* $\mathrm{Arg\,min}\, f_S \neq \emptyset$ *then* $x^k \to \bar{x} \in \arg\min f_S$, *otherwise* $\|x^k\| \to \infty$. *In both cases* $f(x^k) \downarrow \inf f_S$.

The proof is rather technical and we refer the reader to the full version of the paper for details. To sketch the main idea only, assume that test point y^k has caused a null step. Then y^k is the (unconstrained) minimizer of the strongly convex function $\bar{f}^k(\cdot) + \frac{u}{2}\|\cdot - x^k\|^2$ with modulus u and therefore, since $W^k \in \hat{\mathcal{W}}^k$,

$$L(y_{W^{k+1}, \eta^k}; W^{k+1}, \eta^k) \geq L(y^k; W^k, \eta^k) + \frac{u}{2}\|y_{W^{k+1}, \eta^k} - y^k\|^2.$$

Likewise we obtain

$$L(y^{k+1}; W^{k+1}, \eta^{k+1}) \geq L(y_{W^{k+1}, \eta^k}; W^{k+1}, \eta^k) + \frac{u}{2}\|y^{k+1} - y_{W^{k+1}, \eta^k}\|^2.$$

Using these relations and the fact that the subgradients generated by the algorithm remain locally bounded one can arrive at Theorem 1 along a similar line of arguments as given in [7].

5 $\hat{\mathcal{W}}^k$ and the Quadratic Semidefinite Subproblem

Whereas the convergence analysis of the algorithm imposes only very few requirements on $\hat{\mathcal{W}}^k$ and its update to $\hat{\mathcal{W}}^{k+1}$, its actual realization is of utmost importance for the efficiency of the algorithm. The spectral bundle method uses

$$\hat{\mathcal{W}}^k = \left\{W = P^k V (P^k)^T + \alpha \overline{W}^k : \mathrm{tr}(V) + \alpha = 1, V \in S_r^+, \alpha \geq 0\right\},$$

where $P^k \in \mathbb{R}^{n \times r}$ is some fixed matrix with orthonormal columns and $\overline{W}^k \in S_n^+$ satisfies $\operatorname{tr}(\overline{W}^k) = 1$. P^k should span at least partially the eigenspace belonging to the largest eigenvalues of matrices $C - \mathcal{A}^T(y)$ for y in the vicinity of x^k. $\overline{W}^k$ serves to aggregate subgradient information that cannot be represented within the set $\{P^k V (P^k)^T : \operatorname{tr}(V) = 1, V \in S_r^+\}$. The use of $\overline{W}^k$ ensures convergence of the spectral bundle method even if the number of columns of P^k is restricted to $r = 1$. The quality of the semidefinite model as well as the computation time for solving (6) depend heavily on r.

The special structure of $\hat{\mathcal{W}}^k$ allows us to evaluate $\hat{f}^k$ directly:

$$\hat{f}^k(y) = \max \left\{ \lambda_{\max}((P^k)^T (C - \mathcal{A}^T(y)) P^k), \left\langle \overline{W}^k, C - \mathcal{A}^T(y) \right\rangle \right\} + \langle b, y \rangle .$$

Observe that the eigenvalue computation involved is cheap since the argument is an $r \times r$ symmetric matrix.

By the choice of $\hat{\mathcal{W}}^k$, for a fixed $\eta = \eta^k$, (6) is equivalent to the quadratic semidefinite program

$$
\begin{aligned}
\min \ & \tfrac{1}{2u} \| b - \eta - \mathcal{A}(P^k V (P^k)^T + \alpha \overline{W}^k) \|^2 \\
& - \left\langle P^k V (P^k)^T + \alpha \overline{W}^k, C - \mathcal{A}^T(x^k) \right\rangle - \langle b - \eta, x^k \rangle \\
\text{s.t.} \ & \operatorname{tr}(V) + \alpha = 1 \\
& V \succeq 0, \alpha \geq 0.
\end{aligned}
\tag{8}
$$

Its optimal solution (V^*, α^*) gives rise to $W^{k+1} = P^k V^* (P^k)^T + \alpha^* \overline{W}^k$ and is also used to update P^k and $\overline{W}^k$ in Step 4 of the algorithm. Let $Q \Lambda Q^T$ be an eigenvalue decomposition of V^*. Then the 'important' part of the spectrum of W^{k+1} is spanned by the eigenvectors associated with the 'large' eigenvalues of V^*. Thus the eigenvectors of Q are split into two parts $Q = [Q_1 Q_2]$ (with corresponding spectra Λ_1 and Λ_2), Q_1 containing as columns the eigenvectors associated to 'large' eigenvalues of V^* and Q_2 containing the remaining columns. P^{k+1} then contains an orthonormal basis of the columns of $P^k Q_1$ and at least one eigenvector to the maximal eigenvalue of $C - \mathcal{A}^T(y^{k+1})$ computed in Step 3 of the algorithm. The next aggregate matrix is

$$\overline{W}^{k+1} = \alpha^* \overline{W}^k + P^k Q_2 \Lambda_2 Q_2^T (P^k)^T / (\alpha^* + \operatorname{tr}(\Lambda_2)).
\tag{9}$$

This ensures $W^{k+1} \in \hat{\mathcal{W}}^{k+1}$. For $\hat{\mathcal{W}}^0$ we choose P^0 to contain a set of orthonormalized eigenvectors to large eigenvalues of $C - \mathcal{A}(y^0)$. We do not use the aggregate in the first iteration.

(8) has to be solved for each update of η in the inner loop or, if $\varepsilon_M = \infty$, at least once per outer iteration. Thus the efficiency of the whole algorithm hinges on the speed of this computation. We briefly explain the most important issues that arise in this context.

Using the svec-operator described in [12] to expand symmetric matrices from S_r into column vectors of length $\binom{r+1}{2}$ and by ignoring all constants, (8) can be brought into the following form (recall that, for $A, B \in \mathcal{S}_r$, $\langle A, B \rangle = \text{svec}(A)^T \text{svec}(B)$ and that $\text{tr}(V) = \langle I, V \rangle$)

$$\begin{aligned} \min \; & \tfrac{1}{2}\text{svec}(V)^T Q_{11}\text{svec}(V) + \alpha q_{12}^T \text{svec}(V) + \tfrac{1}{2}q_{22}\alpha^2 + c_1^T \text{svec}(V) + c_2\alpha \\ \text{s.t. } & \alpha + s_I^T \text{svec}(V) = 1 \\ & \alpha \geq 0, V \succeq 0, \end{aligned} \qquad (10)$$

where

$$\begin{aligned} Q_{11} &= \tfrac{1}{u}\sum_{i=1}^m \text{svec}(P^T A_i P)\text{svec}(P^T A_i P)^T \\ q_{12} &= \tfrac{1}{u}\text{svec}(P^T \mathcal{A}^T(\mathcal{A}(\overline{W}))P) \\ q_{22} &= \tfrac{1}{u}\langle \mathcal{A}(\overline{W}), \mathcal{A}(\overline{W})\rangle \\ c_1 &= -a\,\text{svec}(P^T(\tfrac{1}{u}\mathcal{A}^T(b-\eta) + C - \mathcal{A}^T(y))P) \\ c_2 &= -a(\langle \tfrac{1}{u}(b-\eta) - y, \mathcal{A}(\overline{W})\rangle + \langle C, \overline{W}\rangle) \\ s_I &= \text{svec}(I). \end{aligned}$$

This problem has $\binom{r+1}{2}+1$ variables and can be solved quite efficiently by interior point methods if r is not too large, smaller than 25, say. Since convergence is guaranteed even for $r = 1$, it is possible to run this algorithm for problems with a huge number of constraints m. Several remarks are in order.

First, it is not necessary to have $\overline{W}$ available as a matrix, it suffices to store the m-vector $\mathcal{A}(\overline{W})$ and the scalar $\langle C, \overline{W}\rangle$. These values are easily updated whenever $\overline{W}$ is changed.

Second, almost all computations involve the projected matrices $P^T A_i P$. These have to be computed only once for each evaluation of (10), they are symmetric of size $r \times r$, and only one such projected matrix has to be kept in memory if the values of Q_{11} to c_2 are accumulated.

Third, the most expensive operation in computing the cost coefficients is the accumulation of Q_{11}, which involves the summation of m dyadic products of vectors of size $\binom{r+1}{2}$ for a total of $O(mr^4)$ operations. Even for rather small r but sufficiently large m, this operation takes longer than solving the reduced quadratic semidefinite program.

Finally, changes in η do not affect the quadratic cost matrices Q_{11} to q_{22}, but only the linear cost coefficients c_1 and c_2. Within the inner loop it suffices to update the linear cost coefficients. The dominating cost of computing Q_{11} can be avoided. If the inner iteration yields only small changes in η then the optimal solution will also change only slightly. This can be exploited in restarting strategies.

We solve (10) by the primal-dual interior point code used in [6]. It maintains primal and dual feasibility throughout (U is the dual variable of V, β is the dual variable of α, and t is the dual variable to the primal equality constraint),

$$Q_{11}\text{svec}(V) + \alpha q_{12} + c_1 - t s_I - \text{svec}(U) = 0, \qquad (11)$$
$$\alpha q_{22} + q_{12}^T\text{svec}(V) + c_2 - t - \beta = 0, \qquad (12)$$
$$1 - \alpha - s_I^T\text{svec}(V) = 0. \qquad (13)$$

We employ a restart procedure for reoptimizing the quadratic program in the inner loop. Let $(U^*, V^*, \alpha^*, \beta^*, t^*)$ denote the optimal solution of the latest problem, and let Δc_1 and Δc_2 denote the update (caused by the change in η) to be added to c_1 and c_2. We determine a starting point $(U^0, V^0, \alpha^0, \beta^0, t^0)$ by the following steps. First, note that $(V, \alpha) = (I, 1)/(r + 1)$ is the analytic center of the primal feasible set. Therefore any point on the straight line segment $\xi(V^*, \alpha^*) + (1 - \xi)(I, 1)/(r + 1)$ with $\xi \in [0, 1]$ is again primal feasible. Furthermore it can be expected to be a point close to the central path if the new optimal solution is relatively close to the old optimal solution. We determine ξ by $\xi = \min\{.99999, \max[.9, \sqrt{1 - \frac{\|(\Delta c_1, \Delta c_2)\|}{\|(c_1, c_2)\|}}]\}$.

In (11) and (12) the change from (V^*, α^*) to the starting point (V^0, α^0) determines the changes in the dual variables up to a diagonal shift that can be applied through t. This diagonal shift is chosen so that the changes in U and β are positive definite. Thus U^0 (β^0) is strictly positive definite but not too far from U^* (β^*). Experience shows that this restarting heuristic reduces the number of iterations to two thirds down to one half depending on the size of the changes in η.

6   Implementation and Combinatorial Applications

The algorithm has been implemented as a direct extension of the algorithm in [6] and uses essentially the same parameter settings. In particular, $m_L = 0.1$, $r = 25$, $\varepsilon_{opt} = 5 \cdot 10^{-4}$, and the weight u is updated dynamically by safeguarded quadratic interpolation. We present some computational results for $\varepsilon_M = .6$ and $\varepsilon_M = \infty$. The choice of $\varepsilon_M = .6$ should keep the number of inner iterations small, at the same time avoiding null steps enforced by the bad quality of the relaxation. Our computational experiments were carried out on a Sun Sparc Ultra 1 with a Model 140 UltraSPARC CPU and 64 MB RAM. Computation times are given in hours:minutes:seconds.

We apply the code to the max-cut instances G_1 to G_{21} on 800 vertices from [6]. Graphs G_1 to G_5 are unweighted random graphs with a density of 6% (approx. 19000 edges). G_6 to G_{10} are the same graphs with random edge weights from $\{-1, 1\}$. G_{11} to G_{13} are toroidal grids with random edge weights from $\{-1, 1\}$ (approx. 1200 edges). G_{14} to G_{17} are unweighted 'almost' planar graphs having as edge set the union of two (almost maximal) planar graphs (approx. 4500 edges). G_{18} to G_{21} are the same almost planar graphs with random edge weights from $\{-1, 1\}$. In all cases the cost matrix C is the Laplace matrix of the graph divided by four, i.e., let A denote the (weighted) adjacency matrix of G, then $C = (\text{Diag}(Ae) - A)/4$.

To the basic semidefinite relaxation ($\text{diag}(X) = e$) we add 1600 triangle inequalities. These were determined by our code from [5], which is a primal-dual interior point code. We applied the separation routine of this code to the optimal solution of the semidefinite relaxation without inequality constraints. We also used the code to solve the relaxation with the same triangle constraints and will compare these results to the spectral bundle method.

Table 1. Comparison of the spectral bundle and interior point approaches.

Problem	IP-sol	spectral bundle		interior point	
		1%	0.1%	1%	0.1%
G_1	12049.33	19	3:51	2:16:14	4:05:13
G_2	12053.88	20	5:37	2:15:57	4:04:42
G_3	12048.56	17	4:13	2:44:06	4:06:09
G_4	12071.59	17	3:11	2:42:53	4:04:20
G_5	12059.68	15	3:21	3:39:56	5:02:24
G_6	2615.55	1:24	14:47	3:37:52	4:59:34
G_7	2451.61	1:28	11:55	3:37:48	4:59:28
G_8	2463.68	1:42	19:19	3:37:47	4:59:27
G_9	2493.51	1:38	16:44	3:38:45	5:00:46
G_{10}	2447.79	1:22	12:17	3:35:53	4:57:31
G_{11}	623.49	9:25	54:55	4:55:20	6:10:01
G_{12}	613.02	9:32	47:01	4:57:09	6:43:04
G_{13}	636.45	8:52	39:00	4:31:17	6:46:56
G_{14}	3181.35	1:19	10:13	3:36:13	5:24:18
G_{15}	3161.89	1:09	11:51	3:35:00	5:22:31
G_{16}	3164.95	1:24	12:55	4:07:35	5:57:37
G_{17}	3161.78	1:20	12:32	3:35:22	5:23:04
G_{18}	1147.61	4:10	15:45	4:30:04	5:51:05
G_{19}	1064.57	4:23	17:17	4:59:07	6:47:54
G_{20}	1095.46	4:31	14:58	4:30:13	5:51:18
G_{21}	1087.89	3:22	12:40	4:56:09	6:16:55

For the bundle algorithm the diagonal of C is removed. This does not change the problem because the diagonal elements of X are fixed to one. The offset $e^T(Ae - \text{diag}(A))/4$ is not used internally in the spectral bundle code. However, after each serious step we check externally whether the stopping criterion is satisfied if the offset is added. If so, we terminate the algorithm. As starting vector y^0 we choose the zero vector.

For inequality constrained problems our spectral bundle code exhibits an even stronger tailing off effect than for equality constrained problems. To solve the relaxation exactly seems to be out of reach. However, within the first few iterations the objective value gets already very close to the optimal value. In cutting plane approaches it does not make sense to solve relaxations to optimality if much faster progress can be achieved by adding new inequalities. Therefore the quick initial convergence is a desirable property for these applications.

In this spirit we give the time needed to get within 1% and 0.1% of the optimum in Table 1 for both the spectral bundle code with $\varepsilon_M = .6$ and the interior point code. The first column specifies the problem instance, the second (IP-sol) gives the optimal value of the relaxation with triangle inequalities. Columns three and five (four and six) display the times needed by both codes to get within 1% (0.1%, resp.) of the optimum. (The entries for the interior point code are estimates, since its output only gives the overall computation time.)

Table 2. Performance of the spectral bundle code for $\varepsilon_{\mathrm{opt}} = 5 \cdot 10^{-4}$ and $\varepsilon_{\mathrm{M}} = .6$.

Problem	f^*	$f(x^k)$	$\frac{f(x^k)-f^*}{f^*}$	total time	eig_time[%]	# calls	# serious
G_1	12049.33	12055.79	$5.4 \cdot 10^{-4}$	5:10	62.9	15	11
G_2	12053.88	12057.64	$3.1 \cdot 10^{-4}$	12:06	53.99	25	16
G_3	12048.56	12051.59	$2.5 \cdot 10^{-4}$	13:15	53.21	27	18
G_4	12071.59	12078.58	$5.8 \cdot 10^{-4}$	4:26	61.28	15	11
G_5	12059.68	12065.88	$5.1 \cdot 10^{-4}$	6:28	56.96	18	13
G_6	2615.55	2617.04	$5.7 \cdot 10^{-4}$	26:17	45.21	49	22
G_7	2451.61	2453.27	$6.8 \cdot 10^{-4}$	15:38	45.95	34	21
G_8	2463.68	2464.86	$4.8 \cdot 10^{-4}$	31:33	46.01	59	22
G_9	2493.51	2494.79	$4.8 \cdot 10^{-4}$	28:06	45.37	54	22
G_{10}	2447.79	2449.71	$7.8 \cdot 10^{-4}$	15:20	43.8	35	22
G_{11}	623.49	623.95	$7.3 \cdot 10^{-4}$	1:11:04	68.64	94	30
G_{12}	613.02	613.61	$9.6 \cdot 10^{-4}$	47:42	78.13	90	35
G_{13}	636.45	637.03	$9.1 \cdot 10^{-4}$	39:00	79.74	82	35
G_{14}	3181.35	3182.53	$3.7 \cdot 10^{-4}$	19:46	64.67	43	29
G_{15}	3161.89	3163.24	$4.3 \cdot 10^{-4}$	19:51	66.16	40	27
G_{16}	3164.95	3165.98	$3.3 \cdot 10^{-4}$	29:32	64.73	50	28
G_{17}	3161.78	3163.52	$5.5 \cdot 10^{-4}$	18:48	67.55	41	28
G_{18}	1147.61	1148.25	$5.6 \cdot 10^{-4}$	21:54	55.48	48	31
G_{19}	1064.57	1065.06	$4.6 \cdot 10^{-4}$	22:33	54.99	48	28
G_{20}	1095.46	1095.94	$4.4 \cdot 10^{-4}$	20:05	56.02	46	29
G_{21}	1087.89	1088.72	$7.6 \cdot 10^{-4}$	14:48	59.57	40	29

The fast progress of the bundle code on examples G_1 to G_5 can be explained by the large offset value (9500) that is added externally. For the other classes the offset is around 0 for G_6 to G_{13} and G_{18} to G_{21}, being about 2300 for G_{14} to G_{17}. The eigenvalue structure of the solutions of G_{10} to G_{13} suggests that the rather poor performance of the algorithm on these problems is caused by a very flat objective.

Table 2 provides more detailed information on the overall performance of the spectral bundle code for $\varepsilon_{\mathrm{M}} = .6$. Column f^* refers to the value of the interior point solution, $f(x^k)$ gives the objective value at termination, $\frac{f(x^k)-f^*}{f^*}$ is the relative accuracy with respect to f^*, *total time* gives the computation time, *eig_time* is the percentage of time spent in the Lanczos code (eigenvalue and eigenvector computation), *# calls* counts the number of objective evaluations and *# serious* the number of serious steps.

Somewhat surprisingly the spectral bundle method also yields reasonable results for $\varepsilon_{\mathrm{M}} = \infty$. The corresponding values are displayed in Table 3.

Table 3. Performance of the spectral bundle code for $\varepsilon_{\mathrm{opt}} = 5 \cdot 10^{-4}$ and $\varepsilon_{\mathrm{M}} = \infty$.

Problem	f^*	$f(x^k)$	$\frac{f(x^k)-f^*}{f^*}$	total time	eig_time[%]	# calls	# serious
G_1	12049.33	12055.79	$5.4 \cdot 10^{-4}$	5:14	62.42	15	11
G_2	12053.88	12057.31	$5.4 \cdot 10^{-4}$	13:13	54.22	27	16
G_3	12048.56	12053.27	$3.9 \cdot 10^{-4}$	11:54	54.06	25	17
G_4	12071.59	12078.58	$5.8 \cdot 10^{-4}$	4:23	61.98	15	11
G_5	12059.68	12065.88	$5.1 \cdot 10^{-4}$	6:26	57.51	18	13
G_6	2615.55	2616.89	$5.1 \cdot 10^{-4}$	26:32	45.79	52	24
G_7	2451.61	2453.27	$6.8 \cdot 10^{-4}$	15:14	46.50	34	21
G_8	2463.68	2464.86	$4.7 \cdot 10^{-4}$	32:02	44.59	59	22
G_9	2493.51	2494.69	$4.7 \cdot 10^{-4}$	30:14	46.14	58	23
G_{10}	2447.79	2449.51	$7.0 \cdot 10^{-4}$	14:32	44.84	34	22
G_{11}	623.49	623.89	$6.4 \cdot 10^{-4}$	1:30:05	71.3	113	33
G_{12}	613.02	613.69	$1.1 \cdot 10^{-3}$	53:17	74.8	95	33
G_{13}	636.45	637.05	$9.4 \cdot 10^{-4}$	43:31	79.1	87	36
G_{14}	3181.35	3182.68	$4.1 \cdot 10^{-4}$	19:23	64.49	43	28
G_{15}	3161.89	3163.18	$4.1 \cdot 10^{-4}$	19:29	66.47	40	27
G_{16}	3164.95	3165.98	$3.3 \cdot 10^{-4}$	28:55	65.42	50	28
G_{17}	3161.78	3163.46	$5.3 \cdot 10^{-4}$	18:35	67.89	41	28
G_{18}	1147.61	1148.28	$5.8 \cdot 10^{-4}$	20:16	58.31	47	30
G_{19}	1064.57	1065.04	$4.4 \cdot 10^{-4}$	23:24	59.69	51	28
G_{20}	1095.46	1095.94	$4.3 \cdot 10^{-4}$	18:39	59.34	46	28
G_{21}	1087.89	1088.63	$6.8 \cdot 10^{-4}$	15:09	61.17	42	29

References

1. S. Benson, Y. Ye, and X. Zhang. Solving large-scale sparse semidefinite programs for combinatorial optimization. Working paper, Department of Management Science, University of Iowa, IA, 52242, USA, Sept. 1997.
2. M. X. Goemans and D. P. Williamson. Improved approximation algorithms for maximum cut and satisfiability problems using semidefinite programming. *J. ACM*, 42:1115–1145, 1995.
3. G. H. Golub and C. F. van Loan. *Matrix Computations.* The Johns Hopkins University Press, 2nd edition, 1989.
4. M. Grötschel, L. Lovász, and A. Schrijver. Polynomial algorithms for perfect graphs. *Annals of Discrete Mathematics*, 21:325–356, 1984.
5. C. Helmberg and F. Rendl. Solving quadratic (0,1)-problems by semidefinite programs and cutting planes. ZIB Preprint SC-95-35, Konrad-Zuse-Zentrum für Informationstechnik Berlin, Nov. 1995. To appear in *Math. Programming*.
6. C. Helmberg and F. Rendl. A spectral bundle method for semidefinite programming. ZIB Preprint SC-97-37, Konrad-Zuse-Zentrum für Informationstechnik Berlin, Aug. 1997.
7. K. C. Kiwiel. Proximity control in bundle methods for convex nondifferentiable minimization. *Math. Programming*, 46:105–122, 1990.
8. A. S. Lewis and M. L. Overton. Eigenvalue optimization. *Acta Numerica*, 149–190, 1996.

9. L. Lovász. On the Shannon capacity of a graph. *IEEE Transactions on Information Theory*, IT-25(1):1–7, 1979.

10. L. Lovász and A. Schrijver. Cones of matrices and set-functions and 0-1 optimization. *SIAM J. Optim.*, 1(2):166–190, 1991.

11. H. Schramm and J. Zowe. A version of the bundle idea for minimizing a nonsmooth function: Conceptual idea, convergence analysis, numerical results. *SIAM J. Optim.*, 2:121–152, 1992.

12. M. J. Todd, K. C. Toh, and R. H. Tütüncü. On the Nesterov-Todd direction in semidefinite programming. Technical Report TR 1154, School of Operations Research and Industrial Engineering, Cornell University, Ithaca, New York 14853, Mar. 1996.

Author Index